Introduction to Functions of a Complex Variable

MONOGRAPHS AND TEXTBOOKS IN PURE AND APPLIED MATHEMATICS

1. *K. Yano,* Integral Formulas in Riemannian Geometry (1970)
2. *S. Kobayashi,* Hyperbolic Manifolds and Holomorphic Mappings (1970)
3. *V. S. Vladimirov,* Equations of Mathematical Physics (A. Jeffrey, editor; A. Littlewood, translator) (1970)
4. *B. N. Pshenichnyi,* Necessary Conditions for an Extremum (L. Neustadt, translation editor; K. Makowski, translator) (1971)
5. *L. Narici, E. Beckenstein, and G. Bachman,* Functional Analysis and Valuation Theory (1971)
6. *D. S. Passman,* Infinite Group Rings (1971)
7. *L. Dornhoff,* Group Representation Theory (in two parts). Part A: Ordinary Representation Theory. Part B: Modular Representation Theory (1971, 1972)
8. *W. Boothby and G. L. Weiss (eds.),* Symmetric Spaces: Short Courses Presented at Washington University (1972)
9. *Y. Matsushima,* Differentiable Manifolds (E. T. Kobayashi, translator) (1972)
10. *L. E. Ward, Jr.,* Topology: An Outline for a First Course (1972) *out of print*
11. *A. Babakhanian,* Cohomological Methods in Group Theory (1972)
12. *R. Gilmer,* Multiplicative Ideal Theory (1972)
13. *J. Yeh,* Stochastic Processes and the Wiener Integral (1973) *out of print*
14. *J. Barros-Neto,* Introduction to the Theory of Distributions (1973) *out of print*
15. *R. Larsen,* Functional Analysis: An Introduction (1973)
16. *K. Yano and S. Ishihara,* Tangent and Cotangent Bundles: Differential Geometry (1973)
17. *C. Procesi,* Rings with Polynomial Identities (1973)
18. *R. Hermann,* Geometry, Physics, and Systems (1973)
19. *N. R. Wallach,* Harmonic Analysis on Homogeneous Spaces (1973)
20. *J. Dieudonné,* Introduction to the Theory of Formal Groups (1973)

21. *I. Vaisman,* Cohomology and Differential Forms (1973)
22. *B.-Y. Chen,* Geometry of Submanifolds (1973)
23. *M. Marcus,* Finite Dimensional Multilinear Algebra (in two parts) (1973, 1975)
24. *R. Larsen,* Banach Algebras: An Introduction (1973)
25. *R. O. Kujala and A. L. Vitter (eds.),* Value Distribution Theory: Part A; Part B. Deficit and Bezout Estimates by Wilhelm Stoll (1973)
26. *K. B. Stolarsky,* Algebraic Numbers and Diophantine Approximation (1974)
27. *A. R. Magid,* The Separable Galois Theory of Commutative Rings (1974)
28. *B. R. McDonald,* Finite Rings with Identity (1974)
29. *J. Satake,* Linear Algebra (S. Koh, T. Akiba, and S. Ihara, translators) (1975)
30. *J. S. Golan,* Localization of Noncommutative Rings (1975)
31. *G. Klambauer,* Mathematical Analysis (1975)
32. *M. K. Agoston,* Algebraic Topology: A First Course (1976)
33. *K. R. Goodearl,* Ring Theory: Nonsingular Rings and Modules (1976)
34. *L. E. Mansfield,* Linear Algebra with Geometric Applications (1976)
35. *N. J. Pullman,* Matrix Theory and its Applications: Selected Topics (1976)
36. *B. R. McDonald,* Geometric Algebra Over Local Rings (1976)
37. *C. W. Groetsch,* Generalized Inverses of Linear Operators: Representation and Approximation (1977)
38. *J. E. Kuczkowski and J. L. Gersting,* Abstract Algebra: A First Look (1977)
39. *C. O. Christenson and W. L. Voxman,* Aspects of Topology (1977)
40. *M. Nagata,* Field Theory (1977)
41. *R. L. Long,* Algebraic Number Theory (1977)
42. *W. F. Pfeffer,* Integrals and Measures (1977)
43. *R. L. Wheeden and A. Zygmund,* Measure and Integral: An Introduction to Real Analysis (1977)
44. *J. H. Curtiss,* Introduction to Functions of a Complex Variable (1978)

Introduction to Functions of a Complex Variable

J.H. Curtiss

Department of Mathematics
University of Florida
Coral Gables, Florida

CRC Press
Taylor & Francis Group
Boca Raton London New York

CRC Press is an imprint of the
Taylor & Francis Group, an informa business

First published 1978 by Marcel Dekker, Inc.

Published 2019 by CRC Press
Taylor & Francis Group
6000 Broken Sound Parkway NW, Suite 300
Boca Raton, FL 33487-2742

First issued in paperback 2019

ISBN 13: 978-0-367-45206-3 (pbk)
ISBN 13: 978-0-8247-6501-9 (hbk)

Visit the Taylor & Francis Web site at
http://www.taylorandfrancis.com

and the CRC Press Web site at
http://www.crcpress.com

Library of Congress Cataloging in Publication Data

Curtiss, John Hamilton, 1909-1977.
Introduction to functions of a complex variable.

(Pure and applied mathematics ; v. 44)
Includes bibliographical references and index.
1. Functions of complex variables. I. Title.
QA331.C84 1978 515'.9 77-25846
ISBN 0-8247-6501-X

To Joseph Leonard Walsh,
in memorium

Foreword

On several occasions, John Curtiss told me of his plans and aspirations for the present volume. It was to be thoroughly modern in approach, featuring a balance of abstract concepts on the one hand and practical applications on the other; relaxed and friendly, but carefully logical, in style; analytically rigorous and independent of geometric intuition, yet with many geometric illustrations and applications; and with a great wealth of examples and exercises. Though intended for an introductory course at the advanced undergraduate and first-year graduate level, it was also to provide doctoral students at any university with a sound preparation for qualifying examinations in complex analysis. For a final, sentimental objective, the book was to expand on his father's brief but still popular Carus Monograph on analytic functions of a complex variable.

John Curtiss was well qualified to write this book. Besides serving terms as Chief of the Applied Mathematics Division of the National Bureau of Standards and as Executive Director of the American Mathematical Society, he held teaching positions at Johns Hopkins University, Cornell University, the Courant Institute of Mathematical Sciences of New York University, and the University of Miami. Much of his original research work involved complex-variable theory; he taught courses on the subject quite often, and his notes for the book developed during many years of experience.

In reviewing the manuscript and in reading proof sheets of the book for typographical errors, I have been pleased to observe that, in my opinion, in each of his aims for the book he has succeeded admirably.

E. F. Beckenbach

Preface

This book is intended for an introductory course in complex analysis at the first-year graduate and advanced undergraduate level. The program of the book will be revealed to some extent to the instructor experienced in the subject by a mere scanning of the table of contents, and no amount of prefatory remarks would be illuminating to most beginners. However, certain highlights of the treatment will be examined here.

It has often been observed that students who enroll in such a course have diverse mathematical backgrounds. Partly for this reason, it is traditional for instructional materials for the course to be self-contained to a considerable degree. This tradition has been observed in the present book. Only some knowledge of the structure of the real number system and some familiarity with elementary calculus are assumed.

Perhaps no other course at a comparable level in the standard mathematics curriculum has been presented with such wide and erratic variations in logical completeness or "rigor." This book falls at the rigorous end of the spectrum. In fact, the intention was to achieve (in some cases to exceed) the same high level of logical completeness which characterizes, for example, the excellent texts on the subject by Ahlfors [2], Heins [8], and Rudin [18]; but with a more relaxed, detailed, and accessible treatment. It goes without saying that to keep the book within reasonable physical dimensions, such a program entailed some sacrifices in the extent of coverage.

The author taught complex analysis for many years, using a variety of successful textbooks. This experience resulted in the compilation of a rather extensive set of lecture notes, from which this book was distilled. Naturally the lectures were influenced by the textbooks, especially those of Ahlfors [2], Hille [9, 10], Rudin [18], and Saks and Zygmund [19]. It is a pleasure to acknowledge the author's debt to these distinguished mathematicians. Some of the proofs in the last two chapters seem to be new. This applies particularly to the Appendix of Chapter 13.

It has been the author's impression that even some of the more competent mathematics majors find their first rigorous course in complex analysis to be difficult, at least at the outset. Therefore the exposition in this book has been paced so as to start out quite slowly and to gather momentum and depth gradually. In fact, the first five chapters, which include a chapter on elementary general topology, will be viewed by some instructors and readers as only preparatory, to be covered rapidly in the course if at all. But even the well-prepared student may find the frequent appearance of back-references to these earlier chapters helpful.

The central theorem of complex analysis is the Cauchy Integral Theorem, which appears first in Chapter 9. In its most general forms the theorem is deep and difficult to prove. However, for a function analytic in a convex or starlike region a straightforward proof can easily be given. Furthermore, even with the validation of the theorem so restricted, a large number of the famous classical results of complex function theory can be usefully derived. This fact has been exploited in this book, partly with a view, as mentioned above, to keeping things simple as long as possible. All of the theorems and formulas in Chapters 9, 10, and 11 (and these are often considered to be the core of classical complex analysis) are obtained from the Cauchy Integral Theorem for a starlike region. Generalizations of the Cauchy theory to cases in which the paths of integration lie in arbitrary open sets are achieved in Chapter 12 by means of Runge's theorem on approximation by rational functions. The concepts of homology and homotopy in their application to the Cauchy theory are deferred until after the Cauchy Integral Theorem has been established in a very general form. Important as these topological concepts are for a full understanding of complex integration theory, the author has found that it tends to be diversionary to introduce them before the student has become acquainted with some of the basic classical results.

Nowadays in the undergraduate mathematics curriculum a considerable amount of emphasis is given to abstract concepts. This has been duly recognized here by presenting a number of the elementary definitions in reasonably abstract contexts. Also, the reader will find various passages in the earlier chapters where deference is shown to the basics of functional analysis.

A secondary theme relating to the important role of approximation theory in analysis runs through the book, as exemplified by the early introduction of power series in Chapter 4, the extensive exposure given to the Runge theorem and its polynomial specializations, and the applications of conformal mapping to polynomial approximations in the last chapter.

For better or for worse, the intuitive geometric aspects of the subject are not strongly emphasized. However, the geometric nature of mappings by complex-valued functions is given an early presentation in exercises in Chapter 1; it is touched on in various ways in the sequel (e.g., the Open Mapping Theorem, Section 10.4); and it is studied at some length theoretically in the concluding chapter on conformal mapping.

The principal statements in this book are numbered consecutively in each section with a triple numbering system which indicates at once the particular chapter, section, and statement in the section. The reader should have no difficulty in understanding the code. The statements are variously labeled lemma, proposition, theorem, or corollary. A theorem is a statement of primary importance which occupies a central position in the theory. A proposition is a statement of some interest for itself alone, but which occupies a more peripheral position than a theorem in the mainstream of the exposition. A lemma is a portion of a proof which for convenience in reference has been set out formally.

There are several hundred exercises which are placed at the ends of the sections. Some of them are meant merely for drill, but many of them contain important extensions and developments of the theory. Exercises which contain propositions cited elsewhere as authorities are indicated by boldface numbers.

It is the author's pleasant duty to give thanks to two friends who were particularly helpful in bringing this book into final form. The first of these is Professor E. F. Beckenbach, who was requested by the publisher to review the first draft of the manuscript. He made many constructive suggestions bearing on both substance and style, and almost all of these were used. The other friend is my colleague Professor Edwin Duda, whose expertise in point-set topology contributed in a significant way to the accuracy of Chapter 5 and of the topological parts of Chapter 13. It goes almost without saying that any errors which may remain are the author's own responsibility. Finally, thanks are due to Mrs. Mamie Cummings of the Mathematics Department of the University of Miami, who overcame the distraction of a multitude of revisions to produce a skillfully typed manuscript.

J. H. Curtiss

Contents

Introduction to Functions of a Complex Variable

1

The Real and Complex Number Fields

1.1 Introduction: Set Theoretic Notation

The basic language of mathematics nowadays is that of set theory. It is safe to assume that the reader is acquainted with the fundamental intuitive ideas of set theory. The purpose here is to summarize the basic notation and terminology to be used in this book.

Given a set E, we write $x \in E$ to express the statement "x is an element (or member) of E." If x is not an element of E, we write $x \notin E$. A set is completely determined by its members, so if sets E and F have the property that $x \in E$ if and only if $x \in F$, then E and F are the same set, and we write $E = F$.

If the sets E and F have the property that $x \in E$ implies $x \in F$, then E is called a *subset* of F, and we write $E \subset F$ or $F \supset E$. In this situation, we shall also say that E is *contained* in F, or more colloquially, *E lies in F.*

There are essentially two ways of specifying a set. The first consists of listing its elements or displaying a partial list which indicates a pattern; the notation used is typified by $\{a, b, c\}$ and $\{1, 2, 3, \ldots\}$. This is sometimes called the census method. The other way, and the one most frequently employed in mathematics, consists of announcing a condition which the elements must satisfy, or a property or properties which each element must have. A set of elements with property P is denoted by $\{x: P(x)\}$. For example, if X is the set of positive real integers, then the subset $\{1, 2, \ldots, 10\}$ can be denoted by $\{x: x \in X, x \leq 10\}$.

In using the census method for specifying a set, it is understood that any permutation of the elements in the list still represents the same set; for example, $\{a, b\} = \{b, a\}$. Notations other than the brace notation, such as (a, b), will be used to indicate an ordering of the elements a, b.

The empty set is the set with no member and is denoted by $\varnothing$. It is a subset of every set.

In any application of the theory of sets, all sets under consideration are regarded as subsets of a fixed set X. This set is given various names in the literature, such as the *universal set*. In mathematical applications, it is often called a *space*. Other synonyms for sets which are used in various contexts (partly to avoid monotonous repetition of "set") are "collection" (for example, "a collection of subsets of X") and "family" (for example, "a family of collections of subsets of X").

If X and Y are two sets, the *Cartesian product set* of X and Y, written $X \times Y$, is the set of all ordered pairs $\{(x,y): x \in X, y \in Y\}$.

Let X be a given set. The *union* of the subsets E and F is the set $E \cup F = \{x: x \in E \text{ or } x \in F\}$. The *intersection* of E and F is the set $E \cap F = \{x: x \in E \text{ and } x \in F\}$. The sets E and F are *disjoint* if $E \cap F = \varnothing$.

More generally, if Λ is a set and E_λ is a subset of X for each $\lambda \in \Lambda$, we use $\cup\{E_\lambda: \lambda \in \Lambda\} = \cup_\lambda E_\lambda$ to mean $\{x: x \in E_\lambda \text{ for some } \lambda \in \Lambda\}$, and we use $\cap\{E_\lambda: \lambda \in \Lambda\} = \cap_\lambda E_\lambda$ for $\{x: x \in E_\lambda \text{ for every } \lambda \in \Lambda\}$. The *difference* of E and F is the set $E \sim F = \{x: x \in E \text{ and } x \notin F\}$. The *complement* of E (with respect to X) is the set $E = X \sim E$. We use the notation E^c for the complement of E.

De Morgan's Laws are as follows:

$$[\cup_\lambda E_\lambda]^c = \cap_\lambda E_\lambda^c,$$
$$[\cap_\lambda E_\lambda]^c = \cup_\lambda E_\lambda^c.$$

1.2 Fields

Let $\mathscr{F}$ be a set containing more than one element, and let there be given two operations $+$ and $\cdot$ in $\mathscr{F}$, called, respectively, *addition* and *multiplication*. The set $\mathscr{F}$ is a *field* with respect to these two operations provided that the following postulates are satisfied:

F_1. *Closure.* If a and b are elements of $\mathscr{F}$, then for each such a and b, there is one and only one element $a + b$ and one and only one element $a \cdot b$.

F_2. *Commutative.* If $a \in \mathscr{F}$, $b \in \mathscr{F}$, then $a + b = b + a$, and $a \cdot b = b \cdot a$.

F_3. *Associative.* If $a \in \mathscr{F}$, $b \in \mathscr{F}$, $c \in \mathscr{F}$, then $a + (b + c) = (a + b) + c$ and $a \cdot (b \cdot c) = (a \cdot b) \cdot c$.

F_4. *Distributive.* If $a \in \mathscr{F}$, $b \in \mathscr{F}$, $c \in \mathscr{F}$, then $a \cdot (b + c) = a \cdot b + a \cdot c$.

F_5. *Existence of neutral elements.* There are elements of $\mathscr{F}$, denoted by 0 and 1, such that for any $a \in \mathscr{F}$, $a + 0 = a$ and $a \cdot 1 = a$.

F_6. *Inverses.* For each $a \in \mathscr{F}$, there is an element $-a \in \mathscr{F}$ (the "additive inverse") such that $a + (-a) = 0$. For each element $a \in \mathscr{F}$ other than 0, there is an element $a^{-1} \in \mathscr{F}$ (the "multiplicative inverse") such that $a \cdot a^{-1} = 1$.

Remark (i). There can be at most one element 0 satisfying F_5, and the following proof is more or less typical of that of many other simple direct consequences of the foregoing axioms. If $x + 0' = x$ for every $x \in \mathscr{F}$, then in particular $0 + 0' = 0 = 0' + 0$ [F_2]†, but $0' + 0 = 0'$ [F_5], so $0' = 0$. Similarly, 1 is unique.

Remark (ii). In his early youth, the reader was undoubtedly introduced to a system of numbers which is the basis of ordinary practical arithmetic. This is called the real number system. It is assumed here that the reader is familiar with the usual arithmetic operations in this system. It is, therefore, obvious that with the field operations $+$ and $\cdot$ interpreted as ordinary real number addition and multiplication, the real number system is a field. We return to this point in Sec. 1.5. For the present, our use of the real number system is confined to employing it for indexing purposes (as in Remark (iii)) and for suggesting some standard conventions and notations for use with general fields. The field of real numbers henceforth is denoted by **R**.

Remark (iii). Let a, b, c be elements of a general field. The indicated continued sum $a + b + c$ is defined to be the element $(a + b) + c$. The indicated continued product $a \cdot b \cdot c$ is defined to mean $(a \cdot b) \cdot c$. The extension of these conventions by induction provides unambiguous interpretations for finite continued sums and products such as $a_1 + a_2 + \cdots + a_n$ and $a_1 \cdot a_2 \cdots a_n$.

Remark (iv). An element $x \in \mathscr{F}$, where $\mathscr{F}$ is a general field, is called a *positive integer* if it is one of the elements $1, 1 + 1, 1 + 1 + 1, \ldots$. The traditional real number symbols $1, 2, 3, \ldots, 9, 10, 11, \ldots$ are often used for the respective terms of this progression.

Remark (v). If n is a positive integer in **R** and a is an element of a field $\mathscr{F}$, we write na for $a + a + \cdots + a$ and a^n for $a \cdot a \cdots a$, where in each case a is repeated n times.

Remark (vi). A *polynomial* in a field $\mathscr{F}$ is an expression of the form $a_0 + (a_1 \cdot x) + (a_2 \cdot x^2) + \cdots + (a_n \cdot x^n)$, where $a_0, a_1, \ldots, a_n, x \in \mathscr{F}, a_n \neq 0$. (The subscripts are, of course, in **R**.) The *degree of the polynomial is n.*

Remark (vii). The inverse elements in Postulate F_6 are uniquely determined by their respective equations. To prove this for the additive inverse, suppose that for some element $a \in \mathscr{F}$, there existed another element $b \in \mathscr{F}, b \neq -a$, which like $-a$ satisfied the equation $a + x = 0$. Then $b = b + 0 = b + (a + (-a)) = (b + a) + (-a) = 0 + (-a)$. (Here we used in succession F_5, the equation for $-a$, then F_3, the equation for b, and finally F_5 and F_2.) A similar argument establishes the uniqueness of a^{-1}.

Remark (viii). The *subtraction operation* is defined by $b + (-a)$ and written $b - a$, and the *division operation* is defined by $b \cdot a^{-1}$ and written b/a. The latter is called the quotient of b and a. Note that $1/a = 1 \cdot a^{-1} = a^{-1}$.

Remark (ix). The notation a^{-n} is used for $(a^{-1})^n, a \in \mathscr{F}$, with n a positive integer in **R**.

Remark (x). The summation convention borrowed from **R** is useful in dealing with repeated addition. Let $a_m, a_{m+1}, \ldots, a_{m+p}$ be elements of a field $\mathscr{F}$ with m a real integer and p a nonnegative real integer. Then according to the summation convention $a_m + a_{m+1} + \cdots + a_{m+p}$ is written $\sum_{k=m}^{m+p} a_k$.

†Throughout this book, cited authorities are displayed in square brackets.

Certain familiar-looking rules for field operations now follow from the field postulates and these conventions. Some of them appear in the exercises that follow.

Remark (xi). As in **R**, the multiplication operation in a general field $\mathscr{F}$ is often indicated by mere juxtaposition of the elements to be multiplied when no confusion will result; then $a \cdot b \cdot c$ becomes abc. Again as in **R**, when an algebraic expression contains both indicated multiplications and additions, it is assumed that the multiplications take precedence over the additions; for example, $ab + cd$ is to be computed as $(a \cdot b) + (c \cdot d)$.

Remark (xii). In further conformance with the conventions in **R**, in displayed formulas, the division operation in a general field is often indicated by a horizontal bar. It is understood that field operations indicated above the bar, and also those below the bar, precede the division operation. Thus for example, if a, b, c, d, e, f are elements of a general field,

$$\frac{a + bc}{d + ef}$$

signifies $[a + (b \cdot c)]/[d + (e \cdot f)]$.

Exercises 1.2

Prove the following statements relating to elements in a field $\mathscr{F}$.

1. $-0 = 0$, $1^{-1} = 1$.
2.† $(-a) \cdot (-b) = a \cdot b$.
3. $a \cdot 0 = 0$; $a \cdot b = 0$ only if $a = 0$ or $b = 0$.
4. $(a/b)(c/d) = (a \cdot c)/(b \cdot d)$.
5. $(a/b) + (c/d) = (a \cdot d + b \cdot c)/(b \cdot d)$.
6. If $a + b = a + c$, then $b = c$.
7. $-a = (-1)a$; $-(-a) = a$; $(-1)(-1) = 1$.
8. If $a \neq 0$ and $ab = ac$, then $b = c$.
9. (Abel's identity.) Let $a_0, a_1, a_2, \ldots$ and $b_0, b_1, b_2, \ldots$ be elements of a field $\mathscr{F}$; let $s_k = a_0 + a_1 + \cdots + a_k$, $k = 0, 1, 2, \ldots$, and $s_{-1} = 0$. Then for any positive real integer n and for $m = 0, 1, 2, \ldots, n - 1$,

$$\begin{aligned} a_m b_m + a_{m+1} b_{m+1} + \cdots + a_n b_n = {} & (b_m - b_{m+1}) s_m + (b_{m+1} - b_{m+2}) s_{m+1} \\ & + \cdots + (b_{n-1} - b_n) s_{n-1} + b_n s_n - b_m s_{m-1}, \end{aligned}$$

or with the summation convention,

$$\sum_{k=m}^{n} a_k b_k = \sum_{k=m}^{n-1} (b_k - b_{k+1}) s_k + b_n s_n - b_m s_{m-1}.$$

The equation remains true with $n = m$, if the summation term on the right-hand side is omitted.

†As stated in the Preface, boldface numbers indicate exercises which are cited elsewhere in this book as authorities.

10. Prove the binomial theorem:

$$(a + b)^n = a^n + na^{n-1}b + \frac{n(n-1)}{1 \cdot 2}a^{n-2}b^2 + \cdots + \frac{n(n-1)(n-2)\cdots(n-k+1)}{1 \cdot 2 \cdot 3 \cdots k}a^{n-k}b^k + \cdots + b^n,$$

where n is a positive integer in **R**.

1.3 Definitions Relating to Ordered Fields

(i) A *positive part* $\mathscr{P}$ of a field $\mathscr{F}$ is a nonempty subset of $\mathscr{F}$ with these properties: (1) if $a \in \mathscr{P}$, $b \in \mathscr{P}$, then $a + b \in \mathscr{P}$ and $a \cdot b \in \mathscr{P}$; (2) the three subsets $\{0\}$, $\mathscr{P}$, and $\{x: -x \in \mathscr{P}\}$ are disjoint (that is, no two of them have an element in common), but their union is $\mathscr{F}$.

(ii) A *field with a simple order*, or briefly, an *ordered field*, is a field together with a positive part of the field.

(iii) Given an ordered field $(\mathscr{F}, \mathscr{P})$, an element $a \in \mathscr{F}$ is "positive" if $a \in \mathscr{P}$, "negative" if $-a \in \mathscr{P}$.

To translate these definitions into more familiar terms, we define certain relations on $\mathscr{F}$, which are customarily called inequalities. The inequality $>$ ("greater than") may be taken as basic and is defined by the statement $b > a$ if and only if $b - a \in \mathscr{P}$. Then the notation $a < b$ (a "is less than" b) means $b > a$; $b \geq a$ means $b > a$ or $b = a$; $a \leq b$ means $a < b$ or $a = b$. The inequalities containing the alternative $=$ are sometimes called weak inequalities, and the others, strong inequalities. The usual properties of inequalities as they are used traditionally in connection with real numbers now follow from Definitions (i), (ii), and (iii). Some of the more important properties are given in the following exercises.

Exercises 1.3

Prove the following assertions about elements in a general ordered field.

1. The multiplicative neutral element 1 is positive. (Hence $1 + 1$, $1 + 1 + 1, \ldots,$ are positive, as was suggested by the name "positive integers.")
2. $a > 0$ if and only if a is positive; $a < 0$ if and only if a is negative.
3. $b > a$ if and only if $-b < -a$.
4. $a > b$ and $b > c$ imply that $a > c$.
5. If $a \geq b$, $b \geq c$, and $a = c$, then $a = b$.
6. If $a \geq 0$, $b \geq 0$, and $a + b = 0$, then $a = b = 0$.
7. If $a > b$, and $c > d$, then $a + c > b + d$.
8. If $a > b$, and $c > 0$, then $a \cdot c > b \cdot c$.

9. If $0 < a < b$, then $0 < b^{-1} < a^{-1}$.
10. $a_1^2 + a_2^2 + \cdots + a_n^2$ is always positive unless $a_1 = a_2 = \cdots = a_n = 0$.

1.4 Definitions Relating to Boundedness

(i) Let E be a subset of an ordered field $(\mathscr{F}, \mathscr{P})$. An *upper bound* for E is an element $b \in \mathscr{F}$ such that $a \leq b$ for every $a \in E$. A *lower bound* for E is an element $c \in \mathscr{F}$ such that $a \geq c$ for every $a \in E$. It may happen that E has no upper bounds or no lower bounds or no bounds of either type. But if E has an upper bound, then E is said to be *bounded above*, and if it has a lower bound, then E is said to be *bounded below*. If bounds of both types exist, the E is said to be *bounded*.

(ii) Let E be bounded above. If there exists an upper bound b_0 such that $b \geq b_0$ for every upper bound b, then b_0 is called the *least upper bound* of E. Similarly if E is bounded below, the *greatest lower bound*, if it exists, is a lower bound c_0 such that $c_0 \geq c$ for every lower bound c.

It is clear that the elements b_0 and c_0 as so defined are unique. They are often called the "supremum" and "infimum" of E, respectively, and this gives rise to the notation $b_0 = \sup E$ or $\sup\{x: x \in E\}$ and $c_0 = \inf E$ or $\inf\{x: x \in E\}$. Note that $\sup E$ and $\inf E$ may or may not be elements of E.

(iii) An ordered field is *complete* if every subset E with an upper bound has a least upper bound.

There are various equivalent formulations of completeness in field theory. One of them is suggested by the following assertions.

Proposition *Let $\mathscr{F}$ be an ordered field. If every set in $\mathscr{F}$ bounded above has a least upper bound, then every set bounded below has a greatest lower bound. Conversely, if every set bounded below has a greatest lower bound, then every set bounded above has a least upper bound.*

The proof, which we leave to the reader, can be given by using Exercise 1.3.3.

Exercises 1.4

1. Let E be a subset of an ordered field $\mathscr{F}$, and suppose that $\sup E = b_0$ exists. Prove that for every $a \in \mathscr{F}$ with $a < b_0$, there must exist at least one element $a' \in E$ such that $a < a' \leq b_0$. (A similar statement applies to $\inf E$.) Also prove that if E is bounded above and $b_1 \in \mathscr{F}$ is an upper bound for E with the property that for every $a \in \mathscr{F}$ with $a < b_1$, there exists $a' \in E$ with $a < a' \leq b_1$, then b_1 is the least upper bound of E. (There is a similar characterization of the greatest lower bound.)

2. Let E_1 and E_2 be subsets of an ordered field with $E_1 \subset E_2$. On the assumption that the indicated least upper bounds and greatest lower bounds exist, prove that $\sup E_1 \leq \sup E_2$ and $\inf E_1 \geq \inf E_2$. (*Hint*: Assume the contrary and use Exercise 1.4.1.)
3. Let E be a subset of a complete ordered field $\mathscr{F}$, and let a be an element of $\mathscr{F}$. Let $E_1 = \{y : y = x + a,\ x \in E\}$ and let $E_2 = \{y : y = a \cdot x,\ x \in E\}$. Prove that if $\sup E$ exists, then $\sup E_1 = a + \sup E$ and that if $a \geq 0$, then $\sup E_2 = a \sup E$.

1.5 The Real Number System

Classical mathematical analysis is based on the fact that the real number system **R** is a complete ordered field with respect to the usual arithmetic operations of addition and multiplication. It is known that this statement can be taken as a postulate which in itself totally describes what is commonly known as the real number system. (See [11, Chapter 5]). We assume that the reader is familiar with one of the constructive processes by which the real numbers can be introduced. Here we pause only briefly to develop some notation and to mention the Axiom of Archimedes.

The positive part of **R** of course consists of the set $\{x : x > 0\}$ (Exercise 1.3.2). It was brought out in Sec. 1.2 that an element $x \in \mathbf{R}$ is called a positive integer if it belongs to the set $\{1, 1 + 1, 1 + 1 + 1, \ldots\}$, where 1 is the neutral element for multiplication in **R**. An element in **R** is called a *whole number* or *nonnegative integer*, if it is a positive integer or zero. The *rational numbers* are the elements of **R** which are of the form m/n, where m and n are positive or negative integers, $n \neq 0$.

The completeness condition implies the Axiom of Archimedes: *Given any real number x, there is an integer n such that $x < n$.* This has the corollary that between any two distinct real numbers there lies a rational number: If $x < y$, there exists a rational n with $x < n < y$. For proofs, see [11, p. 177].

1.6 The Extended Real Numbers

In dealing with unbounded sets of real numbers, it is convenient to extend **R** by adjoining two elements ∞ and $-\infty$ called *plus infinity* and *minus infinity*. The enlarged set is denoted by $\hat{\mathbf{R}}$. We extend the definition of $<$ to the new system by postulating that $-\infty < x < \infty$ for every $x \in \mathbf{R}$.

We define, for all $x \in \mathbf{R}$,

$$x + \infty = \infty + x = \infty,$$
$$x - \infty = -\infty + x = -\infty,$$

$$x \cdot \infty = \infty \cdot x = \infty \text{ if } x > 0,$$
$$x \cdot (-\infty) = (-\infty) \cdot x = -\infty \text{ if } x > 0,$$
$$0 \cdot (\pm\infty) = (\pm\infty) \cdot 0 = 0;$$

and furthermore,

$$\infty + \infty = \infty,$$
$$-\infty - \infty = -\infty,$$
$$\infty \cdot (\pm\infty) = \pm\infty,$$
$$-\infty \cdot (\pm\infty) = \mp\infty.$$

The operations $\infty - \infty$ and $-\infty + \infty$ are not defined.

The extended real numbers $\hat{\mathbf{R}}$ provide an intuitively appealing extension of the definition of the greatest lower bound and least upper bound of a set E in $\mathbf{R}$. If E has no upper bound in $\mathbf{R}$, we write $\sup E = \infty$, with a similar interpretation for $\inf E = -\infty$. Thus, $\sup E$ is now defined for all nonempty sets E in $\mathbf{R}$, and it is the least number in $\hat{\mathbf{R}}$ which is greater than or equal to each element in E. A similar remark applies to $\inf E$.

Let α and β be elements of $\hat{\mathbf{R}}$ with $\alpha < \beta$. The set $\{x: \alpha < x < \beta\}$ is called an *open interval* in $\mathbf{R}$, and the abbreviated notation (α, β) is used for it. However, the reader is warned that a symbol such as (a, b) is also used on occasion to denote an ordered pair of elements of a given set. It should be clear from the context what is meant, even when a and b are real numbers. A *closed interval* in $\mathbf{R}$ is a set of this type: $\{x: \alpha \leq x \leq \beta\}$, where α and β are finite; the usual abbreviated notation is $[\alpha, \beta]$. "Half-open" intervals $[\alpha, \beta) = \{x: \alpha \leq x < \beta \leq \infty\}$ and $(\alpha, \beta] = \{x: -\infty < \alpha < x \leq \beta\}$ will also be used.

1.7 Functions

We interrupt the algebraic train of thought here to review the concept of function, which is fundamental to many parts of mathematics. Given two nonempty set X and Y, a *function* from X into Y is a set f of ordered pairs (x, y), with $x \in X$ and $y \in Y$, such that distinct ordered pairs belonging to it have distinct first components. In symbols, the statements $(x, y) \in f$, $(x, z) \in f$ imply $y = z$. Given $(x, y) \in f$, it is customary to denote y by $f(x)$; this is called the *value* of f at x.

The set $A \subset X$ of first components x such that $(x, y) \in f$ is called the *domain* of f, and the set B of values $f(x)$ is called the *range* or *image* of f. Given a function f with domain $A \subset X$ and range $B \subset Y$, we say that f maps A "into" Y and write $f: A \to Y$. If $B = Y$, we say that f maps A "onto" Y.

Given a function f and a subset C of the domain of f, the *restriction* of f to C, written $f|C$, is the set of ordered pairs belonging to f with first component in C. The *image* of C with respect to f, written $f(C)$, is the range of $f|C$. Given $f: A \to Y$, the *preimage* or *inverse image* of a set $D \subset Y$ with respect to f, written $f^{-1}(D)$, is the set $\{x: f(x) \in D\}$.

A function f with domain A and range B is called a *constant* function if B contains only one element, say b. In this case, we write $f(x) \equiv b$, $x \in A$.

Given a function f with domain A and range B, we say that the function is *univalent* (or the mapping is *one-to-one*) if the preimage of each point in B is only a single point of A. An equivalent formulation of univalence is that for $x_1, x_2 \in A$, the equation $f(x_1) = f(x_2)$ implies that $x_1 = x_2$. In the case of univalence, an *inverse function* f^{-1} exists, which is defined by $f^{-1}(y) = x$ if $y = f(x)$.

Now let X, Y, and Z be nonempty sets; let f be a function having domain $A \subset X$ and range $B \subset Y$; and let g be a function having domain $B \subset Y$ and range $C \subset Z$. The *composition* of g with f, often written $g \circ f$, is defined by the value equation $(g \circ f)(x) = g(f(x))$, $x \in A$. It gives a mapping from A onto C.

It is traditional to specify a function by giving its domain A and the value $f(x)$ at each point of A. Thus, it is within this tradition to use language such as "consider the function x^2 on the interval $[0, 1]$" or "let f be the constant 17 on the real axis." We often follow this tradition when no confusion can result. If $f(x) \equiv b$, $x \in A$, we often say that f is identically equal to b on A.

Many examples of functions appear in the sequel. Two examples are especially relevant to the algebraic content of this chapter, and it is worthwhile to point them out here.

In the first place, consider the field operations $+$ and $\cdot$ introduced in Sec. 1.2. Postulate F_1 implies that each of these operations is a function of which the domain is the set of all pairs a, b with $a \in \mathscr{F}$ and $b \in \mathscr{F}$ (this is the Cartesian product space $\mathscr{F} \times \mathscr{F}$) and the range is in $\mathscr{F}$. In fact, Postulate F_5 implies that the functions are onto $\mathscr{F}$.

The second type of function is known as an isomorphism between two fields. Let $\mathscr{F}_1$ and $\mathscr{F}_2$ be fields. An *isomorphism* between $\mathscr{F}_1$ and $\mathscr{F}_2$ is a univalent function f mapping $\mathscr{F}_1$ onto $\mathscr{F}_2$, which preserves the field operations; that is, if $a \in \mathscr{F}_1$, $b \in \mathscr{F}_1$, then $f(a + b) = f(a) + f(b)$ and $f(a \cdot b) = f(a) \cdot f(b)$. A fundamental theorem in field theory asserts that any two complete ordered fields are isomorphic [11, p. 178, Exercise 12]. It is customary to regard a pair of isomorphic fields as undistinguishable. Thus in this sense, the real number system is the unique complete ordered field.

We return to the definition of function given at the beginning of this section. Let f_1 and f_2 be functions from X into Y with the same domain A. Suppose further that Y is a field. Then $f_1 + f_2$, the sum of f_1 and f_2, is defined to be the function with domain A and with values given by $(f_1 + f_2)(x) = f_1(x) + f_2(x)$; $f_1 \cdot f_2$, the product of f_1 and f_2, is given by $(f_1 \cdot f_2)x =$

$f_1(x)f_2(x)$; and f_1/f_2, the quotient of f_1 and f_2, is the function with domain consisting of all $x \in A$ for which $f_2(x) \neq 0$, and with values given by $(f_1/f_2)x = f_1(x)/f_2(x)$.

If $\mathscr{A}$ is a field, a function $p: \mathscr{A} \to \mathscr{A}$ with values $p(x) = a_0 + a_1x + a_2x^2 + \cdots + a_nx^n$ is called a *polynomial function.* (It is implied that $a_0, a_1, \ldots, a_n$, and x are elements of $\mathscr{A}$.) If the coefficients $a_0, a_1, \ldots, a_n$ are each 0, p is called the *zero polynomial function.*

1.8 The Complex Number Field

It is well known that not every quadratic equation with real number coefficients can be solved in terms of real numbers; for example, $x^2 + 1 = 0$. The complex numbers historically were introduced as a suitable extension of the reals in which such equations would have solutions. Such an extension of **R** can be obtained by taking the linear vector space of two-component vectors with real components and endowing it with a suitable multiplication operation which causes it to be a field.

Definition. The system **C** of complex numbers is the set of all two-vectors (x, y) with real components, together with two operations, addition, $\oplus$, and multiplication, $\odot$. The operation $\oplus$ is the usual vector addition defined by $(x, y) \oplus (u, v) = (x + u, y + v)$; the operation $\odot$ is defined by $(x, y) \odot (u, v) = (x \cdot u - y \cdot v, x \cdot v + y \cdot u)$.

Theorem 1.8.1 *The complex number system* **C** *is a field. The additive neutral element is* $0 = (0, 0)$ *and the multiplicative neutral element is* $1 = (1, 0)$.

Proof. Closure under $\oplus$ and $\odot$ is obvious. The remaining field postulates are satisfied because **R** is a field with respect to + and $\cdot$. We leave most of the verification of this fact to the exercises. For the additive inverse of (x, y), we take $(-x, -y)$ and for the multiplicative inverse of (x, y), $(x, y) \neq 0$, we take

$$(x, y)^{-1} = \left(\frac{x}{x^2 + y^2}, \frac{-y}{x^2 + y^2}\right). \tag{1.8.1}$$

Exercises 1.8

1. Prove that **C** satisfies field postulates F_2, F_3, and F_4.
2. Prove that the complex number $(x, y)^{-1}$ given above in (1.8.1) satisfies the equation $(x, y) \odot (x, y)^{-1} = (1, 0)$.
3. Prove that

$$\frac{(u, v)}{(x, y)} = \left(\frac{ux + vy}{x^2 + y^2}, \frac{vx - yu}{x^2 + y^2}\right).$$

4. Let **K** be the set of all 2 × 2 matrices of the form

$$\begin{bmatrix} x & y \\ -y & x \end{bmatrix},$$

with x and y real. Let binary operations ⊕ and ⊙ be associated with this set, where ⊕ is the usual matrix addition and ⊙ is the usual matrix multiplication. Show that **K** is a field with respect to these two operations and is isomorphic to the complex field. (You can save yourself some work by first noting that there is a one-to-one correspondence between **K** and **C**, which preserves addition and multiplication; then refer to the Theorem 1.8.1.)

1.9 Relationships of C with $\mathbf{R}^2$ and R

By $\mathbf{R}^2$ we mean the linear vector space $\mathbf{R} \times \mathbf{R}$ of all ordered pairs (x, y) with $x, y \in \mathbf{R}$. The *set* **C**, considered as an unstructured set of elements, is identical with $\mathbf{R}^2$, again considered as an unstructured set. When we are emphasizing the complex field structure on this set, as defined in Sec. 1.8, we use the symbol **C**, and when we are emphasizing only the linear vector space structure, we use $\mathbf{R}^2$.

Definition. A *subfield* of a field $\mathscr{F}$ is a subset of $\mathscr{F}$, which is itself a field with respect to the operations of addition and multiplication of $\mathscr{F}$.

Theorem 1.9.1 *The field* **C** *contains a subfield which is isomorphic to* **R**.

Proof. Let $\mathbf{R}^* = \{(x, 0)\colon (x, 0) \in \mathbf{C}\}$. It is easy to verify that $\mathbf{R}^*$ is closed under the operations ⊕ and ⊙ of **C** and that with these operations, $\mathbf{R}^*$ satisfies the field postulates F_2, F_3, and F_4 of Sec. 1.2. The elements $(0, 0)$ and $(1, 0)$ satisfy Postulate F_5 for neutral elements of $\mathbf{R}^*$. There exists an additive inverse of $(a, 0)$ in $\mathbf{R}^*$, given by $(-a, 0)$, and a multiplicative inverse of $(a, 0)$, $a \neq 0$, given by $(1/a, 0)$. Thus, $\mathbf{R}^*$ is a subfield of **C**. To show that it is isomorphic to **R**, define the function $f\colon \mathbf{R} \to \mathbf{R}^*$ by $f(x) = (x, 0)$. The mapping given by f is one-to-one from **R** onto $\mathbf{R}^*$. Furthermore, $f(x + y) = (x + y, 0) = (x, 0) \oplus (y, 0)$ and $f(xy) = (xy, 0) = (x, 0) \odot (y, 0)$. Thus, f preserves the field operations. It follows from the definition of isomorphism that $\mathbf{R}^*$ together with ⊕ and ⊙ is isomorphic to the real number system **R**. Q.E.D.†

We henceforth identify this subfield with **R** and call a complex number of the form $(x, 0)$ a real number. We use the abbreviation x for $(x, 0)$. With these understandings, it no longer serves a useful purpose to employ the symbols ⊕ and ⊙ for the binary operations in the complex field to distinguish them from operations in the real field, so we henceforth use + instead of ⊕ and · (or simply juxtaposition) for ⊙.

†The letters Q.E.D. stand for the Latin words Quod erat demonstrandum (that which was was to be proved). Their presence is a classical indication that a proof is complete.

1.10 The Imaginary Unit

The element (0, 1) is denoted by i or by $\sqrt{-1}$ (or by j in engineering sciences). It has the property that $i^2 = (0, 1)(0, 1) = (-1, 0) = -1$. Thus, the equation $z^2 + 1 = 0$ has the solutions $\pm i$ in the complex field [Exercise 1.2.2].

It is to be noticed that for $z = (x, y)$, we have $z = (x, y) = (x, 0) + (0, y) = (x, 0) + (0, 1)(y, 0) = x + iy$. The right-hand member of this equation is the classical representation of a complex number. Algebraic manipulations are generally easier to handle with the classical notation, using real-number algebra and $i^2 = -1$, than with the vector notation, largely because the classical notation provides a simple way to remember the multiplication operation: $(x + iy)(u + iv) = xu + iyu + ixv + i^2yv = xu - yv + i(uy + xv)$.

1.11 Real and Imaginary Parts; Conjugation

Given $z = (x, y) = x + iy$, x and $y \in \mathbf{R}$, x is called the *real part* of z, written $\operatorname{Re} z$, and y is called the *imaginary part* of z, written $\operatorname{Im} z$. Traditionally, a complex number of the form $0 + iy$ is called a "pure imaginary." (Notice that $\operatorname{Im} z$ is a real number, not a pure imaginary.)

The conjugate complex number for $z = x + iy$ is $\bar{z} = x - iy$. (Sometimes z^* is used for $\bar{z}$.) The mapping given by conjugation is an isomorphism of $\mathbf{C}$ onto itself; in particular,

$$\overline{z + w} = \bar{z} + \bar{w}, \qquad \overline{zw} = \bar{z} \cdot \bar{w}. \tag{1.11.1}$$

We defer the verification to the exercises. Notice also that $\bar{\bar{z}} = z$.

Exercises 1.11

1. Prove that conjugation is an isomorphism of $\mathbf{C}$ onto itself.
2. Show that $\operatorname{Re} z = (z + \bar{z})/2 = \operatorname{Re} z$, $\operatorname{Im} z = (z - \bar{z})/2i = -\operatorname{Re}(iz)$.
3. Show that for any positive integer n, $\bar{z}^n = \overline{z^n}$.
4. With $z \neq 0$, show that $\overline{1/z} = 1/\bar{z}$. (*Hint*: Start with $zz^{-1} = 1$, take conjugates, use (1.11.1).) Also show that $\bar{w}/\bar{z} = \overline{w/z}$.
5. Show that $\operatorname{Re} z\bar{w} = \operatorname{Re} \bar{z}w$, $\operatorname{Im} z\bar{w} = -\operatorname{Im} \bar{z}w$.
6. Show that $z\bar{w}$ is real and ≥ 0 if and only if there exist real numbers $\lambda, \mu \geq 0$, but not both zero, such that $\lambda z = \mu w$. (*Hint*: If $w \neq 0$, $z\bar{w} = (z/w)w\bar{w}$.)
7. Let

$$z = \frac{(4 + 3i)(7 - 2i)}{i(3 - i)}$$

Find $\bar{z}$.

8. Reduce z in Exercise 1.11.7 to the form $x + iy$, $x, y \in \mathbf{R}$.
9. A *rational function* R is defined to be the quotient of two polynomial functions in which the denominator is not the zero polynomial function. Prove that if the polynomials in a rational function R have real coefficients, then $\overline{R(z)} = R(\bar{z})$.
10. Show that if $z + w$ and zw are both real, then either z and w are both real or else $z = \bar{w}$.
11. Let $w = u + iv = z^2$, u and v real. Find z explicitly in terms of u and v.

   Answer. If $v \neq 0$, then

$$z = \pm\left[\left(\frac{u + \sqrt{u^2 + v^2}}{2}\right)^{1/2} + i\frac{v}{|v|}\left(\frac{-u + \sqrt{u^2 + v^2}}{2}\right)^{1/2}\right]$$

   If $v = 0$, the solutions are $z = \pm\sqrt{u}$ if $u \geq 0$, or $z = \pm i\sqrt{-u}$ if $u < 0$. If particular, if $w = -1$, $z = \pm i$.

   Remark. Exercise 1.11.11 establishes the existence of two and only two distinct square roots of the complex number w for each $w \neq 0$.
12. Solve the quadratic equation $z^2 + bz + c = 0$ for z, where $b, c \in \mathbf{C}$ are given.

   Answer. $z = -b/2 \pm \zeta$, where ζ is a solution of $\zeta^2 = (b^2 - 4c)/4$ and is, therefore, given by the answer to Exercise 1.11.11 with $w = (b^2 - 4c)/4$.

   Remark. Exercise 1.11.12 shows that if $b^2 - 4c \neq 0$, there are two and only two distinct solutions of this quadratic equation, and if $b^2 - 4c = 0$, the only solution is $z = -b/2$.
13. Let f be a function which maps a set X, consisting of more than one element, into $\mathbf{C}$. For each $x \in X$, let $u(x) = \operatorname{Re} f(x)$, $v(x) = \operatorname{Im} f(x)$, so $f = u + iv$. Prove that if either u or v is known to be univalent, then f must be univalent. (*Hint*: If f were not univalent, then there would exist $x_1, x_2 \in X$, $x_1 \neq x_2$, such that $f(x_1) = f(x_2)$. What does this equation imply for u and v?)

1.12 Absolute Value

The *absolute value* or *modulus* of $z = x + iy$, written $|z|$, is defined by $|z| = \sqrt{z\bar{z}} = \sqrt{x^2 + y^2}$. The operation of forming the absolute value is a mapping of $\mathbf{C}$ onto the nonnegative reals which preserves multiplication; the proof will be an exercise.

Notice that $|z| = x$ if and only if $y = 0$ (which implies that z is real) and $x \geq 0$. Also $|z| = y$ if and only if $x = 0$ (z is a pure imaginary) and $y \geq 0$. When z is real, say $z = x + 0i$, the definition of absolute value in $\mathbf{C}$ coincides with the usual definition in $\mathbf{R}$: $|z| = \sqrt{x^2}$, where the nonnegative square root is implied. Finally, notice that $|\bar{z}| = \sqrt{\bar{z}\bar{\bar{z}}} = \sqrt{z\bar{z}} = |z|$.

Exercises 1.12

1. Find the absolute value of $(z - 3i)/(1 + 3iz)$ in terms of the real and imaginary

parts of z. (*Hint*: In finding absolute values of quotients, it is often helpful to multiply numerator and denominator by the conjugate of the denominator.)

2. Prove that $|zw| = |z|\,|w|$; and with $z \neq 0$, prove that $|z^{-1}| = 1/|z|$, and $|w/z| = |w|/|z|$. (*Hint*: $|zw|^2 = zw\ \bar{z}\bar{w}$.)
3. Show that $|z \pm w|^2 = |z|^2 + |w|^2 \pm 2\text{Re}\ z\bar{w}$.
4. Show that $|z + w|^2 + |z - w|^2 = 2(|z|^2 + |w|^2)$.
5. For $|z| = 1$, $|z_0| \neq 1$, show that $|z - z_0|/|1 - \bar{z}_0 z| = 1$. (*Hint*: Square and apply the hint in Exercise 1.12.2).
6. Prove Lagrange's Identity

$$\left|\sum_{j=0}^{n} z_j w_j\right|^2 = \sum_{j=0}^{n} |z_j|^2 \sum_{j=0}^{n} |w_j|^2 - \sum_{0 \le j < k \le n} \sum |z_j \bar{w}_k - z_k \bar{w}_j|^2.$$

(*Hint*: Start with the last summation).

1.13 Inequalities Involving Absolute Value

The most fundamental of the inequalities involving complex numbers stem from the vector space structure of **C** and are, therefore, familiar to most readers. We start with

$$-|z| \le \text{Re}\ z \le |z|, \qquad -|z| \le \text{Im}\ z < |z|. \tag{1.13.1}$$

These inequalities are apparent from the definition $|z| = \sqrt{x^2 + y^2}$, where $z = x + iy$. Next comes the triangle inequality,

$$|z + w| \le |z| + |w|, \tag{1.13.2}$$

and its companion,

$$|z - w| \ge |z| - |w|. \tag{1.13.3}$$

Proof. By Exercise 1.12.3, we have $|z \pm w|^2 = |z|^2 + |w|^2 \pm 2\text{Re}\ z\bar{w}$. Now according to (1.13.1) and Exercise 1.12.2, $\text{Re}\ z\bar{w} \le |z\bar{w}| = |z|\,|\bar{w}| = |z|\,|w|$. Thus, $|z + w|^2 \le |z|^2 + |w|^2 + 2|z|\,|w| = (|z| + |w|)^2$, and $|z - w|^2 \ge |z|^2 + |w|^2 - 2|z|\,|w| = (|z| - |w|)^2$. Q.E.D.

The special case of equality in (1.13.2) is worth examining. Clearly from the proof a necessary and sufficient condition for equality is that $\text{Re}\ z\bar{w} = |z\bar{w}|$, so by a remark at the beginning of Sec. 1.12, $z\bar{w}$ is real and ≥ 0. Then from Exercise 1.11.6, we find that a necessary and sufficient condition is that there exist real numbers $\lambda \ge 0$, $\mu \ge 0$, but not both zero, such that $\lambda z = \mu w$.

Some immediate consequences of (1.13.2) and (1.13.3) are worth stating:

$$|z - w| \le |z| + |w|, \qquad |z \pm w| \ge \big||z| - |w|\big|. \tag{1.13.4}$$

$$\left|\sum_{j=0}^{n} z_j\right| \le \sum_{j=0}^{n} |z_j|, \qquad z_0, z_1, \ldots, z_n \in \mathbf{C}. \tag{1.13.5}$$

$$|x + iy| \le |x| + |y|, \qquad x, y \in \mathbf{R}. \tag{1.13.6}$$

We now come to the Cauchy Inequality:

$$\left|\sum_{j=0}^{n} z_j w_j\right|^2 \le \sum_{j=0}^{n} |z_j|^2 \sum_{j=0}^{n} |w_j|^2, \tag{1.13.7}$$

where $z_0, z_1, \ldots, z_n, w_0, w_1, \ldots, w_n$ are complex numbers. The proof is immediate from Lagrange's Identity, Exercise 1.12.6. A necessary and sufficient condition for equality is clearly that $z_j\bar{w}_k = z_k\bar{w}_j$ for all j, $k = 0, 1, 2, \ldots, n$ with $j < k$. An equivalent condition involving fewer equations is developed in Exercise 1.13.5.

The Cauchy Inequality is true for reals, and this fact together with (1.13.5) leads to the continued inequality:

$$\left|\sum_{j=0}^{n} z_j w_j\right|^2 \le \left(\sum_{j=0}^{n} |z_j w_j|\right)^2 \le \sum_{j=0}^{n} |z_j|^2 \sum_{j=0}^{n} |w_j|^2. \tag{1.13.8}$$

The following extension of the triangle inequality is sometimes useful:

$$|z + w|^p \le (|z| + |w|)^p \le \begin{cases} 2^{p-1}(|z|^p + |w|^p), & p > 1; \\ |z|^p + |w|^p, & 0 < p \le 1. \end{cases} \tag{1.13.9}$$

The first inequality is an obvious consequence of the triangle inequality and of the fact (borrowed from real number theory) that if $\alpha \ge 0$ and $\beta \ge 0$ satisfy $\alpha \le \beta$, then for any $p > 0$, $\alpha^p \le \beta^p$. To establish the second inequality in the case $p > 1$, notice that the graph of the equation $y = x^p$, $p > 1$, is concave upward for $x \ge 0$, since the second derivative of x^p is positive for $x > 0$. Therefore, the ordinate of a point on the curve with abscissa $x = (\alpha + \beta)/2$ is less than the average of the ordinates at $x = \alpha$ and $x = \beta$, $0 \le \alpha < \beta$. This means that $[(\alpha + \beta)/2]^p < (\alpha^p + \beta^p)/2$, or $(\alpha + \beta)^p < 2^{p-1}(\alpha^p + \beta^p)$. The case $0 < p < 1$ is an exercise.

Notice that (1.13.9) with $p \ge 1$ is an equality when $z = w = 0$ and when $z = w = 1$.

Exercises 1.13

1. With $z = x + iy$, prove that $|z| \le |x| + |y| \le \sqrt{2}|z|$.
2. Prove the following special case of the so-called Minkowski Inequality:

$$\left(\sum_{j=0}^{n} |z_j + w_j|^2\right)^{1/2} \le \left(\sum_{j=0}^{n} |z_j|^2\right)^{1/2} + \left(\sum_{j=0}^{n} |w_j|^2\right)^{1/2}.$$

(*Hint*: Use the identity $(z_j + w_j)\overline{(z_j + w_j)} = z_j\overline{(z_j + w_j)} + w_j\overline{(z_j + w_j)}$, $j = 0, \ldots, n$; apply the triangle inequality; then sum on j and use the Cauchy Inequality.)

3. Show that the function $w = (z - z_0)/(1 - \bar{z}_0 z)$, with $|z_0| < 1$, gives a one-to-one mapping of $\{z : |z| < 1\}$ onto $\{w : |w| < 1\}$ and of $\{z : |z| = 1\}$ onto $\{w : |w| = 1\}$. (*Hint*: By solving for z, first show that an inverse function exists and is of the same

general form as the orignal function. Then consider the value of $|w|^2$ for $|z| < 1$. The case $|z| = 1$ was covered in Exercise 1.12.15.)

4. Prove (1.13.9) with $0 < p < 1$. (*Hint*: First show that for all $x > 0$, $y = 1 + x^p - (1 + x)^p > 0$, by looking at the derivative.)
5. In considering the Cauchy Inequality, clearly we may discard any pair z_j, w_j, both of which are zero. Suppose then that there are no zero pairs. Show that a necessary and sufficient condition for equality in the Cauchy Inequality is that either (i) all z_j's are zero or all w_j's are zeros or (ii) none of these numbers is a zero and there exists a complex constant $c \neq 0$ such that $z_j = c\bar{w}_j$, $j = 0, \ldots, n$. (*Hint for (ii)*: for the necessity, only the equations $z_0\bar{w}_k = z_k\bar{w}_0$, $k = 1, \ldots, n$ need to be considered.)
6. Prove that for k a positive integer, $|\text{Im } z^k| \leq k|\text{Im } z|\,|z|^{k-1}$. Can you replace Im by Re? (*Hint*: If $\text{Im } z \neq 0$, $|\text{Im } z^k/\text{Im } z| = |(z^k - \bar{z}^k)/2i|/|(z - \bar{z})/2i|$; carry out the division.)
7. Students of complex analysis have been known to accept the following false assertion. Given $z_0, z_1, \ldots, z_n$ and $w_0, w_1, \ldots, w_n$, all in **C**, with $|z_j| < |w_j|$, $j = 0, \ldots, n$; also given $a_0, a_1, \ldots, a_n$ in **C** with $|a_j| < 1$, $j = 0, \ldots, n$; then $|\sum_{j=0}^n a_j z_j| \leq |\sum_{j=0}^n w_j|$. Find a counterexample.
8. Prove that if $z_0, z_1, \ldots, z_n \in \mathbf{C}$ are such that $|z_j| < 1$, $j = 0, \ldots, n$, and $\lambda_0, \lambda_1, \ldots, \lambda_n \in \mathbf{R}$ satisfy the condition $\lambda_j \geq 0$, $j = 0, \ldots, n$, and $\sum_0^n \lambda_j = 1$, then $|\sum_0^n \lambda_j z_j| < 1$.

1.14 Geometric Representation of Complex Numbers

The vector aspect of the complex number $z = (x, y) = x + iy$ leads naturally to the geometric representation on a two-dimensional plane, which is doubtless familiar to the reader as the "picture" of $\mathbf{R}^2$. The picture is obtained by introducing a rectangular coordinate system on the plane and plotting the point $z = (x, y)$ with x as abscissa and y as ordinate. A geometric picture of **C** is thereby obtained, which is called the *complex plane* (or *Gaussian plane*, or *Argand diagram*). The x-axis is called the *real axis*, and the y-axis, the *imaginary axis*.

This book is intended to be a rigorous introduction to complex analysis, and with this in mind, we take the point of view expressed by Ahlfors in [2]: "All conclusions in analysis should be derived from properties of the real numbers and not from the axioms of geometry." However, the geometric picture of **C** is a most valuable visual aid for understanding relations in the complex field and, in particular, mappings. Here we give only a brief introduction to the geometry of **C** in which all the descriptions are derived only from elementary analytic geometry. Even the polar representation of a point in the plane is deferred to Chapter 7, in which an analytic concept of angle is developed.

The addition operation in the field **C** is, of course, vector addition. As described in the elementary texts on vectors and matrices, the point $z_3 = z_1 + z_2$ is found by "completing the parallelogram," as shown in Fig. 1. The difference $z_3 - z_1$ is the located vector represented by the dashed line with initial point z_1 and end point at z_3. Multiplication is most conveniently described in terms of polar coordinates, which we postpone. (For a construction, see [9, p. 22].)

The Euclidean distance between $z_1 = (x_1, y_1)$ and $z_2 = (x_2, y_2)$ in $\mathbf{R}^2$ is given by $\sqrt{(x_1 - x_2)^2 + (y_1 - y_2)^2}$, which is the absolute value $|z_1 - z_2|$. In particular, $|z|$ is the distance between the point z and the origin.

The triangle inequality $|z_1 + z_2| \leq |z_1| + |z_2|$ expresses the fact that the length of the side $[0, z_3]$ of the triangle with vertices $0, z_1, z_3$ in Fig. 1 is less than the sum of the lengths of the sides $[0, z_1]$ and $[z_1, z_3]$. (The latter has the same length as $[0, z_2]$, by construction.) The identity in Exercise 1.12.4 becomes a standard theorem in geometry when interpreted geometrically.

The point $\bar{z}$ is the reflection of z in the real axis. The point $-z$ is the reflection of z in the origin. The points $z, \bar{z}, -z, -\bar{z}$ are the vertices of a rectangle with sides parallel to the coordinate axes and center at 0.

From time to time, it is convenient to use the expression "graph of a function" or "graph of an equation" when the equation defines a function. In the case of functions from $\mathbf{R}$ into $\mathbf{R}$, the traditional meaning of such an expression is the graph of the set of ordered pairs which constitute the function, plotted on the geometric picture of $\mathbf{R}^2$. However, in the case of functions from $\mathbf{R}$ into $\mathbf{C}$ or from $\mathbf{C}$ into $\mathbf{C}$, the expression is used to mean the range of the function, plotted as a geometric configuration on the picture of $\mathbf{R}^2$. Thus, the graph of the function given by $y = x^2$, $x, y \in \mathbf{R}$, is the familiar parabola going through the origin of the (x, y)-plane, but the graph of the

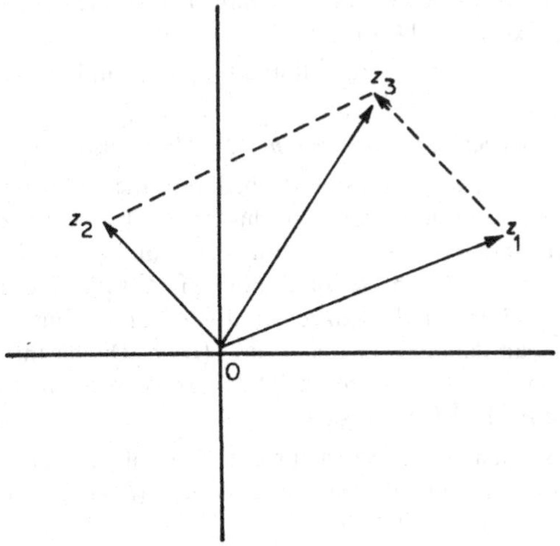

Fig. 1

function given by $w = z^2$, say with z restricted to some set $E \subset \mathbf{C}$, means the image of E under this function, regarded as a point set in $\mathbf{R}^2$.

Exercises 1.14

1. Identify the sets of points which satisfy the following relations:
   (a) $z^2 = 2(z - 1)$,
   (b) $\text{Im } z > k$, k real,
   (c) $0 < \text{Re } z < k$, k real.
2. In vector analytic geometry, vectors (x_1, y_1) and (x_2, y_2) are said to be perpendicular if their *inner product*, defined by $x_1x_2 + y_1y_2$, is zero. (Inner product is also called *scalar product* and *dot product*.) Show that the condition for vectors z_1 and z_2 to be perpendicular is that $\text{Re } z_1\bar{z}_2 = 0$.
3. The straight line through a given point z_0 and perpendicular to a given vector $\zeta \neq 0$ may be defined to be the set of all points z such that the vector $z - z_0$ is perpendicular to ζ. Show that the corresponding equation is of the form $\zeta\bar{z} + \bar{\zeta}z - (z_0\bar{\zeta} + \bar{z}_0\zeta) = 0$ and that this is equivalent to $\text{Re } \zeta\bar{z} - \text{Re } \zeta\bar{z}_0 = 0$. Also show that the set of all points satisfying any equation of the form $a\bar{z} + \bar{a}z + k = 0$, with a complex, but not zero, and k real, forms a straight line (perpendicular to the vector a) in the sense of this definition.
4. Let p be a complex number not zero. Show that according to the definition in Exercise 1.14.3, the graph of the "parametric equation" $z = z_1 + pt$, where t is real, $-\infty < t < \infty$, is a straight line through z_1 perpendicular to the vector ip. (*Hint*: Write $z - z_1 = pt$, $\bar{z} - \bar{z}_1 = \bar{p}t$ and eliminate t by division.)
5. Show using Exercise 1.14.4 that the graph of the equation $z = z_1 + t(z_2 - z_1)$, $-\infty < t < \infty$, $z_1 \neq z_2$, is a straight line through the points z_1 and z_2.
6. Let A, B, C, x, and y all be real number. Show that the equation $Ax + By + C = 0$ represents a straight line, perpendicular to the vector $A + iB$, with "straight line" defined as in Exercise 1.14.3. (*Hint*: Let $x = (z + \bar{z})/2$, $y = (z - \bar{z})/2i$; substitute; use Exercise 1.14.3.)
7. Show that $|z_1 \pm z_2|^2 = |z_1|^2 + |z_2|^2$ if and only if z_1 and z_2 are perpendicular.
8. Show algebraically that $|z_1 - z_3|^p \leq |z_1 - z_2|^p + |z_2 - z_3|^p$, $0 < p \leq 1$, and interpret this geometrically for the case $p = 1$. (*Hint*: use (1.13.9).)
9. A circle with center z_0 and radius r is defined to be the set of all points at distance r from z_0. Thus, the basic equation for this circle is $|z - z_0| = r$. Show that if A and C are real and b is complex, with $A \neq 0$ and $\rho^2 = (|b|/A)^2 - (C/A) > 0$ (equivalent to $|b|^2 > AC$), the set of all points in $\mathbf{C}$ satisfying the equation $Az\bar{z} + b\bar{z} + \bar{b}z + C = 0$ is a circle with center at $-b/A$ and radius ρ; and further that every circle has an equation of this form. (*Hint*: Divide through by A. The equation in Exercise 1.12.3 is then helpful; transpose the first two terms in the right-hand side to the left-hand side.)
10. Show that the set of all points satisfying the equation $|z - z_1| = \alpha|z - z_2|$, α real, $\alpha \neq 0, 1$, is a circle, and find its center and radius. (*Hint*: Put the equation in the form displayed in Exercise 1.14.9.)

11. Let z_1, z_2, z_3 be distinct complex numbers such that $|z_1| = |z_2| = |z_3| = 1$ and $z_1 + z_2 + z_3 = 0$. Show that z_1, z_2, z_3 are the vertices of an equilateral triangle inscribed in the circle with center at origin and radius 1. (*Hint*: To show that, say, $|z_1 - z_2|^2 = |z_2 - z_3|^2$, use the equation $z_1 + z_2 + z_3 = 0$ to replace z_2, which appears on both sides.)
12. Let $z_1 = x_1 + iy_1$ and $z_2 = x_2 + iy_2$ be distinct points of the complex plane, and to fix ideas assume that $x_2 \geq x_1$ and $y_2 \geq y_1$. Show that the set of all points z satisfying the relation

$$\operatorname{Im}\frac{z - z_1}{z_2 - z_3} > 0 \tag{1.14.1}$$

is (a) the half-plane lying to the left of the line with equation $x = x_1$ if $x_2 = x_1$;
(b) the half-plane lying above and to the left of the line with equation $y = y + [(y_2 - y_1)/(x_2 - x_1)](x - x_1)$ if $x_2 > x_1$, $y_2 > y_1$;
and (c) the half-plane lying above the line with equation $y = y_1$ if $y_2 = y_1$. (*Note*: in each of cases (a) and (b), the locus is called the *left half-plane* for the given line, and in case (c) it is called the *upper half-plane* for the given line. The *right half-planes* and *lower half-plane* for the given lines are represented by the relation (1.14.1) with the inequality reversed.)
13. Identify the point sets in **C** satisfying the following relations:
(a) $|z - 1| = |z - i|$,
(*b*) $|z - 1| \geq 2$,
(c) $\operatorname{Re}(z - 2) = |z|$.
14. It is shown in analytic geometry textbooks that the equation in $\mathbf{R}^2$ of an ellipse with center at the origin and major axis along the x-axis of length 2α and minor axis along the y-axis of length 2β is $(x/\alpha)^2 + (y/\beta)^2 = 1$. With $\alpha > 0$, $\beta > 0$, identify the set of all points $z = (x, y)$ satisfying $(\beta^2 - \alpha^2)\operatorname{Re}(z^2) + (\beta^2 + \alpha^2)|z|^2 = 2\alpha^2\beta^2$.
15. Identify the point set satisfying the relation $|z - 1| + |z + 1| \leq 3$.
16. Show that if $R > 1$, the graph of the function defined by $w = (1/2)(Rz + 1/Rz)$, $z \in \mathbf{C}$ with $|z| = 1$, is an ellipse E with major axis of length $R + (1/R)$ and minor axis of length $R - (1/R)$, and that the mapping is one-to-one. Also, show that if $R = 1$, the graph is the interval $[-1, 1]$. (*Hint*: Let $w = u + iv$, $z = x + iy$, and take real and imaginary parts in the equation. Remember that when $|z| = 1$, $x^2 + y^2 = 1$. Be sure to show that the mapping from the unit circle is *onto* the ellipse.)
17. (Continuation of 16). Given an ellipse with the equation displayed in Exercise 1.14.14; the set $\{(x, y): (x/\alpha)^2 + (y/\beta)^2 > 1\}$ is the subset of $\mathbf{R}^2$ which is intuitively "outside" the ellipse, and the points satisfying the opposite inequality are the subset "inside" the ellipse. In the special case in which $\alpha = \beta > 0$, these inequalities define the outside and inside of a circle. Show that the function (*) $w = (1/2)(Rz + 1/Rz)$ with $R > 1$ gives a one-to-one mapping of $D_1 = \{z: |z| > 1\}$ onto the outside of the ellipse E of Exercise 1.14.16, and also gives a one-to-one mapping of $D_2 = \{z: 0 < |z| < 1/R^2\}$ onto the outside of E. (*Hint*: Let $w = u$

$+ iv$, $z = x + iy$, $r^2 = x^2 + y^2$. Show that if z, w satisfy (*), then $x = u/[(1/2)(R + 1/Rr^2)]$, $y = v/[(1/2)(R - 1/Rr^2)]$. Then establish the equation

$$1 = \frac{u^2}{(1/4)(Rr + 1/Rr)^2} + \frac{v^2}{(1/4)(Rr - 1/Rr)^2}. \tag{1.14.2}$$

Then show that if $r > 1$, or $r < 1/R^2$,

$$\frac{u^2}{(1/4)(R + 1/R)^2} + \frac{v^2}{(1/4)(R - 1/R)^2} > 1. \tag{1.14.3}$$

You can do this by examining denominators in (1.14.2) and (1.14.3). This proves that the image of any point in D_1 or D_2 lies outside the ellipse E. Next show that (*) is equivalent to the quadratic equation $z^2 - 2wzR^{-1} + R^{-2} = 0$, and so by Exercise 1.11.12, for any given value of w, there are exactly two solutions z_1 and z_2, which might coincide. Then by substitution in (*) show that if z_1 is a solution, the other solution must be $z_2 = 1/R^2 z_1$. Finally, take $w_0 = u_0 + iv_0$ satisfying (1.14.3) and show that (1.14.2) becomes an inequality like (1.14.3), if it is assumed that $(1/R^2) \leq r \leq 1$, so that it must be true that the two preimages of w_0 lie, respectively, in D_1 and D_2.)

18. The points $w = (u, v)$, u, v real, satisfying an equation of the form $v^2 = 4k(u - c)$, $k > 0$, c real, fulfill the geometric conditions which define a parabola. The set $\{(u, v): u > c, v^2 < 4k(u - c)\}$ constitutes what would be intuitively thought of as the "inside" of the parabola. Show that the mapping given by $w = u + iv = -z^2$ maps the line $\operatorname{Re} z = p > 0$ in a one-to-one manner onto the parabola with equation $v^2 = 4p^2(u + p^2)$, and also maps the strip $\{z: 0 \leq \operatorname{Re} z < p, -\infty < \operatorname{Im} z < \infty\}$ onto the inside of the parabola with the imaginary axis in the z-plane going onto the nonnegative real axis in the w-plane, taken twice, and the interval $\{z: 0 \leq z < p\}$ going onto the interval $\{w: -p^2 < u \leq 0\}$. The mapping of the strip $\{z: 0 < \operatorname{Re} z < p\}$ is one-to-one onto the inside of the parabola with the nonnegative real axis deleted. (*Hint*: For the onto and one-to-one parts of the problem, Exercise 1.11.11 can be used. Since here we have $z^2 = -w$, the u and v in the answer to that exercise must be replaced by $-u$ and $-v$. Only the real part of that answer is used here. In locating the abscissa of a point in the z-plane corresponding to a point (u, v) with $u > -p^2$, $v^2 < 4p^2(u_0 + p^2)$ (inequalities which put the point (u, v) inside the parabola), only the second of these inequalities is needed.)

19. The points $z = x + iy$ which satisfy the equation $\operatorname{Re} z^2 = x^2 - y^2 = p$, $p > 0$, form a rectangular hyperbola. We define the "inside" of each branch by the inequality $y^2 < x^2 - p$. Show that the function given by $w = z^2$ maps the inside of either branch of this hyperbola in a one-to-one manner onto the half-plane $\operatorname{Re} w > p$.

1.15 The Extended Complex Plane and Stereographic Projection

In Sec. 1.6, we extended the real number system by adjoining two elements ∞ and $-\infty$. A motivation for this is that it simplifies certain statements involving unbounded sets of real numbers. Similar considerations lead to an extension of the complex number system **C** by adjoining to it a single ideal element called the *point at infinity*, and written ∞. The notation $\hat{\mathbf{C}}$ is used to designate $\mathbf{C} \cup \{\infty\}$; $\hat{\mathbf{C}}$ is called the *extended complex plane*.

The new element ∞ is, of course, not to be thought of as a member of the complex number field as defined in Sec. 1.8, but certain algebraic connections with numbers in **C** can be set up without danger of running into inconsistencies. These are as follows. For any $z \in \mathbf{C}$, $z \pm \infty = \infty \pm z = \infty$ and $z/\infty = 0$. For $z \in \hat{\mathbf{C}}$, $z \neq 0$, $z \cdot \infty = \infty \cdot z$ and $z/0 = \infty$. We do not define $\infty \pm \infty$, $-\infty + \infty$, $0 \cdot \infty$, and $\infty \cdot 0$.

An example of the usefulness of this extension of **C** is given by the transformation $w = 1/z$. If w and z are restricted to **C**, the transformation is generally one-to-one from **C** onto **C** with the unfortunate exceptions that $z = 0$ has no image and $w = 0$ has no preimage. However, for $w, z \in \hat{\mathbf{C}}$, these exceptions are removed: $z = 0$ goes into $w = \infty$ and $w = 0$ goes into $z = \infty$. This example is generalized in Sec. 1.16.

There is an illuminating geometric model for $\hat{\mathbf{C}}$ in which ∞ has a concrete representation. The details of the model can be arranged in various ways, of which we choose the following one. Consider the sphere S in Euclidean 3-space $\mathbf{R}^3$ given by the equation $x_1^2 + x_2^2 + x_3^2 = x_3$. (See Fig. 2.) The center of S is at $(0, 0, 1/2)$ and the radius is $1/2$. The line segment joining

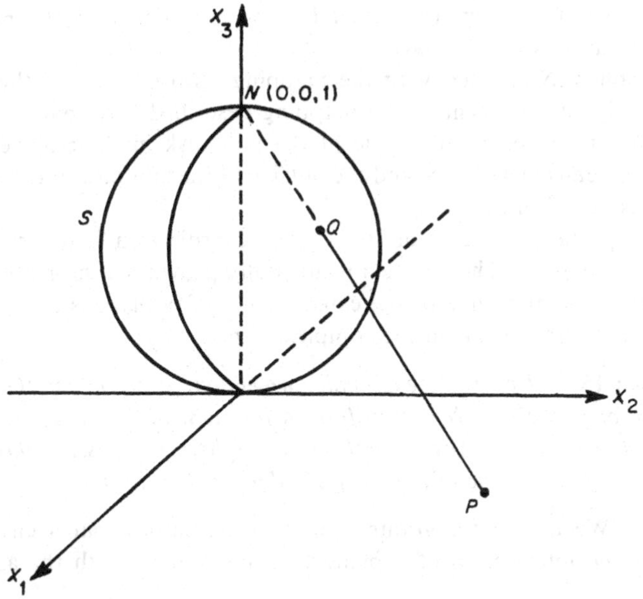

Fig. 2

the "north pole" $N(0, 0, 1)$ to a point $P(x, y, 0)$ in the (x_1, x_2)-plane has parametric equations $x_1 = tx$, $x_2 = ty$, $x_3 = 1 - t$, $0 \leq t \leq 1$. It is not difficult to show that this line segment intersects S at the point Q with coordinates

$$x_1 = \frac{x}{x^2 + y^2 + 1}, \qquad x_2 = \frac{y}{x^2 + y^2 + 1}, \qquad x_3 = \frac{x^2 + y^2}{x^2 + y^2 + 1}, \quad (1.15.1)$$

and the points N and Q are the only points of intersection with S. Conversely, given any point $Q(x_1, x_2, x_3) \in S$ other than N, there is a unique point P at which the line through N and Q intersects the (x_1, x_2)-plane. The coordinates of P in terms of those of Q are

$$\left(\frac{x_1}{1 - x_3}, \frac{x_2}{1 - x_3}, 0\right). \quad (1.15.2)$$

The verifications are deferred to the exercises.

We now identify the (x_1, x_2)-plane with the complex z-plane, $z = x + iy$, oriented so that the real axis lies on the x_1-axis, and the imaginary axis, on the x_2-axis. With this identification, the one-to-one correspondence given by (1.15.1) and (1.15.2) is called a *stereographic projection* of the z-plane onto S. The point $N(0, 0, 1) \in S$ has no corresponding point in the complex plane under this transformation. In view of the mechanism of stereographic projection, it is intuitively appealing to assign the point at infinity in the extended complex plane to N. With this assignment, we have completed the promised construction of a geometric model for the extended complex plane $\hat{\mathbf{C}}$ in which ∞ has a "visible" image.

The sphere S together with the mapping onto $\hat{\mathbf{C}}$ is called the *Riemann sphere*. The arrangement of the mapping described here sends $z = 0$ onto the "south pole" of S, the inside of the unit disk in the z-plane onto the "southern hemisphere" of S, and the outside of the unit disk onto the "northern hemisphere" of S.

An important property of stereographic projection is described in the following theorem. The theorem incidentally provides a motivation for this definition: A straight line of the extended complex plane is a set $L \cup \{\infty\}$, where L is a straight line in the complex plane.

Theorem 1.15.1 *Under stereographic projection, a circle on the Riemann sphere through the pole N is transformed into a straight line of the extended plane and vice versa. Also a circle on the sphere not passing through N is transformed into a circle of the complex plane and vice versa.*

Proof. We take it for granted (or as a definition) that a circle on the sphere is the intersection of a plane with the sphere, with the assumption that the plane intersects the sphere in more than one point. Now assume that the plane with the equation

$$Ax_1 + Bx_2 + Cx_3 = D, \qquad A, B, C, D \in R, \tag{1.15.3}$$

intersects S in a circle Γ on S. We substitute into (1.15.3) with (1.15.1), and after clearing of fractions, we obtain the equation

$$(C - D)(x^2 + y^2) + Ax + By = D. \tag{1.15.4}$$

This is the equation of a certain locus in the z-plane ($z = x + iy$). The locus is a straight line if $C = D$, a circle if $C \neq D$ and $A^2 + B^2 > 4D(C - D)$, or a single point or empty. The latter two cases are ruled out by the assumption that the preimage Γ is nonempty. In the straight line case, Equation (1.15.3) must be of the form $Ax_1 + Bx_2 + C(x_3 - 1) = 0$. The plane represented by this equation passes through $N(0, 0, 1)$, which is mapped into $\{\infty\}$ in $\hat{\mathbf{C}}$, and by definition ∞ lies on any straight line in $\hat{\mathbf{C}}$.

So far, we have shown only that Γ is mapped onto a subset of the locus represented by (1.15.4). Now with the same A, B, C, D, we take *any* point $z = x + iy$ which satisfies (1.15.4). The steps whereby we pass from (1.15.3) to (1.15.4) are reversible, and we find that the image of z under stereographic projection satisfies (1.15.3), with $C = D$ in the straight line case. This shows that no points of the locus represented by (1.15.4) are omitted in the mapping of Γ. The mapping is one-to-one with N going into $\{\infty\}$ in the straight line case.

Conversely, any given straight line in the z-plane, $z = x + iy$, can be represented by an equation of the form $\alpha x + \beta y = \delta$ and any circle by an equation of the form $\gamma(x^2 + y^2) + \alpha x + \beta y = \delta$; here α, β, γ, δ are real numbers. Each of these equations can be thrown into the form (1.15.4) by letting $A = \alpha$, $B = \beta$, $D = \delta$, and $C = \gamma + \delta$ ($\gamma = 0$ in the straight line case). By reversing the steps which led from (1.15.3) to (1.15.4), we find that the stereographic projection of the given line or circle is at least a subset of a circle Γ on S which in the straight line case passes through N. Then by taking any point on this Γ and using (1.15.1) we find that the corresponding point in the z-plane lies on the given line or circle, with the line viewed as a straight line in $\hat{\mathbf{C}}$. The mapping of the line or circle in $\hat{\mathbf{C}}$ is again one-to-one and onto Γ. Q.E.D.

Since both straight lines and circles in $\hat{\mathbf{C}}$ become circles in the geometric model for $\hat{\mathbf{C}}$, it is convenient in certain discussions relating to geometry in $\hat{\mathbf{C}}$ not to distinguish between lines and circles.

The *chordal distance* between two points $z, z' \in \hat{\mathbf{C}}$ is defined to be the distance in $\mathbf{R}^3 = \mathbf{R} \times \mathbf{R} \times \mathbf{R}$ of the images of z and z' on the Riemann sphere, expressed in terms of z and z'. The formulas are:

$$\rho(z, z') = \frac{|z - z'|}{\sqrt{(1 + |z|^2)(1 + |z'|^2)}}, \qquad z, z' \in \mathbf{C}, \tag{1.15.5}$$

$$\rho(z, \infty) = \frac{1}{\sqrt{1 + |z|^2}}, \qquad z \in \mathbf{C}, \qquad \text{and } \rho(\infty, \infty) = 0. \tag{1.15.6}$$

(See Exercise 1.15.3.)

Exercises 1.15

1. Describe the relative positions of the projections of z, $\bar{z}, -z$, and $1/z$ on the Riemann sphere.
2. Verify (1.15.1) and (1.15.2). (*Hint*: Find the values of t for which the line segment with parametric equations $x_1 = tx$, $x_2 = ty$, $x_3 = 1 - t$, $0 \le t \le 1$, satisfies the equation $x_1^2 + x_2^2 + x_3^2 = x_3$.)
3. Derive the chordal distance formulas (1.15.5) and (1.15.6). (*Hint*: Let $z = x + iy$, $z' = x' + iy'$. The formulas (1.15.1) for the projection of z can be written in the form $x_1 = x/(1 + |z|^2)$, $x_2 = y/(1 + |z|^2)$, $x_3 = |z|^2/(1 + |z|^2)$, with similar formulas for the projection (x_1', x_2', x_3') of z'. Use the equation of S to show that $(x_1 - x_1')^2 + (x_2 - x_2')^2 + (x_3 - x_3')^2 = x_3 + x_3' - 2(x_1x_1' + x_2x_2' + x_3x_3')$. Recall from Exercise 1.12.3 that $-2(xx' + yy') = |z - z'|^2 - |z|^2 - |z'|^2$.)

1.16 The Linear Fractional Transformations

We conclude this chapter with a brief introduction to the theory of an important type of mapping known as a *linear fractional transformation*, or *Möbius transformation*. Continuity of thought with the preceding section is provided by the fact, established in this section, that such a transformation maps the family of straight lines and circles in the extended plane onto itself.

A linear fractional transformation is a function from $\hat{\mathbf{C}}$ onto $\hat{\mathbf{C}}$ with values given by an equation of the form $w = (az + b)/(cz + d)$, in which a, b, c, d are complex numbers for which the determinant

$$D = \begin{vmatrix} a & b \\ c & d \end{vmatrix} = ad - bc$$

is not zero. (This condition implies that c and d cannot both be zero.) The point $w = \infty$ corresponds to the point $z = -d/c$, and $w = a/c$ is assigned to $z = \infty$. The reason for the restriction $D \neq 0$ is that otherwise the mapping reduces to that given by a constant function. (See Exercise 1.16.1.)

An example of a linear fractional transformation is given by $w = 1/z$, considered briefly in Section 1.15. Another example of a linear fractional transformation appears in Exercise 1.13.3. The coefficients are $a = 1$, $b = -z_0$, $c = -\bar{z}_0$, $d = 1$ with $|z_0| < 1$. Note that if $|z_0| = 1$, the condition

$ad - bc \neq 0$ is violated. The transformation gives a one-to-one mapping of the disk $\{z: |z| \leq 1\}$ onto itself, with the boundary $\{z: |z| = 1\}$ mapped onto itself.

Proposition 1.16.1

(i) *A linear fractional transformation is one-to-one from* $\hat{\mathbf{C}}$ *onto* $\hat{\mathbf{C}}$, *and the inverse is a linear fractional transformation.*
(ii) *If* f_1 *and* f_2 *are linear fractional transformations, then so is their composition,* $f_1 \circ f_2$.

Proof. For part (i), we have merely to solve for z in the equation $w = (az + b)/(cz + d)$, obtaining $z = (dw - b)/(-cw + a)$. The coefficient restriction $D \neq 0$ in the definition of linear fractional transformation, when expressed in the coefficients of the inverse function, is $da - (-b)(-c) \neq 0$, and, of course, this is true because $ad - bc \neq 0$.

We defer the proof of part (ii) to the exercises.

The effect of a linear fractional transformation on lines and circles is now studied by looking at special types of transformations. It is subsequently shown that every linear fractional transformation can be expressed as a composition of functions of these special types.

We first note that according to Exercises 1.14.3 and 1.14.9, every straight line and every circle in $\mathbf{C}$ can be represented by an equation of the type $Az\bar{z} + b\bar{z} + \bar{b}z + C = 0$, where A and C are real and $|b|^2 > AC$. Conversely, the graph of every such equation with coefficients so restricted is either a straight line or a circle in $\mathbf{C}$. It is a straight line if and only if $A = 0$. Recall that a straight line in $\hat{\mathbf{C}}$ is the union of a straight line in $\mathbf{C}$ with $\{\infty\}$.

In the following discussion, for brevity we follow the usual custom of identifying the set of points satisfying an equation with the equation itself; thus, for example, we speak of "the straight line $b\bar{z} + \bar{b}z + C = 0$."

Type 1. Translation (or shift): $w = z + k$, $z = w - k$, $k \in \mathbf{C}$. The line $b\bar{z} + \bar{b}z + C = 0$, C real, becomes the line $b\bar{w} + \bar{b}w - 2\text{Re}\, b\bar{k} + C = 0$. The circle $|z - z_0| = r$ becomes the circle $|w - (k + z_0)| = r$.

Type 2. Rotation and dilation or contraction: $w = kz$, $z = k^{-1}w$, $k \neq 0$, $k \in \mathbf{C}$. Lines go into lines and circles into circles, as in (1).

Type 3. Reciprocation: $w = 1/z$, $z = 1/w$. The equation $Az\bar{z} + b\bar{z} + \bar{b}z + C = 0$ becomes $Cw\bar{w} + \overline{bw} + bw + A = 0$. The coefficient condition $|b|^2 > AC$ becomes $|\bar{b}|^2 > CA$ for the transformed equation, but, of course, these are equivalent inequalities. The transformation takes a line in $\hat{\mathbf{C}}$ (the case $A = 0$) into a line (if $C = 0$), or into a circle through the origin (if $C \neq 0$). It takes a circle (the case $A \neq 0$) into a circle if $C \neq 0$, or, if $C = 0$, into a line through the origin in $\hat{\mathbf{C}}$.

Theorem 1.16.1 *Let F be the family of straight lines and circles in the extended comples plane. Any linear fractional transformation maps F onto itself.*

Proof. In view of part (ii) of Proposition 1.16.1 and the foregoing discussion of the action of special types of linear transformations, it remains only to show that any linear fractional transformation can be expressed as the composition of translations of the three types. Let the transformation be $w = (az + b)/(cz + d)$, with $ad - bc \neq 0$. If $c = 0$, the transformation is of Type 2 composed with Type 1, so we are through. If $c \neq 0$, we write the equation in the form

$$w = \frac{(bc - ad)/c^2}{z + d/c} + \frac{a}{c} = f(z).$$

Define mappings by $\zeta = f_1(z) = z + d/c$, a translation; $\xi = f_2(\zeta) = 1/\zeta$, a reciprocation; $\tau = f_3(\xi) = (bc - ad)\xi/c^2$, a rotation and dilation; and $f_4(\tau) = \tau + a/c$, a translation. Then $w = f_4(f_3(f_2(f_1(z))))$. Q.E.D.

Exercises 1.16

1. Show that if in the equation $w = (az + b)/(cz + d)$, the restriction $ad - bc \neq 0$ is violated, then the function is a constant function (w is not defined if $c = d = 0$).
2. Prove that if f_1 and f_2 are linear fractional transformations, then so is their composition $f_1 \circ f_2$. In fact, if $f_1(z) = (a_1z + b_1)/(c_1z + d_1)$ and $f_2(\zeta) = (a_2\zeta + b_2)/(c_2\zeta + d_2)$ then $f_1(f_2(\zeta)) = (a_3\zeta + b_3)/(c_3\zeta + d_3)$ where a_3, b_3, c_3, d_3 are given by the matrix equation
$$\begin{bmatrix} a_3 & c_3 \\ c_3 & d_3 \end{bmatrix} = \begin{bmatrix} a_1 & b_1 \\ c_1 & d_1 \end{bmatrix}\begin{bmatrix} a_2 & b_2 \\ c_2 & d_2 \end{bmatrix}.$$
3. Show that the transformation $w = (z + i)/(z - 1)$ carries the line in $\hat{\mathbf{C}}$ with equation $(3 + i)\bar{z} + (3 - i)z + 2 = 0$ into a circle with center at $(12) + i(1/8)$ and radius $\sqrt{17/8}$. What point on the circle would be missing if z were restricted to $\mathbf{C}$?
4. Prove that $w = 1/z$ gives a one-to-one mapping of $\{z:0 \leq |z| \leq 1\}$ onto $\{w:w \geq 1\}$ in which the unit circles in the z- and w-planes are mapped onto each other.
5. Show that $w = z + k$, k real, maps the half-plane $\{z:\text{Re } z \geq 0\}$ in a one-to-one manner onto the half-plane $\{w:\text{Re } w \geq k\}$, so that the imaginary axis is mapped onto the line $\text{Re } w = k$.
6. Prove that $w = (1 + z)/(1 - z)$ gives a one-to-one mapping of the set $\{z:|z| \leq 1, z \neq 1\}$ in the z-plane onto $\{\infty\} \cup \{w:\text{Re } w \geq 0\}$, in which the unit circle in the z-plane with $\{1 + 0i\}$ deleted is mapped onto the imaginary axis in the w-plane and $\{1 + 0i\}$ is mapped onto $\{\infty\}$. (*Hint*: The "onto" is perhaps most easily proved by looking at the inverse function.)
7. Let z_0 be complex and r real, $r > 0$. Show that $w = (z - z_0 - r)^{-1}$ gives a one-

to-one mapping of the disk $\{z:|z - z_0| < r\}$ onto $\{w:\text{Re } w < -1/2r\}$, and that the function maps the boundary $\{z:|z - z_0| = r$ onto the line in $\hat{\mathbf{C}}$ with equation $\text{Re } w = -1/2r$. (*Hint*: First, find Re w and look at the z-set which corresponds to the inequality $\text{Re } w < -1/2r$.)

8. Show that the mapping in Exercise 1.16.6 is one-to-one from $\{z:|z| > 1\} \cup \{\infty\}$ onto $\{w: \text{Re } w < 0\}$, and the mapping in Exercise 1.16.7 is one-to-one from $\{z:|z - z_0| > r\} \cup \{\infty\}$ onto $\{w: \text{Re } w > -1/2r\}$. (*Hint*: Make use of Proposition 1.16.1.)
9. Show that a linear fractional transformation has either no fixed point, or one fixed point, or two fixed points, or else every point in $\hat{\mathbf{C}}$ is a fixed point. (*Hint*: A fixed point is a solution of the equation $z = (az + b)/(cz + d)$. Exercise 1.11.12 can be used.)
10. By compositions of functions previously introduced in Sec. 1.7, find a function which gives a one-to-one mapping of the ellipse with equations $(x^2/\alpha^2) + (y^2/\beta^2) = 1$, $\alpha > \beta$, onto the circle $|w| = 1$ and which also gives a one-to-one mapping of the outside of the ellipse onto the outside of the circle. (*Hint*: According to Exercises 1.14.16 and 1.14.17, if $R > 0$, the function

$$\zeta = \frac{1}{2}\left(Rw + \frac{1}{Rw}\right)$$

maps $|w| \geq 1$ in a one-to-one manner onto

$$\left\{\zeta:\frac{(\text{Re } \zeta)^2}{(1/4)(R + 1/R)^2} + \frac{(\text{Im } \zeta)^2}{(1/4)(R - 1/R)^2} \geq 1\right\} \tag{1.15.1}$$

in such a way that the boundary curves are mapped onto each other. Let k be real, $k > 0$, and let $z = k\zeta$, $\zeta = z/k$. This is a one-to-one transformation of the complex plane onto itself, and by substitution into (1.15.1), it is seen to map the ζ-set there indicated onto the set

$$\left\{z = x + iy:\frac{x^2}{(k/2)^2(R + 1/R)^2} + \frac{y^2}{(k/2)^2(R - 1/R)^2} \geq 1\right\}, \tag{1.15.2}$$

with boundaries mapped onto each other. Thus, $z = (k/2)(Rw + 1/Rw)$ maps $|w| \geq 1$ onto the set (1.15.2) with boundary correspondence. (Now let $R = [(\alpha + \beta)/(\alpha - \beta)]^{1/2}$ and $k = (\alpha^2 - \beta^2)^{1/2}$ in (1.15.2) and see what happens.)
11. Use the mapping in Exercise 1.16.9 in composition with that in Exercise 1.16.10 and a shift to obtain a function which maps the outside of the ellipse $(x^2/4) + (y^2/4) = 1$ in the z-plane onto $\{w:\text{Re } w > 2.5\}$.

Answer:

$$z = \frac{\sqrt{5}}{2}\left(\sqrt{5}\,\frac{w - 2}{w - 3} + \frac{1}{\sqrt{5}}\,\frac{w - 3}{w - 2}\right).$$

12. Find a function which gives a one-to-one mapping of the point set lying between the circles $\{z:|z| = 1\}$ and $\{z:|z - 1/2| = 1/2\}$ onto $\{w:\text{Re } w > 1\}$.

2

Sequences and Series

2.1 Basic Definitions

Let X be a nonempty set. A *sequence* in X (or from X) is a function from the set $\mathbf{Z}$ of real integers into X with domain of the form $\{k: k \geq n\}$ or $\{k: k \leq n\}$ or $\{k: m \leq k \leq n\}$, where m, n, k denote real integers. In the last case, the sequence is called a finite sequence. The ordering in $\mathbf{Z}$ automatically induces an ordering on the values of the sequence. Thus, a finite sequence with, say, N values is an ordered N-tuple.

In specifying a sequence by its values, it is customary to use a subscript notation such as $\langle x_k \rangle_{k=n}^{\infty}$. If the intended domain is $\mathbf{N} = \{k: k = 0, 1, 2, \ldots\}$, we abbreviate the notation to $\langle x_k \rangle$. This notation is used in the case of other domains when no confusion results. When the pattern is clear, a sequence is often informally specified by writing down the first few terms; thus, $x_0, x_1, x_2, \ldots$. We indicate the range of the sequence $\langle x_k \rangle$ by $\{x_k\}$, with appropriate elaboration if the exact domain of the sequence must be emphasized. The range of a sequence is sometimes called the *trace*.

A subsequence of $\langle x_k \rangle_n^{\infty}$ is a sequence of the form $\langle x_{\phi(k)} \rangle$, where ϕ is a function from $[n, \infty)$ into $[n, \infty)$ satisfying $\phi(h) < \phi(k)$, $h < k$.

If X is a field, it is consistent with the definitions in Sec. 1.7 to define the sum of two sequences $\langle x_k \rangle$, $\langle y_k \rangle$, written $\langle x_k \rangle + \langle y_k \rangle$, by $\langle x_k + y_k \rangle$. The product $\langle x_k \rangle \cdot \langle y_k \rangle$ is defined by $\langle x_k \cdot y_k \rangle$ and if $y_k \neq 0$, $k = 0, 1, 2, \ldots$, the quotient $\langle x_k \rangle / \langle y_k \rangle$ is defined by $\langle x_k / y_k \rangle$.

The spaces X in which the sequences in this book lie are $\mathbf{C}$, $\mathbf{R}$, and certain spaces of complex-valued functions.

2.2 Metric Spaces

The concept of convergence and divergence is central in the theory of sequences. It can be treated at various levels of abstraction, but since we specialize in sequences in the spaces mentioned previously, an appropriate general setting for the basic definitions is given by the concept of a *metric space*.

A metric space is a pair (X, d), where X is a nonempty set and d is a real-valued function on $X \times X$ satisfying:

(i) $d(x_1, x_2) \geq 0$, $x_1, x_2 \in X$, and $d(x_1, x_2) = 0$ if and only if $x_1 = x_2$;
(ii) $d(x_1, x_2) = d(x_2, x_1)$, $x_1, x_2 \in X$;
(iii) $d(x_1, x_3) \leq d(x_1, x_2) + d(x_2, x_3)$.

Such a function d is called a *metric* on X.

In the case $X = \mathbf{C}$, the function $d(z_1, z_2) = |z_1 - z_2|$ clearly satisfies (i) and (ii), and according to Exercise 1.14.8 it satisfies (iii). It is the "usual metric" on $\mathbf{C}$. The restriction to $\mathbf{R}$ is the usual metric on $\mathbf{R}$. In each case the geometric interpretation of $d(z_1, z_2)$ is the distance from z_1 to z_2. A geometrically equivalent metric in $\mathbf{R}^2$ is the Euclidean distance $[(x_1 - x_2)^2 + (y_1 - y_2)^2]^{1/2}$, $(x_1, y_1), (x_2, y_2) \in \mathbf{R}^2$.

We note in passing that there are other metrics on $\mathbf{C}$. For example, Exercise 1.14.8 with $0 < p < 1$ gives a family of such metrics. Another example is the chordal distance (1.15.5):

$$d(z_1, z_2) = \frac{|z_1 - z_2|}{(1 + |z_1|^2)^{1/2}(1 + |z_2|^2)^{1/2}}.$$

(A detailed proof that this function satisfies (iii) is found in [9, p. 43].) In this book, the metrics used for $\mathbf{C}$ and $\mathbf{R}$ are always the usual metrics.

A subset of a metric space (X, d) of the form $\{x: x \in X, d(a, x) < r\}$, $r > 0$, is called an open ball with center a and radius r. In $\mathbf{C}$, this becomes $\{z: |z - a| < r\}$, which is geometrically a circular disk with center a and radius r. We denote such a disk in $\mathbf{C}$ by $\Delta(a, r)$. Also $C(a, r)$ is used to denote the circle $\{z: |z - a| = r\}$.

A set in a metric space is said to be *bounded* if it is a subset of some open ball with finite radius. A sequence $\langle x_k \rangle$ from a metric space is said to be bounded if its range $\{x_k\}$ is bounded.

An element x of a metric space is often called a "point," by analogy with $\mathbf{R}$ and $\mathbf{R}^2$.

Exercises 2.2

1. Given points a and b in a metric space (X, d), and a real number $r \geq 0$, suppose that it is known that for every real $\varepsilon > 0$, $d(a, b) \leq r + \varepsilon$. Show that $d(a, b) \leq r$,

and if $r = 0$, then $a = b$.

2. Let X be a nonempty set. Show that the function from $X \times X$ into $\mathbf{R}$ given by $d(x_1, x_2) = 1$ if $x_1 \neq x_2$, $d(x_1, x_2) = 0$ if $x_1 = x_2$, is a metric on X.

2.3 Convergence of Sequences

We start with a review of the classical definition of the convergence of a sequence $\langle \rho_k \rangle_n^\infty$ in the real number system. Such a sequence is said to converge if there is a real ρ with the property that for every real $\varepsilon > 0$, there exists $K_\varepsilon \geq n$ such that if $k \geq K_\varepsilon$, then $|\rho_k - \rho| < \varepsilon$. If no such ρ exists, the sequence is said to diverge. If such a ρ does exist, the sequence is said to converge to ρ, which is called the *limit* of the sequence, and we write $\lim_{k \to \infty} \rho_k = \rho$ or $\lim \rho_k = \rho$ or $\rho_k \to \rho$.

Now let $\langle x_k \rangle_n^\infty$ be a sequence in an arbitrary metric space (X, d). The sequence is said to converge if there is an element $x \in X$ such that the sequence $\langle d(x_k, x) \rangle_n^\infty$ converges to zero. The notations $\lim_{k \to \infty} x_k = x$, $\lim x_k = x$, $x_k \to x$ are used. The sequence is said to converge to x, which is called the limit of the sequence. If no such x exists, the sequence is said to diverge.

Remark (i). The concept of convergence of sequences in $\mathbf{R}$ is thus taken to be fundamental. In arithmetic terms, $\langle x_k \rangle$ converges to x if and only if for every $\varepsilon > 0$ there exists $K_\varepsilon \in \mathbf{N}$ such that $k \geq K_\varepsilon$ implies that $d(x_k, x) < \varepsilon$.

Remark (ii). From the arithmetic formulation of convergence in (i), it is apparent that if a sequence $\langle x_k \rangle_m^\infty$ converges to a limit x, then so does any sequence $\langle x_k \rangle_n^\infty$, $n > m$, and if a sequence $\langle x_k \rangle_n^\infty$, $n > m$, converges to a limit x, then so does $\langle x_k \rangle_m^\infty$.

Remark (iii). A convergent sequence has a unique limit. This is proved as follows. If x and x' were two limits of $\langle x_k \rangle$, then $d(x, x') \leq d(x, x_k) + d(k_k, x')$ by Postulate (iii) of metric spaces. But then for every $\varepsilon > 0$, there exists $K_\varepsilon > 0$ and $K'_\varepsilon > 0$ such that for any $k \geq \max(K_\varepsilon, K'_\varepsilon)$, $d(x, x') \leq \varepsilon + \varepsilon$, which, according to Exercise 2.2.1, implies that $d(x, x') = 0$. By Postulate (i) of metric space, $d(x, x') = 0$ implies that $x = x'$.

A *Cauchy sequence* $\langle x_k \rangle_m^\infty$ in a metric space (X, d) is a sequence with the property that for every $\varepsilon > 0$ there exists, $K_\varepsilon \geq m$ such that if $k > h \geq K_\varepsilon$, then $d(x_k, x_h) < \varepsilon$. Thus, a Cauchy sequence can be loosely described as one in which the tail elements become more and more closely bunched together.

In some metric spaces, a Cauchy sequence does not necessarily have a limit. For example, take X to be the open interval $(0, 1)$ in $\mathbf{R}$ and d to be the usual metric for $\mathbf{R}$, restricted to this X. Look at the sequence $1/2$, $1/3, \ldots, 1/k, \ldots$. The limit is clearly 0, but this is not in X. Yet with $k > h$, $d(x_h, x_k) = |(1/h - (1/k)| \leq (1/h) + (1/k) \leq 2/h \to 0$, so the sequence is Cauchy in X.

A metric space in which every Cauchy sequence *does* have a limit is called *Cauchy complete*, or just *complete*.

Exercises 2.3

1. Show that a convergent sequence in a metric space is a Cauchy sequence. (*Hint*: The device used in Remark (iii) is useful here.)
2. Let $\langle x_k \rangle$ be a sequence in a metric space (X, d) which converges to $x \in X$; let y be an element of X; let $r > 0$ be real; and suppose that for all k sufficiently large, say $k \geq K > 0$, $d(x_k, y) \leq r$. Show that $d(x, y) \leq r$. (*Hint*: See previous hint and Exercise 2.2.1.)
3. Show by an example that if the hypothesis $d(x_k, y) \leq r, k \geq K$, in Exercise 2.3.2 is changed to $d(x_k, y) < r, k \geq K$, then the conclusion is the same as before, and not $d(x, y) < r$.
4. Prove that (a) in a metric space any convergent sequence is bounded and (b) any Cauchy sequence is bounded.
5. Prove that every subsequence of a convergent sequence with limit x in a metric space converges to x.

2.4 Sequences of Reals

The definition of convergence for sequences in **R**, with the usual metric, was presented explicitly at the beginning of Sec. 2.3. A sequence $\langle x_k \rangle$ in **R** is a Cauchy sequence if for every $\varepsilon > 0$, there exists $K_\varepsilon > 0$ such that if $k > h \geq K_\varepsilon$, then $|x_k - x_h| < \varepsilon$.

It is assumed here that the reader has already encountered the elementary theory of sequences of reals as given in calculus textbooks. The purpose here is merely to review certain topics of particular interest in complex analysis. Especial attention is given to the idea of the upper and lower limits of a sequence, since this may be new to some readers.

The following elementary theorem is accepted without proof.

Theorem 2.4.1 *Let the sequences $\langle x_k \rangle$ and $\langle y_k \rangle$ in* **R** *have the limits x and y, respectively. Then:*

(i) *$\langle x_k \pm y_k \rangle$ has the limit $x \pm y$.*
(ii) *$\langle x_k y_k \rangle$ has the limit xy.*
(iii) *If $y_k \neq 0$, $k = 0, 1, 2, \ldots$ and $y \neq 0$, then $\langle x_k / y_k \rangle$ has the limit x/y.*

A deeper result is the following one.

Theorem 2.4.2 *The real number system* **R** *is Cauchy complete.*

This means, of course, that every Cauchy sequence in **R** has a limit. The statement is a consequence of the fact that **R** is an ordered field which is complete in the sense of the least upper bound property. The proof is an exercise [Exercise 2.4.13].

It is convenient to make use of the extended real number system when dealing with limits of sequences of reals. Unless a statement is made to the contrary, only the limits and not the terms of the sequences are allowed to be $\pm\infty$. Given a real sequence $\langle x_k \rangle$, the meaning to be taken for $\lim_{k\to\infty} x_k = \infty$ is that for every real number α, there exists K_α such that $x_k > \alpha$ when $k > K_\alpha$. A similar meaning is ascribed to $\lim_{k\to\infty} x_k = -\infty$. Theorem 2.4.1 remains valid for infinite limits if the cases $\infty - \infty$, $-\infty + \infty$, $0\cdot\infty$, and $\infty\cdot 0$ are ruled out.

A sequence of reals $\langle x_k \rangle$ is said to be *increasing* if $x_k \le x_{k+1}$, $k = 0, 1, 2, \ldots$. It is *decreasing* if $x_{k+1} \le x_k$, $k = 0, 1, 2, \ldots$. In either case, the sequence is said to be monotonic. If the inequality is a strong inequality for all k, the sequence is called strictly monotonic.

Theorem 2.4.3 *Let $\langle x_k \rangle$ be increasing and let $x = \sup\{x_k\}$. Then $\langle x_k \rangle$ has the limit x. If $\langle x_k \rangle$ is decreasing and $x = \inf\{x_k\}$, then $\langle x_k \rangle$ has the limit x.*

Proof. We discuss only the increasing case. If $\sup\{x_k\} = \infty$, then $\{x_k\}$ is unbounded above in **R**, and for an increasing sequence this is equivalent to the definition of $\lim x_k = \infty$. If $\sup\{x_k\} = \beta < \infty$ then for every $\varepsilon > 0$, according to Exercise 1.4.1 there must exist a term $x_{k(\varepsilon)}$ of the sequence in the interval $[\beta - \varepsilon, \beta]$. But then by monotonicity, for all $k > k(\varepsilon)$ we have $\beta - \varepsilon < x_{k(\varepsilon)} \le x_k \le \beta$, which implies that $|x_k - \beta| < \varepsilon$. Thus, β satisfies the definition of limit. Q.E.D.

Examples

(i) $\langle 1/k^p \rangle_{k=1}^{\infty}$. If $p > 0$, the sequence is decreasing and $\lim(1/k^p) = 0$. If $p = 0$, the sequence is both decreasing and increasing according to the foregoing definition and $\lim(1/k^0) = 1$. If $p < 0$, the sequence is unbounded and increasing and $\lim(1/k^p) = \infty$.

(ii) $\langle \rho^k \rangle_0^{\infty}$, $\rho > 0$. If $\rho \in (0, 1)$ the sequence is decreasing and $\lim \rho^k = 0$. If $\rho = 1$, the sequence is both increasing and decreasing according to our definitions, and $\lim \rho_k = 1$. If $\rho > 1$, the sequence is unbounded and increasing, and $\lim \rho^k = \infty$.

(iii) $\langle \rho^{1/k} \rangle_{k=1}^{\infty}$, $\rho > 0$. If $\rho \in (0, 1)$, the sequence is increasing and the limit is 1. If $\rho = 1$, the limit is again 1 trivially. If $\rho > 1$, the sequence is decreasing and the limit is again 1.

(iv) $\langle k^{1/k} \rangle_{k=3}^{\infty}$. The sequence is decreasing and has the limit 1 [6, p. 68, Exercise 10].

We now discuss the concept of the upper limit and lower limit of a sequence. Let $\langle x_k \rangle_0^\infty$ be a sequence of reals, and define the sequence $\langle b_n \rangle_0^\infty$ in the extended real number system by

$$b_n = \sup\{x_n, x_{n+1}, x_{n+2}, \ldots\}, \qquad n = 0, 1, 2, \ldots.$$

According to Exercise 1.4.2, $\langle b_n \rangle$ is decreasing and so [Theorem 2.4.3] has a limit, say β. This limit is called the *upper limit* (or *limit superior*) of $\langle x_k \rangle$ and is written $\overline{\lim}\, x_k$ or $\limsup x_k$. The *lower* limit (or *limit inferior*) of $\langle x_k \rangle$ is similarly defined as the limit, say γ, of the increasing sequence $\langle c_n \rangle_n^\infty$, where $c_n = \inf\{x_n, x_{n+1}, x_{n+2}, \ldots\}$.

It should be noted that whether or not a sequence of reals has a limit, finite or infinite, its upper and lower limits always exist in $\hat{\mathbf{R}}$. The concept of upper and lower limits provides a valuable extension of that of limit.

Proposition 2.4.1

(i) *If the sequence* $\langle x_k \rangle$ *is bounded and* $\beta = \overline{\lim}\, x_k$, *then* $-\infty < \beta < \infty$. *For every* $\varepsilon > 0$ *there exists* K_ε *such that* $x_k < \beta + \varepsilon$ *for every* $k \geq K_\varepsilon$, *and also given any integer* $K > 0$, *there exists an integer* $k \geq K$ *such that* $x_k > \beta - \varepsilon$. *Furthermore, a finite real number* β' *which satisfies these conditions must be equal to* $\overline{\lim}\, x_k$.

(ii) *If the sequence* $\langle x_k \rangle$ *is bounded and* $\gamma = \underline{\lim}\, x_k$, *then* $-\infty < \gamma < \infty$. *For every* $\varepsilon > 0$, *there exists* K_ε *such that* $x_k > \gamma - \varepsilon$ *for every* $k \geq K_\varepsilon$, *and also given any* $K > 0$, *there exists an integer* $k \geq K$ *such that* $x_k < \gamma + \varepsilon$. *Furthermore, a finite real number* γ' *which satisfies these conditions must be* $\underline{\lim}\, x_k$.

Proof. We give a proof only for part (i), since the proof for part (ii) is essentially the same with inequalities reversed.

Let $b > 0$, $b \neq \infty$, be such that $-b < x_k < b$, $k = 0, 1, 2, \ldots$. Then by inspection of the definition of b_n, it is clear that $-b \leq b_n \leq b$ for $n = 0, 1, 2, \ldots$. This inequality is the same as $|b_n| \leq b$. Then by Exercise 2.3.2 (with $y = 0$) it follows that $|\beta| \leq b < \infty$. This proves the first sentence in (i).

Given $\varepsilon > 0$, there exists N_ε such that for $n \geq N_\varepsilon$, $\beta \leq b_n \leq \beta + \varepsilon$. The inequality $b_n \leq \beta + \varepsilon$ implies that $x_n \leq \beta + \varepsilon$, $x_{n+1} \leq \beta + \varepsilon, \ldots$, so $x_k \leq \beta + \varepsilon$ for all $k \geq N_\varepsilon$. Now take any integer $K > 0$. Since b_n is the *least* upper bound of $\{x_n, x_{n+1}, \ldots\}$ for each n, and in particular for $n = K$, there must be some $k \geq K$ such that $x_k > b_K - \varepsilon \geq \beta - \varepsilon$. This proves the second sentence in (i).

Now suppose β' is such that the inequalities in part (i) of the proposition are satisfied. Let $\beta = \overline{\lim}\, x_k = \lim b_n$, where $b_n = \sup\{x_n, x_{n+1}, \ldots\}$. Take $\varepsilon > 0$. By hypothesis, for $n \geq K_\varepsilon$, $x_n \leq \beta' + \varepsilon$, $x_{n+1} \leq \beta' + \varepsilon, \ldots$. Yet in every set $\{x_n, x_{n+1}, \ldots\}$ by hypothesis there is at least one element, say x_k, with $x_k > \beta' - \varepsilon$. The implication of these inequalities for b_n is that $\beta' - \varepsilon < b_n < \beta' + \varepsilon$, which is equivalent to $|b_n - \beta'| < \varepsilon$. But then, by Exercise 2.3.2, $|\beta - \beta'| \leq \varepsilon$. Since $\varepsilon > 0$ is arbitrary, it follows from Exercise 2.2.1 that $\beta = \beta'$. Q.E.D.

Proposition 2.4.2 *Given the sequence* $\langle x_k \rangle$, $\overline{\lim}\, x_k = \infty$ *if and only if there is a subsequence* $\langle x_{k(n)} \rangle$ *with* $\lim_{n\to\infty} x_{k(n)} = \infty$; *and* $\overline{\lim}\, x_k = -\infty$ *if and only if* $\lim x_k = -\infty$. *Similarly,* $\underline{\lim}\, x_k = -\infty$ *if and only if there is a subsequence* $x_{k(n)} \to -\infty$, *and* $\underline{\lim}\, x_k = \infty$ *if and only if* $\lim x_k = \infty$.

The proposition is a direct consequence of the definitions.

Corollary If $\overline{\lim}\, x_k < \infty$, then the range of $\langle x_n \rangle$ is bounded above. If $\underline{\lim}\, x_k > -\infty$, then the range of $\langle x_n \rangle$ is bounded below.

Proposition 2.4.3 *The sequence $\langle x_k \rangle$ has a limit x in the extended reals if and only if $\overline{\lim}\, x_k = \underline{\lim}\, x_k = x$.*

Proof. The cases in which $x = \pm\infty$ are contained in Proposition 2.4.2. If $\overline{\lim}\, x_k = \underline{\lim}\, x_k = x$, and $-\infty < x < \infty$, then by the definitions, the sequence $\langle x_k \rangle$ is bounded. Parts (i) and (ii) of Proposition 2.4.1 imply the assertion that for every ε there exists K_ε such that for every $k \geq K_\varepsilon$, $x - \varepsilon < x_k < x + \varepsilon$ (which is the same as $|x_k - x| < \varepsilon$, and this is the definition of convergence to x).

If the sequence converges to the limit x, then for every $\varepsilon > 0$ there exists K_ε such that for every $k \geq K_\varepsilon$, $x - \varepsilon < x_k < x + \varepsilon$. The third sentence in Part (i) of Proposition 2.4.1 and the third sentence in Part (ii) together imply that $x = \overline{\lim}\, x_k = \underline{\lim}\, x_k$. Q.E.D.

Exercises 2.4 (All sequences here are real valued.)

1. Let $x_k = (-1)^k[2 + (3/k)]$, $k = 1,2,\ldots$. Find $\overline{\lim}\, x_k$ and $\underline{\lim}\, x_k$.
2. If $x_k = (-1)^{k+1}k - k$, $k = 0,1,2,\ldots$, show that $\overline{\lim}\, x_k = 0$ and $\underline{\lim}\, x_k = -\infty$.
3. Show that if $\langle x_k \rangle$ is a sequence with nonnegative terms which converges to x, then $x \geq 0$. Thereafter, show that if $\langle x_k \rangle$ is any sequence of nonnegative terms, then $\underline{\lim}\, x_k \geq 0$. (*Hint*: Use the definition of lower limit.)
4. Let $\langle x_k \rangle$ and $\langle y_k \rangle$ be sequences which converge, respectively, to x and y. Show that if $x_k \leq y_k$, $k = 0,1,2,\ldots$, then $x \leq y$. (*Hint*: Exercise 2.4.3 and Theorem 2.4.1.)
5. Show that for any sequences $\langle x_k \rangle$, $\underline{\lim}\, x_k \leq \overline{\lim}\, x_k$. (*Hint*: Use Exercise 2.4.4 in conjunction with the definition of upper and lower limits.)
6. Show that if in the sequences $\langle x_k \rangle$ and $\langle y_k \rangle$, $x_k \leq y_k$, $k = 0,1,2,\ldots$, then $\overline{\lim}\, x_k \leq \overline{\lim}\, y_k$. (*Hint*: See hint for Exercise 2.4.5.) Would it also be true that $\overline{\lim}\, x_k \leq \underline{\lim}\, y_k$?
7. Let $\langle x_k \rangle$ be a sequence of nonnegative terms. Show that if $\overline{\lim}\, x_k = 0$, then $\langle x_k \rangle$ converges to the limit 0.
8. Prove that if $\langle x_k \rangle$ and $\langle y_k \rangle$ are sequences with nonnegative terms, then $\overline{\lim}\, x_k y_k \leq \overline{\lim}\, x_k \;\overline{\lim}\, y_k$ provided the right-hand side is not of the form $0 \cdot \infty$ or $\infty \cdot 0$. (*Hint*: Let $b_n = \sup\{y_n, y_{n+1}, y_{n+2}, \ldots\}$. Then $b_n \geq y_{n+k}$, $k = 0,1,2,\ldots$, and $\sup\{x_n y_n, x_{n+1}y_{n+1}, x_{n+2}y_{n+2}, \ldots\} \leq \sup\{x_n b_n, x_{n+1}b_n, x_{n+2}b_n, \ldots\}$. Use Exercises 1.4.3 and 2.4.4.)
9. For arbitary real sequences $\langle x_k \rangle$ and $\langle y_k \rangle$, prove that $\overline{\lim}\, x_k + \underline{\lim}\, y_k \leq \overline{\lim}(x_k + y_k) \leq \overline{\lim}\, x_k + \overline{\lim}\, y_k$, provided that the left-hand side is not of the form $\infty - \infty$ or $-\infty + \infty$.
10. Let $\langle x_k \rangle$ be a sequence of nonnegative terms with limit x, $0 < x < \infty$, and let $\langle y_k \rangle$ be another sequence of nonnegative terms. (The case $\overline{\lim}\, y_k = \infty$ is not ex-

cluded.) Prove that $\overline{\lim}\, x_k y_k = x\, \overline{\lim}\, y_k$. (*Hint*: Exercises 2.4.7 and 2.4.8 show that $\overline{\lim}\, x_k y_k \le x\, \overline{\lim}\, y_k$. Take any $\varepsilon > 0$ such that $x - \varepsilon > 0$. There is an N such that for $n \ge N, |x_n - x| < \varepsilon$, which implies that for $n \ge N$, $x_n > x - \varepsilon$. Use this appropriately in $\sup\{x_n y_n, x_{n+1} y_{n+1}, \ldots\}$, as suggested by the hint in Exercise 2.4.8, to obtain another inequality for $\overline{\lim}\, x_k y_k$.)

11. Let $\langle \alpha_k \rangle_1^\infty$ be a sequence of nonnegative reals, and suppose that $\overline{\lim}\, \alpha_k^{1/k} = \alpha \le \infty$. Show that $\overline{\lim}\, (k\alpha_k)^{1/k} = \alpha$. (*Hint*: See Example (iv) of monotonic sequences in this section.)
12. Let $\langle x_k \rangle$ be a sequence of nonnegative reals having the limit β, $0 \le \beta \le \infty$. Prove that the sequence $\langle x_k^p \rangle, p > 0$, has the limit β^p if $\beta \ne \infty$ and the limit ∞ if $\beta = \infty$. (*Hint*: Given any $\eta_1 > 0$, $\eta_2 > 0$, for all sufficiently large values of k, $\beta - \eta_1 < x_k < \beta + \eta_2$. In the case $0 < \beta < \infty$, given $\varepsilon > 0$, $\varepsilon < \beta$, take $\eta_1 = \beta - (\beta^p - \varepsilon)^{1/p}$, $\eta_2 = (\beta^p + \varepsilon)^{1/p} - \beta$. Verify that these numbers are positive.)
13. Prove Theorem 2.4.2. (*Hint*: If the sequence $\langle x_k \rangle$ is assumed to converge, then the Cauchy property can easily be established by the triangle inequality. Conversely, if $\langle x_k \rangle$ is a Cauchy sequence, then it is bounded [Exercise 2.2.4]. Use Proposition 2.4.1 to show that the assumption $\underline{\lim}\, x_k < \overline{\lim}\, x_k$ leads to a contradiction of the Cauchy property.)

2.5 Sequences of Complex Numbers

Let $\langle z_k \rangle$ denote such a sequence of complex numbers. To recapitulate from earlier sections, it is bounded if and only if the range $\langle z_k \rangle$ is contained in some disk $\Delta(a, r)$, $r > 0$. It converges in the usual metric to the limit z if and only if for every $\varepsilon > 0$, there exists K_ε such that $k \ge K_\varepsilon$ implies $|z_k - z| < \varepsilon$. It is a Cauchy sequence if and only if, for every $\varepsilon > 0$, there exists K_ε such that $|z_k - z_h| < \varepsilon$ for every $h, k, k > h \ge K_\varepsilon$.

Many of the basic facts about complex sequences can be deduced from the parallel facts for real sequences by means of the following simple theorem.

Theorem 2.5.1 *The sequence $\langle z_k = x_k + iy_k \rangle$, x_k, y_k real, converges to the limit $z = x + iy$, where x and y are finite, if and only if the sequences $\langle x_k \rangle$ and $\langle y_k \rangle$ converge, respectively, to x and y. The sequence $\langle z_k = x_k + iy_k \rangle$ is a Cauchy sequence if and only if the sequences $\langle x_k \rangle$ and $\langle y_k \rangle$ are both Cauchy sequences.*

Proof. Recall Exercise 1.13.1: If $a = \alpha + i\beta$, α, β real, then $|a| \le |\alpha| + |\beta| \le \sqrt{2}|a|$. The statement in the first sentence of the theorem is established by applying the definition of convergence to the inequality $|z_k - z| = |x_k - x + i(y_k - y)| \le |x_k - x| + |y_k - y| \le \sqrt{2}|z_k - z|$. For the assertion in the second sentence, we use $|z_k - z_h| = |x_k - x_h + i(y_k - y_h)| \le |x_k - x_h| + |y_k - y_h| \le \sqrt{2}|z_k - z_h|$. The arithmetic details are deferred to the exercises.

Thus, if $\langle z_k \rangle$ is a Cauchy sequence, then so are the sequences of real and imaginary parts. But then by Theorem 2.4.2, the latter two sequences both have limits, say x and y, respectively. It then follows from the first part of Theorem 2.5.1 that $\langle z_k \rangle$ converges to $x + iy$. In other words, we have the following result.

Corollary 1 The complex number system with the usual metric is Cauchy complete.

Corollary 2 (Comparison Test for Convergence). Let $\langle z_k \rangle$ and $\langle w_k \rangle$ be such that there exists $K > 0$ such that $|z_k - z_h| \leq |w_k - w_h|$ for every h, k, $k > h \geq K$. Then if $\langle w_k \rangle$ converges, so does $\langle z_k \rangle$.

Proof. If $\langle w_k \rangle$ converges, it is a Cauchy sequence. The hypothesis then implies that $\langle z_k \rangle$ is a Cauchy sequence, so by Corollary 1 it converges. Q.E.D.

A sequence $\langle z_k \rangle$ in $\mathbf{C}$ is said to have the limit $\infty \in \hat{\mathbf{C}}$ if and only if $\lim_{k\to\infty} |z_k| = \infty$, and we then write $\lim_{k\to\infty} z_k = \infty$.

Exercises 2.5

1. Show that if the sequence $\langle z_k \rangle$ of complex numbers has the limit z, then $\langle |z_k| \rangle$ has the limit $|z|$. Also show by example that the converse is not in general true, but that if $\langle |x_k| \rangle$ converges to 0, then so does $\langle z_k \rangle$.
2. Show that $\langle z_k \rangle$ has the limit z if and only if $\langle \bar{z}_k \rangle$ has the limit $\bar{z}$.
3. Prove that $\lim_{k\to\infty} z^k = 0$ if $|z| < 1$, and the sequence $\langle z^k \rangle$ has the limit infinity if $|z| > 1$. (Reference may be made to examples in Sec. 2.4.)
4. Fill in the details of the proof of Theorem 2.5.1.
5. Prove that if the sequences $\langle z_k \rangle$ and $\langle w_k \rangle$ have respective limits z and w, with z, $w \in \hat{\mathbf{C}}$, then provided that the indicated limits fall within the algebraic restrictions imposed on ∞ in Sec. 1.15, the following statements are true:
   (i) $\langle z_k \pm w_k \rangle$ has the limit $z \pm w$;
   (ii) $\langle z_k w_k \rangle$ has the limit zw;
   (iii) if $w_k \neq 0$, $k = 0,1,2,\ldots$, and $w \neq 0$, then $\langle z_k/w_k \rangle$ has the limit z/w.
   (Use Theorems 2.4.1 and 2.5.1 as needed.)
6. Suppose that the sequences $\langle z_k \rangle_m^\infty$ and $\langle w_k \rangle_m^\infty$ are such that $\langle w_k \rangle$ converges to $w \in \mathbf{C}$ and $\langle z_k + w_k \rangle_m^\infty$ converges to $s \in \mathbf{C}$. Show that $\langle z_k \rangle$ converges to $s - w$.
7. Prove that if $\langle z_k \rangle_1^\infty$ has the limit 0, then so does the sequence $\langle (z_1 + z_2 + \cdots + z_n)/n \rangle_1^\infty$. Thereafter, prove that if $\lim z_k = z \in \mathbf{C}$ then $\lim[(z_1 + z_2 + \cdots + z_n)/n] = z$. (*Hint*: Write $(z_1 + z_2 + \cdots + z_n)/n = (z_1 + z_2 + \cdots + z_{K-1})/n + (z_K + z_{K+1} + \cdots + z_n)/n$. For every $\varepsilon > 0$, there exists K such that $|z_k| < \varepsilon/2$, $k \geq K$. Thereby show that the absolute value of the second fraction on the right-hand side is less than $\varepsilon/2$ for any $n \geq K$. Then adjust n to control the first fraction.)

8. Prove that if $\langle z_k\rangle_1^\infty$ converges to $z \in \mathbf{C}$ and $\langle w_k\rangle_1^\infty$ converges to $w \in \mathbf{C}$, then the sequence $C = \langle (z_1 w_n + z_2 w_{n-1} + \cdots + z_n w_1)/n\rangle_1^\infty$ converges to zw. (*Hint*: First, prove the result for $z = 0$ and $\langle w_k\rangle$ merely bounded. Then write $z_k = z + \varepsilon_k$, $w_k = w + \eta_k$, where $\lim \varepsilon_k = 0 = \lim \eta_k$. Substitute into C and use Exercise 2.5.7).

2.6 Infinite Series in C

Let $\langle z_k\rangle_{k=m}^\infty$ be a sequence in $\mathbf{C}$, where m is any integer, and let $s_n = z_m + z_{m+1} + \cdots + z_n$, $n \geq m$. The *infinite series* associated with $\langle z_k\rangle_m^\infty$ is defined to be the sequence $\langle s_n\rangle_{n=m}^\infty$. If the latter sequence converges to a limit $s \in \mathbf{C}$, then s is called the *sum* of the infinite series, and the series is said to *converge to the sum s*. If the sequence $\langle s_n\rangle$ has no limit in $\mathbf{C}$, the infinite series is said to diverge. If $\lim s_n = \infty$, the infinite series is said to diverge to infinity. We often abbreviate "infinite series" to just "series."

It is traditional to denote both the series associated with a sequence $\langle z_k\rangle_m^\infty$ and its sum (when it exists) by $\sum_m^\infty z_k$. When the pattern is clear, a symbol such as $z_m + z_{m+1} + \cdots$ is often used to denote both the sum and the series. When $m = 0$, or the value of m is immaterial for a statement, we indicate the series by $\sum z_k$. In the cases $m = 0, 1$, s_n is called the nth partial sum of the series.

Proposition 2.6.1 *Let $\langle z_k\rangle_m^\infty$ be a sequence in $\mathbf{C}$, and let m_1 and m_2 be integers, $m_2 > m_1 \geq m$.*

(i) *If $\sum_{m_2}^\infty z_k$ converges to the sum s, then $\sum_{m_1}^\infty z_k$ converges to $s + (z_{m_1} + z_{m_1+1} + \cdots + z_{m_2-1})$.*
(ii) *If $\sum_{m_1}^\infty z_k$ converges to the sum s', then $\sum_{m_2}^\infty z_k$ converges to $s' - (z_{m_1} + \cdots + z_{m_2-1})$.*

Proof. Let $s_n = z_{m_1} + z_{m_1+1} + \cdots + z_n$, and let $s_n' = z_{m_2} + z_{m_2+1} + \cdots + z_n$. Let $\sigma = z_{m_1} + z_{m_1+1} + \cdots + z_{m_2-1}$. By Remark (ii) in Sec. 2.3, the sequence $\langle s_n\rangle_{m_1}^\infty$ converges if and only if $\langle s_n\rangle_{m_2}^\infty$ does, and the limits are the same. Notice that if $n > m_2$, $s_n = \sigma + s_n'$.

Now suppose that $\sum_{m_2}^\infty z_k$ converges to the sum s. This means that $\langle s_n'\rangle_{m_2}^\infty$ converges to s. It follows from Exercise 2.5.5(i) that $\langle s_n = \sigma + s_n'\rangle_{m_2}$ converges to $\sigma + s$. But then $\langle s_n\rangle_{m_1}^\infty$ also converges to $\sigma + s$, which proves statement (i) in the proposition. The proof of (ii) is similar. Q.E.D.

More generally, it is true that convergence or divergence of a series $\sum z_k$ is not affected by omission or insertion of a finite number of terms of the sequence $\langle z_k\rangle$. Of course, the sum may be altered by such manipulations.

Proposition 2.6.2 *If $\sum_m^\infty z_k$ converges to $s_1 \in \hat{\mathbf{C}}$ and $\sum_m^\infty w_k$ converges to $s_2 \in \mathbf{C}$ then $\sum_m^\infty (z_k + w_k)$ converges to $s_1 + s_2$, $\sum_m^\infty (z_k - w_k)$ converges to $s_1 - s_2$, and $\sum_m^\infty z z_k$ converges to $z s_1$ for any $z \in \mathbf{C}$. If $z \neq 0$ and $\sum_m^\infty z_k$ diverges, then so does $\sum_m^\infty z z_k$.*

The proof is an exercise.

The literature of infinite series contains many tests for convergence. A comprehensive treatment will be found in [3]. Here only a few tests useful in complex analysis are discussed.

The only nontrivial general *necessary* and *sufficient* convergence test seems to be the one given by the Cauchy completeness of **C**, which can be restated for series as follows.

Proposition 2.6.3 (*Cauchy Criterion*). *The series* $\sum_m^\infty z_k$ *converges if and only if for any* $\varepsilon > 0$ *there is an integer* $K_\varepsilon \geq m$ *such that if* $k > h \geq K_\varepsilon$, *then* $|z_{h+1} + z_{h+2} + \cdots + z_k| < \varepsilon$.

Proof. Let $s_k = z_m + z_{m+1} + \cdots + z_k$, $k \geq m$. Then for $k > h$, $s_k - s_h = z_{h+1} + z_{h+2} + \cdots + z_k$. The statement in the proposition is equivalent to asserting that $\langle s_k \rangle_m^\infty$ converges if and only if it is a Cauchy sequence. The fact that a convergent sequence in a metric space must be a Cauchy sequence was brought out in Exercise 2.3.1. The assertion that a Cauchy sequence in **C** must converge is Corollary 1 of Theorem 2.5.1. Q.E.D.

Corollary If $\sum z_k$ converges, then $\lim z_k = 0$.

Proof. This is obtained from the special case of the Cauchy criterion in which we take $h + 1 = k$.

Remark. The converse of the corollary is not true. For example, it is shown later in this section that the series $\sum_1^\infty (1/k)$ diverges, although $\lim(1/k) = 0$. The corollary is often used in a contrapositive mode: If $\lim z_k \neq 0$, then $\sum z_k$ cannot converge.

A series $\sum z_k$ is said to *converge absolutely* if and only if the series of nonnegative real numbers $\sum |z_k|$ has a finite sum.

Proposition 2.6.4 *If a series in* **C** *converges absolutely, then it converges.*

Proof. Let the series be $\sum_0^\infty z_k$. If $\sum |z_k|$ converges, then it satisfies the Cauchy criterion in Proposition 2.6.3. But for any n, m, $n > m \geq 0$, we have $|z_{m+1} + z_{m+2} + \cdots + z_n| \leq |z_{m+1}| + |z_{m+2}| + \cdots + |z_n|$. Therefore, $\sum z_k$ also satisfies the Cauchy criterion, so the series must converge. Q.E.D.

Remark. Again the converse is not true. It is shown later that the series $1 - (1/2) + (1/3) - (1/4) + \cdots$ converges; but $1 + (1/2) + (1/3) + \cdots$ does not converge.

Proposition 2.6.5 *The geometric series* $1 + z + z^2 + z^3 + \cdots$ *converges to the sum* $1/(1 - z)$ *and converges absolutely for* $|z| < 1$. *It diverges for* $|z| \geq 1$.

Proof. It is readily verified that for every $z \neq 1$, $s_n = 1 + z + z^2 + \cdots + z^n = (1 - z^{n+1})/(1 - z)$. Convergence and absolute convergence for

$|z| < 1$ follow from Exercises 2.5.3 and 2.5.5. Divergence for $|z| \geq 1$ follows from the corollary of Proposition 2.6.3, because then the sequence $\langle z^k \rangle_0^\infty$ does not have the limit 0.

Proposition 2.6.6 (*Comparison Test*). *If the series $\sum_0^\infty z_k$ and $\sum_0^\infty w_k$ are such that $\sum_0^\infty w_k$ converges absolutely, and $K \geq 0$ exists such that for every $k \geq K$, $|z_k| \leq |w_k|$, then $\sum_0^\infty z_k$ converges absolutely and so converges.*

The proof is almost the same as that of Proposition 2.6.4 and is left as an exercise.

Corollary Let $\sum_0^\infty w_k$ converge absolutely, and suppose that $\sum_0^\infty z_k$ is such that $|z_k| \leq |w_k|$ for $k = 0, 1, 2, \ldots$. Let s be the sum of $\sum_0^\infty z_k$, S that of $\sum_0^\infty |z_k|$, and T that of $\sum_0^\infty |w_k|$. Then $|s| \leq S \leq T$.

Theorem 2.6.1 (*Cauchy kth Root Test*). *Given the series $\sum_m^\infty z_k$, $m \geq 1$; let $r = \overline{\lim}\, |z_k|^{1/k}$. The series converges absolutely if $0 \leq r < 1$ and diverges if $1 < r \leq \infty$. When $r = 1$, no conclusion is indicated.*

Proof. If $0 \leq r < 1$ the sequence $\langle |z_k|^{1/k} \rangle$ must be bounded. (See the corollary to Proposition 2.4.2.) Take any ρ, $0 \leq r < \rho < 1$. By Proposition 2.4.1(i) there exists $K > 0$ such that $|z_k|^{1/k} < \rho$ for every $k \geq K$. This inequality is equivalent to $|z_k| < \rho^k$. By Proposition 2.6.5 the series $\sum_0^\infty \rho^k$ converges, so by Proposition 2.6.6 the series $0 + z_1 + z_2 + \cdots$ converges absolutely.

If $1 < r < \infty$, again by Proposition 2.4.1(i), for every real K there exists some integer $k \geq K$ such that $|z_k|^{1/k} > 1$, which implies that $|z_k| > 1$. But then it is impossible for $\langle z_k \rangle$ to have the limit 0, so by the corollary of Proposition 2.6.3 the associated series must diverge. If $r = \infty$, then $\langle |z_k|^{1/k} \rangle$ is unbounded, and the same conclusion is reached. Q.E.D.

Theorem 2.6.2 (*Ratio Test*). *Given the series $\sum_m^\infty z_k$, in which $z_k \neq 0$, $k = m, m+1, \ldots$, let $\mu = \overline{\lim}\, |z_{k+1}/z_k|$ and $\nu = \underline{\lim}\, |z_{k+1}/z_k|$. If $0 \leq \mu < 1$, the series converges absolutely; if $\nu > 1$, it diverges. When $\nu \leq 1 \leq \mu$, no conclusion is indicated.*

Proof. If $\mu < 1$, the sequence $\langle z_{k+1}/z_k \rangle$ is bounded. Take any ρ, $\mu < \rho < 1$. By Proposition 2.4.1(i) there exists $K > 0$ such that for every $k \geq K$, $|z_{k+1}/z_k| < \rho$. This implies that $|z_{K+1}| < \rho|z_K|$, $|z_{K+2}| < \rho|z_{K+1}| < \rho^2|z_K|, \ldots, |z_{K+n}| < \rho^n|z_K|, \ldots$. The series $\sum_0^\infty \rho^k$ converges, and therefore by Proposition 2.6.2, the series $\sum_{k=m}^\infty |z_K|\rho^k$ also converges. Thus, by the comparison test the series $\sum z_k$ converges absolutely. Q.E.D.

We defer the case $\nu > 1$ to the exercises.

Remark (i). Often the ratio test is easier to apply than the kth root test. In many cases, $\lim|z_{k+1}/z_k|$ exists and can easily be found.

Remark (ii). If convergence is indicated by the ratio test, then it is also indicated by the kth root test, but the converse statement is not true. (See Exercise 2.6.11.)

Remark (iii). Let μ and ν be defined for $\sum_0^\infty z_k$ as in Theorem 2.6.2. It is possible for the series to converge even if $\mu = \infty$ and to diverge even if $\nu = 0$. (Exercise 2.6.12.)

The tests for convergence of series in **C** which proceed via the absolute convergence route, of course, all belong to the theory of series of nonnegative terms in real analysis. A detailed treatment would be inappropriate here; but we go just a little further in this direction to develop a standard set of comparison series.

Proposition 2.6.7 *Let $\sum_m^\infty \alpha_k$ be a series of nonnegative real numbers, and let $S = \{\sum_m^n \alpha_k, n = m, m + 1, \ldots\}$. If $\sup S = \infty$, the series diverges to infinity, and if $\sup S < \infty$, the series converges to the sum $\sup S$.*

Proof. The sequence $\langle s_n = \sum_m^n \alpha_k \rangle_{n=m}^\infty$ is clearly increasing, so Theorem 2.4.3 is applicable. Q.E.D.

Notice that the only possible mode of divergence of such a series is divergence to infinity.

In connection with the following proposition, we assume here and elsewhere in this book that the reader is familiar with the definition and elementary properties of the Riemann integral as presented in elementary calculus textbooks.

Proposition 2.6.8 *Let $\sum_m^\infty \alpha_k$ be a series of nonnegative real numbers and let the sequence $\langle \alpha_k \rangle$ be decreasing. Let f be a function from the nonnegative reals into the nonnegative reals such that $f(x) \leq f(x')$ if $x > x'$ and such that $f(k) = \alpha_k$, $k = m, m + 1, \ldots$. Then either the series $\sum_m^\infty \alpha_k$ and the sequence $\langle \int_m^{n+1} f(x)\,dx \rangle_{n=m}^\infty$ are both convergent, or else the series diverges to infinity and the sequence has the limit ∞.*

Proof. By the definition of the Riemann integral and the decreasing character of $f(x)$, we have

$$\sum_{k=m}^{n} f(k)(k + 1 - k) \geq \int_m^{n+1} f(x)\,dx \geq \sum_{k=m}^{n} f(k + 1)(k + 1 - k),$$

which is equivalent to

$$\sum_m^n \alpha_k \geq \int_m^{n+1} f(x)\,dx \geq \sum_m^n \alpha_{k+1}.$$

The conclusions of the proposition follows at once from this inequality. For instance, if the sequence in the theorem converges to a finite limit, then the partial sums $\alpha_{m+1} + \alpha_{m+2} + \cdots + \alpha_{n+1}$, $n = m, m + 1, \ldots$, are bounded, and convergence of the series follows from Proposition 2.6.7. The reader can easily supply the remaining details.

Corollary The series $\sum_1^\infty (1/k)^p$ and $\sum_2^\infty 1/[k(\log k)^p]$ converge for $p > 1$ but diverge for $p \leq 1$.

The proof is left as an exercise. A special case encountered frequently is the harmonic series $1 + (1/2) + (1/3) + \cdots$, which is thus seen to diverge to infinity.

To complement the information in the corollary in the divergence cases, we conclude the text of this section with the simple convergence test for *alternating series.*

Theorem 2.6.3 *Let $\langle \alpha_k \rangle_0^\infty$ be a decreasing sequence of nonnegative reals such that $\lim \alpha_k = 0$. The series $\alpha_0 - \alpha_1 + \alpha_2 - \alpha_3 + \cdots$ converges to a sum s, and $0 \le s \le \alpha_0$.*

Proof. As usual, let $s_n = \sum_0^n (-1)^k \alpha_k$. Then $s_{2m} = \alpha_0 - (\alpha_1 - \alpha_2) - (\alpha_3 - \alpha_4) - \cdots - (\alpha_{2m-1} - \alpha_{2m})$. The terms in parentheses are nonnegative, so the subsequence $\langle s_{2m} \rangle$ is decreasing and, therefore, has a limit, say $\sigma \ge -\infty$. Also it is clear that $s_{2m} \le \alpha_0$ for every m.

Now $s_{2m+1} = (\alpha_0 - \alpha_1) + (\alpha_2 - \alpha_3) + \cdots + (\alpha_{2m} - \alpha_{2m+1})$. The terms in parentheses are nonnegative, so $\langle s_{2m+1} \rangle$ is increasing and must have a limit, say $\tau \le \infty$. Also $s_{2m+1} \ge 0$. It is to be noted that $s_{2m+1} = s_{2m} - \alpha_{2m+1}$, so $s_{2m+1} \le s_{2m}$.

By combining these observations, we obtain two inequalities:

$$0 \le s_{2m+1} \le s_{2m} \le \alpha_0,$$

$$0 < s_{2m} - s_{2m+1} = \alpha_{2m+1}.$$

The first one shows that σ and τ are finite and $0 \le \sigma, \tau \le \alpha_0$ [Exercise 2.3.2 with $y = 0$, $r = \alpha_0$]. By Theorem 2.4.1, $\lim (s_{2m} - s_{2m+1}) = \sigma - \tau$. But by hypothesis, $\lim \alpha_{2m+1} = 0$, so by Exercises 2.4.3 and 2.4.4, $\sigma - \tau = 0$ or $\sigma = \tau$. We write $s = \sigma = \tau$.

Take any $\varepsilon > 0$ and let K_1 be such that for every $m \ge K_1$, $|s_{2m} - s| < \varepsilon$. Let K_2 be such that for every $m \ge K_2$, $|s_{2m+1} - s| < \varepsilon$. Then for every $n \ge \max(2K_1, 2K_2 + 1)$, $|s_n - s| < \varepsilon$. Q.E.D.

For example, the theorem assures the convergence of $1 - 1/2^p + 1/3^p - 1/4^p + \cdots$, $p > 0$, to a sum s, $0 \le s \le 1$.

Further information can be obtained by combining the theorem with Proposition 2.6.1. We obtain the following result.

Corollary The sum s in Theorem 2.6.3 satisfies $0 \le s - (\alpha_0 - \alpha_1 + \cdots - \alpha_{2m-1}) \le \alpha_{2m}$ and $0 \le \alpha_0 - \alpha_1 + \cdots + \alpha_{2m} - s \le \alpha_{2m+1}$.

Example. As a numerical illustration which is used later in Chapter 6, consider the series $1 - (2^2/2!) + (2^4/4!) - (2^6/6!) + \cdots$, where $k! = k(k-1)\cdot \cdots \cdot 3 \cdot 2 \cdot 1$. Recognize that this series represents the cosine of 2 radians. The terms are decreasing after the first one, because for $k \ge 1$, we have $1 > 4/[(2k+2)(2k+1)]$, which implies that $2^{2k}/(2k)! > 2^{2k+2}/(2k+2)!$. Hence, $(\cos 2) - 1 + 2^2/2! < 2^4/4!$, which is equivalent to $\cos 2 < -1/3$.

Exercises 2.6

1. Given a sequence $\langle w_n \rangle_0^\infty$ in **C**, construct an infinite series for which the numbers w_n are the partial sums.
2. Prove Proposition 2.6.2 by using relevant results for sequences of reals and Theorem 2.5.1.
3. Prove that if the series $\sum_0^\infty z_k$ and $\sum_0^\infty w_k$ converge absolutely, then so do the series $\sum_0^\infty (z_k + w_k)$ and the series $\sum zz_k$, $z \in \mathbf{C}$.
4. Prove the comparison test, Proposition 2.6.6.
5. Describe the set of values of z for which $\sum_{n=1}^\infty [(z-1)/(z+1)]^n$ converges absolutely.
6. The series $1 + \sum_1^\infty [a(a-1)(a-2)\cdots(a-k+1)/k!]z^k$, a, $z \in \mathbf{C}$, is called the binomial series. Show that if a is fixed, the series converges absolutely for every z for which $|z| < 1$. Also show that if z is fixed, $|z| = r < 1$, the series converges absolutely for all $a \in \mathbf{C}$.
7. Test the series $\sum_1^\infty i^k/k$ for convergence and for absolute convergence. (*Hint*: The kth root and ratio tests fail; take real and imaginary parts.)
8. Test the series $\sum_1^\infty (k!i^k/k^k)$ for convergence and absolute convergence.
9. Test the series

$$\sum_1^\infty \left(\frac{k^2-1}{k^2+5k+4}\right)^k i^k$$

for convergence. (*Hint*: Use the fact that $\lim_{n\to\infty} [1 + (\alpha/n)]^n = e^\alpha$.)

10. Given the series $\sum_0^\infty z_k$ in which $z_k \neq 0$, $k = 0,1,2,\ldots$. Prove that if $\underline{\lim} |z_{k+1}/z_k| > 1$, then the series diverges. (*Hint*: The proof resembles that of the divergence case in the kth root test.)
11. Show that if $\sum_1^\infty z_k$ satisfies the ratio test for convergence, then it also satisfies the kth root test for convergence. (*Hint*: Look at the inequalities in the proof of the ratio test which precede the comparison with the series $\sum \rho^k$.)
12. Find a convergent series $\sum_0^\infty z_k$ with $z_k \neq 0$, $k = 0,1,2,\ldots$, such that $\mu = \overline{\lim} |z_{k+1}/z_k| = \infty$.
13. Show that for the series $1/4 + 1/2 + 1/8 + 1/4 + 1/16 + 1/8 + 1/32 + \cdots$ of which the terms are given by $z_{2k} = 1/2^k$, $z_{2k-1} = 1/2^{k+1}$, $K = 1,2,\ldots$, the ratio test in inconclusive, but the kth root test indicates convergence.
14. Verify the corollary of Proposition 2.6.8. (*Hint*: You can use the fact that $\lim_{n\to\infty} \log n = \infty$.)
15. Let l^2 be the set of all sequences $\langle z_k \rangle_0^\infty$ in **C** such that $\sum |z_k|^2$ converges. Recall (See. 2.1) that the sum $\langle z_k \rangle + \langle w_k \rangle$ of two sequences $\langle z_k \rangle$ and $\langle w_k \rangle$ is the sequence $\langle z_k + w_k \rangle$, and the product $\langle z_k \rangle \cdot \langle w_k \rangle$ is $\langle z_k w_k \rangle$. Multiplication by a constant c is defined to be $\langle cz_k \rangle$. Prove that l^2 is closed under the operations $+$, $\cdot$, and multiplication by constants. (*Hint*: Use the Cauchy and Minkowski inequalities which appear in Sec. 1.13.)

16. Given $\sum_0^\infty z_k$. As usual let $s_n = z_0 + z_1 + \cdots + z_n$. The series is said to be summable $(C, 1)$ to the sum s if $\lim_{n\to\infty} [(s_0 + s_1 + \cdots + s_n)/(n + 1)] = s$. (The C refers to Cesàro, who generalized this averaging process.) Prove that if the series converges to s, it is summable $(C, 1)$ to s. (*Hint*: Exercise 2.5.7.) Show by example that a series can be summable $(C, 1)$ without converging.

2.7 Algebraic Operations on Infinite Series

We review briefly here the definition of a vector space or linear space. A vector space V over a field $\mathscr{F}$ (in which the elements are called scalars) is a set of elements, called vectors, and an operation $+$, called addition, such that (i) to every ordered pair (u, v) of vectors in V there corresponds a unique vector $u + v$ in V (called their sum); (ii) $u + v = v + u$; (iii) $(u + v) + w = u + (v + w)$; (iv) there is a neutral element in V, denoted by $\mathbf{0}$ and called the zero vector, such that $\mathbf{0} + u = u + \mathbf{0} = u$; (v) for every ordered pair (u, v) of vectors there is a vector x such that $u + x = v$; (vi) for every scalar $c \in \mathscr{F}$ and every $u \in V$ there is a unique vector $c \cdot u \in V$, called the scalar product, with the properties $(c_1, c_2 \in \mathscr{F})$ (1) $(c_1 + c_2)\cdot u = c_1 \cdot u + c_2 \cdot u$, (2) $c \cdot (u + v) = c\cdot u + c \cdot v$, (3) $(c_1 c_2)\cdot u = c_1(c_2 \cdot u)$, and (4) $1 \cdot u = u$.

Consider now the family S of all series in $\mathbf{C}$ of the type $\sum_0^\infty z_k$. The operation of addition for two such series and of multiplication by a complex number z are defined as follows (regardless of convergence): $(\sum z_k) + (\sum w_k) = \sum (z_k + w_k)$; $z \cdot \sum z_k = \sum z z_k$. It is clear from these definitions and from the properties of the complex field that with the structure provided by these operations, the set S is a vector space over the complex field $\mathbf{C}$. The neutral element $\mathbf{0}$ is the series $0 + 0 + 0 + \cdots$, a convergent series.

Proposition 2.6.2 shows that the subset of S consisting of the convergent series is closed under the operations in question, so it is a subspace of the vector space of series. The set of absolutely convergent series is a subspace of this subspace.

This brings up the question of whether a useful multiplication of series can be defined under which the family of convergent series would again be closed. The key word here is "useful." There are two types of multiplication of series in general use, both of which preserve absolute convergence of the factor series and neither of which necessarily preserves convergence.

The more simple type is the *Dirichlet product*. Let $\sum_0^\infty z_k$ and $\sum_0^\infty w_k$ be any two series; the Dirichlet product series is $\sum_0^\infty z_k w_k$.

Proposition 2.7.1 *If $\sum_0^\infty z_k$ and $\sum_0^\infty w_k$ converge absolutely, then so does $\sum_0^\infty z_k w_k$.*

The proof is an exercise. The Dirichlet product series does not in general converge to the product of the sums. For example, with $|z| < 1$, the product series for $1 + z + z^2 + \cdots = 1/(1 - z)$ with itself is $1 + z^2 + z^4 + \cdots = 1/(1 - z^2) \neq [1/(1 - z)]^2$.

A more elaborate criterion for the convergence of $\sum z_k w_k$, in which neither of the original series needs to converge, is found in Proposition 3.4.7.

The most commonly used type of multiplication operation for series in complex analysis results in the *Cauchy product series.* The Cauchy product series for $\sum_0^\infty z_k$ and $\sum_0^\infty w_k$ is the series $\sum_0^\infty c_n$ where $c_n = z_0 w_n + z_1 w_{n-1} + \cdots + z_n w_0$, $n = 0, 1, 2, \ldots$. It is brought out in the following propositions that if the Cauchy product series does converge, it converges to the "right" sum, unlike the Dirichlet product. The Cauchy product series is interesting in complex analysis chiefly because if $\sum_0^\infty z_k$ and $\sum_0^\infty w_k$ are of the type $\sum_0^\infty a_n z^k$, $\sum_0^\infty b_n z^k$, then the product series is of the same type. Series of this type are known as *power series* and are of central importance in complex function theory.

Proposition 2.7.2 *If $\sum_0^\infty z_k$ and $\sum_0^\infty w_k$ converge absolutely, then so does the Cauchy product series.*

Proof. With the notation in the definition,

$$\begin{aligned} 0 &\le \sum_0^m |c_n| \le |z_0|\,|w_0| + (|z_1|\,|w_0| + |z_0|\,|w_1|) + \cdots \\ &\quad + (|z_m|\,|w_0| + \cdots + |z_0|\,|w_m|) \le \sum_0^m |z_k| \sum_0^m |w_k| \\ &\le \sum_0^\infty |z_k| \sum_0^\infty |w_k| < \infty. \end{aligned}$$

But $\langle \sum_{n=0}^m |c_n| \rangle_{m=0}^\infty$ is increasing, so since it is bounded, it has a finite limit. (Theorem 2.4.3.) Q.E.D.

Proposition 2.7.3 *Let $\sum_0^\infty z_k$ converge to z and $\sum_0^\infty w_k$ converge to w. Then if the Cauchy product series converges, it must converge to zw.*

Proof. The proof depends upon Exercises 2.5.7 and 2.5.8. Let $s_n = \sum_0^n z_k$, $t_n = \sum_0^n w_k$, and $u_n = \sum_0^n c_k$, where $\sum_0^\infty c_k$ is the Cauchy product series. It can readily be verified that $u_k = w_0 s_k + w_1 s_{k-1} + w_2 s_{k-2} + \cdots + w_k s_0$, and then that

$$\sum_{k=0}^n u_k = t_0 s_n + t_1 s_{n-1} + t_2 s_{n-2} + \cdots + t_n s_0.$$

Now by Exercise 2.5.8 the sequence

$$\left\langle \frac{t_0 s_n + t_1 s_{n-1} + \cdots + t_n s_0}{n + 1} \right\rangle_0^\infty \tag{2.7.1}$$

converges to zw. Thus, the sequence $S = \langle \sum_0^n u_k/(n+1) \rangle_0^\infty$, which is identical with the sequence in (2.7.1), must converge to zw. But by Exercise 2.5.7, if the sequence $\langle u_k \rangle_0^\infty$ converges, then so does S, and to the same limit. Q.E.D.

The method of proof establishes at the same time a more general statement, which we phrase in the language introduced in Exercise 2.6.16.

Theorem 2.7.1 *The Cauchy product series of two convergent series with respective sums z and w is summable* (C, 1) *to zw.*

In the literature of infinite series there are numerous results of a type in which convergence of the Cauchy product is derived from weaker hypotheses than the absolute convergence of the factor series. Perhaps the best known is the following theorem.

Theorem 2.7.2 (*Mertens' Theorem*). *The Cauchy product series converges if one of the factor series converges absolutely and the other merely converges.*

For a proof, see [3, pp. 92–93].

It is possible for neither of the factor series to converge absolutely and for the Cauchy product series to converge. A result along these lines which provides a link with Dirichlet multiplication to given by Bromwich [3, p. 94] and is ascribed to Pringsheim. It reads as follows.

Proposition 2.7.4 *Let* $\langle \alpha_k \rangle_0^\infty$ *and* $\langle \beta_k \rangle_0^\infty$ *be strictly decreasing sequences such that* $\lim \alpha_k = \lim \beta_k = 0$. *If* $\sum^\infty \alpha_k \beta_k$ *converges, then so does the Cauchy product of* $\sum_0^\infty (-1)^{k+1}\alpha_k$ *and* $\sum_0^\infty (-1)^{k+1}\beta_k$.

The reader is referred to Bromwich [3, p. 94] for a proof. The alternating series in the conclusion of the proposition both converge, by the alternating series test.

Exercises 2.7

1. Show by an example that the Dirichlet product of two convergent series need not converge.
2. Prove Proposition 2.7.1. (*Hint*: An argument something like the proof of Proposition 2.7.2 suffices.)
3. Prove that the Cauchy product of the convergent series $1 - (1/2)^p + (1/3)^p - \cdots + (-1)^{n+1}(1/n)^p + \cdots$ with itself diverges if $0 < p \le 1/2$. (*Hint*: In the notation of the definition of Cauchy product, with $z_0 = 0$, $c_{n+1} = 1^p \cdot (-1)^{n+1}(1/n)^p - (1/2)^p(-1)^n(1/(n-1))^p + (1/3)^p(-1)^{n-1}(1/(n-2))^p - \cdots + (-1)^{n+1}(1/n)^p \cdot 1^p = (-1)^{n+1}[1^p(1/n)^p + (1/2)^p(1/n-1)^p + \cdots + (1/n)^p \cdot 1^p]$. Notice that each term in the square bracket is greater than $(1/n)^p(1/n)^p = (1/n)^{2p}$, and thereby show that $\langle c_n \rangle$ cannot converge to zero.)

4. Show that the Cauchy product in Exercise 2.7.3 does converge for $p > 1/2$, although the original series does not converge absolutely for $1/2 < p \leq 1$. (*Hint*: Proposition 2.7.4.)
5. Show that the Cauchy product of two power series $\sum_0^\infty a_k(z - a)^k$ and $\sum_0^\infty b_k(z - a)^k$ is another power series of the same type.
6. Let $\mathscr{A}$ be a set containing more than one element, and let there be given two functions $+$ and $\cdot$ from $\mathscr{A} \times \mathscr{A}$ into $\mathscr{A}$, called addition and multiplication. Furthermore, let $\mathscr{F}$ be a given field, and let there be a third operation, called scalar multiplication, which is a function from $\mathscr{F} \times \mathscr{A}$ into $\mathscr{A}$. The set $\mathscr{A}$ with this structure is called an *algebra over* $\mathscr{F}$ if
   (i) $\mathscr{A}$ with $+$ and scalar multiplication is a vector space over $\mathscr{F}$;
   (ii) The following further postulates are satisfied:
   (a) $(x \cdot y) \cdot z = x \cdot (y \cdot z)$;
   (b) $x \cdot (y + z) = x \cdot y + x \cdot z$; $(y + z) \cdot x = y \cdot x + z \cdot x$;
   (c) $(ax) \cdot (by) = (ab)x \cdot y$ where $x, y \in \mathscr{A}$, $a, b \in \mathscr{F}$, and ax denotes scalar multiplication.

   In the second and third paragraphs of this section, it was observed that the infinite series of type $\sum_0^\infty z_k$ in **C** with the usual addition and multiplication by numbers (here to be regarded as scalar multiplication) form a linear vector space over **C**, and the absolutely convergent series are a subspace. Starting with this, prove the following.

   (A) With Dirichlet multiplication used as $\cdot$, the vector space of series of the previously discussed type form an algebra over **C**.

   (B) The absolutely convergent series with Dirichlet multiplication are again an algebra over **C**.

   (C) With the Cauchy product used for $\cdot$, the conclusion in (A) follows.

   (D) With the Cauchy product used for $\cdot$, the conclusion in (B) follows.

3

Sequences and Series of Complex-valued Functions

3.1 Pointwise Convergence

Let X be a nonempty set of functions from **C** into **C** which have a common domain E. A sequence of these functions in X is formally defined exactly as a sequence is defined in Sec. 2.1: It is a function from the real integers **Z** into X with a domain of the type described in that section. The usual symbolism for a sequence of functions is typified by $\langle f_k \rangle$, $\langle f_k \rangle_m^\infty$.

But something new is present in the case of sequences of functions. Associated with a sequence of functions $\langle f_k \rangle$, there is a family of sequences of complex numbers $\langle f_k(z) \rangle$, one sequence for each point $z \in E$. It is by means of these associated sequences of functional values that algebraic operations on sequences of functions and various concepts of convergence are defined.

The sequence $\langle f_k \rangle$ is said to converge at a point $z_0 \in E$, if there is a $w_0 \in \mathbf{C}$ such that $\langle f_k(z_0) \rangle$ converges to w_0. The sequence is said to *converge pointwise* on E, if it converges at each point of E. (The word "pointwise" is often dropped when no confusion can result.) Since the limit, when it exists, of a sequence in **C** is unique (Sec. 2.3, Remark (iii)), in the case of pointwise convergence the limits define a function from **C** into **C** with domain E. This function is called the pointwise limit or limit function of the sequence.

The sequence $\langle f_k \rangle$ is called a Cauchy sequence at a point $z_0 \in E$ if and only if $\langle f_k(z_0) \rangle$ is a Cauchy sequence in **C**. The sequence $\langle f_k \rangle$ is said to have the *Cauchy property pointwise* on E (or just the Cauchy property on E) if and only if $\langle f_k(z) \rangle$ is a Cauchy sequence for every $z \in E$.

With these definitions, many facts about convergence of numerical sequences in **C** carry over immediately to convergence of sequences of functions at a point and pointwise. We summarize some of the more important ones in the following propositions, using the language for the pointwise case. The references in square brackets provide instant proofs. The functions f, g, f_k, g_k, which appear in these propositions, are all complex-valued functions with a common domain $E \subset \mathbf{C}$.

Proposition 3.1.1 *Let $f_k = u_k + iv_k$, $k = 0, 1, 2, \ldots$, be functions with common domain $E \subset \mathbf{C}$, and let u_k and v_k be real valued for each k. The sequence $\langle f_k \rangle$ converges pointwise on E to the function $f = u + iv$ if and only if the sequences $\langle u_k \rangle$ and $\langle v_k \rangle$ converge pointwise on E, respectively, to u and v. The sequence $\langle f_k \rangle$ has the Cauchy property pointwise on E if and only if the sequences $\langle u_k \rangle$ and $\langle v_k \rangle$ each have the Cauchy property pointwise on E* [*Theorem 2.5.1*].

Proposition 3.1.2 *The sequence $\langle f_k \rangle$ converges pointwise on E if and only if it has the Cauchy property pointwise on E* [*Corollary 1 of Theorem 2.5.1*].

Proposition 3.1.3 *If $\langle f_k \rangle$ converges to the pointwise limit f on E, then $\langle |f_k| \rangle$ converges pointwise to the limit $|f|$, where $|f_k|$ and $|f|$ denote, respective, the functions with values $|f_k(z)|$ and $|f(z)|$. If $\langle |f_k| \rangle$ converges pointwise on E to zero, then so does $\langle f_k \rangle$* [*Exercise 2.5.1*].

Proposition 3.1.4 *If $\langle f_k \rangle_0^\infty$ and $\langle g_k \rangle_0^\infty$ converge pointwise on E to f and g, respectively, then $\langle f_k \pm g_k \rangle$ and $\langle f_k \cdot g_k \rangle$ converge pointwise on E, respectively, to $f \pm g$ and $f \cdot g$; and if $g(z) \neq 0$, $g_k(z) \neq 0$ for every $z \in E$, $k = 0, 1, 2, \ldots$, then $\langle f_k/g_k \rangle$ converges pointwise on E to f/g* [*Exercise 2.5.5*].

It is consistent with the definition of addition and multiplication of sequences in Sec. 2.1 and addition and multiplication of functions in Sec. 1.7 to make the following definitions for sum, difference, and product of sequences of functions. Let $\langle f_k \rangle_m^\infty$ and $\langle g_k \rangle_m^\infty$ be two sequences of functions into **C** with common domain $E \subset \mathbf{C}$. Their sum, $\langle f_k \rangle_m^\infty + \langle g_k \rangle_m^\infty$, is the sequence $\langle f_k + g_k \rangle_m^\infty$, in which function addition as defined in Sec. 1.7 appears. Their product, $\langle f_k \rangle_m^\infty \cdot \langle g_k \rangle_m^\infty$ is the sequence $\langle f_k \cdot g_k \rangle_m^\infty$, in which function multiplication as defined in Sec. 1.7 appears.

We further define the scalar product $a\langle f_k \rangle_m^\infty$ of $\langle f_k \rangle_m^\infty$ with a complex number a by the value sequence $\langle af_k(z) \rangle_m^\infty$. The difference $\langle f_k \rangle_m^\infty - \langle g_k \rangle_m^\infty$ is interpreted to mean $\langle f_k \rangle_m^\infty + (-1)\langle f_k \rangle_m^\infty$ and has the value sequence $\langle f_k(z) - g_k(z) \rangle_m^\infty$.

It follows from these definitions that the collection of all sequences of complex-valued functions of the type $\langle f_k \rangle_m^\infty$ with common domain E, taken with the operations just now defined, forms an algebra over **C**. The zero element in this algebra is the sequence $\langle f_k \rangle_m^\infty$ in which $f_k(z) \equiv 0$, $z \in E$,

$k = m, m + 1, \ldots$. (See Exercise 2.7.6 for the definition of "algebra" in this sense.) It further follows from Proposition 3.1.4 that the subcollection of all such sequences which converge pointwise on E again forms an algebra over **C**. (The null sequence already defined belongs to this subcollection. The fact that the scalar product of a pointwise convergent sequence with a number a is also pointwise convergent can be validated by choosing $f_k(z) \equiv a$, $z \in E$, in Proposition 3.1.4.)

3.2 Bounded and Continuous Functions

A function $f: E \to \mathbf{C}$, $E \subset \mathbf{C}$ is said to be *bounded* on E if there is a real $M > 0$ such that $|f(z)| \leq M$ for every $z \in E$. This amounts to saying that the condition for boundedness is the existence of a disk $\Delta(0, M)$ such that $\{f(z): z \in E\} \subset \Delta(0, M)$. The definition is consistent with the definition of a bounded sequence $\langle z_k \rangle$ given at the beginning of Sec. 2.5. In that case, E is a subset of the integers.

A collection of functions $f: E \to \mathbf{C}$ is said to be a *set of bounded functions* if each one has its own private bound M_f for z on E; that is, the functions are individually bounded but not necessarily collectively bounded. But a *uniformly bounded collection* or set of functions with domain E is a collection with the property that there is a real $M > 0$ such that every function f in the family satisfies $|f(z)| \leq M$ for every $z \in E$. To illustrate, $1, z, z^2, z^3, \ldots$ is a set of bounded functions on $E \in \Delta(0, 2) = \{z: |z| < 2\}$, but the sequence is not uniformly bounded on E.

A function $f: E \to \mathbf{C}$, $E \subset \mathbf{C}$, is said to be *continuous* at a point $z_0 \in E$ if for every $\varepsilon > 0$ there exists $\delta_{\varepsilon,z_0} > 0$ such that $|f(z) - f(z_0)| < \varepsilon$ for every $z \in \Delta(z_0, \delta_{\varepsilon,z_0}) \cap E$. It is said to be *continuous* on E when it is continuous at every point of E. It is said to be *uniformly continuous* on E if it is continuous on E and if δ_{ε,z_0} can be replaced by a single δ_ε valid for every $z_0 \in E$.

It might be noted that the definition of continuity at a point z_0 given here includes the case in which z_0 is an isolated point of E, that is, the case in which there exists a disk $\Delta(z_0, r)$, $r > 0$, with no other point of E in it. Thus, with this definition, it is correct to say that the function given by $f(z) = z^2$ is continuous on $E = \{1, 1/2, 1/3, \ldots, 0\}$.

Theorem 3.2.1 *Let the function $f: E \to \mathbf{C}$ have values $f(z) = u(z) + iv(z)$, $z = x + iy \in E$, where $u(z)$ and $v(z)$ are real. The function f is continuous at $z_0 \in E$ if and only if u and v are continuous at z_0; it is continuous on E if and only if u and v are continuous on E; it is uniformly continuous on E if and only if u and v are uniformly continuous on E.*

Proof. The technique used for complex sequences in the proof of Theorem

2.5.1 is applicable. In fact, for each z, $z_0 \in E$, we have the inequality:

$$\begin{aligned} |f(z) - f(z_0)| &= |u(z) - u(z_0) + i(v(z) - v(z_0))| \\ &\leq |u(z) - u(z_0)| + |v(z) - v(z_0)| \\ &\leq \sqrt{2}|f(z) - f(z_0)|. \end{aligned}$$

The remaining details are deferred to the exercises.

The complex-valued functions continuous on a set $E \subset \mathbf{C}$, taken with the usual operations of addition, multiplication, and multiplication by a complex constant, form an algebra over **C**. The proof of this is contained in Exercise 3.2.8. A notation such as $C(E)$ is traditional for this algebra.

Exercises 3.2

1. Let $\langle f_k \rangle_0^\infty$ be a sequence of complex-valued functions with common domain $E \subset \mathbf{C}$; let $\langle f_k \rangle_0^\infty$ converge pointwise on E to the limit f; and let there exist $M > 0$ such that $|f_k(z)| \leq M$, $k = 0,1,2, \ldots$, every $z \in E$. Prove that $|f(z)| \leq M$, $z \in E$.
2. Notice that the set of function $f_0, f_1, f_2, \ldots$ in Exercise 3.2.1 is uniformly bounded. Exhibit a sequence of bounded continuous functions which converge pointwise on a set E to a function which is not bounded on E. (*Hint*: Start with a limit function such as that given by $f(x) = 1/x$, x real, $0 < x \leq 1$. Let $f_k(x) = 1/x$ part of the way from 1 to 0.)
3. Let $f: E \to \mathbf{C}$, $E \subset \mathbf{C}$, be continuous at $z_0 \in E$. Prove the following statements:
   (a) Given any real number $M > |f(z_0)|$, there exists a disk $\Delta(z_0, r)$, $r > 0$, such that $|f(z)| < M$ for every $z \in \Delta(z_0, r) \cap E$.
   (b) If $f(z_0) \neq 0$, for any real number m, $0 \leq m < |f(z_0)|$, there exists a disk $\Delta(z_0, r)$, $r > 0$, such that $|f(z)| > m$ for every $z \in \Delta(z_0, r) \cap E$.
   (c) If f is *real valued* and $f(z_0) > 0$, then for any real number m, $0 \leq m < f(z_0)$, there exists $\Delta(z_0, r)$, $r > 0$, such that $f(z) > m$ for $z \in \Delta(z_0 r) \cap E$.
   (*Hint*: For (a), in the definition of continuity take $\varepsilon = \mathrm{M} - |f(z_0)|$; recall 1.13.4. For (b), take $\varepsilon = |f(z_0)| - m$.)
4. Let f be a complex-valued function with domain a bounded set $E \subset \mathbf{R}$. Let f be continuous at each point of E, and suppose that $f(x) = 0$ for some $x \in E$. Let $x_0 = \inf\{x : f(x) = 0, x \in E\}$. Prove that if $x_0 \in E$, then $f(x_0) = 0$. (*Hint*: By Exercise 1.4.1, for every $h > 0$, there must exist $x_h \in E$, $x_0 \leq x_h < x_0 + h$, with $f(x_h) = 0$. Now use Exercise 3.2.3(b).)
5. Let $f: \mathrm{E} \to \mathbf{C}$, $E \subset \mathbf{C}$ be continuous at $z_0 \in E$. Prove that the functions given by $\operatorname{Re} f(z)$, $\overline{f(z)}$, and $|f(z)|$, are also continuous at z_0. Also show that the identity mapping $f(z) = z$ is continuous on **C**.
6. Fill in the details of the proof of Theorem 3.2.1.
7. Let F_b be the collection of complex-valued functions which are bounded on a common domain $E \subset \mathbf{C}$. Show that F_b, taken with the usual operations of addition and multiplication of functions, forms an algebra over **C**. (See Exercise 2.7.6 for the definition of "algebra" in this sense.)

8. Let f and g be complex-valued functions with a common domain $E \subset \mathbf{C}$, and let f and g be continuous at a point $z_0 \in E$. Prove that $f \pm g$ and $f \cdot g$ are also continuous at z_0. Also show that f/g (domain $E \sim \{z : g(z) = 0\}$) is continuous at z_0 if $g(z_0) \neq 0$. (*Hint*: The proofs go through exactly as given for functions from $\mathbf{R}$ into $\mathbf{R}$ in the more rigorous calculus textbooks. For the product and quotient functions, Exercise 3.2.3(a) and (b) are needed, respectively.)
9. Starting with the assumption that the identity mapping given by $f(z) = z$ is continuous on $\mathbf{C}$ (Exercise 3.2.5), use Exercise 3.2.8 to show that the polynomial $\sum_0^n a_k z^k$ is continuous on $\mathbf{C}$. Then show that the rational function $r(z) = p(z)/q(z)$, where $p(z)$ and $q(z)$ are polynomials and $q(z)$ is not the zero polynomial, is continuous everywhere on $\mathbf{C}$ except at points z where $q(z) = 0$.
10. Let g have domain $E \subset \mathbf{C}$ and range $G \subset \mathbf{C}$; let f be a complex-valued function with domain G. Let g be continuous at the point $z_0 \in E$, and let f be continuous at the point $g(z_0) \in G$. Prove that the composition function $f \circ g$ defined by the values $f(g(z))$ is continuous at z_0. (*Hint*: Let $w = g(z)$. Write out the arithmetic criterion for $f(w)$ to be continuous at $w_0 = g(z_0)$ and then show that for any $\delta > 0$, there is a restriction on z which yields the inequality $|w - w_0| < \delta$.)

    Remark. The exercise implies that if g is continuous on E and f is continuous on G, then $f \circ g$ is continuous on E. A more elegant proof in a more general context is encountered in Exercise 5.2.2.
11. Let V be a vector space over the complex field $\mathbf{C}$. (See Sec. 2.7 for the definition.) A *norm* for V is a real-valued function p with domain V such that for all $u, v \in V$: (i) $p(u) \geq 0$ and $p(u) = 0$ if and only if $u = \mathbf{0}$; (ii) $p(a \cdot u) = |a| p(u)$ for all $a \in \mathbf{C}$; (iii) $p(u + v) \leq p(u) + p(v)$. It is easy to see that $p(u - v)$ satisfies the postulate for a metric given in Sec. 2.2. (Here $u - v$ means $u + (-v)$, where $(-v)$ is the solution of $v + x = \mathbf{0}$.)

    Now let V be the vector space over $\mathbf{C}$ of bounded functions $f: E \to \mathbf{C}$, $E \subset \mathbf{C}$, with the usual function addition, and with scalar multiplication defined by the usual multiplication of functions by constants. Show that $\|f\|_E = \sup\{|f(z)| : z \in E\}$ is a norm for this V. (*Hint*: For Postulate (ii) use Exercise 1.4.3. Fro (iii), pass to the "sup" one term at a time in the relevant inequality for absolute values.)
12. A normed algebra $\mathscr{A}$ is an algebra wich is a normed vector space, and for which the norm $p(u)$ further satisfies (iv) $p(u \cdot v) \leq p(u)p(v)$, $u, v \in \mathscr{A}$. Continue Exercise 3.2.11 by showing that with the usual multiplication of functions the particular V described there with the norm $\|f\|_E$ is a normed algebra.

3.3 Uniform Convergence of Sequences of Functions

Let $\langle f_k \rangle_m^\infty$ be a sequence of complex-valued functions, each with domain $E \subset \mathbf{C}$. The sequence is said to have the *uniform Cauchy property* on E if for every $\varepsilon > 0$ there exists $K_\varepsilon \geq m$ such that for every k, h, with $k > h \geq K_\varepsilon$, $\sup\{|f_k(z) - f_h(z)| : z \in E\} \leq \varepsilon$. The sequence is said to *converge uniformly* on E to the limit function f if it converges pointwise on E to f and for every

$\varepsilon > 0$ there exists $K_\varepsilon \geq m$ such that for $k \geq K_\varepsilon$, $\sup\{|f_k(z) - f(z)|: z \in E\} \leq \varepsilon$.

Remark (i). It is clear that if the sequence of functions has the Cauchy property uniformly on E, then it has the Cauchy property pointwise on E.

Remark (ii). It is also obvious that if $\langle f_k \rangle_m^\infty$ converges to f uniformly on E, then for every $n > m$ the sequence $\langle f_k \rangle_{k=n}^\infty$ converges uniformly on E to f; and conversely, if for some $n > m$ the sequence $\langle f_k \rangle_n^\infty$ converges uniformly on E to f, then so does $\langle f_k \rangle_m^\infty$.

Examples. The sequence $\langle z^k/K \rangle_1^\infty$, converges uniformly to the constant function $f(z) \equiv 0$ on $\{z:|z| \leq 1\}$. The sequence $\langle z^k \rangle_0^\infty$ converges uniformly to $f(z) = 0$ on $\Delta(0, 1/2)$.

Theorem 3.3.1 *If $\langle f_k \rangle_m^\infty$ converges uniformly on E, then it has the uniform Cauchy property on E. If $\langle f_k \rangle_m^\infty$ has the uniform Cauchy property on E, then it converges uniformly on E.*

Proof. The first sentence is proved by applying the definitions to the inequality $|f_k(z) - f_h(z)| \leq |f_k(z) - f(z)| + |f(z) - f_h(z)|$, where f is the limit function implied by the hypothesis. We concentrate on the second sentence in which the uniform Cauchy property is the hypothesis.

This hypothesis implies that for every $\varepsilon > 0$ there exists $K_\varepsilon \geq m$ such that for every $z \in E$ and for $k > h \geq K_\varepsilon$, $|f_k(z) - f_h(z)| \leq \varepsilon$. According to Proposition 3.1.2 and Remark (i) in this section, the sequence $\langle f_k \rangle_m^\infty$ converges pointwise on E to some function f. Hold h temporarily fixed. By Proposition 3.1.4, the sequence $\langle f_k - f_h \rangle_{k=m}^\infty$ converges pointwise on E to $f - f_h$. But then by Exercise 3.2.1, for every $z \in E$, $|f(z) - f_h(z)| \leq \varepsilon$.

Now notice that this last inequality is valid for every $h \geq K$. Thus, the definition of uniform convergence of $\langle f_h \rangle_{k=m}^\infty$ to f on E is satisfied. Q.E.D.

Some further insight into uniform convergence may be provided by the following negative statement.

Proposition 3.3.1 *Let $\langle f_k \rangle_0^\infty$ be a sequence of complex-valued functions with common domain E, converging pointwise on E to a function f. Suppose that there is an infinite subsequence $\langle f_{\phi(k)} \rangle$ with the property that for some $\alpha > 0$ and for each k, $k = 0, 1, 2, \ldots$, there is a point $z = z_{\phi(k)} \in E$ such that $|f_{\phi(k)}(z_{\phi(k)}) - f(z_{\phi(k)})| \geq \alpha$. Then $\langle f_k \rangle_0^\infty$ does not converge uniformly on E.*

Proof. The hypothesis implies that $\sup\{|f_{\phi(k)}(z) - f(z)|: z \in E\} \geq \alpha$, $k = 0, 1, 2, \ldots$. But if there were uniform convergence, then there would be a number K such that $|f_{h'}(z) - f(z)| \leq \alpha/2$ for every $h' \geq K$ and every $z \in E$. Now the sequence $\langle \phi(k) \rangle_0^\infty$ is strictly increasing, by definition of subsequence, so eventually $\phi(k) \geq K$, and a contradiction is obtained. Q.E.D.

Theorem 3.3.2 *Let $\langle f_h \rangle_m^\infty$ and $\langle g_h \rangle_m^\infty$ be two sequences of complex-valued functions with common domain $E \subset \mathbf{C}$. If the sequences converge uniformly on*

E, respectively to functions f and g, then the sequences $\langle f_h \pm g_h\rangle_m^\infty$ *converge uniformly on E to* $f \pm g$. *Also if* $a \in \mathbf{C}$, *the sequence with values* $\langle af_h(z)\rangle_m^\infty$ *converges uniformly on E to* $a\cdot f$.

The proof is deferred to the exercises.

The conclusion of uniform convergence in Theorem 3.3.2 is not true in general for the product sequence, although, of course, pointwise convergence takes place [Proposition 3.1.4]. For example, let $f_k(z) = g_k(z) = (1/z) + (1/k)$, $k = 1, 2, \ldots$. The sequences $\langle f_k\rangle$ and $\langle g_k\rangle$ converge uniformly to $1/z$ on $E = \mathbf{C} \sim \{0\}$. Now

$$(f_k\cdot g_k)(z) = \frac{1}{z^2} + \frac{2}{z\cdot k} + \frac{1}{k^2},$$

and $\langle f_k\cdot g_k\rangle$ converges pointwise to $1/z^2$ on E. But it does not converge uniformly to $1/z^2$, because for every k, $k = 1, 2, \ldots$, it is possible to find $z_k \in E$ such that

$$\left|(f_k\cdot g_k)(z_k) - \frac{1}{z_k^2}\right| = \left|\frac{2}{kz_k} + \frac{1}{k^2}\right| \geq 1.$$

(For example, take $z_k = 1/k$.) Therefore, by Proposition 3.3.1, the convergence is not uniform.

Theorem 3.3.3 *Let* $\langle f_k\rangle_m^\infty$ *be a sequence of complex-valued functions continuous on* $E \subset \mathbf{C}$. *If the sequence converges uniformly on E to the limit function f, then f is continuous on E.*

Proof. For any z, $z_0 \in E$ and any k, $k = m, m+1, \ldots$, we have the inequality

$$|f(z) - f(z_0)| \leq |f(z) - f_k(z)| + |f_k(z) - f_k(z_0)| + |f_k(z_0) - f(z_0)|.$$

By uniform convergence, given any $\varepsilon > 0$, there exists an integer k independent of z, $z_0 \in E$ such that the first and last terms in the sum on the right-hand side do not exceed $\varepsilon/3$. Hold this k fixed. By the continuity of f_k at z_0, there exists a disk $\Delta(z_0, \delta)$ such that $|f_k(z) - f_k(z_0)| < \varepsilon/3$ for every $z \in \Delta(z_0, \delta) \cap E$. This shows that $|f(z) - f(z_0)| < (\varepsilon/3) + (\varepsilon/3) + (\varepsilon/3) = \varepsilon$ for every $z \in \Delta(z_0, \delta) \cap E$. Q.E.D.

Example. Theorem 3.3.3 can be used in the contrapositive to detect the absence of uniform convergence. Look at the sequence $\langle f_k(x)\rangle = \langle(kx^2 - 1)/(kx^2 + 1)\rangle_0^\infty$ with domain $E = \{x : x \geq 0\}$. It is easily seen that $\lim f_k(x) = 1$, $x > 0$, and $f_k(0) = -1$, $k = 0, 1, 2, \ldots$. The limit function is not continuous on E, so the sequence cannot converge uniformly on E.

Proposition 3.3.2 (*Comparison Test for Uniform Convergence of Sequences*). *Let* $\langle f_k\rangle_m^\infty$ *be a sequence of complex-valued functions, each with domain* $E \subset \mathbf{C}$.

If there exists a convergent sequence of complex constants $\langle a_k \rangle_m^\infty$ *and an integer* K *such that for* $k > h \geq K$, $|f_k(z) - f_h(z)| \leq |a_k - a_h|$ *for every* $z \in E$, *then* $\langle f_k \rangle$ *converges uniformly on* E.

Proof. The hypothesis implies that $\langle f_k \rangle$ has the uniform Cauchy property on E, since by the corollary of Theorem 2.5.1, $\langle a_k \rangle$ is a Cauchy sequence. Q.E.D.

We conclude the text of this section with some remarks of an elementary function-theoretical nature concerning bounded functions and bounded continuous functions. Let F_b denote the set of all bounded complex-valued functions which have a common domain $E \subset \mathbf{C}$, and let F_{bc} denote the subset of F_b consisting of the bounded functions continuous on E. According to Exercises 3.2.7 and 3.2.8, F_b and F_{bc} taken with the usual operations of functional addition, multiplication, and multiplication by constants, are algebras over $\mathbf{C}$. In Exercises 3.2.11 and 3.2.12 it was brought out that $\|f\|_E = \sup\{|f(z)| : z \in E\}$ provides a norm for these algebras. The norm also sets up a metric $d^*(f, g) = \|f - g\|_E$ on the sets F_b and F_{bc}.

Now let $\langle f_h \rangle_m^\infty$ denote a sequence in either F_b or F_{bc}. The definition of the uniform Cauchy property can be rephrased to read: For every $\varepsilon > 0$, there exists $K_\varepsilon > 0$ such that for $k > h \geq K_\varepsilon$, $\|f_k - f_h\|_E \leq \varepsilon$. This is equivalent to saying that with the "sup norm" metric d^*, a sequence with the uniform Cauchy property is a Cauchy sequence in the metric space (F_b, d^*) or (F_{bc}, d^*), as the case may be. The definition of uniform convergence to a limit function f can be written: The sequence $\langle f_h \rangle$ converges pointwise on E to f and $\lim_{k\to\infty} \|f_k - f\|_E = 0$. This is equivalent to saying that the sequence $\langle f_k \rangle$ of elements in the metric space (F_b, d^*) or (F_{bc}, d^*) converges to f.

It can be shown (Exercise 3.3.5) that the limit function of a convergent sequence $\langle f_k \rangle$ in (F_b, d^*) is itself an element of F_b. Theorem 3.3.3 with the result just cited implies in the present context that the limit function of a convergent sequence in (F_{bc}, d^*) is itself an element of F_{bc}. Theorem 3.3.1 establishes that every Cauchy sequence in (F_b, d^*) or (F_{bc}, d^*) does converge to a limit function. Recall that a metric space in which every Cauchy sequence has a limit in that space is called Cauchy complete. In short, $(F_b, \|\cdot\|_E)$ and $(F_{bc}, \|\cdot\|_E)$ are Cauchy complete normed algebras.

A Cauchy complete normed vector space is called a *Banach space*, and a Cauchy complete normed algebra is called a *Banach algebra*. A more elementary example of a Banach algebra is the complex number system with absolute value as norm.

Exercises 3.3

1. Show that $\langle z^k \rangle_0^\infty$ converges but does not converge uniformly on $\Delta(0, 1) \cup \{1\}$. (*Hint*: Theorem 3.3.3.)
2. Show that $\langle (1 - z^k)/(1 - z) \rangle_0^\infty$ converges uniformly on any $\Delta(0, r)$, $0 < r < 1$, but

does not converge uniformly on $\Delta(0, 1)$. (*Hint*: For the nonuniformity in Proposition 3.3.1, take $\alpha = 1$ and $\phi(k) = k$. It suffices to let $z_{\phi(k)}$ be real.)

3. Let r be real, $0 < r < 1$, and let $E_1 = \{z : |z| \leq r\}$ and $E_2 = \{z : |z| \geq 1/r\}$. Show that $\langle (1 - z^k)/(1 + z^k) \rangle_0^\infty$ converges uniformly on both E_1 and E_2. What is the limit function?
4. Let $\langle f_k = u_k + iv_k \rangle_m^\infty$ be a sequence of functions, each with domain $E \subset \mathbf{C}$ and with u_k and v_k real valued. Prove that the sequence has the uniform Cauchy property on E if and only if the sequences $\langle u_k \rangle_m^\infty$ and $\langle v_k \rangle_m^\infty$ have the uniform Cauchy property on E. Also prove that the sequence converges uniformly on E to a function $f = u + iv$, u, v real valued, if and only if $\langle u_k \rangle_m^\infty$ converges uniformly on E to u and $\langle v_k \rangle_m^\infty$ converges uniformly on E to v.
5. Show that the sequence $\langle (1/z) + (1/k) \rangle_1^\infty$ converges uniformly on $E = \mathbf{C} \sim \{0\}$ and that $\sup\{|(1/z) + (1/k)| : z \in E\}$ does not fulfill all the requirements for a norm. (The example shows that the concept of uniform convergence is applicable to a wider class of functions with a given domain E than the algebras which can be normed by the sup norm.)
6. Let $\langle f_k \rangle_m^\infty$ be a sequence of complex-valued functions with common domain $E \subset \mathbf{C}$. Prove that if f_k is bounded on E, $k = m, m + 1 \ldots$, and if $\langle f_k \rangle$ converges uniformly on E to the limit function f, then f is also bounded on E. (Do not assume that the collection $\{f_k\}$ is uniformly bounded on E. *Hint*: Take $\varepsilon = 17$ in the definition of uniform convergence, and do not let k get away from you.)
7. Let $\langle f_k \rangle_m^\infty$ be a sequence of complex-valued functions with a common domain $E \subset \mathbf{C}$. Show that if the sequence converges uniformly on E to a limit function f which is bounded on E, then there exists $N > m$ such that the collection of functions $\{f_k, k \geq N\}$ is uniformly bounded on E.
8. Prove Theorem 3.3.2.
9. Let $\langle f_k \rangle_m^\infty$ and $\langle g_k \rangle_m^\infty$ be sequences of complex-valued functions with a common domain $E \subset \mathbf{C}$. Suppose that these sequences, respectively, converge uniformly to functions f and g, which are bounded on E. Prove that the product sequence $\langle f_k \cdot g_k \rangle_m^\infty$ converges uniformly on E to $f \cdot g$. (*Hint*: $|f_k \cdot g_k - f \cdot g| \leq |f_k| |g_k - g| + |g| |f_k - f|$. Use Exercise 3.3.7 with this inequality.)

3.4 Infinite Series of Functions

Let $\langle f_k \rangle_m^\infty$ be a sequence of functions from $\mathbf{C}$ into $\mathbf{C}$ with a common domain E, and let $S_n = f_m + f_{m+1} + \cdots + f_n$, $n \geq m$. The *infinite series of functions* associated with $\langle f_k \rangle_m^\infty$ is the sequence of functions $\langle S_n \rangle_m^\infty$. If the latter sequence converges to w_0 at a point $z_0 \in E$, then w_0 is called the sum of the series at z_0. If the sequence $\langle S_n \rangle$ converges pointwise (or uniformly) on E to a function f, then the series is said to converge pointwise (or uniformly) on E to the sum function f (or just to the "sum f"). If the sequence $\langle S_n \rangle$ has no limit at a point $z_0 \in E$, then the series is said to diverge at z_0; and if $\lim_{n \to \infty} S_n(z_0) = \infty$, the series is said to diverge to infinity at z_0.

It is customary to denote both the series associated with a sequence of functions $\langle f_k \rangle_m^\infty$ and (if it exists) the sum function by $\sum_m^\infty f_k$, or by $f_m + f_{m+1} + \cdots$. As in the case of sequences of functions, a series is frequently specified through its series of values $\sum_m^\infty f_k(z)$; for example, $\sum_0^\infty (z^2 - 1)^k$. If $m = 0$, or if the value of m is immaterial for a statement, we often indicate the series by $\sum f_k$. In the case $m = 0$ or 1, S_n is often called the nth partial sum of the series.

An infinite series of complex-valued functions $\sum_m^\infty f_k$ with common domain E is said to converge absolutely at a point $z_0 \in E$, if the series of real numbers $\sum_m^\infty |f_k(z_0)|$ converges. The series $\sum_m^\infty f_k$ is called absolutely convergent on E, if it converges absolutely at each point of E.

With these definitions, the study of infinite series of functions becomes largely a matter of translating relevant results for series of numbers and sequences of functions into a different terminology. We first describe the customary algebraic operations on series of functions.

Let $\sum_m^\infty f_k$ and $\sum_m^\infty g_k$ be series of complex-valued functions with common domain E. Addition and subtraction of the series are defined by the two series $\sum_m^\infty (f_k \pm g_k)$. The Dirichlet product of the series is the series $\sum_m^\infty f_k g_k$. The Cauchy product in the case $m = 0$ is the series $\sum_0^\infty C_k$, where $C_k = f_k g_0 + f_{k-1} g_1 + f_{k-2} g_2 + \cdots + f_0 g_k$. The product of the series $\sum_m^\infty f_k$ with the complex number $a \in \mathbf{C}$ is defined by the value series $\sum_m^\infty a f_k(z)$. With these operations, and with either type of multiplication, the set of all series $\sum_0^\infty f_k, f_k \colon E \to \mathbf{C}$, forms an algebra over $\mathbf{C}$ in which the zero element is the series $0 + 0 + \cdots$, where 0 here denotes the function with the constant value zero on E.

In the list of propositions which follows, the functions f_k and g_k are complex valued with common domain $E \subset \mathbf{C}$. The references in square brackets provide instant proofs.

Proposition 3.4.1 *Let $f_k = u_k + iv_k$, with u_k and v_k real valued, $k = m, m + 1, \ldots$. The series $\sum_m^\infty f_k$ converges pointwise on E to $f = u + iv$ if and only if the series $\sum_m^\infty u_k$, and $\sum_m^\infty v_k$ converge pointwise on E to u and v, respectively* [*Proposition 3.1.1*]. *The statement remains valid with uniform convergence substituted for pointwise convergence* [*Exercise 3.3.4*].

Proposition 3.4.2 *The series $\sum f_k$ converges pointwise on E, if it converges absolutely on E* [*Proposition 2.6.4*].

Proposition 3.4.3

(i) *If $\sum_m^\infty f_h$ and $\sum_m^\infty g_h$ converge pointwise on E to f and g, respectively, then the two series $\sum_m^\infty (f_h \pm g_h)$ converge pointwise on E to $f \pm g$; and for any $a \in \mathbf{C}$, $\sum_m^\infty a f_h$ converges pointwise on E to $a \cdot f$* [*Proposition 3.1.4*].

(ii) *The statements in (i) are true with uniform convergence substituted for pointwise convergence* [*Theorem 3.3.2*].

(iii) *If* $\sum_m^\infty f_h$ *and* $\sum_m^\infty g_h$ *converge absolutely on E, then so do* $\sum_m^\infty (f_h \pm g_h)$ *and* $\sum_m^\infty af_h$ *for any* $a \in \mathbf{C}$ [*Exercise 2.6.3*].

Proposition 3.4.4

(i) *If* $\sum_0^\infty f_k$ *converges absolutely on E, and thus converges pointwise to a sum function f, and if* $\sum_0^\infty g_k$ *converges pointwise on E to g, then the Cauchy product series converges to fg pointwise on E* [*Theorem 2.7.2 and Proposition 2.7.3*].

(ii) *If* $\sum_0^\infty f_k$ *and* $\sum_0^\infty g_k$ *converge absolutely on E, so does the Dirichlet product series* $\sum_0^\infty f_k g_k$ [*Proposition 2.7.1*].

Proposition 3.4.5 *Let* $\sum_m^\infty f_k$ *be a uniformly convergent series with sum function f.*

(i) *If* $\{f_k\}$ *is a collection of bounded functions, then f is bounded on E* [*Exercises 3.2.7 and 3.3.6*].

(ii) *If each function* f_k *is continuous on E, then f is continuous on E* [*Exercise 3.2.8 and Theorem 3.3.3*].

The series $\sum_m^\infty f_k$ of complex-valued functions with common domain E is said to satisfy the *Cauchy criterion* for convergence at a point $z_0 \in E$ if the sequence $\langle S_n(z_0)\rangle_m^\infty$ is a Cauchy sequence. It is said to satisfy the *Cauchy criterion pointwise* on E if $\langle S_n\rangle_m^\infty$ has the Cauchy property pointwise on E, and it satisfies the *uniform Cauchy criterion* on E if $\langle S_n\rangle$ has the uniform Cauchy property on E.

The arithmetical formulation of the uniform Cauchy criterion in terms of the functions f_k reads as follows: For every $\varepsilon > 0$, there exists $K_\varepsilon \geq m$ such that if $k > h \geq K_\varepsilon$, then $|f_{h+1}(z) + f_{h+2}(z) + \cdots + f_k(z)| < \varepsilon$ for every $z \in E$. In the case of pointwise convergence, this formulation becomes: For every $\varepsilon > 0$ and each $z \in E$, there exists $K_{\varepsilon,z}$ such that if $k > h \geq K_{\varepsilon,z}$, then $|f_{h+1}(z) + f_{h+2}(z) + \cdots + f_k(z)| < \varepsilon$.

Proposition 3.4.6

(i) *A necessary and sufficient condition for* $\sum_m^\infty f_k$ *to converge at* $z_0 \in E$ *is that it satisfies the Cauchy criterion at* z_0. *A necessary and sufficient condition for* $\sum_m^\infty f_k$ *to converge pointwise on E is that it satisfies the Cauchy criterion pointwise on E* [*Proposition 3.1.2*].

(ii) *A necessary and sufficient condition for* $\sum_m^\infty f_k$ *to converge uniformly on E is that it satisfies the uniform Cauchy criterion on E* [*Theorem 3.3.1*].

Corollary If $\sum_m^\infty f_k$ converges at z_0, then the sequence $\langle f_k(z_0)\rangle$ converges to zero. If the series converges pointwise on E, then $\langle f_k \rangle$ converges pointwise to zero on E. If the series converges uniformly on E, then $\langle f_k \rangle$ converges uniformly to zero on E.

Proof of the Corollary. If convergence of one of the types described in the corollary takes place, then by Proposition 3.4.6 the series satisfies the relevant Cauchy criterion. The corollary is obtained by taking $k = h + 1$ in the arithmetic formulation of the Cauchy criterion, as in the proof of the corollary of Proposition 2.6.3. Q.E.D.

Remark. It is possible for $\langle f_k \rangle$ to converge uniformly to 0 on a set E, without the series $\sum f_k$ converging pointwise on E. A simple example is given by the series $\sum_1^\infty z^k/k$, $E = C(0, 1) = \{z : |z| = 1\}$.

Theorem 3.4.1 ("*Weierstrass M-test*"). *The series* $\sum_m^\infty f_k$ *converges absolutely and uniformly on* E *if there exists a convergent series* $\sum_n^\infty M_k$, $n \geq m$, *in which* M_k *is a nonnegative real number such that* $|f_k(z)| \leq M_k$ *for every* $z \in E$, $k = n, n + 1, \ldots$.

The test amounts to requiring that $\{f_k\}$ be a collection of bounded functions at least for $k \geq n$, and that the infinite series associated with the sequence of bounds converges.

Proof. Given $\varepsilon > 0$, there exists K_ε such that if $k > h \geq K_\varepsilon \geq n$, then $M_{h+1} + M_{h+2} + \cdots + M_k < \varepsilon$, since $\sum M_k$ must satisfy the Cauchy criterion. But then for $k > h \geq K_\varepsilon$, we have for every $z \in E$,

$$\begin{aligned}|f_{h+1}(z) + f_{h+2}(z) + \cdots + f_k(z)| \\ \leq |f_{h+1}(z)| + |f_{h+2}(z)| + \cdots + |f_k(z)| \\ \leq M_{h+1} + M_{h+2} + \cdots + M_k < \varepsilon.\end{aligned}$$

This implies that $\sum_n^\infty f_k$ satisfies the Cauchy criterion for uniform convergence on E and therefore the series converges uniformly on E [Proposition 3.4.6(ii)]. Q.E.D.

Notice that if the M-test is satisfied, something more than plain absolute convergence on E is implied. It is the *uniform convergence* of the series $\sum |f_k|$, where $|f_k|$ is the function with value $|f_k(z)|$.

Another test for uniform convergence which is sometimes useful is as follows.

Proposition 3.4.7 (*Dirichlet's Test Adapted to Uniform Convergence*). *Let* $\sum_0^\infty a_k$ *be a series in* **C**, *let* $\sum_0^\infty f_k$ *be a series of complex-valued functions with common domain* E, *and let* $S_n = f_0 + f_1 + \cdots + f_n$. *If the following conditions are all satisfied*:

(a) *there exists* $M > 0$ *such that* $|S_n(z)| \leq M$ *for every* $z \in E$, $n = 0, 1, 2, \ldots$,
(b) $\sum_0^\infty |a_k - a_{k+1}| < \infty$, *and*

(c) $\lim_{k\to\infty} a_k = 0$,

then $\sum_0^\infty a_k f_k$ *converges uniformly on E. If in addition the sequence* $\langle a_k \rangle$ *is real and decreasing, then (b) is redundant and for every* $z \in E$, $|\sum_0^\infty a_k f_k(z)| \le a_0 M$.

Proof. By Abel's identity, Exercise 1.2.9, if $n > m + 1$,

$$\sum_{k=m+1}^{n} a_k f_k(z) = \sum_{k=m+1}^{n-1} (a_k - a_{k+1})S_k(z) + a_n S_n(z) - a_{m+1}S_m(z).$$

Then by part (a) of the hypothesis, still with $n > m + 1$,

$$\left|\sum_{k=m+1}^{n} a_k f_k(z)\right| \le M \sum_{m+1}^{n-1} |a_k - a_{k+1}| + |a_n| + |a_{m+1}|. \tag{3.4.1}$$

If $n = m + 1$, then

$$\begin{aligned}\left|\sum_{k=m+1}^{n} a_k f_k(z)\right| &= |a_{m+1} f_{m+1}(z)| \\ &= |a_{m+1}(S_{m+1}(z) - S_m(z))| \le 2M|a_{m+1}|. \end{aligned}\tag{3.4.2}$$

The inequalities (3.4.1) and (3.4.2) are valid for every $z \in E$. Since the series in part (b) of the hypothesis converges, it must satisfy the Cauchy criterion. Thus, given any $\varepsilon > 0$, there exists K_ε such that for $n > m + 1 \ge K_\varepsilon$, $\sum_{m+1}^{n-1} |a_k - a_{k+1}| < \varepsilon/3M$. But by part (c) of the hypothesis, there also exists K'_ε such that for $m + 1 \ge K'_\varepsilon$, $|a_{m+1}| < \varepsilon/3M$. Take $K''_\varepsilon = \max(K_\varepsilon, K'_\varepsilon)$. Then (3.4.2) implies that for $n = m + 1 \ge K''_\varepsilon$,

$$\left|\sum_{k=m+1}^{n} a_k f_k(z)\right| < \frac{2M\varepsilon}{3M} < \varepsilon,$$

and (3.4.1) implies that for $n > m + 1 \ge K''$,

$$\left|\sum_{k=m+1}^{n} a_k f_k(z)\right| < M\left[\frac{\varepsilon}{3M} + \frac{\varepsilon}{3M} + \frac{\varepsilon}{3M}\right] = \varepsilon.$$

Thus, $\sum_0^\infty a_k f_k$ satisfies the uniform Cauchy criterion on E, and the conclusion of the theorem as to uniform convergence follows from Proposition 3.4.6(ii).

If $\langle a_k \rangle$ is real and decreasing, to obtain the indicated bound for the series we take $m = -1$ in Abel's identity and define $S_{-1}(z)$ to be zero. Then (3.4.1) becomes

$$\begin{aligned}\left|\sum_{k=0}^{n} a_k f_k(z)\right| &\le M[(a_0 - a) + (a - a_2) + \cdots + (a_{n-1} - a_n) + a_n] \\ &= Ma_0. \end{aligned}$$

Q.E.D.

Exercises 3.4

1. Show that the series $\sum_1^\infty z^k/k^p$, $p > 1$, converges uniformly and absolutely on $E = \{1 : |z| \leq 1\}$. (*Hint*: Find an M-test.)
2. The curve Γ_μ with equation $|z^2 - 1| = \mu > 0$ is a special case of what is known as a lemniscate, and its interior is defined to be the set $E_\mu = \{z : |z^2 - 1| < \mu\}$. Thus, the lemniscate $\Gamma_{\mu'} = \{z : |z^2 - 1| = \mu' < \mu\}$ and its interior $E_{\mu'}$ lie in the interior of Γ_μ. Prove that the series

$$\sum_{k=0}^{\infty} \frac{(z^2 - 1)^k}{\mu^k}(-1)^k$$

converges uniformly on $\Gamma_{\mu'} \cup E_{\mu'}$, for any fixed $\mu' < \mu$, $\mu' > 0$. (*Hint*: An M-text is easily visible.) What is the sum of the series?
3. Let f and g be complex-valued functions with domain **C**. Let S be an infinite series with the value series

$$S(z) = g(z) + [f(z) - g(z)]\left\{\frac{1}{1+z} + \left[\frac{1}{1+z^2} - \frac{1}{1+z}\right] + \left[\frac{1}{1+z^3} - \frac{1}{1+z^2}\right] + \left[\frac{1}{1+z^4} - \frac{1}{1+z^3}\right] + \cdots\right\}.$$

Show that $S(z) = f(z)$ if $|z| < 1$ and $S(z) = g(z)$ if $|z| > 1$.
4. Let $\sum_0^\infty f_k$ be a series of complex-valued functions with common domain E, and suppose that there exists $M > 0$, such that $\sum_0^n |f_k(z)| \leq M$ for every $z \in E$ and for $n = 0, 1, 2, \ldots$. Then $\sum_0^\infty f_k$ converges absolutely on E [Proposition 2.6.7]. It may then be tempting to conclude that $\sum_0^\infty |f_k|$ converges uniformly on E, in which case so also would $\sum_0^\infty f_k$, by the Cauchy criterion. Show that this conclusion is false by examining the series $(1 - x) + (x - x^2) + (x^2 - x^3) + \cdots$, x real, $0 \leq x \leq 1$.
5. Let $\sum_0^\infty f_k$ be a series of complex-valued functions with common domain E. Prove that the series converges uniformly on E to the sum function f if for some $m > 0$, the series $\sum_m^\infty f_k$ converges uniformly on E to the sum $f - f_1 - f_2 - \cdots - f_{m-1}$. Also prove that if $\sum_0^\infty f_k$ converges uniformly to f on E, then for each $m > 0$ the series $\sum_m^\infty f_k$ converges uniformly on E to $f - f_1 - f_2 - \cdots - f_{m-1}$. (*Hint*: Remark (ii) in Sec. 3.3 and Theorem 3.3.2.)
6. Show that the series $\sum_1^\infty z^k(1/k)^p$, $0 < p \leq 1$, and $\sum_2^\infty (z^k/\log k)$ converge uniformly on the set $E = \{z : |z| \leq 1\} \cap \{z : |z - 1| \geq \rho\}$, where $0 < \rho < 2$. (*Hint*: This is the classical application of Proposition 3.4.7 in the case in which $\langle a_k \rangle$ is real and decreasing.)
7. Let $\sum_0^\infty f_k$ be a series of complex-valued functions with common domain $E \cup E_0$, where $E_0 \cap E = \phi$ and E_0 is a *finite* point set. Prove that if the series converges uniformly on E and also converges at each point of E_0, then it converges uniformly on $E \cup E_0$. (*Hint*: Set up the arithmetic characterizations of the uniform convergence on E and the convergence at each point of E_0, and combine into a single characterization of convergence on $E_0 \cup E$.)

4

Introduction to Power Series

4.1 Radius of Convergence

Special power series were encountered in Sec. 2.7 in connection with the Cauchy product of series. In general, a power series in $\mathbf{C}$ is a series of functions $\sum_m^\infty f_k$, $m \ge 0$, where $f_k(z) = a_k(z-a)^k$ with $a, a_k \in \mathbf{C}$, $k = m, m+1, m+2, \ldots$. The numbers a_k, $k = m, m+1, \ldots$, are called the coefficients of the series. In the case $m = 0$, we use the convention $(z-a)^0 = 1$ even when $z = a$.

It turns out that the natural convergence set for a power series is a circular disk of some sort. Recall the notation $\Delta(a, r) = \{z : |z-a| < r\}$, $r > 0$; $C(a, r) = \{z : |z-a| = r\}$, $r \ge 0$. We let $\bar{\Delta}(a, r) = \Delta(a, r) \cup C(a, r)$.

Our starting point for the theory of convergence is the following simple result.

Theorem 4.1.1 *If there exist $z_0 \ne a$ and $M > 0$ such that $|a_k(z_0 - a)^k| \le M$, $k = m, m+1, \ldots$, $m \ge 0$, then*

(i) *The series $\sum_m^\infty a_k(z-a)^k$ converges uniformly on $\bar{\Delta}(a, r)$ for any r, $0 < r < |z_0 - a|$, and*
(ii) *This power series converges absolutely on $\Delta(a, |z_0 - a|)$.*

Proof. We construct an M-test [Theorem 3.4.1]. Take any r, $0 < r < |z_0 - a|$. For every $z \in \bar{\Delta}(a, r)$, we have ($k \ge m$)

$$|a_k(z-a)^k| = |a_k(z_0-a)^k| \left|\frac{z-a}{z_0-a}\right|^k \le M\left(\frac{r}{|z_0-a|}\right)^k = M\rho^k,$$

where $0 < \rho < 1$. The series $\sum_m^\infty M\rho^k$ converges, so we have set up a valid M-test with $M_k = M\rho^k$. This proves part (i) of the theorem.

Now let z_1 be any point in $\Delta(a, |z_0 - a|)$, and choose $r = |z_1 - a|$. The M-test indicates absolute, as well as uniform, convergence for every $z \in \bar{\Delta}(a, r)$. Q.E.D.

Example. Look at the power series $\sum_0^\infty z^k = 1/(1 - z)$, $|z| < 1$. Take $z_0 = 1$; then $|1 \cdot 1^k| = 1$ and the hypothesis of the theorem is satisfied with this z_0 and $M = 1$. The series converges absolutely on $\Delta(0, 1)$ and uniformly on any $\bar{\Delta}(0, r)$, $0 < r < 1$. The nth partial sum is $s_n(z) = \sum_0^n z^k = (1 - z^{n+1})/(1 - z)$. According to Exercise 3.3.2, the sequence $\langle s_n \rangle$ (and, therefore, the series) does not converge uniformly on $\Delta(0, 1)$.

Corollary If $\sum_m^\infty a_k(z - a)^k$ converges for some $z_0 \neq a$, then the conclusions of Theorem 4.1.1 are valid with this z_0.

Proof. The hypothesis of the corollary implies that the sequence $\langle a_k(z - a)^k \rangle_m^\infty$ converges to zero (corollary of Proposition 2.6.3), so, of course, the sequence is bounded and the hypothesis of the theorem is satisfied. Q.E.D.

Notice that in this example, the hypothesis of the theorem is satisfied for the chosen z_0, but not the hypothesis of the corollary, because the series obviously diverges at $z = 1$.

The *radius of convergence* of the power series $\sum_m^\infty a_k(z - a)^k$ is defined to be the extended real number $R = 1/\overline{\lim}_{k\to\infty} |a_k|^{1/k}$. When $\overline{\lim} |a_k|^{1/k} = 0$, we set $R = \infty$, and when $\overline{\lim} |a_k|^{1/k} = \infty$, we set $R = 0$.

The justification for the term radius of convergence lies in the following theorem.

Theorem 4.1.2 *Let R be the radius of convergence of $\sum_m^\infty a_k(z - a)^k$.*

(i) *If $0 < R < \infty$, the series converges absolutely on $\Delta(a, R)$ and uniformly on $\bar{\Delta}(a, r)$ for any r, $0 < r < R$. It diverges at each point of $\{z: |z - a| > R\}$.*

(ii) *If $R = 0$, the series converges only at $z = a$. If $R = \infty$, the series converges absolutely for every $z \in \mathbf{C}$ and converges uniformly on each $\bar{\Delta}(a, r)$, $0 < r < \infty$.*

Proof. The proof stems from the Cauchy kth root test [Theorem 2.6.1]. First, we need a computation. Take any $z_0 \in \mathbf{C}$, but with $z_0 \neq a$. Then

$$\begin{aligned} \mu &= \overline{\lim} |a_k(z_0 - a)^k|^{1/k} \\ &= \overline{\lim} [|z_0 - a| \, |a_k|^{1/k}] \\ &= |z_0 - a| \, \overline{\lim} |a_k|^{1/k}. \end{aligned} \tag{4.1.1}$$

The authority for the third equation is Exercise 2.4.10.

Now suppose as in part (*i*) of the theorem that $0 < R < \infty$. Then $\mu =$

$|z_0 - a|/R$. If $z_0 \in \Delta(a, R)$, $z_0 \neq a$, then $(|z_0 - a|/R) < 1$ and according to the kth root test, the series $\sum a_k(z_0 - a)^k$ converges absolutely. But then by the corollary of Theorem 4.1.1, it converges uniformly on any disk $\bar{\Delta}(a, r)$, $0 < r < |z_0 - a| < R$. Since $z_0 \in \Delta(a, R)$ is arbitrary, we have proved the assertion in the first sentence of part (i).

If on the other hand $|z_0 - a| > R$, then $\mu > 1$, and divergence follows from the kth root test. This completes the proof of (i).

Now $R = 0$ means $\overline{\lim}\, |a_k|^{1/k} = \infty$. Since $z_0 - a \neq 0$, the computation in (4.1.1) is valid (Exercise 2.4.10 still is applicable) and $\mu = \infty$. But then the kth root test indicates divergence. The series converges trivially if $z = a$.

Now suppose $R = \infty$. This means that $\overline{\lim}\, |a_k|^{1/k} = 0$. Again (4.1.1) is valid, and $\mu = 0$ for every $z_0 \in \mathbf{C}$ (including $z_0 = a$). The kth root test indicates absolute convergence for every such z_0. Thereupon, since z_0 can be chosen so that $|z_0 - a|$ is arbitrarily large, the corollary of Theorem 4.1.1 again becomes applicable to assure uniform convergence on any disk $\bar{\Delta}(a, r)$, $0 < r < \infty$. Q.E.D.

We call $\Delta(a, R)$ the disk of convergence of $\sum a_k(z - a)^k$ and $C(a, R)$ the circle of convergence.

Corollary 1 If the power series $\sum a_k(z - a)^k$ is known to converge at each point of the disk $\Delta(a, r)$, $0 \leq r \leq \infty$, and is known to diverge at each point of $\{z: |z - a| > r\}$ (an empty set if $r = \infty$), then r must be the radius of convergence.

Proof. If the radius of convergence R were not equal to r, and if, say, $R > r$, then according to the theorem there would be convergence at some points outside of $\Delta(a, r)$, a contradiction. Similarly, $R < r$ leads to a contradiction.

Corollary 2 Let $\sum_0^\infty a_k(z - a)^k$ have radius of convergence R. Then $\sum_{k=0}^\infty a_{k+m}(z - a)^k$, $m > 0$, and $\sum_{k=0}^\infty a_k(z - a)^{m+k}$, $m > 0$, each has radius of convergence R.

Proof. The series $\sum_{k=0}^\infty a_{k+m}(z - a)^{k+m} = \sum_{k=m}^\infty a_k(z - a)^k$ has radius of convergence $1/\overline{\lim}\, |a_k|^{1/k} = R$, by the definition of radius of convergence. Therefore, it converges on $\Delta(a, R)$ and diverges for $|z - a| > R$.

Now suppose that $z \neq a$. By Proposition 2.6.2, the series

$$(z - a)^{-m} \sum_{k=0}^{\infty} a_{k+m}(z - a)^{k+m} = \sum_{k=0}^{\infty} a_{k+m}(z - a)^k \tag{4.1.2}$$

converges for $0 < |z - a| < R$ and diverges for $|z - a| > R$. The series $\sum_{k=0}^\infty a_{k+m}(z - a)^k$ converges trivially to a_m at $z = a$.

If $R = 0$, the series $\sum_0^\infty a_k(z - a)^k$ converges at $z = a$. The foregoing argument shows that $\sum_{k=0}^{\infty} a_{k+m}(z-a)^k$ diverges for every $z \neq a$. There is the usual trivial convergence at $z = a$.

If $R = \infty$, the argument proves convergence of $\sum_{k=0}^{\infty} a_{k+m}(z-a)^k$ for every $z \neq a$, and there is trivial convergence at $z = a$.

We have now shown that for any R, $0 \leq R \leq \infty$, $\sum_{k=0}^{\infty} a_{k+m}(z-a)^k$ converges pointwise on $\Delta(a, R)$ (or just at $z = a$ if $R = 0$) and diverges at every point of $\{z: |z-a| > R\}$. (The latter set is empty if $R = \infty$.) Thus, by Corollary 1, R is the radius of convergence of this series.

A similar argument applies to $\sum_0^{\infty} a_k(z-a)^{m+k}$. Q.E.D.

Remark. The sum function of the series $\sum_0^{\infty} a_{k+m}(z-a)^k$ is easily computed from that of $\sum_0^{\infty} a_k(z-a)^k$ by using (4.1.2) and Exercise 3.4.5 (or Proposition 2.6.1). The result is $(z-a)^{-m}[\sum_0^{\infty} a_k(z-a)^k - \sum_0^{n-1} a_k(z-a)^k]$, if $0 < |z-a| < R$ and a_m if $z = a$. Similarly, $\sum_0^{\infty} a_k(z-a)^{m+k} = (z-a)^m \sum_0^{\infty} a_k(z-a)^k$.

Proposition 4.1.1 (*Ratio Test*). *Let $\sum_m^{\infty} a_k(z-a)^k$ be a power series in which $a_k \neq 0, k = m+1, m+2, \ldots$, and for which the sequence $\langle |a_k/a_{k+1}| \rangle_m^{\infty}$ converges to a number r, $0 \leq r \leq \infty$. Then the radius of convergence is r.*

Proof. The ratio test for series of numbers, Theorem 2.6.2, applied in the present context, concerns the behavior of the sequence

$$\left\langle \frac{|a_{k+1}(z-a)^{k+1}|}{|a_k(z-a)^k|} \right\rangle = \left\langle |z-a| \left| \frac{a_{k+1}}{a_k} \right| \right\rangle. \tag{4.1.3}$$

If $r = 0$, $\lim |a_{k+1}/a_k| = \infty$ and for every $z \neq a$, the sequence (4.1.3) has the limit ∞. The ratio test then indicates divergence for every $z \neq a$. If $r = \infty$, then $\lim |a_{k+1}/a_n| = 0$, and the sequence (4.1.3) converges to zero for every z. The ratio test indicates convergence for every z. If $0 < r < \infty$, the sequence (4.1.3) has the limit $|z-a|/r$. If $|z-a| < r$, then $|z-a|/r < 1$ and the ratio test indicates convergence. Similarly, it indicates divergence if $|z-a| > r$. Corollary 1 of Theorem 4.1.2 completes the proof that r is the radius of convergence. Q.E.D.

Power series with finite radii of convergence exhibit a variety of convergence and divergence phenomena on their circles of convergence. The study of this behavior occupies a large literature, mainly developed during the last 100 years. We pause here only to indicate some of the possibilities.

Proposition 4.1.2 *If a power series $\sum_m^{\infty} a_k(z-a)^k$ with finite radius of convergence R converges absolutely at a single point of the circle of convergence, then it converges absolutely and uniformly at every point of $\bar{\Delta}(a, R)$.*

The proof is an easy exercise.

Examples

(i) The series $\sum_0^{\infty} z^k$ has $R = 1$, but it converges nowhere on $C(0, 1)$ because the terms of the series do not approach zero if $|z| = 1$.

(ii) Consider the series $\sum_1^\infty (z^k/k^p)$, $p > 0$. Here by the ratio test for power series we find that $R = \lim(k+1)^p/k^p = \lim(1 + 1/k)^p = 1$ [Exercise 2.4.12]. If $p > 1$, according to the corollary of Proposition 2.6.8 the series converges absolutely at $z = 1$. Therefore, by Proposition 4.1.2 it converges absolutely and uniformly on $\bar{\Delta}(0, 1)$. If $0 < p \leq 1$, by the same authority it diverges at $z = 1$. However, according to Exercise 3.4.6, this power series converges uniformly (but not absolutely) on $\{z: |z| \leq 1\} \sim \Delta(1, \rho)$, where $\Delta(1, \rho)$ is an arbitrarily small disk centered at $z = 1$.

(iii) L. Fejér gave an example of a power series which converges uniformly but not absolutely on the cricle of convergence. See [9, p. 122].

Some remarks about the behavior of a power series outside the circle of convergence are in order. With radius of convergence R finite, $\sum_0^\infty a_k(z-a)^k$ diverges for every z in $\{z: |z - a| > R\}$. Therefore, for every such z, $\sum_0^\infty |a_k(z-a)^k| = \infty$. This raises the question as to whether for every such z, $\sum_0^\infty a_k(z-a)^k = \infty$. (This equation is interpreted to mean that the sequence of partial sums $s_n(z) = \sum_0^n a_k(z-a)^k$, $n = 0, 1, 2, \ldots$, has the limit ∞ [Sec. 2.6], which in turn is characterized by $\lim_{n\to\infty} |s_n(z)| = \infty$.) The answer is in the negative. It was shown by M. B. Porter in 1906–07 that there exists a power series with disk of convergence $\Delta(0, 1)$, with the property that a properly chosen infinite subsequence of the partial sums converges *uniformly* on an infinite point set in $\{z: |z| > 1\}$. His example is found in Ref. 8 [p. 116], where references to the research literature are given.

Exercises 4.1

1. Prove Proposition 4.1.2 by showing that the hypothesis supplies an M-test.
2. Find the radius of convergence of each of the following series.

(a) $1 + \sum_1^\infty (z^k/k!)$. *Answer*: $R = \infty$.

(b) $\sum_1^\infty (\sqrt{n})^n z^n$. *Answer*: $R = 0$.

(c) $1 + \sum_1^\infty \dfrac{a(a-1)(a-2)\cdots(a-k+1)}{k!} z^k$, a complex.

Answer: $R = 1$ unless $a \in \mathbf{N}$.

(d) $\sum_1^\infty a_k z^k$ with $a_{2m-1} = 1/m$, $a_{2m} = 1$, $m = 1,2,\ldots$. (Ratio test fails.)
Answer: $R = 1$.

(e) $\sum_0^\infty 17^n (z-a)^n$.

3. Let the radii of convergence of $\sum_0^\infty a_k z^k$ and $\sum_0^\infty b_k z^k$ each be not less than $\rho > 0$. Prove that the Cauchy product series $\sum_0^\infty c_k z^k$ has radius of convergence at least as great as ρ. (*Hint*: The problem is solved by showing that the Cauchy product converges pointwise on $\Delta(0, \rho)$. Recall Proposition 2.7.2.)

Remark. Exercise 4.1.3 shows that the collection of power series of type $\sum_0^\infty a_k z^k$ with radius of convergence at least as great as $\rho > 0$ forms an algebra over $\mathbf{C}$, with the usual addition of series and scalar product, and the Cauchy product operation used for multiplication. See Exercise 2.7.6.

4. Let the radii of convergence of $\sum_0^\infty a_k(z-a)^k$ and $\sum_0^\infty b_k(z-a)^k$ be, respectively, R_a and R_b. Let the radii of convergence of $\sum_0^\infty (a_k+b_k)(z-a)^k$ and $\sum_0^\infty a_h b_h(z-a)^h$ be, respectively, R_s and R_p.
   (i) Prove that $R_s \geq \min(R_a, R_b)$. (*Hint*: Proposition 2.6.2.)
   (ii) Give an example of equality in (i).
   (iii) Give an example with $R_a < \infty$, $R_b < \infty$, but $R_s = \infty$.
   (iv) Prove that if $R_a \neq 0$, $R_b \neq 0$, then $R_p \geq R_a R_b$. (*Hint*: Exercise 2.4.8.)
   (v) Give an example of equality in (iv.)
5. Given the series $\sum_0^\infty a_k(z-a)^k$ with radius of convergence $R > 0$, show that the series $\sum_1^\infty k a_k(z-a)^{k-1}$ has the same radius of convergence. (*Hint*: The second series can be written as $\sum_0^\infty (k+1)a_{k+1}(z-a)^k$. Find the radius of convergence of $\sum_0^\infty k a_k(z-a)^k$ by using the definition of radius of convergence and the information in Sec. 2.4; then apply Corollary 2 of Theorem 4.1.2.)
6. If $\sum a_k z^k$ has radius of convergence R, show that $\sum a_k z^{2k}$ has radius $\sqrt{R}$. (*Hint*: Corollary 1 of Theorem 4.1.2.)

4.2 Continuity Properties of the Sum Function of a Power Series

The words "converges uniformly" appeared frequently in the preceding section, so the reader may have anticipated the following assertion.

Theorem 4.2.1

(i) *A power series with disk of convergence* $\Delta(a, R)$, $R > 0$, *converges to a continuous sum function on* $\Delta(a, R)$.
(ii) *If a power series with disk of convergence* $\Delta(a, R)$, $R > 0$, *converges absolutely at a point of the circle of convergence, then it converges to a continuous function on* $\bar{\Delta}(a, R)$.

Proof.

(i) Let the series be $\sum a_k(z-a)^k$. By Theorem 4.1.2 the series converges uniformly on any disk $\bar{\Delta}(a, r) \subset \Delta(a, R)$. The terms $a_k(z-a)^k$ are continuous on **C** [Exercise 3.2.9], so by Proposition 3.4.5(ii) the sum function is continuous on $\bar{\Delta}(a, r)$. But r, $0 < r < R$, is arbitrary, so given any $z \in \Delta(a, R)$, by taking $r > |z-a|$ we have $z \in \Delta(a, r)$. Thus, the point set on which the sum function is continuous includes all of $\Delta(a, R)$.
(ii) This part of the theorem follows at once from Propositions 4.1.2 and 3.4.5(ii) and the continuity of the terms of the series. Q.E.D.

Part (ii) of the theorem contains a sufficient condition for the sum function of a power series to have a continuous extension onto the entire circle of convergence. Various conditions are known under which the sum function is continuous on only a subset of the circle of convergence. Take, for example, the following proposition.

Proposition 4.2.1 *Let $\sum_0^\infty a_k z^k$ be a power series in which the coefficients satisfy the conditions $\sum_{k=0}^\infty |a_k - a_{k+1}| < \infty$ and $\lim a_k = 0$. This power series converges uniformly on $E_\rho = \{z: |z| \leq 1\} \sim \Delta(1, \rho)$, $0 < \rho < 2$, and the sum function f is continuous on E_ρ. The series converges pointwise on $E_0 = \{z: |z| \leq 1\} \sim \{1\}$ and represents a continuous function there. If in addition the sequence of coefficients $\langle a_k \rangle$ is real and decreasing, then $|f(z)| \leq 2a_0/(1 - \rho)$ for $z \in E_\rho$.*

Proof. The proposition is the adaptation of Proposition 3.4.7 to the power series. Special cases appeared in Exercise 3.4.6. Let $S_n(z)$ in Proposition 3.4.7 be $1 + z + z^2 + \cdots + z^n, n = 0, 1, 2, \ldots$. Then $S_n(z) = (1 - z^{n+1})/(1 - z)$, and for every $z \in E_\rho$ we have $|z - 1| \geq \rho$ and $|z| \leq 1$. Thus,

$$|S_n(z)| \leq \frac{1 + |z|^{n+1}}{|1 - z|} \leq \frac{2}{1 - \rho}.$$

The last member of this inequality is taken to be the "M" in Proposition 3.4.7.

If z_0 is an arbitrary point on E_0, it lies in E_ρ if $\rho \leq |z_0 - 1|$, and the series then converges at z_0. Also it converges uniformly on this E_ρ, and so the

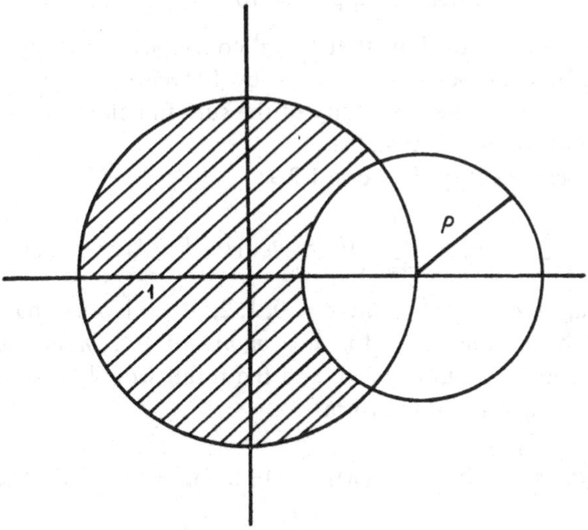

Fig. 1

sum function is continuous at z_0 [Proposition 3.4.5(ii)]. The last sentence in the proposition is the translation of the last sentence of Proposition 3.4.7. Q.E.D.

Example. (See Example (ii) in Sec. 4.1 and Exercise 3.4.6.) If $p > 1$, the sum function of $\sum_1^\infty z^k/k^p$ is continuous on the circle of convergence $C(0, 1)$ and on $\bar{\Delta}(0, 1)$. If $p \leq 1$, the sum function is continuous on $E_0 = \bar{\Delta}(0, 1) \sim \{1\}$. The series with $p \leq 1$ does not converge absolutely at any point of $C(0, 1)$.

In Theorem 4.2.1, continuity of the sum function everywhere on the circle of convergence is deduced from the hypothesis of absolute convergence at a point of the circle of convergence. This suggests the following question. Suppose that there is ordinary convergence but not necessarily absolute convergence at a point of the circle of convergence; what does this imply for the sum function, if anything? An answer is given by a result known as Abel's limit theorem.

Theorem 4.2.2 (*Stolz' Form of Abel's Limit Theorem*). *Let the power series $f(z) = \sum_0^\infty a_k z^k$ have radius of convergence 1. If the series converges for $z = 1$ (thus defining $f(1)$), then f is not only continuous on $\Delta(0, 1)$, but is also continuous on the set*

$$B_K = \{z: |1 - z| \leq K(1 - |z|), K \geq 1\} \cap \bar{\Delta}(0, 1).$$

Note. Clearly, the interval [0, 1] lies in B_K. Also, since $|z| = 1$ at each point of $C(0, 1)$, the point $z = 1$ is the only point of $C(0, 1)$ which lies in B_K. The purport of the conclusion of the theorem is that the sum function f is continuous at $z = 1$ when the approach to $z = 1$ is suitably restricted.

Proof. The plan is to show that $\sum a_k z^k$ converges uniformly on $B_K \sim \{1\}$. It converges by hypothesis at $z = 1$, so by Exercise 3.4.7 we have uniform convergence on B_K. The continuity of the sum function on B_K then follows as usual from Proposition 3.4.5(ii).

Recall Abel's identity, Exercise 1.2.9:

$$\sum_{k=m+1}^{n} a_k b_k = \sum_{k=m+1}^{n} (b_k - b_{k+1}) s_k + b_n s_n - b_{m+1} s_m,$$

where $s_k = a_0 + a_1 + \cdots + a_k$, $k = 0, 1, 2, \ldots$. Notice that if s_k were to be replaced by a constant s for $k = m, m + 1, \ldots, n$ in the right-hand member, everything would cancel, and the result would be zero. Thus, we can rewrite the identity in the form

$$\sum_{m+1}^{n} a_k b_k = \sum_{m+1}^{n-1} (b_k - b_{k+1})(s_k - s) + b_n(s_n - s) - b_{m+1}(s_m - s).$$

The identity is valid for $n = m + 1, m + 2, \ldots$, if the summation on the right-hand side is omitted for $n = m + 1$.

We now apply this to the series $\sum_0^\infty a_k z^k$, letting $b_k = z^k$, $s_k = a_0 + a_1 + \cdots + a_k$, $k = 0, 1, 2, \ldots$, and $s = \lim s_k = \sum_0^\infty a_k = f(1)$. The result is

$$\sum_{m+1}^{n} a_k z^k = \sum_{m+1}^{n-1} (z^k - z^{k+1})(s_k - s) + z^n(s_n - s) - z^{m+1}(s_m - s)$$

$$= (1 - z)\sum_{m+1}^{n-1} z^k(s_k - s) + z^n(s_n - s) - z^{m+1}(s_m - s).$$

This is valid for $n = m + 1, m + 2, \ldots$, if the first term on the right-hand side is omitted for $n = m + 1$. Now take any $\varepsilon > 0$. The convergence of $\sum a_k$ implies that there exists $N_\varepsilon > 0$ such that for every $k \geq N_\varepsilon$, $|s_k - s| < \varepsilon/(K + 2)$, where K appears in the statement of the theorem. It follows that for $n > m + 1 \geq N_\varepsilon$,

$$\left|\sum_{m+1}^{n} a_k z^k\right| \leq \frac{|1 - z|\varepsilon}{K + 2}\sum_{m-1}^{n-1} |z|^k + \frac{\varepsilon(|z|^n + |z|^{m+1})}{K + 2}. \tag{4.2.1}$$

Now suppose that $z \in B_K \sim \{1\}$. Then $|z| < 1$, and $\sum_{m+1}^{n-1} |z|^k < \sum_0^\infty |z|^k = 1/(1 - |z|)$, and $|1 - z|/(1 - |z|) \leq K$. Substituting into (4.2.1), we have

$$\left|\sum_{m+1}^{n} a_k z^k\right| \leq \frac{|1 - z|\varepsilon}{(1 - |z|)(K + 2)} + \frac{2\varepsilon}{K + 2}$$

$$\leq \frac{K\varepsilon}{K + 2} + \frac{2\varepsilon}{K + 2} = \varepsilon. \tag{4.2.2}$$

If $n = m + 1 \geq N_\varepsilon$, the estimate in the third member of (4.2.2) becomes $2\varepsilon/(K + 2) < \varepsilon$.

The conclusion to be drawn from these inequalities is that $\sum a_k z^k$ satisfies the uniform Cauchy criterion on $B_K \sim \{1\}$ and, therefore, converges uniformly on $B_K \sim \{1\}$. It also converges at $z = 1$, so by Exercise 3.4.7 it converges uniformly on B_K. Q.E.D.

Corollary 1 (Abel's Original Formulation). Let $\sum a_k z^k$, with radius of convergence 1, converge at $z = 1$. Then the sum function is continuous on the interval $[0, 1]$.

Corollary 2 Let $\sum a_k z^k$ with radius of convergence $R > 0$ converge at $z = z_0$, $|z_0| = R$. Then the sum function f is continuous on the line segment [Exercise 1.14.4] with equation $z = tz_0$, $0 \leq t \leq 1$.

Proof. Look at the function $\phi(w) = \sum a_k z_0^k w^k = f(z_0 w)$. The radius of convergence of this power series in w is easily seen to be 1 by using the definition of radius of convergence; and the series converges at $w = 1$, by hypothesis. Thus, by Abel's theorem, $\phi(w)$ is continuous on the line segment with equation $w = t$, $0 \leq t \leq 1$.

Now consider the composition of the function ϕ with the function given by $w = z/z_0$. The latter function is continuous everywhere on **C**, so its restriction to a domain $D = \{z: z = tz_0, 0 \le t \le 1\}$ is, of course, continuous. The range of $w = z/z_0$ with this restriction is $\{w: w = tz_0/z_0 = t, 0 \le t \le 1\}$ and as observed previously, ϕ is continuous on this point set. It follows from Exercise 3.2.9 that the composition function $\phi(z/z_0) = f[z_0(z/z_0)] = f(z)$ is continuous on D. Q.E.D.

Remark (i). By examining the geometry of the set B_K in Theorem 4.2.2 it can be shown [3, p. 254] that the convergence of the series $f(z) = \sum_0^\infty a_k z^k$ at $z = 1$ implies the continuity of $f(z)$ at $z = 1$ when the approach to $z = 1$ is restricted to the subset of $\Delta(0, 1)$ lying between two chords of $C(0, 1)$ intersecting at $z = 1$. (This is Stolz' original formulation of the theorem.)

Remark (ii). It is not possible in general to eliminate all restrictions on the approach to $z = 1$ on $\Delta(0, 1)$. Hardy and Littlewood [*Proceedings of the London Mathematical Society*, **13**, 174–191, 1964] have given an example of a series $\sum^\infty a_k z^k$ with radius of convergence 1, convergent at $z = 1$, such that the sum function is *not* continuous on any circle C which lies in $\Delta(0, 1)$ except for the point $z = 1$, at which point C is tangent to $C(0, 1)$.

It has become customary to say that a theorem about power series is "Abelian" if the hypothesis contains a statement involving convergence or summability at a point of the circle of convergence and the conclusion is a continuity property of the sum function at that point. Thus, Theorem 4.2.2 and the corollaries are of Abelian type. A theorem which goes in the opposite direction, deducing convergence or summability at a point of the circle of convergence from a continuity hypothesis, is said to be of "Tauberian" type. Notice that the series $\sum_0^\infty (-z)^n = 1/(1 + z)$ has a sum function which is continuous everywhere on the circle of convergence $C(0, 1)$ except at $z = -1$, but it converges nowhere on $C(0, 1)$. Thus, some side conditions are necessary to obtain nontrivial Tauberian theorems. We give here only Tauber's original result.

Theorem 4.2.3 *Let the power series* $f(z) = \sum a_k z^k$ *have radius of convergence 1, let the sequence* $\langle ka_k \rangle$ *converge to zero, and let* f *have the property that there exists* $s \in \mathbf{C}$ *such that if* $f(1)$ *is defined by* $f(1) = s$, *then* f *is continuous on the interval* $[0, 1]$. *Then* $\sum a_k 1^k$ *converges to* s.

A proof, and other Tauberian theorems, can be found in [8, pp. 35–37].

4.3 Uniqueness of the Power Series Representation

We now introduce a factorization of the sum function of a power series. This leads to an important result concerning the uniqueness of the power series representation of a function.

Theorem 4.3.1 *Let $\sum_0^\infty a_k(z - a)^k = f(z)$ have radius of convergence $R > 0$, and suppose that $a_0 = a_1 = \cdots = a_{m-1} = 0$, but $a_m \neq 0$. Then there exists a function $g\colon \Delta(a, R) \to \mathbf{C}$ such that $f(z) = (z - a)^m g(z)$ for every $z \in \Delta(a, R)$. The function g is the sum function of a power series $\sum_0^\infty b_k(z - a)^k$ with radius of convergence R, and so is continuous on $\Delta(a, R)$. Furthermore, $g(z) \neq 0$ on some $\Delta(a, \rho) \subset \Delta(a, R)$, $\rho > 0$.*

Proof. By Corollary 2 of Theorem 4.1.2, the power series $g(z) = \sum_{k=0}^\infty a_{k+m}(z - a)^k$ has radius of convergence R. The remark following that corollary shows that $g(z) = (z - a)^{-m}f(z)$, $z \in \Delta(a, R) \sim \{a\}$. Thus, $f(z) = (z - a)^m g(z)$, $0 < |z - a| < R$, and since $f(a) = 0$, the equation remains valid at $z = a$. By Theorem 4.2.1(i), g is continuous on $\Delta(a, R)$. Now $g(a) = a_m \neq 0$, so by Exercise 3.2.3(b) there is a disk $\Delta(a, \rho)$, $\rho > 0$, such that $g(z) \neq 0$ on $\Delta(a, \rho)$. Q.E.D.

Proposition 4.3.1 *Let $f(z) = \sum_0^\infty a_k(z - a)^k$ have radius of convergence $R > 0$. Let $\langle z_n \rangle_0^\infty$ be a sequence of points in $\Delta(a, R)$ converging to a, with $z_n \neq a$, $n = 0, 1, 2, \ldots$. If $f(z_n) = 0$, $n = 0, 1, 2, \ldots$, then $a_k = 0$, $k = 0, 1, 2, \ldots$, and so $f(z) = 0$ for every $z \in \Delta(a, R)$.*

Proof. Suppose that for some $m \geq 0$, $a_0 = a_1 = \cdots = a_{m-1} = 0$ but $a_m \neq 0$. (Nothing is said about a_k for $k > m$.) By Theorem 4.3.1, there exists $g\colon \Delta(a, R) \to \mathbf{C}$ with $f(z) = (z - a)^m g(z)$, and there is a disk $\Delta(a, \rho) \subset \Delta(a, R)$, $\rho > 0$, such that $g(z) \neq 0$ in $\Delta(a, \rho)$. Since $\lim z_n = a$, there exists a number N such that for every $n \geq N$, $z_n \in \Delta(a, \rho)$. Now by hypothesis $z_n \neq a$, $n = 0, 1, 2, \ldots$, so $(z_n - a)^m \neq 0$, $n = 0, 1, 2, \ldots$. Also $g(z_n) \neq 0$, $n = N, N + 1, N + 2, \ldots$, because $z_n \in \Delta(a, \rho)$, $n = N, N + 1, \ldots$. Yet, by hypothesis, $f(z_n) = 0$, $n = 0, 1, 2, \ldots$. This is a contradiction, so it must be true that $a_m = 0$.

The proof is completed by induction on m: With $m = 0$, the foregoing argument shows that $a_0 = 0$, and if for any $m > 0$, $a_0 = a_1 = \cdots = a_{m-1} = 0$, then $a_m = 0$. Q.E.D.

As an illustration, suppose that f is known to be the sum function of a power series $\sum a_k z^k$ with disk of convergence $\Delta(0, 1)$. It would be impossible for $f(z)$ to have the value zero at every point of the interval $(0, 1/2)$ without being identically zero in $\Delta(0, 1)$. (Take for example the sequence $\langle (1/2)^{n+2} \rangle_0^\infty$ as the sequence $\langle z_n \rangle$ in the proposition.)

Theorem 4.3.2 (*Uniqueness Theorem*). *Let $f(z) = \sum_0^\infty a_k(z - a)^k$ have radius of convergence $R_1 > 0$, and let $g(z) = \sum_0^\infty b_k(z - a)^k$ have radius of convergence $R_2 > 0$. Let $\langle z_n \rangle_0^\infty$ be a sequence of points in $\Delta(a, R)$, where $R = \min(R_1, R_2)$, such that $\lim z_n = a$ and $z_n \neq a$, $n = 0, 1, 2, \ldots$. Then if $f(z_n) = g(z_n)$, $n = 0, 1, 2, \ldots$, it follows that $a_k = b_k$, $k = 0, 1, 2, \ldots$, and, consequently, $R_1 = R_2 = R$ and $f(z) = g(z)$ for every $z \in \Delta(a, R)$.*

Proof. The series $\sum_0^\infty a_k(z-a)^k$ converges pointwise to the value $f(z)$ for every $z \in \Delta(a, R)$, and the series $\sum_0^\infty b_k(z-a)^k$ converges pointwise to the value $g(z)$ for every such z. Therefore, by Proposition 3.4.3(i) the series $\sum_0^\infty [a_k(z-a)^k - b_k(z-a)^k] = \sum_0^\infty (a_k - b_k)(z-a)^k$ converge pointwise on $\Delta(a, R)$ to the sum function with values $f(z) - g(z)$, so its radius of convergence must be at least $R > 0$. By hypothesis, $f(z_n) - g(z_n) = 0$, $n = 0, 1, \ldots$, with $z_n \neq a$ and $z_n \to a$. From Proposition 4.3.1, it must follow that $a_k - b_k = 0$, $a_k = b_k$, $k = 0, 1, 2, \ldots$. Consequently, by the definition of radius of convergence, $R_1 = R_2$, and then $R = R_1 = R_2$. The sum functions f and g must have identical values on $\Delta(a, R)$ because the coefficients of their power series are identical. Q.E.D.

Corollary If the function f from **C** into **C** is known to be the sum function of the power series $\sum_0^\infty a_k(z-a)^k$ with radius of convergence R, $0 < R \leq \infty$, then this power series representation of f is unique in the sense that if $\sum_0^\infty b_k(z-a)^k$ is known to converge to the sum function f in some disk $\Delta(a, \rho)$, with $\rho > 0$, then $a_k = b_k$, $k = 0, 1, 2, \ldots$.

Proof. Let $r = \min(R, \rho)$. If $\sum b_k(z-a)^k$ converges pointwise on $\Delta(a, \rho)$, $\rho > 0$ to f, then its sum function has values identical with those of the sum function of $\sum a_k(z-a)^k$ on $\Delta(a, r)$. In particular, the values of the sum functions coincide on any sequence of points in $\Delta(a, r)$ converging to a; to be specific, say the sequence $a + (r/2), a + (r/3), \ldots, a + r/(n+1), \ldots$. But then by the theorem, $a_k = b_k$, $k = 0, 1, 2, \ldots$.

Remark (i). The purport of the corollary is that given a complex-valued function f with domain E which contains a disk $\Delta(a, \rho)$, $\rho > 0$, if by some device (say division or substitution) one can find a power series of the type $\sum a_k(z-a)^k$ which converges to f on $\Delta(a, \rho)$, then this is the only power series representation for f on $\Delta(a, \rho)$ in powers of $(z-a)$.

Remark (ii). The condition $\rho > 0$ is essential in the theorem, because $f(a) + \sum_1^\infty b_k(z-a)^k$ converges trivially at $z = a$ to $f(a)$ for any choice of the coefficients b_k.

Remark (iii). It is shown later on that a function which has a power series representation $\sum a_k(z-a)^k$ valid on some disk $\Delta(a, \rho)$, $\rho > 0$, also has an infinite number of convergent power series representations with different values of a. Here we consider only a simple example. Let $f(z) = 1/(1+z)$. Then (see the example which follows Theorem 4.1.1) for $|z| < 1$, $f(z) = 1 - z + z^2 - z^3 + \cdots$. But also

$$\frac{1}{1+z} = \frac{1}{2(1+(z-1)/2)} = \frac{1}{2} - \frac{1}{2}\left(\frac{z-1}{2}\right) + \frac{1}{2}\left(\frac{z-1}{2}\right)^2 - \frac{1}{2}\left(\frac{z-1}{2}\right)^3 + \cdots,$$

which is convergent for $|(z-1)/2| < 1$ or $|z-1| < 2$. Both power series converge to $f(z)$ in $\Delta(0, 1)$.

Exercises 4.3

1. Find the power series representation for $(z + 1)(z - 1)^{-1}$ with disk of convergence $\Delta(0, 1)$. (*Hint*: Multiply the series for $1/(z - 1)$ by $z + 1$ and collect terms [Exercise 4.1.3].
2. Find the two power series in powers of z of which $1/(1 - z)^2$ and $1/(1 - z)^3$ are, respectively, the sum functions. (*Hint*: Cauchy product again.)
3. Let $f(z) = \sum_0^\infty a_k z^k$ have radius of convergence 1, and suppose $f(z) \neq 0$ on $\Delta(0, 1)$. Suppose further that it is known that $1/f(z)$ has a power series representation $\sum_0^\infty b_k z^k$ on $\Delta(0, 1)$. (It turns out later that the second sentence follows necessarily from the first one.) Set up a sequence of formulas whereby if the sequence $\langle a_k \rangle$ is known, the sequence $\langle b_k \rangle$ can be computed. (*Hint*: $1 = f(z)/f(z)$; use Cauchy multiplication.)
4. Given $f(z) = \sum_0^\infty a_k z^k$, convergent in $\Delta(0, r)$, $r > 0$. Show that if $[f(z)]^2 = f(z)$ for every z on the interval $(0, r)$, then either $f(z) \equiv 1$ or $f(z) \equiv 0$ for every $z \in \Delta(0, r)$. (*Hint*: Theorem 4.3.2; equate coefficients of f^2 and f.)
5. Consider the polynomial $\sum_0^n a_k z^k = p(z)$. Show by power series theory that there cannot exist another linear combination of powers of z, say $b_0 + b_1 z + \cdots + b_m z^m$, having the same value as $p(z)$ at each point of a disk $\Delta(0, r)$, $r > 0$.

5

Some Elementary Topological Concepts

5.1 Basic Definitions

The role of elementary topology in mathematical analysis is to give precision to concepts of "nearness," "inside," "outside," "boundary," "connected," and so on. In the earlier chapters of this book these questions are handled by simple *ad hoc* methods using inequalities in **R**. However, to give a satisfactory treatment of the more sophisticated limit processes involved in complex differentiation and integration, an explicit formulation of some of the basic topological concepts is useful. Most of the topological tools required for complex analysis belong to the theory of general topological spaces, but for this book it suffices to consider only the special case of metric spaces.

Recall the definition of a metric space given in Sec. 2.2: It is a pair (X, d), where X is a nonempty set and d is a real-valued function on $X \times X$ satisfying certain postulates given in Sec. 2.2. Recall also the definition of an open ball in X with center $a \in X$ and radius $r > 0$: $B(a, r) = \{x: x \in X, d(a, x) < r\}$. (If $X = \mathbf{C}$ and d is the usual metric on $\mathbf{C}$, we have been replacing B by Δ.) Our basic definitions are the following three.

(a) A set $E \subset X$ is *open* if and only if for every $x_0 \in E$ there exists $r > 0$ such that $B(x_0, r) \subset E$. (It is not implied that the same r is valid for every x_0.)

(b) Given a set $E \subset X$, $\bar{x} \in X$ is a *point of closure* of E if and only if for every $\delta > 0$, there exists $x_\delta \in E$ such that $x_\delta \in B(\bar{x}, \delta)$. (In particular, all of the points of E are points of closure.) The set of all points of closure of E are denoted by $\bar{E}$. It is called the *closure* of E.

(c) A set $E \subset X$ is *closed* if and only if $E = \bar{E}$.

A number of relatively trivial deductions are made from these definitions, some with the use of the De Morgan's laws for sets. We group some of them together in the following proposition, in which O, O_1, O_2, O_α denote open sets in a metric space (X, d), F, F_1, F_2, F_α denote closed sets, and for any set $E \subset X$, $E^c = X \sim E$, the *complement* of E (with respect to X).

Proposition 5.1.1

(i) *X and $\varnothing$ are both open and closed.*
(ii) *$O_1 \cap O_2$ is open; the intersection of a finite number of open sets is open.*
(iii) *$\bigcap_{\alpha \in A} O_\alpha$ is open, where the index set A is finite or infinite.*
(iv) *Any O is a union of open balls.*
(v) *For any $E \subset X$, $E \subset \bar{E}$, and $\bar{\bar{E}} = \bar{E}$ (so $\bar{E}$ is closed).*
(vi) *For any E_1, E_2 in X with $E_1 \subset E_2$, $\bar{E}_1 \subset \bar{E}_2$.*
(vii) *For any E_1, E_2 in X, $\overline{E_1 \cup E_2} = \bar{E}_1 \cup \bar{E}_2$ and $\overline{E_1 \cap E_2} \subset \bar{E}_1 \cap \bar{E}_2$.*
(viii) *For any $E \subset X$, $\bar{E}$ is contained in each closed set containing E.*
(ix) *$F_1 \cup F_2$ is closed; the union of any finite number of closed sets is closed.*
(x) *$\bigcup_{\alpha \in A} F_\alpha$ is closed, where A is finite or infinite.*
(xi) *O^c is closed, F^c is open. A set $E \subset X$ is open if and only if its complement is closed.*

Proof. Most of the statements are obvious from the definitions. Among the slightly more sophisticated parts, we select the first assertion in (xi) to typify the reasoning needed. Take any $x \in O$. There exists $\delta > 0$ such that $B(x, \delta) \subset O$. Hence x is not a point of closure of O^c, by definition of point of closure. Thus, O^c contains all its points of closure, since a point of closure of O^c not in O^c has to lie in O. Thus, O^c is closed, by definition.

A *neighborhood* of a point x in the metric space X is an open set $V \subset X$ which contains x. Generally, the type of neighborhood we use is an open ball, but it is convenient for some formulations to have the more general definition. If V is a neighborhood of x, we call $V \sim \{x\}$ a *deleted* neighborhood of x.

The *frontier* or *boundary* of a set E in the metric space X is the set $\partial E = \bar{E} \cap \overline{E^c}$.

Example. With $X = \mathbf{C}$, $E = \Delta(0, 1)$, $\partial E = C(0, 1) = \bar{\Delta}(0, 1) \cap \{z : |z| \geq 1\}$.

The *interior* of E is the maximal open set contained in E. A point of the *interior* is called an *interior point* of E. (The interior of E may be empty.)

The *exterior* of E is the interior of E^c.

The point $x_0 \in E$ is an *isolated point* of E if and only if there exists a neighborhood V of x_0 such that $V \cap E = \{x_0\}$. (That is, x_0 is the only point of E in V.)

The point $x_0 \in X$ is an *accumulation point* of E if and only if in every neighborhood of x_0 there exists a point $x \in E, x \neq x_0$. (Thus, an accumulation point of E is a point of the closure of E which is not an isolated point of E. A brief characterization is that it is a point of the closure of $E \sim \{x_0\}$.)

Given a metric space (X, d) and a nonempty set $E \subset X$, it can easily be seen that the restriction $\rho = d|(E \times E)$ of d to E satisfies the postulates for a metric on E. Thus, (E, ρ) is again a metric space. We say that ρ is the relative metric induced on E by d. When no confusion can result, we refer to the "metric space E." In this space, an open ball with center a and radius $r > 0$ is the set $B_E(a, r) = \{x : x \in E, d(a, x) < r\}$. We then start all over again to define the open sets, points of closure, and closed sets as in (a), (b), and (c), using open balls of the type $B_E(a, r)$. It is not difficult to see that since $B_E(a, r) = B(a, r) \cap E$, an open set in the metric space E is of the form $O \cap E$, where O is open in (X, d). Also a closed set in the metric space E is of the form $F \cap E$, where F is closed in (X, d).

Example. Let $X = \mathbf{C}$, with the usual metric, and let $E = \bar{\Delta}(0, 1)$. The set E is closed in $\mathbf{C}$ [Exercise 5.1.7] but both open and closed in E Proposition 5.1.1(i)]. The set $\Delta(1, 1) \cap E$ is an open ball in E and so is open in E [Exercise 5.1.1], but it is neither open nor closed in $\mathbf{C}$. The set $\Delta(1/2, 1/2) \cap E$ is again an open ball in E so is open in E. It is also open in $\mathbf{C}$.

Exercises 5.1

1. Prove that an open ball $B(a, r)$ in a metric space X satisfies the definition of an open set in X. (*Hint*: The problem is to show that for any $x_0 \in B(a, r)$, there exists $B(x_0, r_0) \subset B(a, r)$, $r_0 > 0$. Take $r_0 = r - d(x_0, a)$, and show by the triangle inequality that if $x \in B(x_0, r_0)$ (which means that $d(x_0, x) < r_0$) then $x \in B(a, r)$; i.e., $d(a, x) < r$.)
2. Show that $\bar{x}$ is a point of closure of a set E in the metric space X if and only if there is a sequence $\langle x_k \rangle$ in E with $\lim x_k = \bar{x}$.
3. In Proposition 5.1.1, prove that $\bar{E} = \bar{\bar{E}}$.
4. In Proposition 5.1.1, prove that if $E_1 \subset E_2$, then $\bar{E}_1 \subset \bar{E}_2$.
5. Find an example in which $\overline{E_1 \cap E_2} \neq \bar{E}_1 \cap \bar{E}_2$.
6. In Proposition 5.1.1, prove that $\bar{E}$ is the minimal closed set containing E. (*Hint*: Exercises 5.1.4 and 5.1.3.)
7. Show that the closure of the open ball $\Delta(a, r)$ in the metric space $\mathbf{C}$ is the set $\{z : |z - a| \leq r\}$.
8. Show that the intervals of real numbers $(-\infty, a)$, (a, ∞), (a, b), are open sets in the metric space $\mathbf{R}$.
9. Let O be an open set in $\mathbf{R}^2$, regarded as a metric space with the usual Euclidean metric. Prove that the sets $A = \{x \in \mathbf{R} : (x, y_0) \in O, y_0 \text{ fixed}\}$ and $B = \{y \in \mathbf{R} : (x_0, y) \in O, x_0 \text{ fixed}\}$, are open in the metric space $\mathbf{R}$. (*Hint*: The problem for A is to prove that if $x_1 \in A$, there exists $r > 0$ such that for every $x \in (x_1 - r, x_1 + r)$, the corresponding point $(x, y_0) \in O$. Write out the condition that (x_1, y_0) is an

interior point of O.)

10. Show that the half-planes in **C**, $\{z : \text{Re } z > 0\}$, $\{z : \text{Re } z < 0\}$, $\{z : \text{Im } z > 0\}$, $\{z : \text{Im } z < 0\}$, are open sets in **C**.

5.2 Continuity and Functional Limits

In Sec. 3.2, we define continuity for a function from **C** into **C**. We now generalize this definition. Let f be a function with domain E in a metric space (X, d_X) and with range in another metric space (Y, d_Y). The notation $B_X(a, r)$ is used for an open ball in X and $B_Y(b, r)$ for an open ball in Y. The function f is continuous at $x_0 \in E$ if for every $\varepsilon > 0$ there exists $\delta > 0$ (δ ordinarily depends on x_0 and ε) such that if $x \in B_X(x_0, \delta) \cap E$ then $f(x) \in B_Y(f(x_0), \varepsilon)$. If f is continuous at each point of E, it is said to be continuous on E.

This definition can be rephrased in a useful way as follows.

Theorem 5.2.1 *The function f is continuous on E if and only if for each open set $O \subset Y$ the inverse image $f^{-1}(O)$ is open in the metric space E.*

We remind the reader that the metric space E means the set E with the relative metric $d_X|(E \times E)$.

Proof. Suppose first that for each open $O \subset Y$ the set $f^{-1}(O)$ is open in E. Let $x_0 \in E$ and $\varepsilon > 0$ be given. The open ball $B_Y(f(x_0), \varepsilon)$ is an open set in Y [Exercise 5.1.1] so $O' = f^{-1}[B_Y(f(x_0), \varepsilon)]$ is open in E. Of course, $x_0 \in O'$. Since O' is open in E, there must exist an open ball $B = B_X(x_0, \delta) \cap E$, $\delta > 0$ (open in the metric space E) contained in O', and the image $f(x)$ of every point x in this open ball B must lie in $B_Y(f(x_0), \varepsilon)$ because $B \subset f^{-1}[B_Y(f(x_0), \varepsilon)]$. The definition for continuity of f at x_0 is, therefore, satisfied.

Now conversely suppose that the definition of continuity is satisfied. Let $O \subset Y$ be open, and let x_0 be a point of $f^{-1}(O)$, which is assumed to be nonempty. This places $f(x_0)$ in O. Since O is open, there must be some $\varepsilon > 0$ such that $B_Y(f(x_0), \varepsilon) \subset O$. By the definition of continuity, there exists $\delta > 0$ such that if $x \in B_Y(x_0, \delta) \cap E$, then $f(x) \in B_Y(f(x_0), \varepsilon) \subset O$. This in turn implies that x must lie in $f^{-1}(O)$. We have now shown that any point $x_0 \in f^{-1}(O)$ is the center of an open ball $B_X(x_0, \delta) \cap E$, $\delta > 0$, consisting only of points of $f^{-1}(O)$. This means that $f^{-1}(O)$ is an open set in the metric space E, by definition of open set. Q.E.D.

Example.

(i) Let X and Y be **R** with the usual metric. The open balls are open intervals. Let $E = [0, 1]$ and $f(x) = x^2$, restricted to E. Consider the open set $(1/4, 1)$ in Y. Its inverse image is $(1/2, 1)$, which is the intersection of, say, $(1/2, \infty)$ and E and so is open in E although not open in X.

(ii) Let X and Y be **R** with the usual metric, let $E = \mathbf{R}$, and let $f(x) = x^2/(1 + x^2)$. Then $f[(-\infty, \infty)] = [0, 1)$, which is neither open nor closed in Y, although $(-\infty, \infty)$ is open in X.

The second example shows that in general the image of an open set under a continuous mapping is not necessarily open. There is, however, an important case in which open sets correspond to open sets. This is the case of a *homeomorphism*. A homeomorphism from a set E in a metric space X onto a set F in a metric space Y is a univalent (Sec. 1.7) continuous function, with domain E and range F, possessing a continuous inverse. In the case of such a mapping, Theorem 5.2.1 shows that an open set in the metric space F (with F and its relative metric playing the role of (Y, d_Y) in the theorem) is transformed into an open set in the metric space E, and an open set in the metric space E (with E and its relative metric used for (Y, d_Y) in the theorem) is mapped onto an open set in the metric space F.

We now define the *limit of a function*. Let X and Y be metric spaces, let f from X into Y be a function with domain $E \subset X$, and let a be an accumulation point of E, which may or may not be an element of E. An element $L \in Y$ is said to be the limit on E of f as x approaches a if and only if there exists a function $l: E \cup \{a\} \to Y$, continuous at a, such that $l(x) = f(x)$ for every $x \in E \sim \{a\}$ ($= E$ if $a \notin E$), and $l(a) = L$. Under these circumstances we write $\lim_{x\to a} f(x) = L$, $x \in E$, to indicate both the existence and value of the limit simultaneously. The notation $x \in E$ after the limit symbol is dropped when no confusion can result.

Remark (i). The definition is equivalent to this statement: The element L is such that for every real $\varepsilon > 0$ there exists $\delta > 0$ such that if $x \in B_X(a, \delta) \cap (E \sim \{a\})$, then $f(x) \in B_Y(L, \varepsilon)$.

Remark (ii). Since every neighborhood of a must contain a point of E, the definition corresponds to the intuitive idea of the value of f approaching L as x on E approaches a.

Remark (iii). If $a \in E$, *the function* f is continuous at a if and only if $\lim\limits_{x\to a} f(x) = f(a)$, $x \in E$.

Remark (iv). The definition includes the idea of left- and right-hand limits of real-valued functions of which the domain is an interval in **R**.

Theorem 5.2.2 *Let f and g be functions from the metric space X into the metric space Y with common domain $E \subset X$, and let a be an accumulation point of E. Let Y also be a field. Further let* $\lim_{x\to a} f(x) = L_1$ *and* $\lim_{x\to a} g(x) = L_2$. *Then (i)* $\lim_{x\to a}[f(x) + g(x)] = L_1 + L_2$, *(ii)* $\lim_{x\to a} f(x)g(x) = L_1L_2$, *and (iii) if $g(x) \neq 0$ on E and $L_2 \neq 0$,* $\lim_{x\to a}[f(x)/g(x)] = L_1/L_2$.

Proof. There exist functions l_1 and l_2 with common domain $E \cup \{a\}$

and range in Y such that $l_1(x) = f(x)$, $l_2(x) = g(x)$ for every $x \in E \sim \{a\}$, and $l_1(a) = L_1$, $l_2(a) = L_2$. The proof of the theorem merely involves showing that $l_1 + l_2$, $l_1 \cdot l_2$, and l_1/l_2 are also continuous at a.

This proposition was covered for the case $X = \mathbf{C}$, $Y = \mathbf{C}$ in Exercise 3.2.8, and since that is the only case needed in the sequel, we leave the general case to the reader.

Exercises 5.2

1. Let f be a function from a metric space X into a metric space Y, and let the domain of f be E. Show that f is continuous on E if and only if for each closed set $F \subset Y$, $f^{-1}(F)$ is closed in the metric space E. (*Hint*: Theorem 5.2.1 and set complementation.)
2. Let X, Y, and Z be metric spaces. Let the function g be continuous on X and have range $E \subset Y$. Let $f: E \to Z$ be continuous on E. Then the composition $f \circ g$ is continuous on X. (*Hint*: Theorem 5.2.1. This exercise is a generalization of Exercise 3.2.10.)
3. Let f be a function from a metric space X into a metric space Y, let E be the domain of f, and let a be an accumulation point of E. Prove that $\lim_{x \to a} f(x) = L$ if and only if for every sequence $\langle x_n \rangle_0^\infty$ in $E \sim \{a\}$ converging to a, the sequence $\langle f(x_n) \rangle_0^\infty$ in Y converges to L.
4. Let f be a continuous univalent mapping of a set E in a metric space onto the set $f(E)$ in a metric space. Prove that if $a \in E$ is an accumulation point of E, then $f(a)$ is an accumulation point of $f(E)$.
5. Let X be a metric space, let f be a complex-valued function with domain $E \subset X$, and let a be an accumulation point of E. Let $f(x) = u(x) + iv(x)$, $x \in E$, where u and v are real valued. Prove that f has the limit $L = L_1 + iL_2$ as x approaches a if and only if u has the limit L_1 and v has the limit L_2 as x approaches a. (*Hint*: Theorem 3.2.1.)
6. Let X be $\mathbf{C}$, and let α be real. Prove that the half-planes $\{z = x + iy : x > \alpha\}$, $\{z : x < \alpha\}$, $\{z : y > \alpha\}$, $\{z : y < \alpha\}$, are open in $\mathbf{C}$. (*Hint*: Translations of the types $w = z - \alpha$, $w = z - i\alpha$ give homemorphisms from $\mathbf{C}$ onto $\mathbf{C}$ which reduce the problem to Exercise 5.1.10.)
7. With $m \in \mathbf{R}$, show that the half-planes defined by $\{z = x + iy : y > mx\}$ and $\{z = x + iy : y < mx\}$ are open. (*Hint*: Look at the homeomorphism from $\mathbf{C}$ into $\mathbf{C}$ given by $w = (1 - im)z$.)
8. Show that these circular sectors are open: $\{z = x + iy : x > 0, |z| < r\}$, $\{z : y > 0, |z| < r\}$, $\{z : x < 0, |z| < r\}$, $\{z : y < 0, |z| < r\}$ and $S_k = \{z : -kx < y < kx, x > 0, |z| < r\}$, where $k > 0$, $r > 0$. (*Hint*: S_k is the intersection of three open half-planes and an open disk.)
9. Prove that the half-plane $H = \{z = x + iy : (x/\alpha) + (y/\beta) > 1\}$, $\alpha > 0$, $\beta > 0$, is open. (*Hint*: Look at the homeomorphisms from $\mathbf{C}$ onto $\mathbf{C}$ given by $w = z - \alpha$, $\zeta = (\beta - i\alpha)w$.)
10. Show that the interval $[\alpha, \beta] \subset \mathbf{R}$ is closed in $\mathbf{C}$. (*Hint*: Look at its complement.)

5.3 Connected Sets

Intuitively speaking, a set is "connected" if it consists of one piece. It is important for complex analysis to make this idea precise, which we now do.

Let E be a subset of a metric space (X, d). A *separation* (splitting) of E is a pair of disjoint nonempty sets O_1 and O_2, each open in the metric space E (with the relative metric) and such that $E = O_1 \cup O_2$.

The set E is *connected* if and only if there exists no such separation.

Notice that in a separation of E, O_1 is the complement of O_2 with respect to E, and vice versa. Therefore, by Proposition 5.1.1(xi), O_2 and O_1 are closed as well as open in the metric space E. It follows that a set E is connected if and only if $\varnothing$ and E are the only subsets which are both open and closed in the metric space E.

We call a nonempty open connected set in a metric space a *region*.

Proposition 5.3.1 *Let E be a connected subset of a metric space X, and let f be a function with domain E and range $R = f(E) \subset Y$, where Y is a metric space. If f is continuous on E, then R is connected.*

Proof. Suppose R were not connected. Then there would exists two nonempty disjoint sets O_1 and O_2, each open in the metric space R, and such that $R = O_1 \cup O_2$. By Theorem 5.2.1, $f^{-1}(O_1)$ and $f^{-1}(O_2)$ are open in the metric space E. Neither of the sets $f^{-1}(O_1)$ and $f^{-1}(O_2)$ can be empty, because their images under f are not empty. Furthermore, $f^{-1}(O_1) \cup f^{-1}(O_2) = E$. But then we have a separation of E, which is a contradiction.

Proposition 5.3.2 (*Intermediate Value Theorem*). *Let E be a connected subset of a metric space X and let f be a real-valued function continuous on E. If x and y are two points in E and c is a real number such that $f(x) < c < f(y)$, then there must exist $x_c \in E$ such that $f(x_c) = c$.*

Proof. Suppose that there did not exist such a point x_c. The sets $f^{-1}[(-\infty, c)]$ and $f^{-1}[(c, \infty)]$ are open in the metric space E, by Exercise 5.1.8 and Theorem 5.2.1. Neither can be empty, because x belongs to the first and y, to the second. Finally, since our supposition implies that c is omitted from the range of f, the union of these two sets, which are disjoint (by the definition of function), is E. Thus, E is not connected, a contradiction.

Proposition 5.3.3 *Any interval of* **R** *is connected.*

Proof. Let I be any interval of **R**, which may be open, closed, half-open, half-closed, or infinite (see Sec. 1.6). Suppose that I is not connected. There then exist open subsets $O_1, O_2 \subset \mathbf{R}$ such that $I \cap O_1 \neq \varnothing$, $I \cap O_2 \neq \varnothing$, and $I = (I \cap O_1) \cup (I \cap O_2)$. Take any $x \in I \cap O_1$ and $y \in I \cap O_2$. Without loss of generality, since $x \neq y$, we assume that $x < y$. (This is merely a question of labeling.)

Let $s = \sup \{x' : x' \in O_1 \text{ and } x' < y\}$. Clearly, $x \leq s \leq y$. Since I is an interval, s must lie in I, so either $s \in O_1$ or else $s \in O_2$. We find that either alternative leads to a contradiction.

Suppose $s \in O_1$. Since O_1 is open, there exists $\delta > 0$ such that $[s - 2\delta, s + 2\delta] \subset O_1$, and in particular $s + \delta \in O_1$. If $s + \delta < y$, we have exhibited a number lying in O_1 which is greater than s and less than y, which contradicts the definition of s. If $s + \delta \geq y$, then $s \leq y \leq s + \delta$; but the entire interval $[s, s + \delta]$ is contained in O_1, which contradicts the assumption that $y \in O_2$.

Now suppose that $s \in O_2$. Again there exists $\delta > 0$ (the reasoning is the same as before) such that $[s - \delta, s] \subset O_2$. But there simply must be a point of O_1 in this interval because if not, $\varepsilon - \delta$ would be an upper bound for $\{x' : x' \in O_1, x' < y\}$, and s would not be the least upper bound. Q.E.D.

Proposition 5.3.4 *Let E be a connected subset of a metric space X, let F be another subset of the space, and suppose that $E \subset F \subset \bar{E}$. Then F is connected.*

The proof is an exercise. The proposition implies that if E is connected, so is $\bar{E}$.

Proposition 5.3.5 *A connected set E in* **R** *which consists of more than one point is an interval.*

Lemma If $x, y \in E$, $x < y$, where E is a connected subset of **R**, then $[x, y] \subset E$.

Proof. Suppose $x' \in (x, y)$ but $x' \notin E$. Then $E \cap (-\infty, x')$ and $E \cap (x', \infty)$ form a separation of E [Exercise 5.1.8], which is a contradiction.

Proof of the Proposition. We claim that if $\alpha = \inf E$, $\beta = \sup E$, then $(\alpha, \beta) \subset E$. For if there is $x_1 \in (\alpha, \beta)$, but $x_1 \notin E$, then there must be $x_0 \in E$, $\alpha \leq x_0 < x_1$ and an $x_2 \in E$, $x_0 < x_2 \leq \beta$ [Exercise 1.4.1]. But by the lemma, $[x_0, x_2] \subset E$, so there can be no such x_1.

By the definition of sup and inf, we have $E \subset [\alpha, \beta]$, so $(\alpha, \beta) \subset E \subset [\alpha, \beta]$. The implication is that E must be some sort of interval, finite or infinite; open, closed, or neither. Q.E.D.

An important type of connected subset of a metric space (X, d) is what is known as the *graph of an arc*, a term which by tradition is often abbreviated to just an "arc." Formally, an *arc* in X is a continuous function γ from an interval $[\alpha, \beta] \subset$ **R** into X. We always take α and β to be finite. Its *graph* is the point set in X which is the range of the function. The point $\gamma(\alpha)$ is called the *initial point* and the point $\gamma(\beta)$ is called the *terminal point* or *endpoint*. The arc is said to *join* $\gamma(\alpha)$ to $\gamma(\beta)$. The *parametric equation* of the arc is $x = \gamma(t)$, $\alpha \leq t \leq \beta$.

The connectivity in X of the arc is assured by Propositions 5.3.1 and 5.3.3.

Examples

(i) Let $X = \mathbf{C}$ with the usual metric. In Exercise 1.14.5, it was indicated that the graph of the function $z = z_1 + t(z_2 - z_1), -\infty < t < \infty, z_1 \neq z_2$, is a straight line through z_1 and z_2. The restriction of t to the interval $[0, 1]$ produces a line segment joining z_1 to z_2. The arc given by

$$z = z_1 + \frac{\tau - \alpha}{\beta - \alpha}(z_2 - z_1), \qquad \alpha \leq \tau \leq \beta,$$

has the same graph, as is easily verified. We denote this graph by $[z_1, z_2]$.

(ii) Let $X = \mathbf{C}$ and a, b be complex numbers. A polygonal arc joining a to b is an arc of which the graph is a finite sequence of line segments joining $a = z_0$ to z_1, z_1 to z_2, z_2 to $z_3, \ldots, z_{n-1}$ to $z_n = b$. A parametric equation of this arc is

$$z = \gamma(t) = z_j + \frac{t - t_j}{t_{j+1} - t_j}(z_{j+1} - z_j), \qquad t_j \leq t \leq t_{j+1}, \qquad j = 0, 1, 2, \ldots, n-1,$$

where $t_0 < t_1 < t_2 < \cdots < t_n$. (Clearly γ is continuous on $[t_0, t_n]$.)

Let (X, d) be a metric space. A subset $E \subset X$ is said to be *arcwise connected* if and only if any two points in E can be joined by an arc with its graph contained in E.

Theorem 5.3.1 *If a subset E of a metric space is arcwise connected, then it is connected.*

Proof. We show that there cannot exist a separation of E. Suppose that O_1 and O_2 *is* a separation of E, where O_1 and O_2 are open in the metric space E. Take $x_1 \in O_1$, $x_2 \in O_2$. Let $x = \gamma(t)$, $\alpha \leq t \leq \beta$ be the parametric equation of an arc γ joining x_1 to x_2, and with graph Γ contained in E. Then $O_1 \cap \Gamma$ and $O_2 \cap \Gamma$ are open sets in the metric space Γ (that is, Γ with the metric given by the restriction of the metric of E to Γ). By Theorem 5.2.1, $\gamma^{-1}(O_1)$ and $\gamma^{-1}(O_2)$ are open sets in the metric space $[\alpha, \beta]$ (that is, $[\alpha, \beta]$ with the metric inherited from the usual metric on $\mathbf{R}$). Neither set is empty, because $\alpha \in \gamma^{-1}(O_1)$ and $\beta \in \gamma^{-1}(O_2)$. Their union contains every point in $[\alpha, \beta]$. These sets are disjoint, by the definition of function and the fact that O_1 and O_2 are disjoint. Thus, $\gamma^{-1}(O_1)$ and $\gamma^{-1}(O_2)$ constitute a separation of the interval $[\alpha, \beta]$, which contradicts Proposition 5.3.3. Q.E.D.

Examples

(i) An important class of connected subsets of $\mathbf{C}$ is that of the *convex sets*. A set E containing two or more points is convex if and only if any two points in it can be joined by a line segment lying entirely in E.

(ii) Another important class of connected sets in $\mathbf{C}$ which play a central role in some later chapters of this book is the class of *starlike sets*. A nonempty set S of $\mathbf{C}$ is called starlike if there is some point $a \in S$ such that if b is any other point in S, then the line segment joining a to b is contained in S. The point a is called a *star center* of S, and S is said to be starlike with respect to a. Clearly, any convex set is starlike with respect to any of its points.. Any two points b and c in a set which is starlike with respect to a can be joined by the polygonal arc joining b to a and a to c, so by Theorem 5.3.1 a starlike set is connected. An example of a starlike but nonconvex set is given below in Exercise 5.3.9.

Proposition 5.3.6 *A disk $\Delta(a, r)$, $r > 0$, in* **C** *is a convex set.*

Proof. Let z_1 and z_2 lie in $\Delta(a, r)$, $z_1 \neq z_2$. The line segment joining z_1 to z_2 has a parametric equation

$$z = z_1 + t(z_2 - z_1) = (1 - t)z_1 + tz_2, \qquad 0 \le t \le 1. \quad (5.3.1)$$

Clearly $a = (1 - t)a + ta$, so the parametric equation can be rewritten in the form

$$z - a = (1 - t)(z_1 - a) + t(z_2 - a), \qquad 0 \le t \le 1.$$

By the triangle inequality, using the fact that $|z_1 - a| < r$, $|z_2 - a| < r$, we obtain

$$|z - a| \le |(1 - t)(z_1 - a)| + |t(z_2 - a)| < (1 - t)r + tr = r.$$

Thus, every point z on the line segment joining z_1 to z_2 lies in $\Delta(a, r)$. Q.E.D.

The converse of Theorem 5.3.1 is not true in general, but it is true for regions in **C**.

Theorem 5.3.2 *A region in* **C** *is arcwise connected by polygonal arcs.*

Proof. Let R be the region, and let a be in R. Let B be the union of $\{a\}$ with the set of all points $z \in R$ such that there exists a polygonal arc with graph contained in R joining a to z. The set B obviously is not empty. We prove that it is open.

Suppose that $z_0 \in B$. Since R is open, there exists a disk $\Delta(z_0, r)$, $r > 0$, contained in R. The point a can be joined to any point $z \in \Delta(z_0, r)$ by a polygonal arc joining a to z_0 augmented by a line segment joining z_0 to z and lying in $\Delta(z_0, r)$, thus lying in R [Proposition 5.3.5]. The union of the arc and the line segment is again a polygonal arc, so if $z \in \Delta(z_0, r)$, it follows that $z \in B$. Therefore, B is open. (The polygonal arc joining a to z_0 is to be omitted if $a = z_0$.)

Now look at $R \sim B$. If this set is empty, then there is no more to be proved. But it must be empty, because if $z_1 \in R \sim B$, let $\Delta(z_1, r)$, $r > 0$, be any disk contained in R with center at z_1. If this disk were to contain a point $z_2 \in B$, then a could be joined to z_1 by a polygonal arc joining a to z_2 and a line segment joining z_2 to z_1. This contradicts the assumption that $z_1 \in R \sim B$. So

$\Delta(z_1, r) \subset R \sim B$. But that implies that $R \sim B$ is open. So if $R \sim B$ is not empty, it is open, and B together with $R \sim B$ constitutes a separation of R. This contradicts the hypothesis that R is connected. Q.E.D.

Exercises 5.3

1. In Exercise 1.14.17, the set $E = \{z = x + iy : (x/\alpha)^2 + (y/\beta)^2 < 1\}$ is identified as the "inside" of the ellipse in $\mathbf{R}^2$ with major axis 2α and minor axis 2β. Show that E is connected. (*Hint*: First, show that if $z_1 \in E$, the line segment with equation $z = tz_1$, $0 \le t \le 1$, which joins $z = 0$ to z_1, lies in E.)
2. Is a singleton $\{x\}$ in a metric space connected?
3. Let the real-valued function f be continuous on an interval $I \subset \mathbf{R}$. Prove that $f(I)$ is an interval.
4. A real-valued function f of which the domain is a set E in $\mathbf{R}$ is said to be strictly increasing if $f(x_1) < f(x_2)$ for every $x_1, x_2 \in E$ with $x_1 < x_2$. It is said to be strictly decreasing if $f(x_1) > f(x_2)$ for every $x_1, x_2 \in E$ with $x_1 < x_2$. (The definition is generally used only when E is an interval.) Prove that if the real-valued function f is continuous on an interval $I \subset \mathbf{R}$, then it gives a univalent mapping [Sec. 1.7] of I onto $f(I)$ if and only if it is either strictly decreasing or strictly increasing. (*Hint*: The "if" part is easy and does not require continuity. Prove the other part by proving the contrapositive: If f is not strictly increasing or strictly decreasing, then it is not univalent. The hypothesis here is that there must exist x_1, x_2, x_3 in I with $x_1 < x_2$, $x_1 < x_3$, and such that $f(x_1) < f(x_2)$ and $f(x_1) > f(x_3)$. Thus, $f(x_3) < f(x_1) < f(x_2)$. Use Proposition 5.3.2.)
5. Let f be real valued, continuous, and univalent on an interval $I \subset \mathbf{R}$. Show that the image of a subinterval of I is the same kind of subinterval (closed, open, half-open) of $f(I)$.
6. Let f be real valued, continuous, and univalent on an interval $I \subset \mathbf{R}$. Prove that f must be a homeomorphism. (*Hint*: The problem is to show that f^{-1} is continuous on $f(I)$. Suppose f is strictly increasing. Take $y_0 \in f(I)$, and let $x_0 = f^{-1}(y_0) \in I$. First, suppose that x_0 is an interior point of I. Given $\varepsilon > 0$, the number $h < \varepsilon$ can be chosen so that $I_h = (x_0 - h, x_0 + h) \in I$. By Exercise 5.3.5, $f(I_h) = (f(x_0 - h), f(x_0 + h))$. Let $\delta = \min[y_0 - f(x_0 - h), f(x_0 + h) - y_0]$. Then if $y \in (y_0 - \delta, y_0 + \delta)$, it follows that $f^{-1}(y) \in I_h \subset (x_0 - \varepsilon, x_0 + \varepsilon)$. Thus, by the basic definition of continuity, $f^{-1}(y)$ is continuous at y_0. Complete the proof for other cases.)
7. Prove Proposition 5.3.4. (*Hint*: Given any $\bar{x} \in F$, since by hypothesis $\bar{x}$ is also a point of $\bar{E}$, it follows that in every neighborhood of $\bar{x}$ there must be a point of E. Suppose F is not connected, and let $O_1 \cap F$ and $O_2 \cap F$ be a separation of F, where O_1 and O_2 are open in X. Show that $O_1 \cap E$ and $O_2 \cap E$ is a separation of E. Since E is connected, one of these sets must be empty and the other must be E. Suppose $O_1 \cap E = E$. Look at $O_2 \cap F$ in the light of the first sentence of this hint, and arrive at a contradiction.)

8. Show that the intersection of two convex sets in $\mathbf{C}$ is convex if is consists of two or more points.
9. Show that the set $\mathbf{C} \sim P$, where P is the nonnegative real axis, is starlike but not convex. (*Hint*: Take $z = -1$ as star center and show that the imaginary part of any z lying on the line segment joining -1 to $z_1 \in \mathbf{C} \sim P$, with $\text{Im } z_1 \neq 0$, is never zero.
10. A *weighted average* of the distinct points $z_1, z_2, \ldots, z_n$ in $\mathbf{C}$ is an expression of the type $w = m_1 z_1 + m_2 z_2 + \cdots + m_n z_n$, where $m_1, m_2, \ldots, m_n$, are real, $0 \leq m_k \leq 1$, $k = 1, 2, \ldots, n$, and $\sum_1^n m_k = 1$. The *convex hull* of a set $E \subset \mathbf{C}$ is the intersection of all convex sets containing E. Prove that the convex hull of a set $E \subset \mathbf{C}$ is equal to the set W of all weighted averages of points of E. (*Hint*: First, show that the set W is convex by showing that any point on the line segment joining any two points in W is itself a weighted average of points of E. Then show that if H is any convex set containing E, any weighted average $w = m_1 z_1 + m_2 z_2 + \cdots + m_n z_n$ of points of E lies in H. Start by observing that $w_2 = [m_1/(m_1 + m_2)]z_1 + [m_2/(m_1 + m_2)]z_2$ is on the line segment joining z_1 and z_2 and so lies in H; proceed inductivity.)
11. Let z_1, z_2, and z_3 be distinct points in $\mathbf{C}$. Show that the convex hull of these points is the union of all the line segments joining z_1 to the points of the line segment joining z_2 to z_3. (*Hint*: Use the characterization of convex hull contained in Exercise 5.3.10.)
12. Show that the half-planes $\{z = x + iy : x > \alpha\}$, $\{z : x < \alpha\}$, $\{z : y > \alpha\}$, $\{z : y < \alpha\}$ are convex regions.
13. Show that the circular sectors in Exercise 5.2.8 are convex regions. (*Hint*: Exercise 5.3.8.)
14. Let F be a closed subset of a metric space (X, d). Show that if O_1 and O_2 constitute a separation of F, then O_1 and O_2 are *closed* in X. (*Hint*: Let Ω_1 and Ω_2 be disjoint open sets in X such that $O_1 = \Omega_1 \cap F$, $O_2 = \Omega_2 \cap F$. Now $\overline{\Omega_1 \cap F} \subset \bar{\Omega}_1 \cap F$. Take any $z_1 \in \bar{\Omega}_1 \cap F$; then since $z_1 \in F$, we have $z_1 \in \Omega_1 \cap F$ or $z_1 \in \Omega_2 \cap F$. Show that the latter case is impossible by the definition of openness and closure.)

5.4 Components†

A component of a metric space (X, d) is a nonempty connected subset of X which is not a subset of any other connected subset of X. In other words, a component of X is a maximal connected subset. In the applications later in this book, X is itself a subset of a larger metric space (which usually is $\mathbf{C}$), so connectivity is defined in terms of the relative metric on X.

The following result is useful for discussing the nature of components.

†The material in the remainder of this chapter is not needed until Chapter 8 and reading may, therefore, be postponed.

Proposition 5.4.1 *Let $\{E_\alpha : \alpha \in A\}$ (A is an index set) be a family of connected subsets of a metric space X such that $I = \bigcap_{\alpha \in A} E_\alpha \neq \varnothing$. Then $E = \bigcup_{\alpha \in A} E_\alpha$ is connected.*

Proof. Assume that E is not connected and that O_1 and O_2 are nonempty sets, open in the relative metric on E, and such that $O_1 \cap O_2 = \varnothing$, $O_1 \cup O_2 = E$. Either O_1 or O_2 must contain a point of I; suppose O_1 does. *We assert that for each $\alpha \in A$, $E_\alpha \subset O_1$.* This follows from these facts: (a) $E_\alpha \cap O_1 \neq \varnothing$; (b) $E_\alpha \subset E = O_1 \cup O_2$; (c) therefore, if $E_\alpha \cap O_2 \neq \varnothing$, the sets $E_\alpha \cap O_1$ and $E_\alpha \cap O_2$ would constitute a separation of E_α, contrary to the hypothesis that E_α is connected. Since $E_\alpha \subset O_1$ for each $\alpha \in A$, the union $E \subset O_1$, which implies that $O_2 = \varnothing$—a contradiction. Q.E.D.

Corollary Let $\langle E_n \rangle_0^\infty$ be a sequence of connected subsets of a metric space X, and let $E_n \cap E_{n+1}$ be nonempty, $n = 0, 1, 2, \ldots$. Then $\bigcap_{n=0}^\infty E_n$ is connected.

Proof. According to the proposition, $E_0 \cup E_1$ is connected. It follows from induction on n that $\bigcap_{n=0}^N E_n$ also is connected for any positive integer N.

Suppose now that $E = \bigcup_{n=0}^\infty E_n$ is nevertheless not connected, and that $\{O_1, O_2\}$ is a separation of E. The set E_0 lies entirely in either O_1 or O_2; otherwise $\{E_0 \cap O_1, E_0 \cap O_2\}$ would be a separation of E_0. If $E_0 \subset O_1$, then $\bigcup_0^N E_n \subset O_1$, $N = 1, 2, \ldots$, because if for some N, $(\bigcup_0^N E_n) \cap O_2 \neq \varnothing$, then $\{O_1 \cap (\bigcup_0^N E_n), O_2 \cap (\bigcup_0^N E_n)\}$ is a separation of $\bigcup_0^N E_n$. Similarly, if $E_0 \subset O_2$, then $\bigcup_0^N E_n \subset O_2$, $N = 1, 2, \ldots$.

Now suppose $E_0 \subset O_1$. Since by assumption $O_2 \neq \varnothing$, it follows that there must exists some element x which lies in both E and O_2. This x must be an element of one of the sets E_n; say $x \in E_N$. But we have just shown that $\bigcup_0^N E_n \subset O_1$, which is a contradiction. Similarly, a contradiction is reached if $E_0 \subset O_2$. It follows that no separation of E can exist, so E is connected. Q.E.D.

Theorem 5.4.1 *If C is a nonempty connected subset of a metric space X, then C is contained in one and only one component of X. This component is the union of all connected subsets of X which contain C.*

Notice that C can be a single point in X.

Proof. Let E be the union described in the theorem. By Proposition 5.4.1, E is connected. It is a component of X, because if E_0 is any connected set containing E, by construction $E = E_0 \cup E$. If E_1 were any other component of X containing C, then again by construction of E, since E_1 is connected, $E_1 \subset E$; but unless $E_1 = E$, E_1 would not be a maximal connected subset of X. Q.E.D.

Corollary 1 The components of a metric space X are pairwise disjoint, and their union is X.

Proof. According to the theorem, each point in X lies in one but only one component. Q.E.D.

Corollary 2 Let O be an open set in **C**. The components of O are regions.

Proof. Let C be a component of O, and let a be any point in C. There exists a disk $\Delta(a, r) \subset O$, $r > 0$. But $\Delta(a, r)$ is a connected subset of O [Proposition 5.3.6]. Therefore, by the theorem, $\Delta(a, r) \subset C$. Thus, C is open, and since it is connected, it satisfies the definition of a region.

Remark. An open subset of **C** has at most a countable number of components; that is, the components can be put into one-to-one correspondence with a subset of the integers. The proof, which we do not give in detail here, depends on the fact that integers k, m, n can be chosen so that any component contains at least one member of the set $\{2^{-k}(\mathrm{m} + \mathrm{in})\}$, and this set is countable.

Proposition 5.4.2 *Let E be a bounded set in* **C**. *The complement E^c has one and only one unbounded component. If $E \subset \bar{\Delta}(a, R)$, then the unbounded component contains* $[\bar{\Delta}(a, R)]^c$.

Lemma The punctured disk $\Delta' = \{z: 0 < |z - a| < R\}$, $R > 0$, is a region and $[\bar{\Delta}(a, R)]^c$ is a region.

Proof of the Lemma. The homeomorphism of **C** onto **C** given by $w = z - a$ carries Δ' onto $\{w: 0 < |w| < R\}$, so by Theorem 5.2.1 and Proposition 5.3.1, it suffices to prove the lemma with $a = 0$. In that case, we express Δ' in the form $\Delta' = S_1 \cup S_2 \cup S_3 \cup S_4$, where $S_1 = \{z = x + iy: x > 0, |z| < R\}$, $S_2 = \{z: y > 0, |z| < R\}$, $S_3 = \{z: x < 0, |z| < R\}$, $S_4 = \{z: y < 0, |z| < R\}$. According to Exercise 5.3.13, S_1, S_2, S_3, and S_4 are convex regions, and, therefore, each is connected. According to Exercise 5.2.8, each of these sectors is open, so their union is open. By the corollary of Proposition 5.4.1, the union is connected Therefore, Δ' is a region.

We now examine the complement $\bar{\Delta}^c$ of $\bar{\Delta}(a, R)$. This complement is the open set $\{\zeta: |\zeta - a| > R\}$. It is easily seen that $\bar{\Delta}^c$ is the image of Δ' under the homeomorphism given by $\zeta = a + R^2/(z - a)$ of which $\mathbf{C} \sim \{a\}$ is both the domain and range. It follows from Theorem 5.2.1 and Proposition 5.3.1 that $\bar{\Delta}^c$ is a region. The proof of the lemma is complete.

Proof of the Proposition. Suppose that $E \subset \bar{\Delta}(a, R)$. Then $\bar{\Delta}^c = [\bar{\Delta}(a, R)]^c \subset E^c$. According to the lemma, $\bar{\Delta}^c$ is connected in **C** and, therefore, it is also connected in the metric space E^c. It follows from Theorem 5.4.1 that $\bar{\Delta}^c$ is contained in one and only one component of E^c, say E^c_∞. This component must be unbounded because $\bar{\Delta}^c$ is unbounded. No point belonging to one of the other components of E^c can lie in $\bar{\Delta}^c$, because then it would also lie in E^c_∞ [Corollary 1 of Theorem 5.4.1]. Therefore, the components of E^c other than E^c_∞ are all contained in $\bar{\Delta}(a, R)$, so they are bounded sets. Q.E.D.

5.5 Compactness

Let (X, d) be a metric space, and let E be a subset of X. A family Ω of open sets of X is said to be an *open covering* of E if $E \subset \bigcup_{O_\alpha \in \Omega} O_\alpha$. An *$\Omega$-subcovering* of E is a subcollection of the family Ω which is still an open covering of E. A *finite Ω-subcovering* of E is an Ω-subcovering of E which contains only a finite number of sets.

There are several equivalent definitions of compactness for metric spaces. The most convenient for applications to complex analysis seems to be the following one. A set E in a metric space X is *compact* if it satisfies the Heine-Borel-Lebesgue condition: Each open covering Ω of E contains a finite Ω-subcovering of E. The space X is said to be a compact metric space if X itself is compact.

Since E with its relative metric is itself a metric space, the definition raises a question as to whether we have introduced two concepts of compactness for E: compactness defined in terms of coverings by sets open in X and compactness in terms of coverings by sets open in the metric space E. The answer is that the two concepts are equivalent. The reader should have no difficulty in supplying a formal proof.

Proposition 5.5.1 *A compact subset E of a metric space (X, d) is closed and bounded.*

Proof. We first show that E is closed. If $E = X$, the conclusion follows automatically [Proposition 5.1.1(i)]. If E is a proper subset of X, the plan is to prove that $E^c = X \sim E$ is open. Then it follows at once that E is closed.

Take any $y \in E^c$, and hold it fixed. For each $x \in E$, let $r(x) = d(x, y)/2$. Now for each $x \in E$, $B(x, r(x)) \cap B(y, r(x)) = \varnothing$, because if $x_0 \in X$ were simultaneously in both these balls, we would have the contradiction $d(x, y) \leq d(x, x_0) + d(x_0, y) < r(x) + r(x) = 2[d(x, y)/2] = d(x, y)$. The set of balls $\Omega = \{B(x, r(x)) : x \in E\}$, is an open covering of E, so by our definition of compactness, there is a finite Ω-subcovering, say $\{B(x_j, r(x_j)), j = 1, \ldots, n\}$. Let $\delta = \min r(x_j)$, $j = 1, \ldots, n$. Then $B(y, \delta) \cap E \subset B(y, d) \cap [\bigcup_j B(x_j, r(x_j))] = \bigcup_j [B(y, \delta) \cap B(x_j, r(x_j))] = \varnothing$. So $B(y, \delta) \subset E^c$, and since $y \in E^c$ was arbitrary, this shows that E^c is open. Therefore, E is closed.

Our proof that E is bounded includes the case $E = X$. Take any $r > 0$.

Then $\Omega = \{B(x, r): x \in E\}$ is an open covering of E. Let $\{B(x_j, r): j = 1, \ldots, n\}$ be a finite Ω-subcovering, and let $M = \max d(x_1, x_j), j = 1, \ldots, n$. For any $x \in E$, pick j so that $x \subset B(x_j, r)$. Then $d(x, x_1) \leq d(x, x_j) + d(x_j, x_1) < r + M$. It follows that $E \subset B(x_1, r + M)$. Q.E.D.

Corollary A compact subset of $\mathbf{R}$ has both a greatest and a least element.

Proof. According to the proposition, if $E \subset \mathbf{R}$ is compact, it is bounded and closed. Let $\beta = \sup E$ and $\alpha = \inf E$. By Exercise 1.4.1, α and β are each points of closure of E, and, therefore, $\alpha \in E$, $\beta \in E$. Clearly, $\alpha = \min x, x \in E$, and $\beta = \max x, x \in E$. Q.E.D.

Proposition 5.5.2 *Let (X, d) be a metric space, let $E \subset X$ be compact, and let $F \subset E$ be closed in X. Then F is also compact.*

Proof. First, notice that $F^c = X \sim F$ is open. Let Ω be an open covering of F. Then $\Phi = \{O: O \in \Omega\} \cup \{F^c\}$ is an open covering of E, and so by hypothesis, there is a finite Φ-subcovering of E. If we exclude F^c from this finite Φ-subcovering of E, we obtain a finite open Ω-subcovering of F. This proves that F is compact. Q.E.D.

The following result is needed in Chapter 9.

Proposition 5.5.3 *Let E be a compact nonempty subset of a metric space X, and let $\langle F_n \rangle_{n=1}^{\infty}$ be a sequence of nonempty subsets of E, each closed in X, and such that $E \supset F_1 \supset F_2 \supset \cdots$. Then $I = \bigcup_{n=1}^{\infty} F_n$ is not empty.*

Proof. For each n, $F_n^c = X \sim F_n$ is open in X, so $E \cap F_n^c$ is open in the metric space E. Suppose that $I = \varnothing$, so $I^c = X$. We show that this implies that one of the sets F_n is empty, a contradiction of the hypotheses.

Now $\bigcup_{n=1}^{\infty} (E \cap F_n^c) = E \cap [\bigcup_{n=1}^{\infty} F_n^c] = E \cap [\bigcap_{n=1}^{\infty} F_n]^c = E \cap I^c = E \cap X = E$. Therefore, $\{E \cap F_n^c, n = 1, 2, \ldots\}$ is an open covering of E. Since E is compact, this open covering contains a finite subcovering, say $\{E \cap F_{n_j}^c,\ j = 1, 2, \ldots, m,\ n_1 < n_2 < \cdots < n_m\}$. By hypothesis, $F_{n_1}^c \subset F_{n_2}^c \subset \cdots \subset F_{n_m}^c$, so $E \cap F_{n_m}^c$ itself covers E. Since $F_{n_m} \subset E$, it follows that $F_{n_m} = \varnothing$. Q.E.D.

The next result is a version of the famous Heine-Borel-Lebesgue theorem. We use the symbol $\mathbf{R}^n$ to denote the metric space of all ordered real n-tuples with the usual Euclidean-distance metric.

Theorem 5.5.1 *A bounded closed subset of $\mathbf{R}^n$ is compact.*

Proof. We give a proof only for the case $n = 2$, which is the most important one for complex analysis.

In the proof of this theorem and in Example (i) immediately following it, the symbol (x, y) means an ordered pair of reals x and y, not the open interval with endpoints x and y.

Let $F \subset \mathbf{R}^2$ be closed and bounded. There exists a disk $\Delta(0, R)$, $R > 0$, containing F, and this disk is contained in the square $S = \{(x, y): -R \leq x \leq R, -R \leq y \leq R\}$. We prove that S is a compact subset of $\mathbf{R}^2$. It then follows from Proposition 5.5.2 (with $X = \mathbf{R}^2$, $E = S$) that F is compact.

Lemma The rectangle $Q = \{(x, y): \alpha \leq x \leq \alpha + h, \beta \leq y \leq \beta + k\}$, α, β, h, k, $\in \mathbf{R}$, $h > 0$, $k > 0$, is a closed subset of $\mathbf{R}^2$.

The proof of the lemma is an exercise.

Suppose that S is not compact, and let Ω be an open covering of S which has no finite Ω-subcovering. Join the midpoints of the opposite sides of S by line segments and so divide S into four squares with sides of length R. At least one of these squares has no finite Ω-subcovering. Let S_1 be one such square, and let its lower left-hand vertex be (α_1, β_1). Repeat the process on S_1, and obtain a square S_2 with side-length $R/2$ and with no finite Ω-subcovering. Let the lower left-hand vertex of S_2 be (α_2, β_2). Notice that $-R \leq \alpha_1 \leq \alpha_2 \leq R$ and $-R \leq \beta_1 \leq \beta_2 \leq R$. Proceed by induction to obtain for each positive integer n a square S_n with side-length $R/2^{n-1}$, with lower left-hand vertex (α_n, β_n), and with no finite Ω-subcovering. We have $-R \leq \alpha_1 \leq \cdots \leq \alpha_n \leq R$ and $-R \leq \beta_1 \leq \cdots \leq \beta_n \leq R$. The sequences $\langle \alpha_k \rangle$ and $\langle \beta_k \rangle$ are monotonic and bounded, and so have limits [Theorem 2.4.3] which we denote, respectively, by α and β.

Now $d[(\alpha_n, \beta_n), (\alpha, \beta)] = [(\alpha_n - \alpha)^2 + (\beta_n - \beta)^2]^{1/2} \leq |\alpha_n - \alpha| + |\beta_n - \beta|$ (see, e.g., (1.13.2)), so $\lim (\alpha_n, \beta_n) = (\alpha, \beta)$. According to the lemma, S is closed, so by Exercise 5.1.2, $(\alpha, \beta) \subset S$. There is, therefore, an open set $O \subset \Omega$ for which $(\alpha, \beta) \subset O$. It contains an open ball $B = B[(\alpha, \beta), r]$, $r > 0$. *We now claim that if n is sufficiently large*, $S_n \subset B$, which implies that O alone forms a finite Ω-subcovering of S_n, a contradiction to the assumption that S_n has no finite Ω-subcovering.

To prove the claim, we first observe that every point of a square $\{(x, y): \alpha_0 \leq x \leq \alpha_0 + h, \beta_0 \leq y \leq \beta_0 + h, h > 0\}$ lies in $B[(\alpha_0, \beta_0), 2h]$. With this in mind, we choose N so that $d[(\alpha_N, \beta_N), (\alpha, \beta)] < r/2$ and $2(R/2^{N-1}) < r/2$. (The square S_N is then contained in $B[(\alpha_N, \beta_N), 2(R/2^{N-1})]$.) Now for every $(x, y) \in S_N$, by the triangle inequality, $d[(x, y), (\alpha, \beta)] \leq d[(x, y), (\alpha_N, \beta_N)] + d[(\alpha_N, \beta_N), (\alpha, \beta)] < 2(R/2^{N-1}) + (r/2) < r/2 + r/2 = r$. Thus, $S_N \subset B$, and the claim is proved. Q.E.D.

This proof admits of a direct generalization to Euclidean spaces of any dimensionality.

Example.

(i) The rectangle Q which appears in the lemma for Theorem 5.5.1 is contained in any open disk $B[(\alpha, \beta), R]$, with $R > (h^2 + k^2)^{1/2}$. According to the lemma, it is closed in $\mathbf{R}^2$, so its is compact.

We now examine some connections between compactness and continuity of functions. A basic one is this theorem.

Theorem 5.5.2 *Let X and Y be metric spaces, and let the function $f\colon X \to Y$ be continuous. If X is compact, then $f(X)$ is compact in Y, and if f is univalent on X, then f^{-1} is continuous on $f(X)$.*

Proof. Let Ω_Y be an open covering of $f(X)$. By Theorem 5.2.1, if O is any open set in $\Omega_Y, f^{-1}(O)$ is open in X. The family $\Omega_X = \{f^{-1}(O), O \in \Omega_Y\}$ is an open covering of the compact metric space X, so there is a finite Ω_X-subcovering of X, say $\{f^{-1}(O_1), f^{-1}(O_2), \ldots, f^{-1}(O_n)\}$. Then $\{O_1, O_2, \ldots, O_n\} \subset \Omega_Y$, and this collection of sets forms a finite open Ω_Y-subcovering of $f(X)$. Therefore, $f(X)$ is compact in Y.

Suppose now that f, furthermore, is univalent on the compact space X. Let E be an arbitrary closed subset of X. Then E is compact [Proposition 5.5.2], and by the first part of this theorem, $(f^{-1})^{-1}(E) = f(E)$ is compact. Therefore, $(f^{-1})^{-1}(E)$ is closed [Proposition 5.5.1]. It follows from Exercise 5.2.1 that since E is arbitrary, f^{-1} is continuous on $f(X)$. Q.E.D.

Corollary 1 Let X be a compact metric space, and let $f\colon X \to \mathbf{R}$ be continuous. Then f attains a maximum and a minimum value on X. In particular, if the domain of f is a bounded closed subset of $\mathbf{R}^n$, then $f(x)$ has a maximum and minimum.

Proof. The first sentence of the conclusion follows from the theorem and the corollary of Proposition 5.5.1. The second sentence is a consequence of Theorem 5.5.1 and the first sentence of the conclusion. Q.E.D.

Corollary 2 Let X be a compact metric space, let Y be a metric space, and let the function $f\colon X \to Y$ be continuous. Then $f(X)$ is bounded and closed.

Remark. In Sec. 1.9, it was observed that when $\mathbf{C}$ and $\mathbf{R}^2$ are viewed merely as unstructured sets of elements, they are identical. In discussions of limits and continuity of functions with domains in $\mathbf{R}^2$, we are viewing $\mathbf{R}^2$ as a metric space with the usual Euclidean metric $d[(x_1, y_1), (x_2, y_2)] = [(x_2 - x_1)^2 + (y_2 - y_1)^2]^{1/2}$. Since the metric which we use in $\mathbf{C}$ is $|z_2 - z_1| = [(x_2 - x_1)^2 + (y_2 - y_1)^2]^{1/2}$, where $z_1 = x_1 + iy_1$, $z_2 = x_2 + iy_2$, it follows that considered simply as metric spaces, $\mathbf{R}^2$ and $\mathbf{C}$ are the same. The implication of Theorem 5.5.1 is that any bounded closed subset of $\mathbf{C}$ is compact. Corollary 1 of Theorem 5.5.2 implies that a real-valued function with a bounded closed domain in $\mathbf{C}$ attains a maximum and a minimum value, and the purport of Corollary 2 for complex analysis is that a complex-valued function with a bounded closed domain in $\mathbf{C}$ has a bounded closed range.

Examples

(ii) Let (X, d) be a metric space, and let γ be an arc in X (Sec. 5.3). Recall that γ is a continuous function from an interval $[\alpha, \beta] \subset \mathbf{R}$ into X. Let Γ be its graph; i.e., $\Gamma = \gamma[\alpha, \beta]$, the range of γ. According to Exercise 5.2.10, $[\alpha, \beta]$ is closed in $\mathbf{R}$, and clearly it is bounded. Therefore, by Theorem 5.5.1, $[\alpha, \beta]$ is compact. Hence, by Theorem 5.5.2, Γ is compact in X, and by Proposition 5.5.1, it is bounded and closed in X.

(iii) (Continuation of (ii).) Let f be a continuous function from Γ into the metric space Y. Then $f(\Gamma)$ is compact and, thus, is bounded and closed in Y.

(iv) Let T^* be the convex hull of the set $\{a, b, c\} \subset \mathbf{C}$. Then T^* is compact in $\mathbf{C}$. For according to Exercise 5.3.11, T^* is the union of all line segments joining a to the line segment joining b to c. A parametric equation of the last line segment is $\zeta = tb + (1-t)c, 0 \leq t \leq 1$. A parametric equation of the line segment joining a to ζ is $z = sa + (1-s)\zeta$, $0 \leq s \leq 1$. Therefore, the function given by $z = sa + (1-s)tb + (1-s)(1-t)c, 0 \leq s \leq 1, 0 \leq t \leq 1$, is a mapping of the square in $\mathbf{R}^2$ defined by these inequalities onto T^*. This function is clearly continuous on its domain, which according to Example (i) is compact. Hence, by Theorem 5.5.2, T^* is compact.

In Sec. 3.2, the idea of a *uniformly continuous* function from $\mathbf{C}$ into $\mathbf{C}$ is introduced. In the setting of metric spaces, the definition takes the following form. Let (X, d_X) and (Y, d_Y) be metric spaces. The function $f: X \to Y$ is uniformly continuous on X if and only if for every $\varepsilon > 0$, there exists $\delta_\varepsilon > 0$ such that if x' and x'' are any two elements of X with $d_X(x', x'') < \delta_\varepsilon$, then $d_Y[f(x'), f(x'')] < \varepsilon$. An important connection with the concept of compactness is given by the following statement.

Theorem 5.5.3 *Let (X, d_X) and (Y, d_Y) be metric spaces with X compact. If $f: X \to Y$ is continuous on X, then it is uniformly continuous on X.*

Proof. By hypothesis, given $\varepsilon > 0$, for every $x \in X$, there exists $\delta(x) = \delta(\varepsilon, x)$ such that if $d(x', x) < \delta(x)$, $x' \in X$, then $d_Y[f(x'), f(x)] < \varepsilon/2$. The family of open balls $\Omega = \{B[x, \delta(x)/2], x \in X\}$ is an open covering of X, so by our definition of compactness, there exists a finite Ω-subcovering, say $\{B[x_j, \delta(x_j)/2], j = 1, \ldots, n\}$. Let $\mu = \min_j \delta(x_j)/2$. We show that if x', x'' are any two elements of X with $d(x', x'') < \mu$, then $d_Y[f(x'), f(x'')] < \varepsilon$.

Now x' must be contained in at least one of the balls $B[x_j, \delta(x_j)/2]$; say in the kth one. Therefore, $d_Y[f(x'), f(x_k)] < \varepsilon/2$. By the triangle inequality, $d(x'', x_k) \leq d(x'', x') + d(x', x_k) < \mu + [\delta(x_k)/2] \leq 2[\delta(x_k)/2] = \delta(x_k)$. Therefore, also $d_Y[f(x''), f(x_k)] < \varepsilon/2$. We use the triangle inequality again to show that $d_Y[f(x'), f(x'')] \leq d_Y[f(x'), f(x_k)] + d_Y[f(x_k), f(x'')] < (\varepsilon/2) + (\varepsilon/2) = \varepsilon$. Q.E.D.

Exercise 5.5

1. Prove the lemma of Theorem 5.5.1. (*Hint*: See the remark following Corollary 2 of Theorem 5.5.2, and use Exercise 5.2.6.)
2. Let g be a continuous function from the compact metric space X into the metric space Y, and let f be a continuous function from $g(X)$ into a metric space Z. Show that the composition function $f \circ g: X \to Z$ is uniformly continuous on X.
3. Prove that the set T^* appearing in Example (iv) is compact by first observing that it is the range of a linear function $z = sa + tb + [1 - (s + t)]c$, $s \geq 0$, $t \geq 0$, s

$+ t \leq 1$, and by then proving that the triangle in the (s, t)-plane defined by these inequalities is bounded and closed. (*Hint*: Exercises 5.1.10 and 5.2.9.]

4. (Bolzano–Weierstrass Theorem.) Let E be a compact nonempty subset of a metric space X, and let F be an infinite subset of E. Prove that there is at least one accumulation point of F in E. (*Hint*: If the contrary were true, then for each $x \in E$, there would exist a neighborhood of x containing *no more than a finite number* of points of F. These neighborhoods for every $x \in E$ would form an open covering of E.)
5. Prove that any bounded sequence in a compact metric space has a convergent subsequence.
6. Let $B \subset \mathbf{C}$ be a region which intersects both $H = \{z: \operatorname{Im} z > 0\}$ and $H^* = \{z: \operatorname{Im} z < 0\}$, and such that the interval (α, β) is the intersection of B with the real axis. Prove that $B_1 = B \cap H$ and $B_2 = B \cap H^*$ are regions. (*Hint*: B_1 and B_2 are open in $\mathbf{C}$ so the problem is to prove connectivity. This can be done by showing that these sets are arcwise connected [Theorem 5.3.1]. We concentrate on B_1; the argument for B_2 is similar. Suppose we have $a \in B_1$ and $B \in B_1$. The plan is to show that any arc with graph in B which joins a to b and which intersects $(\alpha, \beta) \cup B_2$ can be replaced by an arc which connects a and b and of which the graph lies entirely in B_1. If an arc with graph in B and with equation $z = \gamma(t) = \gamma_1(t) + i\gamma_2(t)$, γ_1, γ_2 real, $\alpha \leq t \leq \beta$, $\gamma(\alpha) = a$, $\gamma(\beta) = b$, intersects B_2, then γ_2 has negative values on $[\alpha, \beta]$. Therefore, since $\operatorname{Im} a = \gamma_2(\alpha) > 0$ and $\operatorname{Im} b = \gamma_2(\beta) > 0$, and since γ_2 is continuous on $[\alpha, \beta]$, there exist points $t \in [\alpha, \beta]$ such that $\gamma_2(t) = 0$ [Proposition 5.3.2]. Of course, this remains true if the arc only intersects (α, β). Clearly $t = \inf\{t: \gamma(t) = 0\}$ and $t_2 = \sup\{t: \gamma(t) = 0\}$ both lie in (α, β). According to Exercise 3.2.4, $\gamma(t_1) = \gamma(t_2) = 0$ and so t_1 and t_2 are, respectively, the "first" and "last" zeros of γ on (α, β). In other terms, $\gamma(t) > 0$ for $\alpha \leq t \leq t_1$ and for $t_2 < t \leq \beta$. Every point of (α, β) is an interior point of B, so for every point $t' \in [t_1, t_2]$ there exists a disk $\Delta[t', \rho(t')]$, $\rho(t') > 0$, contained in B. The family of such disks forms an open covering of the closed interval $[t_1, t_2]$, which is compact in $\mathbf{C}$, so there exists a finite subcovering of this interval from this family. We assume that the disks with centers at t_1 and t_2 are retained in this finite subcovering. Let O be the union of the disks in the subcovering, and let $d = \operatorname{dist}([t_1, t_2], O^c) > 0$ [Theorem 5.6.1]. We leave it to the reader to show how to construct a line segment L in $O \cap H$ which joins a point in the disk for t_1 with a point in the disk for t_2, and thereby to show that any two points in $B \cap H$ can always be joined by an arc with graph entirely in $B \cap H$.)

5.6 Distance Between Sets; Diameter

Let (X, d) be a metric space. Given $E \neq \varnothing$, $E \subset X$, and $x \in X$, the *distance between* x and E is defined by $\operatorname{dist}(x, E) = \inf\{d(x, y): y \in E\}$. If E_1 and E_2 are nonempty subsets of X, the *distance between* E_1 and E_2 is defined by $\operatorname{dist}(E_1, E_2) = \inf\{\operatorname{dist}(x, E_2): x \in E_1\}$. It is clear from these definitions that if E_1 and E_2 are nonempty subsets of X, then for any $x \in E_1$, $y \in E_2$,

$$d(x, y) \geq \text{dist}\,(x, E_2) \geq d(E_1, E_2) \geq 0. \tag{5.6.1}$$

In later chapters of this book there are many occasions in which the following result is needed.

Theorem 5.6.1 *Let K be a nonempty compact subset of the metric space X, and let F be a closed subset of X. If $K \cap F = \varnothing$, then* $\text{dist}\,(K, F) > 0$.

Proof. We give the proof by proving the contrapositive statement: If $\text{dist}\,(K, F) = 0$, then $K \cap F \neq \varnothing$. We use two lemmas which are of some interest for themselves alone.

Lemma 1 Let E be a nonempty subset of the metric space X. Then the function with values $\text{dist}\,(x, E)$, $x \in X$, is uniformly continuous on X.

Proof. For any x, $y \in X$, the following inequalities are valid. Given $\eta > 0$, there exists $z \in E$ such that $d(y, z) < \text{dist}\,(y, E) + \eta$, and then

$$\begin{aligned} \text{dist}\,(x, E) &\leq d(x, z) \leq d(x, y) + d(y, z) \\ &< d(x, y) + \text{dist}\,(y, E) + \eta. \end{aligned}$$

This implies that

$$\text{dist}\,(x, E) \leq d(x, y) + \text{dist}\,(y, E).$$

By reversing the roles of x and y, we obtain

$$\text{dist}\,(y, E) \leq d(x, y) + \text{dist}\,(x, E).$$

Therefore,

$$-d(x, y) \leq \text{dist}\,(y, E) - \text{dist}\,(x, E) \leq d(x, y).$$

Given $\varepsilon > 0$, for every x, $y \in X$ with $d(x, y) < \varepsilon$, we have $|\text{dist}\,(y, E) - \text{dist}\,(x, E)| < \varepsilon$.

Lemma 2 With E as in Lemma 1, $\text{dist}\,(x, E) = 0$ if and only if $x \in \bar{E}$.

Proof. If $x \in E$, it is clear from the definition that $d(x, E) = 0$, because $d(x, x) = 0$. If $x \in \bar{E}$ but $x \notin E$, then by the definition of $\bar{E}$, for every $\delta > 0$ there exists $x_\delta \in E \cap B(x, \delta)$. Given $\varepsilon > 0$, by Lemma 1 if δ is properly chosen, we have $|\text{dist}\,(x_\delta, E) - \text{dist}\,(x, E)| = |0 - \text{dist}\,(x, E)| < \varepsilon$. This implies that $\text{dist}\,(x, E) = 0$.

Conversely, suppose x is such that $d(x, E) = 0$. For a given $\varepsilon > 0$, we can take $y_\varepsilon \in E$ so that $d(x, y_\varepsilon) \leq \text{dist}\,(x, E) + \varepsilon = 0 + \varepsilon$. This fact shows that for any $\varepsilon > 0$, there is always a $y_\varepsilon \in E \cap B(x, \varepsilon)$, so by definition of closure, $x \in \bar{E}$.

We return to the proof of the theorem. By Lemma 1, $\text{dist}\,(x, F)$ is continuous for x on K, and since K is compact, by Corollary 1 of Theorem 5.5.2, dist (x, F) actually assumes a minimum value for some $x \in K$, say $\bar{x}$. By the definition of inf, dist $(\bar{x}, F) =$ dist (K, F). So if dist $(K, F) = 0$, then dist $(\bar{x}, F) = 0$, and by Lemma 2, this can happen only if $\bar{x} \in \bar{F} = F$ (since F is closed), in which case $K \cap F \neq \varnothing$. Q.E.D.

Corollary Let $X = \mathbf{C}$, and let E be a closed subset of the disk $\Delta(a, R)$, $R > 0$. There exists a closed disk $\bar{\Delta}(a, r)$, $0 < r < R$, which contains E.

Proof. The result is true trivially if $E = \{a\}$. Henceforth, we assume that E contains points other than a. By Theorem 5.5.4, E is compact. Also $\Delta^c = [\Delta(a, R)]^c = \{z: |z - a| \geq R\}$ is closed [Exercise 5.1.1]. Therefore, by Theorem 5.6.1, $m =$ dist $(E, \Delta^c) > 0$. We show simultaneously that $m < R$ and that $E \subset \bar{\Delta}(a, R - m)$.

Take any $z \in E$, $z \neq a$. The point $w = a + (z - a)R|z - a|^{-1}$ is such that $|w - a| = R$, so $w \in \Delta^c$. Then

$$\begin{aligned} m &\leq |z - w| \\ &= \left|z - a - (z - a)R|z - a|^{-1}\right| \\ &= |z - a|\left|1 - \frac{R}{z - a}\right| \\ &= \left||z - a| - R\right| \\ &= R - |z - a|. \end{aligned}$$

Thus, $m < R$ and $|z - a| \leq R - m$. Of course, if $a \in E$, then $a \in \bar{\Delta}(a, R - m)$. Q.E.D.

Example. Let O be an open set in $\mathbf{C}$, and let $\Gamma \subset O$ be the graph of an arc or of a closed curve. Then Γ is compact [Example (i), Sec. 5.5] and O^c is closed. According to Theorem 5.6.1, since $\Gamma \cap O^c = \phi$, $|z - w| \geq d(\Gamma, O^c) > 0$ for every $z \in \Gamma$ and every $w \in O^c$.

The *diameter* of a nonempty set E in a metric space (X, d) is defined by the extended real number diam $(E) = \sup \{d(x, y): x, y \in E\}$. Clearly, if $E_1 \subset E_2$, then diam $(E_1) \leq$ diam (E_2) [Exercise 1.4.2]. We consider here two results involving diameter which are needed in Chapter 9. The first one is a refinement of Proposition 5.5.3.

Proposition 5.6.1 *Let E be a compact nonempty subset of a metric space X, and let $\langle F_n \rangle_{n=1}^{\infty}$ be a sequence of nonempty subsets of E, each closed in X, and such that $E \supset F_1 \supset F_2 \supset \cdots$. If furthermore* $\lim_{n \to \infty}$ diam $(F_n) = 0$, *then (i) the intersection $I = \bigcap_{n=1}^{\infty} F_n$ consists of a single point x_0, and (ii) given any $\varepsilon > 0$, there exists $N_\varepsilon > 0$ such that $F_n \subset B(x_0, \varepsilon)$ if $n \geq N_\varepsilon$.*

Proof. According to Proposition 5.5.3, there certainly exists at least one point in I. Suppose then that $x_0 \in I$. We prove part (ii) of the proposition first and then show that x_0 is unique.

Given $\varepsilon > 0$, by hypothesis there exists N_ε such that if $n \geq N_\varepsilon$, then $\operatorname{diam}(F_n) < \varepsilon$. For any fixed $n \geq N_\varepsilon$, let y_n be any point in F_n. We have $d(y_n, x_0) \leq \operatorname{diam}(F_n) < \varepsilon$, and this shows that $F_n \subset \{y_n : d(y_n, x_0) < \varepsilon\} = B(x_0, \varepsilon)$.

Now suppose $y \in I$. Then y lies in each F_n, and if $n \geq N_\varepsilon$, we have $d(y, x_0) \leq \operatorname{diam}(F_n) < \varepsilon$. Since $\varepsilon > 0$ is arbitrary, this implies that $d(y, x_0) = 0$ and $y = x_0$. Q.E.D.

The second question is a rather special one concerning the diameter of a triangular region. The conclusion is intuitively obvious but does seem to call for a formal proof.

Proposition 5.6.2 *Let T^* be the convex hull of the set $T = \{a, b, c\} \subset \mathbf{C}$. Then* $\operatorname{diam}(T^*) = \max\{|a - b|, |b - c|, |c - a|\}$.

Proof. The proof is facilitated by referring to Exercise 5.3.11, in which it is asserted that T^* is the union of all line segments joining one of the points of T to the points of the line segment joining the other two. Without loss of generality, we can assume that $|a - b| = \max\{|a - b|, |b - c|, |c - a|\}$. Let ζ be any point in the line segment $[b, c]$. Then $\zeta = (1 - t)b + tc$ for some number $t \in [0, 1]$, and

$$\begin{aligned}
|\zeta - a| &= |(1 - t)(b - a) + t(c - a)| \\
&\leq (1 - t)|b - a| + t|c - a| \\
&\leq |b - a|(1 - t + t) \\
&= |b - a|.
\end{aligned}$$

This, of course, accords with intuition.

Now take any z and w in T^*. Suppose $\zeta_1, \zeta_2 \in [b, c]$ are such that $z \in [a, \zeta_1]$, $w \in [a, \zeta_2]$. There exist numbers $t_1, t_2 \in [0, 1]$ such that $z = (1 - t_1)a + t_1\zeta_1$ and $w = (1 - t_2)a + t_2\zeta_2$. We have

$$\begin{aligned}
z - w &= (1 - t_1)a + t_1\zeta_1 - (1 - t_2)a - t_2\zeta_2 \\
&= (t_2 - t_1)a + t_1\zeta_1 - t_2\zeta_2. \qquad (5.6.2)
\end{aligned}$$

If $t_1 \leq t_2$, we add and subtract $t_1\zeta_2$ in the right-hand member of (5.6.2), obtaining

$$\begin{aligned}
|z - w| &= |(t_2 - t_1)a - (t_2 - t_1)\zeta_2 + t_1(\zeta_1 - \zeta_2)| \\
&\leq (t_2 - t_1)|a - \zeta_2| + t_1|\zeta_1 - \zeta_2| \\
&\leq (t_2 - t_1)|a - b| + t_1|b - c| \\
&\leq (t_2 - t_1)|a - b| + t_1|a - b| \\
&= t_2|a - b| \\
&\leq |a - b|.
\end{aligned}$$

If $t_2 \leq t_1$, we add and subtract $t_2\zeta_1$ in the right-hand member of (5.6.2) and proceed as before to find again that $|z - w| \leq |a - b|$. This establishes that $\text{diam}(T^*) = \sup\{|z - w| : z, w \in T^*\} \leq |a - b|$. The sup is attained for the points $z = a$, $w = b$ in T^*. Q.E.D.

Exercises 5.6

1. Given the power series $\sum_{k=0}^{\infty} a_k(z - a)^k$ with radius of convergence $R > 0$, show that the series converges uniformly on any closed subset of $\Delta(a, R)$. (*Hint*: According to Theorem 4.1.2, the series converges uniformly on any closed disk $\bar{\Delta}(a, r) \subset \Delta(a, R)$, $0 < r < R$.)
2. Let $X = \mathbf{C}$. Show that the diameters of $C(a, R)$, $\Delta(a, R)$, and $\Delta(a, R) \cup C(a, R)$, as defined in this section, are all equal to $2R$. (*Hint*: Triangle inequality applied to $|z_1 - z_2| = |z_1 - a + a - z_2|$.)
3. Show that the diameter of the rectangle in $\mathbf{R}^2$, $B = \{(x, y)\colon \alpha \leq x \leq \alpha + h, \beta \leq y \leq \beta + k\}$, $h > 0$, $k > 0$, is $(h^2 + k^2)^{1/2}$. (*Hint*: First, show that the inequalities defining B imply that every point in B lies in a closed circular disk with center at $(\alpha + (h/2)), \beta + (k/2))$ and radius $(h^2 + k^2)^{1/2}/2$.)
4. Subsets A and B of the metric space (X, d) are said to be *mutually separated* if $\bar{A} \cap B = \phi$ and $A \cap \bar{B} = \phi$. Show that if A and B are mutually separated, then there exist disjoint sets O_1 and O_2, each open in X, such that $A \subset O_1$ and $B \subset O_2$. (*Hint*: Define $f\colon X \to \mathbf{R}$ by $f(x) = \text{dist}(x, A) - \text{dist}(x, B)$. Use Lemma 1 of Theorem 5.6.1. Look at $f^{-1}(-\infty, 0)$ and $f^{-1}(0, \infty)$.)
5. Let E_1 and E_2 be nonempty subsets of a metric space. Prove that $\text{dist}(E_1, E_2) = \text{dist}(E_2, E_1)$.

6

Complex Differential Calculus

6.1 The Derivative of a Complex-valued Function

Let f be a complex-valued function with domain $E \subset \mathbf{C}$; and let a be an accumulation point of E, $a \in E$. The *difference quotient* of f at a is defined by

$$\delta_a(z) = \frac{f(z) - f(a)}{z - a}, \qquad z \in E \sim \{a\}.$$

The derivative of f at a, if it exists, is $\lim_{z \to a} \delta_a(z)$, $z \in E \sim \{a\}$. Some standard notations for the derivative are

$$f'(a), \qquad \left.\frac{df}{dz}\right]_{z=a}.$$

If the derivative exists at a point, the function is said to be *differentiable* at the point.

Proposition 6.1.1 *Given $f: E \to \mathbf{C}$, $E \subset \mathbf{C}$; let a be an accumulation point of E, $a \in E$. The derivative $f'(a)$ exists if and only if there exists a function d with domain E such that d is continuous at $z = a$, and such that $f(z) - f(a) = d(z)(z - a)$ for every $z \in E$. When such a function d exists, $d(a) = f'(a)$.*

The proposition is merely the direct adaptation of the definition of functional limit given in Sec. 5.2 to the limit of a difference quotient.

Corollary If the function f described in the proposition has a derivative at the accumulation point $a \in E$, it is continuous at a.

Proof. We have $f(z) = f(a) + d(z)(z - a)$, $z \in E$, where d is continuous at a. The right-hand member of the equation is continuous at a, by Exercises 3.2.8 and 3.2.7. Q.E.D.

Theorem 6.1.1 *Let f_1 and f_2 be complex-valued functions with a common domain $E \subset \mathbf{C}$, and let a be an accumulation point of E, $a \in E$. If $f_1'(a)$ and $f_2'(a)$ exist, then the following formulas are valid, with implication of existence of the left-hand member in each case:*

(i) $(f_1 \pm f_2)'(a) = f_1'(a) \pm f_2'(a)$,
(ii) $(f_1 \cdot f_2)'(a) = f_1(a)f_2'(a) + f_2(a)f_1'(a)$.

Furthermore, if $f_2(a) \neq 0$, then

(iii) $(f_1/f_2)'(a) = (f_2(a)f_1'(a) - f_1(a)f_2'(a))/[f_2(a)]^2$.

These are the analogs for complex differentiation of the familiar rules of elementary calculus. The proofs are facilitated by the use of Proposition 6.1.1 and are left as an exercise. It is to be understood in the case of the quotient f_1/f_2 that this function is restricted to the subset E' of E on which $f_2 \neq 0$. By Exercise 3.2.3 and the corollary to Proposition 6.1.1, there exists a disk $\Delta(a, r)$, $r > 0$, such that $\Delta(a, r) \cap E \subset E'$.

Theorem 6.1.2 (*Chain Law*). *Let g be a complex-valued function with domain $E \subset \mathbf{C}$ and range $G \subset \mathbf{C}$; let f be a complex-valued function with domain G. Let a be an accumulation point of E, $a \in E$, and let $b = g(a)$ be an accumulation point of G. Let $f'(b)$ and $g'(a)$ exist. Then $(f \circ g')(a)$ exists and equals $f'(b)g'(a)$.*

Proof. Let $w = g(z)$, $z \in E$. By Proposition 6.1.1, there is $d_g\colon E \to \mathbf{C}$, continuous at a, with $d_g(a) = g'(a)$, and $g(z) - g(a) = d_g(z)(z - a)$ for every $z \in E$. There also exists $d_f\colon G \to \mathbf{C}$, continuous at b, with $d_f(b) = f'(b)$, and $f(w) = f(b) = d_f(w)(w - b)$ for every $w \in G$. So we have by substitution,

$$\begin{aligned} f(g(z)) - f(g(a)) &= d_f(g(z))[g(z) - g(a)] \\ &= d_f(g(z)) \cdot d_g(z)(z - a), \qquad z \in E. \end{aligned}$$

According to Exercise 3.2.10 and the corollary of Proposition 6.1.1, $d_f(g(z))$ is continuous at $z = a$. Its value at $z = a$ is $d_f(b) = f'(b)$. By Exercise 3.2.8, $d_f(g(z)) \cdot d_g(z)$ is continuous at $z = a$, and its value there is $f'(b)g'(a)$. The proof is now concluded by reference to Proposition 6.1.1. Q.E.D.

Theorem 6.1.3 (*Inverse Function Law*). *Let f be a homeomorphism from a set $E \subset \mathbf{C}$ onto $f(E) \subset \mathbf{C}$, with inverse function f^{-1}. Let a be an accumulation point of E, $a \in E$; let $b = f(a)$, and let $f'(a)$ exist, $f'(a) \neq 0$. Then $(f^{-1})'(b)$ exists and equals $1/f'(a)$.*

Proof. By Exercise 5.2.4, b is an accumulation point of $f(E)$. By Proposition 6.1.1, we have an equation $f(z) - f(a) = d(z)(z - a)$, $z \in E$, where d has domain E, is continuous at $z = a$, and $d(a) = f'(a) \neq 0$. Notice that $d(z) \neq 0$ for every $z \in E$, because if for some $z_0 \in E$, $d(z_0) = 0$, then $f(z_0) = f(a)$, and the univalence of f implies that $z_0 = a$.

Substituting $w = f(z)$, $z = f^{-1}(w)$, we obtain

$$w - b = d[f^{-1}(w)][f^{-1}(w) - f^{-1}(b)],$$

or

$$f^{-1}(w) - f^{-1}(b) = \frac{1}{d[f^{-1}(w)]}(w - b), \qquad w \in f(E). \tag{6.1.1}$$

The function $f^{-1}(w)$ is continuous on $f(E)$, by hypothesis, so by Exercise 3.2.10, $d[f^{-1}(w)]$ is continuous at b. Then by Exercise 3.2.8, $1/d[f^{-1}(w)]$ is continuous at b, where its value is $1/d[f^{-1}(b)] = 1/d(a) = 1/f'(a)$. The sufficient condition in Proposition 6.1.1 for f^{-1} to have a derivative $1/f'(a)$ at b is established by (6.1.1) and these remarks. Q.E.D.

Exercises 6.1

1. Prove Theorem 6.1.1 by using Propostion 6.1.1 and the relevant facts about functions continuous at a point. (For the quotient function, it might simplify matters to prove first that $(1/f_2)'(a) = -f_2'(a)/[f_2(a)]^2$.)
2. Let f be a complex-valued function with domain $E \subset \mathbf{R}$, and let $f = f_1 + if_2$, where f_1 and f_2 are real valued. Let a be an accumulation point of E, $a \in E$. Prove that $f'(a)$ exists if and only if both $f_1'(a)$ and $f_2'(a)$ exist, and if $f'(a)$ exists, it equals $f_1'(a) + if_2'(a)$. (*Hint*: Exercise 5.2.5.)
3. Consider the restriction of each of the following functions to a set $E \subset \mathbf{C}$, and let $a \in E$ be an accumulation point of E. Prove these derivative formulas (which imply existence of the derivatives in question):
   (a) $f(z) \equiv c$, $z \in E$, where c is a complex number; $f'(a) = 0$;
   (b) $f(z) = z$; $f'(a) = 1$;
   (c) $f(z) = z^n$, n a positive integer; $f'(a) = na^{n-1}$ (*Hint*: Theorem 6.1.1(ii).);
   (d) $0 \notin E$; $f(z) = 1/z$; $f'(a) = -1/a^2$;
   (e) $0 \notin E$; $f(z) = z^{-n}$, n a positive integer; $f'(z) = -na^{-n-1}$;
   (f) $f(z) = c_0 + c_1 z + c_2 z^2 + \cdots + c_n z^n$; $f'(z) = c_1 + 2c_2 a + \cdots + nc_n a^{n-1}$.
4. A rational function is the quotient of two polynomials in which the denominator is not the zero polynomial. Let $r(z) = p(z)/q(z)$ be a rational function, where p and q are the polynomials. Show that $r'(z)$ exists at every point z of $\mathbf{C}$ at which $q(z) \neq 0$, and is itself a rational function. (*Hint*: Theorem 6.1.1 and Exercise 6.1.3(f).)
5. Let f be complex valued and have domain E, and let $f'(a)$ exist, where $a \in E$ is an accumulation point. Show that if c is a constant, $(cf)'(a) = cf'(a)$, and if n is a positive integer, then the derivative of $[f(z)]^n$ at $z = a$ exists and is $n[f(a)]^{n-1}f'(a)$.

6. (L'Hospital's rule.) Let f and g be complex valued, each with domain $E \subset \mathbf{C}$, let a be an accumulation point of E, $a \in E$; let $f'(a)$ and $g'(a)$ exist, $g'(a) \neq 0$, and finally let $f(a) = g(a) = 0$. Show that

$$\lim_{z \to a} \frac{f(z)}{g(z)} = \frac{f'(a)}{g'(a)}, \qquad z \in E \sim \{a\}.$$

(*Hint*: Proposition 6.1.1 and Exercise 3.2.8.)

7. Let f be real valued with domain an interval $[\alpha, \beta] \subset \mathbf{R}$, let f be continuous on $[\alpha, \beta]$, and let $f'(x)$ exist for every $x \in (\alpha, \beta)$. Prove by Proposition 6.1.1 that if $f(x)$ has a relative maximum at $\xi \in (\alpha, \beta)$ (that is, if there exists $h > 0$ such that $(\xi - h, \xi + h) \subset (a, b)$ and $f(x) \leq f(\xi)$ for every $x \in (\xi - h, \xi + h)$), then $f'(\xi) = 0$. (*Hint*: Recall Exercise 3.2.3(c).)

Remark (i). By a well-known theorem of Weierstrass, of which Corollary 1 of Theorem 5.5.2 was a generalization, a real-valued function continuous on an interval $[\alpha, \beta]$ must have a maximum value at some point in the interval. This fact, together with Exercise 6.1.7, leads to Rolle's theorem and the mean value theorem in elementary calculus. Rolle's theorem states that if $f = [\alpha, \beta] \to \mathbf{R}$ is continuous on $[\alpha, \beta]$, $f(\alpha) = f(\beta) = 0$, and f' exists everywhere on (α, β) then there must exist a point $\xi \in (\alpha, \beta)$ such that $f'(\xi) = 0$. (This is because there must be a relative maximum of either $f(x)$ or $-f(x)$ somewhere in (α, β).)

Remark (ii). The mean value theorem for differential calculus is this: If $f: [\alpha, \beta] \to \mathbf{R}$ is continuous on $[\alpha, \beta]$, differentiable on (α, β), then there exists $\xi \in (\alpha, \beta)$ such that $f(\beta) - f(\alpha) = f'(\xi)(\beta - \alpha)$. (The proof is immediate when Rolle's theorem is applied to the artificial-looking $\phi(x) = f(x) - f(\alpha) - (x - \alpha)(\beta - \alpha)^{-1}[f(\beta) - f(\alpha)]$.)

Remark (iii). An important consequence of the mean value theorem is that under the hypotheses on f in (ii), if $f'(x) > 0$ for every $x \in (\alpha, \beta)$, then $f(x)$ must be strictly increasing on $[\alpha, \beta]$, and if $f'(x) < 0$ on (α, β), $f(x)$ is strictly decreasing on $[\alpha, \beta]$.

6.2 Analytic Functions

The definition of derivative given in Section 6.1 is more general than that usually found in introductory textbooks in complex analysis. It has the advantage of including concepts such as (i) the right-hand derivative and left-hand derivative of a function f from an interval $[\alpha, \beta] \subset \mathbf{R}$ at a point $x \in [\alpha, \beta]$, and (ii) various formulations of the derivative at a boundary point of a region $B \subset \mathbf{C}$ for a complex-valued function with domain $\bar{B}$.

Given a complex- or real-valued function f with domain E in $\mathbf{C}$ or $\mathbf{R}$, the restriction of f to a subset E_0 of E may have a derivative at an accumulation point $a \in E_0$ without the existence of a derivative of the unrestricted function at a. Also if a is an accumulation point of another subset E_1 of E, the restriction of f to E_1 may have a derivative at a which differs in value from $(f|E_0)'(a)$. An example is given by the function $f(x) = |x|$, $x \in \mathbf{R}$,

$E = [-1, 1]$, $E_0 = [-1, 0]$, $E_1 = [0, 1]$, $a = 0$. Here $(f|E_0)'(0) = -1$, $(f|E_1)'(0) = +1$, and $f'(0)$ does not exist.

It is clear from the definition of the limit process that if the *unrestricted* function f has a derivative $f'(a)$ at the accumulation point a, then any restriction of f to a subset E_0 of E with $a \in E_0$, a an accumulation point of E_0, has a derivative at a, and $(f|E_0)'(a) = f'(a)$. We use this remark in an essential way in deriving the Cauchy-Riemann equations in Section 6.3.

The classical definition of derivative of a complex-valued function specifies an open set O for the domain of the function. In that case, any point of O is an accumulation point and is the center of a disk contained in O. Intuitively in this case, the limiting process can be thought of as an approach "from all sides" of a.

A complex-valued function with domain E in $\mathbf{C}$ containing an open set O is defined to be *analytic* or *holomorphic* on O if and only if it has a derivative at every point of O. A complex-valued function f with domain $E \subset \mathbf{C}$ is said to be *analytic* at a point $a \in E$ if and only if there exists a disk $\Delta(a, r) \subset E$, $r > 0$, such that f is analytic on $\Delta(a, r)$.

The values of the derivatives of a function f, which is analytic on an open set O, define a new function f' with domain O, called the derivative function. Later on it is shown that this function f' is also analytic on O, which leads to a hierarchy $f' = f^{(1)}, f^{(2)}, f^{(3)}, \ldots$, of derivative functions for an analytic function, in which $f^{(n)} = [f^{(n-1)}]'$. The number n is called the *order* of the derivative.

The functions which appear in Exercise 6.1.3(a)–(f) are all examples of functions analytic on $\mathbf{C}$, with the exception of $1/z$ and z^{-n} (with $n > 0$), which are analytic on $\mathbf{C} \sim \{0\}$. The implication of Exercise 6.1.4 is that a rational function $r = p/q$ is analytic on $\mathbf{C} \sim \{z\colon q(z) = 0\}$.

Here is an example of a function which has a derivative at a point but is *not* analytic at the point. Consider the function $f(z) = |z|^2$, with domain $\mathbf{C}$. To compute $f'(0)$, we have the difference quotient $(|z|^2 - 0)/(z - 0) = |z|(|z|/z)$, which has the limit 0 as $z \to 0$. The difference quotient at any point $a \neq 0$ is of the form $(|a + h|^2 - |a|^2)/(a + h - a)$, where $h = z - a$. Using Exercise 1.12.3, we simplify this to $h^{-1}(|h|^2 + 2 \operatorname{Re} \bar{h}a)$. Now let $z = z_n = a + (1/n)$, $n = 1, 2, \ldots$, so $h = h_n = 1/n$, $n = 1, 2, \ldots$. The sequence $\langle z_n \rangle$ converges to a. The corresponding sequence of difference quotients is $\langle (1/n) + 2 \operatorname{Re} a \rangle$ which converges to $2 \operatorname{Re} a$, a real number not equal to zero. Again, let $z = z_n = a + (i/n)$, $n = 1, 2, \ldots$, so $h = h_n = i/n$. The sequence $\langle z_n \rangle$ converges to a as before. But the corresponding sequence of difference quotients is $\langle (-i/n) + 2i \operatorname{Re} ia \rangle$, which converges to $2i \operatorname{Re} ia$, a pure imaginary. So the limits of the difference quotient along the two sequences converging to a are not equal, and by Exercise 5.2.3 the difference quotient has no limit as z approaches a on $\mathbf{C}$. Thus, there is no disk $\Delta(0, r)$, $r > 0$, in which this function is analytic, and it is then differentiable but not analytic at $z = 0$.

Theorem 6.2.1 *Let the complex-valued function f be analytic on a region $B \subset \mathbf{C}$. If $f'(z) = 0$ for every $z \in B$, then f is a constant function on B.*

Lemma Let ϕ be a complex-valued function continuous on the interval $[\alpha, \beta] \subset \mathbf{R}$, and let ϕ' exist with $\phi'(x) = 0$ for every $x \in (\alpha, \beta)$. Then ϕ is a constant function.

Proof. Let $\phi = \phi_1 + i\phi_2$, ϕ_1, ϕ_2 real valued. By Exercise 6.1.2, ϕ_1', ϕ_2' exist and have the value 0 for every $x \in (\alpha, \beta)$. The functions ϕ_1 and ϕ_2 are continuous on $[\alpha, \beta]$, by Theorem 3.2.1. By the mean value theorem (Remark (ii) in Exercises 6.1), for every $x \in [\alpha, \beta]$, $\phi_j(x) - \phi_j(\alpha) = 0\cdot(x - \alpha) = 0$, $j = 1, 2$, so ϕ_1 and ϕ_2 are constant functions on $[\alpha, \beta]$ and, therefore, so is ϕ.

Proof of the Theorem. By the corollary of Proposition 6.1.1, f is continuous on B. Let z_1 and z_2 be any two points on B such that the line segment joining them lies entirely in B. A parametric equation of this segment is $z = z_1 + t(z_2 - z_1)$, $0 \leq t \leq 1$. Let $\phi(t) = f[z_1 + t(z_2 - z_1)]$. By the chain law, Theorem 6.1.2, the derivative $\phi'(t)$ exists for every t, $0 < t < 1$, and is given by $f'[z_1 + t(z_2 - z_1)]\cdot(z_2 - z_1)\cdot 1 = 0$. The function ϕ is the composition of a continuous function f with a continuous function of t, and so is continuous on $[0, 1]$ [Exercise 5.2.2]. Therefore the lemma is applicable and $\phi(t) = \phi(0)$ for every $t \in [0, 1]$. This implies that $f(z) = f(z_1)$ for every z on the line segment, and in particular $f(z_2) = f(z_1)$.

By Theorem 5.3.2, the region B is arcwise connected by polygonal arcs. Let a be a fixed point of B, and let b be any other point of B. There exists a polygonal arc lying in B and joining a to b, of which the graph is a finite sequence of line segments, respectively joining $a = z_1$ to z_2, z_2 to $z_3, \ldots,$ z_{n-1} to $z_n = b$. The argument in the preceding paragraph shows that $f(a) = f(z_2) = f(z_3) = \cdots = f(b)$. Since b is arbitrary, this shows that $f(z) = f(a)$ for every $z \in B$. Q.E.D.

Corollary Let the complex-valued functions f and g be analytic on a region $B \subset \mathbf{C}$, and let $f'(z) = g'(z)$ for every $z \in B$. Then $f(z) = g(z) + c$ for every $z \in B$, where c is a complex number.

Proof. Look at the function $f - g$. By Theorem 6.1.1(i) its derivative at each point z of B is $f'(z) - g'(z) = 0$. Theorem 6.2.1 completes the proof. Q.E.D.

Given a complex-valued function f with domain $E \subset \mathbf{C}$, every point of which is an accumulation point, if there exists a function $F: E \to \mathbf{C}$ differentiable at each point of E and such that $F'(z) = f(z)$ for every $z \in E$, then F is called a *primitive* (or *antiderivative* or *indefinite integral*) of f on E. In practice, E is usually an open set in $\mathbf{C}$ or an interval in R.

If F is a primitive of f on E, then so is $F + c$, where c is any complex number [Exercise 6.1.3(a)]. According to this corollary, if E is a region, any two primitives differ by a constant function.

Exercises 6.2

1. Show that the functions $f(z) = |z|$ and $f(z) = \text{Re } z$ are nowhere differentiable on $\mathbf{C}$, although they are continuous on $\mathbf{C}$.
2. Show that the collection of complex-valued functions analytic on an open set $O \subset \mathbf{C}$ (taken with the usual rules for function addition, multiplication, and multiplication by complex constants) is an algebra $\mathscr{A}(O)$ over $\mathbf{C}$ which is a subalgebra of the algebra of complex-valued function continuous on O [Section 3.2]. Furthermore, show that the complex polynomials form a subalgebra of $\mathscr{A}(O)$.
3. Find a primitive valid in $\mathbf{C}$ for the polynomial function $f(z) = c_0 + c_1 z + c_2 z^2 + \cdots + c_n z^n$.
4. Given the following data about a complex-valued function f: It is analytic on a region $B \subset \mathbf{C}$ and there is a number $c \in \mathbf{C}$ such that $f'(z) = c$ for every $z \in B$. Prove that f must be of the form $cz + d$, where d is some complex number.

6.3 The Cauchy–Riemann Equations

We have shown that if the complex-valued function $f(z) = u(z) + iv(z)$, with u and v real valued, is continuous on a set $E \subset \mathbf{C}$, then both u and v are continuous on E, and conversely [Theorem 3.2.1]. In this section, we establish connections between the differentiability properties of f, u, and v.

In a remark in Sec. 5.5, it was pointed out that $\mathbf{R}^2$ and $\mathbf{C}$, considered solely as metric spaces with the usual metrics, are indistinguishable. In particular, if a function f from $\mathbf{C}$ into a metric space Y is continuous at a point $x_0 = x_0 + iy_0$ in its domain, then when f is regarded as a mapping from $\mathbf{R}^2$ into Y, it remains continuous at (x_0, y_0), by the definition of continuity given at the beginning of Sec. 5.2. The converse statement is also true. These observations are of importance for this section, in which we often conveniently identify a given function $f: E \to \mathbf{C}$, $E \subset \mathbf{C}$ with an associated function $F: E_{\mathbf{R}\times\mathbf{R}} \to \mathbf{C}$, where $E_{\mathbf{R}\times\mathbf{R}} = \{(x, y): (x, y) \in \mathbf{R}^2, z = x + iy \in E\}$, and $F(x, y) = f(x + iy)$.

We now review the definition of partial derivatives. Let F be a complex-valued function with domain an open set O in $\mathbf{R}^2$. Let (x_0, y_0) lie in O.

Form the difference quotients

$$\delta_1(x, y_0) = \frac{F(x, y_0) - F(x_0, y_0)}{x - x_0},$$

$$\delta_2(x_0, y) = \frac{F(x_0, y) - F(x_0, y_0)}{y - y_0},$$

with (x, y_0), $(x_0, y) \in O$, $x \neq x_0$, $y \neq y_0$. By Exercise 5.1.9, the sets $\{x: (x, y_0) \in O\}$ and $\{y: (x_0, y) \in O\}$ are open sets in $\mathbf{R}$, so x_0 and y_0 are accumulation points of these two sets, respectively. The partial derivative $F_x(x_0, y_0)$ of F with respect to x at (x_0, y_0), if it exists, is defined to be $\lim_{x \to x_0} \delta_1(x, y_0)$, and the partial derivative $F_y(x_0, y_0)$ of F with respect to y at (x_0, y_0), if it exists, is defined to be $\lim_{y \to y_0} \delta_2(x_0, y)$.

Another notation for the partial derivatives is the familiar

$$\left.\frac{\partial F}{\partial x}\right]_{(x_0, y_0)}, \quad \left.\frac{\partial F}{\partial y}\right]_{(x_0, y_0)}.$$

As in the case of the complex derivative, it is useful to formulate these limits in the manner of the definition of limit in Sec. 5.2.

Proposition 6.3.1 *Let F be a complex-valued function with domain an open set $O \subset \mathbf{R}^2$. Let (x_0, y_0) be a point of O. The partial derivative $F_x(x_0, y_0)$ exists if and only if there is a complex-valued function $d_1(x)$ with domain $O_x = \{x: (x, y_0) \in O\}$, continuous at x_0 and such that $F(x, y_0) - F(x_0, y_0) = d_1(x)(x - x_0)$ for every $x \in O_x$. If such a function d_1 exists, then $d_1(x_0) = F_x(x_0, y_0)$. The partial derivative $F_y(x_0, y_0)$ exists if and only if there is a function $d_2(y)$ with domain $O_y = \{y: (x_0, y) \in O\}$, continuous at y_0 and such that $F(x_0, y) - F(x_0, y_0) = d_2(y)(y - y_0)$ for every $y \in O_y$. Then $d_2(y_0) = F_y(x_0, y_0)$.*

Proposition 6.3.2 *Let F be a complex-valued function with domain an open set O in $\mathbf{R}^2$. Let (x_0, y_0) lie in O. Let $F = U + iV$, where U and V are real valued. The partial derivative $F_x(x_0, y_0)$ exists if and only if both $U_x(x_0, y_0)$ and $V_x(x_0, y_0)$ exist, and $F_y(x_0, y_0)$ exists if and only if both $U_y(x_0, y_0)$ and $V_y(x_0, y_0)$ exist. In case of existence, $F_x(x_0, y_0) = U_x(x_0, y_0) + iV_x(x_0, y_0)$ and $F_y(x_0, y_0) = U_y(x_0, y_0) + iV_y(x_0, y_0)$.*

Proof. For example,

$$\operatorname{Re} \delta_{x_0}(x, y_0) = \frac{U(x, y_0) - U(x_0, y_0)}{x - x_0},$$

$$\operatorname{Im} \delta_{x_0}(x, y_0) = \frac{V(x, y_0) - V(x_0, y_0)}{x - x_0}.$$

According to Exercise 5.2.5, lim $\delta_{x_0}(x, y_0)$ exists if and only if the limit of its real and imaginary parts exist, and the limits must satisfy the equation in the last sentence of the proposition; we have a similar result for F_y. The proof was anticipated in Exercise 6.1.2. Q.E.D.

Theorem 6.3.1 (*Cauchy-Riemann Equations*). *Let the domain of the complex-valued function* $f(z) = f(x + iy) = F(x, y) = U(x, y) + iV(x, y)$ *(*U, V *real valued) be an open set* $O \in \mathbf{C}$ *and* $\mathbf{R}^2$. *If the derivative* $f'(z_0) = f'(x_0 + iy_0)$ *exists at a point* $z_0 \in O$, *then*

(a) $F_x(x_0, y_0)$, $F_y(x_0, y_0)$ *exist*,
(b) $f'(z_0) = F_x(x_0, y_0) = -iF_y(x_0, y_0)$,
(c) $U_x(x_0, y_0) = V_y(x_0, y_0)$, $U_y(x_0, y_0) = -V_x(x_0, y_0)$.

Remark (i). The existence of the partial derivatives of U and V appearing in (c) is assured by (a) taken with Proposition 6.3.2.

Remark (ii). Both (b) and (c) are referred to as the Cauchy-Riemann equations.

Proof. According to the remarks at the beginning of Sec. 6.2, we can compute the derivative $f'(z_0)$ by restricting f either to $E_x = O \cap \{z = x + iy_0 : -\infty < x < \infty\}$ or to $E_y = O \cap \{z = x_0 + iy : -\infty < y < \infty\}$. By Proposition 6.1.1, there exists a function d continuous at $z_0 = x_0 + iy_0$ such that $f(z) - f(z_0) = d(z)(z - z_0)$ for every $z \in O$ and $d(z_0) = f'(z_0)$. The restriction to E_x yields the equation

$$\begin{aligned} f(x + iy_0) - f(x_0 + iy_0) &= F(x, y_0) - F(x_0, y_0) \\ &= d(x + iy_0)(x + iy_0 - x_0 - iy_0) \\ &= d(x + iy_0)(x - x_0). \end{aligned}$$

The function $d(x + iy_0)$ (with y_0 fixed) is a complex-valued function with domain $O_x = \{x : (x, y_0) \in O\}$, continuous at x_0, and $d(x_0 + iy_0) = f'(z_0)$. Thus, by Proposition 6.3.1 with $d_1(x) = d(x + iy_0)$, $F_x(x_0, y_0)$ exists and equals $f'(z_0)$.

The restriction to E_y yields

$$\begin{aligned} f(x_0 + iy) - f(x_0 + iy_0) &= F(x_0, y) - F(x_0, y_0) \\ &= d(x_0 + iy)(x_0 + iy - x_0 - iy_0) \\ &= id(x_0 + iy)(y - y_0). \end{aligned}$$

Here $id(x_0 + iy)$ plays the role of $d_2(y)$ in Proposition 6.3.1, and the conclusion is that $F_y(x_0, y_0)$ exists and equals $id(x_0 + iy_0) = if'(z_0)$. Thus, $f'(z_0) = -iF_y(x_0, y_0)$. We then have $-iF_y(x_0, y_0) = f'(z_0) = F_x(x_0, y_0)$ as in part (b) of the theorem.

According to Proposition 6.3.2, we have now established that $U_x(x_0, y_0) + iV_x(x_0, y_0) = -i[U_y(x_0, y_0) + iV_y(x_0, y_0)]$. Part (c) of the theorem is obtained by equating real and imaginary parts in this equation. Q.E.D.

Notice the asymmetry in the equations in (c). The equations show that simultaneous differentiability of $f = U + iV$ and $\bar{f} = U - iV$ at $z = z_0$ is impossible unless $f'(z_0) = 0$.

The converse of Theorem 6.3.1 is not valid without further hypotheses. For example, consider the function with domain **C** given by

$$f(z) = \begin{cases} \dfrac{z^5}{|z|^4}, & z \neq 0. \\ 0, & z = 0. \end{cases}$$

The difference quotient at $z = 0$ is $\delta_0(z) = [(z^5/|z|^4) - 0]/(z - 0) = z^4/|z|^4$, which has no limit as $z \to 0$. But if we set $F(x, y) = f(x + iy)$, we find that $F_x(0, 0)$ and $F_y(0, 0)$ both exist and satisfy the Cauchy-Riemann equations. We defer the proof to the exercises. The example was given by Menchoff.

To look further into the question of a converse for Theorem 6.3.1, we introduce the idea of the *differentiability* of a function with domain in $\mathbf{R}^2$. Let F be a complex-valued function with domain an open set O in $\mathbf{R}^2$. The function is said to be *differentiable* at $(x_0, y_0) \in O$ if and only if there exist complex numbers p_1, p_2 and a function $\Phi: O \to \mathbf{C}$ continuous at (x_0, y_0) with $\Phi(x_0, y_0) = 0$, such that

$$\begin{aligned} F(x, y) - F(x_0, y_0) &= (x - x_0)p_1 + (y - y_0)p_2 \\ &\quad + \Phi(x, y)[(x - x_0)^2 + (y - y_0)^2]^{1/2}. \quad (6.3.1) \end{aligned}$$

Clearly, such a function F is continuous at (x_0, y_0). By setting up the difference quotients, it is easily seen that $F_x(x_0, y_0)$ and $F_y(x_0, y_0)$ exist and equal p_1 and p_2, respectively. A sufficient condition for differentiability is that $F_x(x_0, y_0)$ exist, and $F_y(x, y)$ exist for every $(x, y) \in O$ and be continuous at (x_0, y_0). (For a proof, see [8, pp. 80–81].)

Theorem 6.3.2 *Let $f(z) = f(x + iy) = F(x, y)$ be a complex-valued function with domain O an open set in* **C**. *Let $z_0 = x_0 + iy_0$ be a point in O.*

(a) *If $f'(z_0)$ exists, then F is differentiable at (x_0, y_0).*
(b) *If F is differentiable at (x_0, y_0) and satisfies the Cauchy-Riemann equations $F_x(x_0, y_0) = -iF_y(x_0, y_0)$, then $f'(z_0)$ exists.*

Proof of (a). By Proposition 6.1.1, there exists a function $d: O \to \mathbf{C}$ continuous at z_0 satisfying for every $z \in O$ the equation

$$\begin{aligned} f(z) - f(z_0) &= F(x, y) - F(x_0, y_0) \\ &= d(z)(z - z_0) \\ &= d(z_0)(z - z_0) + [d(z) - d(z_0)](z - z_0). \end{aligned}$$

Then for $z \neq z_0$, and with $z = x + iy$, $z_0 = x_0 + iy_0$, we have

$$f(z) - f(z_0) = d(z_0)(x - x_0) + id(z_0)(y - y_0) + [d(z) - d(z_0)]\frac{z - z_0}{|z - z_0|}[(x - x_0)^2 + (y - y_0)^2]^{1/2}.$$

Define

$$\phi(z) = \Phi(x, y) = \begin{cases} \dfrac{[d(z) - d(z_0)](z - z_0)}{|z - z_0|}, & z \in O, \quad z \neq z_0. \\ 0, & z = z_0. \end{cases}$$

Since d is continuous at z_0, given any $\varepsilon > 0$ there exists a disk $\Delta(z_0, r) \subset O$, $r > 0$, such that if $z \in \Delta(z_0, r)$, then $|d(z) - d(z_0)| < \varepsilon$. But

$$\left|\frac{[d(z) - d(z_0)](z - z_0)}{|z - z_0|} - 0\right| = |d(z) - d(z_0)|,$$

so $\phi(z)$ is continuous at z_0, and by definition $\phi(z_0) = 0$. Thus, by a remark made at the beginning of this section, $\Phi(x, y)$ is continuous on $\mathbf{R}^2$ at (x_0, y_0). The criterion (6.3.1) for F to be differentiable at (x_0, y_0) is satisfied.

Proof of (b). Into (6.3.1), we substitute with the Cauchy-Riemann equations written in the form $iF_x(x_0, y_0) = F_y(x_0, y_0)$. As usual, $z = x + iy$, $z_0 = x_0 + iy_0$. The result, valid for every $z \in O$, is

$$\begin{aligned} f(z) - f(z_0) &= (x - x_0)F_x(x_0, y_0) + (y - y_0)F_y(x_0, y_0) + \Phi(x, y)|z - z_0| \\ &= (x - x_0)F_x(x_0, y_0) + i(y - y_0)F_x(x_0, y_0) + \Phi(x, y)|z - z_0| \\ &= F_x(x_0, y_0)(z - z_0) + \Phi(x, y)|z - z_0|. \end{aligned}$$

If $z \neq z_0$, this can be written in the form

$$f(z) - f(z_0) = (z - z_0)\left[F_x(x_0, y_0) + \phi(z)\frac{|z - z_0|}{z - z_0}\right],$$

where $\phi(z) = \Phi(x, y)$. Define

$$d(z) = \begin{cases} F_x(x_0, y_0) + \phi(z)\dfrac{|z - z_0|}{z - z_0}, & z \in O, \quad z \neq z_0. \\ F_x(x_0, y_0), & z = z_0. \end{cases}$$

Then the equation $f(z) - f(z_0) = d(z)(z - z_0)$ is valid for every $z \in O$. Now ϕ is continuous at z_0, and $\phi(z_0) = 0$. Therefore, given $\varepsilon > 0$, there exists a disk $\Delta(z_0, r) \subset O$, $r > 0$, such that if $z \in \Delta(z_0, r)$, then $|\phi(z)| < \varepsilon$. But for every $z \in \Delta(z_0, r)$ (in fact, for every $z \in O$), $|d(z) - F_x(x_0, y_0)| = |\phi(z)|$, since $\big||z - z_0|/(z - z_0)\big| = 1$, $z \neq z_0$. Thus, for every $z \in \Delta(z_0, r)$, $|d(z) - F_x(x_0, y_0)| < \varepsilon$, so by the definition of continuity, d is continuous at z_0. Also by definition, $d(z_0) = F_x(x_0, y_0)$. It follows from Proposition 6.1.1 that $f'(z_0)$ exists and equals $F_x(x_0, y_0) = -iF_y(x_0, y_0)$. Q.E.D.

Proposition 6.3.3 *Let the complex-valued function $f = u + iv$ (u, v, real valued) be analytic on a region $B \subset \mathbf{C}$. Each of the following conditions by itself implies that f is a constant function on B:*

(a) *u is a constant function on B,*
(b) *v is a constant function on B,*
(c) *$|f|$ is a constant function on B, where $|f|$ denotes the function of which the values are the absolute values of the values of f.*

Corollary A nonconstant complex-valued function analytic on a region B cannot have only real or only imaginary values on B.

Remark (i). The corollary is the special case of Proposition 6.3.3(a) and (b) in which $v(z) = 0$ or $u(z) = 0$ for every $z \in B$.

Remark (ii). The corollary provides a quick proof of the fact that $|z|^2$ and $|z|$ cannot be analytic on any region $B \subset \mathbf{C}$. This was established in a more cumbersome manner in Sec. 6.2. (Of course, the idea there was to demonstrate directly the limit process involved in complex differentiation.)

Proof of the Proposition. The proof of (a) and (b) is left as an exercise. For (c), first notice that if $f(z) = 0$ at a point of B, then $|f(z)| = 0$ at that point, and then by hypothesis $|f(z)| = 0 = f(z)$ at every point of z. Suppose then that $f(z) \neq 0$ for every $z \in B$. Let $c = |f(z)|$, and as usual, let $f(x + iy) = U(x, y) + iV(x, y)$. We have the equation $c^2 = [U(x, y)]^2 + [V(x, y)]^2$. By Theorem 6.3.1, the partial derivatives U_x, V_x, U_y, V_y exist at every point (x_0, y_0) of B and satisfy the Cauchy-Riemann equations. The chain law, Theorem 6.1.2, is applicable to the differentiation of U^2 and V^2 restricted to the set $\{x: (x, y_0) \in B\}$ and to the set $\{y: (x_0, y) \in B\}$, and we obtain

$$0 = 2U(x_0, y_0)U_x(x_0, y_0) + 2V(x_0, y_0)V_x(x_0, y_0),$$
$$0 = 2U(x_0, y_0)U_y(x_0, y_0) + 2V(x_0, y_0)V_y(x_0, y_0).$$

We substitute with $V_x = -U_y$ in the first equation and $V_y = U_x$ in the second. The equations become

$$U(x_0, y_0)U_x(x_0, y_0) - V(x_0, y_0)U_y(x_0, y_0) = 0,$$
$$V(x_0, y_0)U_x(x_0, y_0) + U(x_0, y_0)U_y(x_0, y_0) = 0.$$

These equations can be viewed as simultaneous linear equations in $U_x(x_0, y_0)$, $U_y(x_0, y_0)$. The determinant is $[U(x_0, y_0)]^2 + [V(x_0, y_0)]^2 \neq 0$. Thus, the only solution is $U_x(x_0, y_0) = U_y(x_0, y_0) = 0$. But then by the Cauchy-Riemann equations, $V_y(x_0, y_0) = -V_x(x_0, y_0) = 0$. It follows that $f'(x_0 + iy_0) = 0$. This holds true for every $z_0 = x_0 + iy_0 \in B$, so by Theorem 6.2.1, f must be a constant function on B. Q.E.D.

Exercises 6.3

1. Show by partial differentiation that $f(z) = z^3$ satisfies the Cauchy-Riemann equations at every $z \in \mathbf{C}$.
2. Show that the example of Menchoff satisfies the Cauchy-Riemann equations at the origin. (*Hint*: Use the definition of $F_x(0, 0)$, $F_y(0, 0)$.)
3. Let the function f be analytic on a region B. Show that if $\bar{f}$ is also analytic on B, then f is a constant function. (*Hint*: Cauchy-Riemann equations and Theorem 6.2.1.)
4. Prove Proposition 6.3.3(a) and (b).
5. Let F be a complex-valued function with domain an open set $O \subset \mathbf{R}^2$, and let $(x_0, y_0) \in O$. Show that if $F_x(x_0, y_0)$ exists, then $F(x, y_0)$ as a function of x is continuous at x_0, and that if $F_y(x_0, y_0)$ exists, then $F(x_0, y)$ as a function of y is continuous at y_0.
6. Show that the function

$$F(x, y) = \begin{cases} \dfrac{xy}{x_2 + y^2}, & (x, y) \neq (0, 0) \\ 0, & (x, y) = (0, 0) \end{cases}$$

has partial derivatives $F_x(0, 0)$ and $F_y(0, 0)$ but is not continuous at $(0, 0)$ (as a function with domain $\mathbf{R}^2$).

7. Let B be a region which is symmetric with respect to the real axis; that is, when $z \in B$, also $\bar{z} \in B$. Prove that if f is analytic on $B_1 = B \cap \{z : \text{Im } z > 0\}$, then the function ϕ with values $\phi(z) = \bar{f}(\bar{z})$ is analytic on $B_2 = B \cap \{z : \text{Im } z < 0\}$. (*Hint*: First, show that if $G(x, y)$ is any function which is differentiable on B_1, then $H(x, y) = G(x, -y)$ is differentiable on B_2. Then show that ϕ satisfies the Cauchy-Riemann equations on B_2, and use Theorem 6.3.2(b).)
8. Let $U(x, y) = x/(x^2 + y^2)$, $(x, y) \in \mathbf{R}^2 \sim \{(0, 0)\}$. By solving the Cauchy-Riemann equations, find a function $V(x, y)$ such that if $u(z) = u(x + iy) = U(x, y)$ and $v(z) = v(x + iy) = V(x, y)$, the function $f = u + iv$ is analytic on $\mathbf{C} \sim \{0\}$. (*Hint*: If $F(x, y)$ is an antiderivative of $f(x, y)$ with respect to x, so $F_x(x, y) = f(x, y)$, then also $F(x, y) + c(y)$ is an antiderivative with respect to x, where $c(y)$ is a function of y only.)
9. Show that there cannot exist a function analytic on an open set $O \in \mathbf{C}$ with real part $x - 2y^2$.
10. Find a function f analytic on $\mathbf{C}$ such that $\text{Re } f'(z) = 3x^2 - 4y - 3y^2$, $\text{Re } f(0) = 6$, and $f(1 + i) = 0$. (*Answer*: $f(z) = z^3 + 2iz^2 + (6 - 2i)$. Prove by integrating the Cauchy-Riemann equations, not by working backward from the answer.)

6.4 Analyticity of Power Series

Exercise 4.1.5 shows that if the power series $f(z) = \sum_0^\infty a_k(z - a)^k$ has radius of convergence $R > 0$, then the series $f_1(z) = \sum^\infty ka_k(z - a)^{k-1}$ has the same radius of convergence. The latter series is the series of derivatives of the terms of the former one. We now prove that the series formed by term-by-term differentiation of a power series represents the authentic derivative of the sum function of the series.

Theorem 6.4.1 *Let $f(z) = \sum_0^\infty a_k(z - a)^k$ have radius of convergence $R > 0$. Then $f'(z)$ exists at every point of $\Delta(a, R)$ and $f'(z) = \sum_1^\infty ka_k(z - a)^{k-1}$, $z \in \Delta(a, R)$. The radius of convergence of the derivative series is R.*

Proof. For simplicity, we first take $a = 0$.

Lemma 1 If $z \neq z_0$, then

$$\frac{z^k - z_0^k}{z - z_0} - kz_0^{k-1}$$
$$= \begin{cases} 0, & k = 1. \\ (z - z_0)[1 \cdot z^{k-2} + 2z_0 z^{k-3} + \cdots + hz_0^{h-1}z^{k-h-1} + \cdots \\ \qquad + (k-1)z_0^{k-2}], & k = 2, 3, \ldots. \end{cases}$$

The proof of this identity is an exercise.

Lemma 2 The radius of convergence of $\sum_0^\infty k^2 a_k z^k$ is the same as that of $\sum_0^\infty a_k z^k$.

Proof. According to Sec. 2.4, Example (iv), $\lim_{k \to \infty} k^{1/k} = 1$. By Exercise 2.4.12, $\lim (k^{1/k})^2 = 1^2 = 1$. By Exercise 2.4.10,

$$\frac{1}{\overline{\lim}\, |k^2 a_k|^{1/k}} = \frac{1}{\lim (k^{1/k})^2\, \overline{\lim}\, |a_k|^{1/k}} = \frac{1}{\overline{\lim}\, |a_k|^{1/k}}.$$

Q.E.D.

Proof of the Theorem. Let $f_1(z) = \sum_0^\infty ka_k z^{k-1}$. By Exercise 4.1.5, the radius of convergence of this series is R. For every $z \in \Delta(0, R)$, $z_0 \in \Delta(0, R)$, $z \neq z_0$, according to Proposition 2.6.2, we have

$$D(z) = \frac{f(z) - f(z_0)}{z - z_0} - f_1(z_0) = \sum_{k=2}^{\infty} a_k \left[\frac{z^k - z_0^k}{z - z_0} - kz_0^{k-1}\right].$$

Take any r, $0 < r < R$. For $|z| < r$, $|z_0| < r$, we use Lemma 1 to obtain

$$|D(z)| \leq \sum_{k=2}^{\infty} \{|a_k|\, |z - z_0|[|z|^{k-2} + \cdots + h|z_0|^{h-1}|z|^{k-h-1} + \cdots$$
$$+ (k-1)|z_0|^{k=-2}]\}$$
$$\leq |z - z_0| \sum_{k=2}^{\infty} [|a_k| r^{k-2}(1 + 2 + \cdots + k - 1)]$$

$$= |z - z_0| \sum_{k=2}^{\infty} \frac{(k-1)k}{2} |a_k| r^{k-2}$$

$$\leq \frac{|z - z_0|}{r^2} \sum_{k=2}^{\infty} k^2 |a_k| r^k.$$

By Lemma 2 and Theorem 4.1.2(i), the power series in the last member of this inequality converges; let the sum be s. Given any $\varepsilon > 0$, for every z, $z \neq z_0$, satisfying the inequality $|z - z_0| < \min(r - |z_0|, \varepsilon r^2/s)$, we have

$$\left| \frac{f(z) - f(z_0)}{z - z_0} - f_1(z_0) \right| < \varepsilon.$$

Thus, the limit of the difference quotient of f at z_0 exists and equals $f_1(z_0)$.

Now suppose $a \neq 0$. We could simply go through the proof again with $z - a$ replacing z, but a simple transformation saves work. Let $w = z - a$, $z = a + w$. Then $\phi(w) = f(a + w) = \sum_0^\infty a_k w^k$, with radius of convergence still equal to R. Thus, by the first part of our proof, $\phi'(w) = \sum_1^\infty k a_k w^k$, valid for every $w \in \Delta(0, R)$. By the chain law, $\phi'(w) = f'(z) \cdot (a + w)' = f'(z)$, $z = a + w$, $z \in \Delta(a, R)$, $w \in \Delta(0, R)$, and, thus, $f'(z) = \phi'(z - a) = \sum_1^\infty k a_k (z - a)^k$, $z \in \Delta(a, R)$. Q.E.D.

And so the sum function of a power series $\sum_0^\infty a_k(z - a)^k$ with positive radius of convergence R is analytic on the disk of convergence. Since the derivative series $f'(z) = \sum_1^\infty k a_k (z - a)^{k-1}$ itself has radius of convergence R, it follows that the second derivative $f''(z)$ exist in $\Delta(a, R)$ and is represented by $\sum_2^\infty k(k - 1)(z - a)^{k-2}$; and so on. We formulate the implications as follows.

Corollary Let $f(z) = \sum_0^\infty a_k(z - a)^k$ have radius of convergence $R > 0$. The function f has derivatives of all orders on $\Delta(a, R)$, and the nth derivative is the sum of the power series

$$f^{(n)}(z) = \sum_{k=n}^{\infty} k(k-1) \cdots (k - n + 1) a_k (z - a)^{k-n},$$

which has radius of convergence R. The first coefficient ($k = n$) is the sum of the series for $z = a$ and, thus, satisfies the equation

$$f^{(n)}(a) = n!\, a_n, \qquad a_n = \frac{f^{(n)}(a)}{n!}. \tag{6.4.1}$$

When the power series $f(z) = \sum_0^\infty a_k(z - a)^k$ is written in the form $\sum_0^\infty [f^{(k)}(a)/k!](z - a)^k$,it is often called the Taylor series for f. In the special case $a = 0$, it is called the Maclaurin series. It is to be understood here that $f^{(0)} = f$ and $f^{(1)} = f'$.

Exercises 6.4

1. Let $f(z) = \sum_0^\infty a_k(z-a)^k$ have radius of convergence $R > 0$. Recall that a primitive of f on $\Delta(a, R)$ is a function F such that $F'(z) = f(z)$ for every $z \in \Delta(a, R)$. Prove that a primitive for f on $\Delta(a, R)$ is given by $F(z) = \sum_0^\infty a_k(k+1)^{-1}(z-a)^{k+1}$. What is the radius of convergence of this power series? (*Hint*: By Corollary 2 of Theorem 4.1.2, the radius of convergence of the series for F is the same as that of $\sum_0^\infty a_k(k+1)^{-1}(z-a)^k$. Notice that $(k+1)^{-1/k} < k^{-1/k}$ so $\lim (k+1)^{-1/k} \le \lim k^{-1/k}$ [Exercise 2.4.6]. Use this to show that the radius of convergence of the series for F is $\ge R$. Apply Theorem 6.4.1 to F.)
2. Prove Lemma 1.
3. A real-valued function ϕ of which the domain is an interval $(\alpha, \beta) \in \mathbf{R}$ is said to be analytic on (α, β) if at each point $x_0 \in (\alpha, \beta)$ there exists a power series $\sum c_k(x_0)(x-x_0)^k$, convergent to $\phi(x)$ in some interval $(x_0 - h, x_0 + h)$, $h > 0$, where h depends on x_0. Show that the complex-valued power series $\sum c_k(x_0)(z-x_0)^k$, $z \in \mathbf{C}$, converges on the disk $\Delta(x_0, h) \subset \mathbf{C}$ and, thus, defines an analytic function $f : \Delta(x_0, h) \to \mathbf{C}$, which coincides with ϕ on $(x_0 - h, x_0 + h)$. Use this to show that ϕ has derivatives of all orders on (α, β) and that the local power series for ϕ at x_0 has the form $\sum [\phi^{(k)}(x_0)/k!](x-x_0)^k$.
4. There exists a real-valued function ϕ with domain $\mathbf{R}$ such that ϕ is analytic on $\mathbf{R} \sim \{0\}$, $\phi(x) > 0$, $x \neq 0$, and $\phi(0) = 0$, and all the derivatives of ϕ at $x = 0$ exist and have the value zero. Assuming this to be true, show why ϕ cannot be analytic on any open interval containing $x = 0$.

   Remark. Notice the contrast between the use of "analytic" in the case of functions from $\mathbf{R}$ into $\mathbf{R}$ on the one hand and functions from $\mathbf{C}$ into $\mathbf{C}$ on the other hand. Exercise 6.4.4 shows that in the real case, a function can have derivatives of all orders at every point of an open domain in $\mathbf{R}$ and not be analytic on that domain. But in the complex case, if a function has just a first derivative at every point of its domain (assumed to be an open set), then it is called analytic. The apparent divergence in the usage of the term "analytic" in the two cases is reconciled partly by the first sentence of the next section.
5. Suppose that $f(z) = \sum_0^\infty a_k z^k$ has radius of convergence $R = \infty$, and suppose that $f^{(n)}$ is a constant function. Show that f is a polynomial function of degree n.

6.5 Functions Locally Representable by Power Series

It is shown later in this book that any complex-valued function analytic on an open set $O \subset \mathbf{C}$ has the property that at each point $b \in O$ there exists a power series $\sum a_k(b)(z-b)^k$ which converges to $f(z)$ at each point of a disk $\Delta(b, r_b)$, $r_b > 0$. We call this the LPS property until we establish its equivalence with analyticity. (LPS stands for local power series.) The standard proof of the equivalence with analyticity depends on establishing a certain representation of f in terms of a complex line integral. (See Sec. 9.4.)

It is the purpose of this section to take note of certain elementary but important features of functions having the LPS property on an open set or a region which can be established easily without recourse to complex integration theory.

The first result is a fairly obvious consequence of Theorem 6.4.1 and its corollary.

Proposition 6.5.1 *Let the complex-valued function f have the LPS property on the open set $O \subset \mathbf{C}$. Then f is analytic on O and has analytic derivative functions of all orders on O, each of which has the LPS property.*

Proof. Let b be an arbitrary point in O. There exists a power series $\sum_0^\infty a_k(b)(z-b)^k$, which converges to $f(z)$ at each point of some disk $\Delta(b, r_b)$, $r_b > 0$. Thus, the radius of convergence of this power series is at least as large as r_b. The sum function of the series is analytic on its disk of convergence and has analytic derivatives of all orders on this disk [Theorem 6.4.1 and corollary]. Each derivative of the sum function is itself the sum of a series of powers of $(z-b)$ converging on this disk [Theorem 6.4.1 and corollary]. The fact that $f(z)$ is identically equal to this sum function on $\Delta(b, r_b)$ completes the proof. Q.E.D.

In Theorem 4.3.2, it was brought out that if the sum functions of two power series in powers of $(z-a)$, with positive radii of convergence, coincide on a sequence $\langle z_n \rangle \to a$, then the two series are identical. We now extend this result to LPS functions with common domains. Here is the basic theorem.

Theorem 6.5.1 *Let $f: B \to \mathbf{C}$, B a region, have the LPS property on B. Suppose that at some point $a \in B$, $f(a) = f'(a) = f''(a) = \cdots = f^{(n)}(a) = \cdots = 0$. Then $f(z)$ and all the derivatives are equal to zero at every point $z \in B$.*

Proof. Let B_1 be the subset of B with the property that at every $b \in B_1$, $f^{(k)}(b) = 0$, $k = 0, 1, 2, \ldots$ ($f^{(0)}$ means f itself). For $b \in B_1$, let $f(z) = \sum a_k(b)(z-b)^k$ be the local power series representation of $f(z)$ in the disk $\Delta(b, r_b)$. According to the corollary of Theorem 6.4.1, the coefficient $a_k(b)$ is the kth derivative of the sum function of this series divided by $k!$, $k = 0, 1, 2, \ldots$, so $a_k(b) = 0$ and, therefore, $f(z) = 0$ for every $z \in \Delta(b, r_b)$. It follows from this (or from further applications of the corollary) that $f^{(k)}(z) = 0$ for every $z \in \Delta(b, r_b)$, $k = 1, 2, \ldots$. Therefore, B_1 is open, and since $a \in B_1$, $B_1 \neq \varnothing$.

Now let $B_2 = B \sim B_1$. If B_2 is empty, there is nothing more to prove. If B_2 were not empty, at a typical point $b \in B_2$, an inequality $f^{(n)}(b) \neq 0$ would hold true for some n, $n = 0, 1, 2, \ldots$; say $n = m$. By Proposition 6.5.1, $f^{(m)}$ has the LPS property on B, so by Theorem 4.2.1(i), the function

$f^{(m)}$ is continuous at b. By Exercise 3.2.3(b), there would exists a disk $\Delta(b, r) \subset B$, $r > 0$, such that $f^{(m)}(z) \neq 0$ for every $z \in \Delta(b, r)$. This disk would have to lie in B_2, by the definition of B_1. But this would mean that B_2 is a nonempty open set. Then $\{B_1, B_2\}$ would form a separation of B, and B would not be connected, a contradiction to the definition of region. Q.E.D.

Corollary 1 Let B be a region in **C**, and let $f: B \to \mathbf{C}$ and $g: B \to \mathbf{C}$ each have the LPS property on B. If at some point $a \in B$, $f(a) = g(a)$, $f^{(n)}(a) = g^{(n)}(a)$, $n = 1, 2, \ldots$, then $f(z) = g(z)$ on B.

Proof. The LPS property for the function $f - g$ is guaranteed by Proposition 3.4.3(i). The corollary thus follows directly when f in the theorem is replaced by $f - g$. Q.E.D.

Corollary 2 Let B be a region in **C**, and let $f: B \to \mathbf{C}$ and $g: B \to \mathbf{C}$ each have the LPS property on B. Let $\langle z_n \rangle_0^\infty$ be a sequence of points in B converging to $a \in B$, with $z_n \neq a$, $n = 0, 1, 2, \ldots$. Finally, let $f(z_n) = g(z_n)$, $n = 0, 1, 2, \ldots$. Then $f(z) = g(z)$ for every z on B.

Proof. The LPS property assures the existence of a disk $\Delta(a, R) \subset B$, $R > 0$, and power series $\sum_0^\infty a_k(z - a)^k$ and $\sum_0^\infty b_k(z - a)^k$ converging, respectively, to $f(z)$ and $g(z)$ on $\Delta(a, R)$. There exists a number N such that for every $n \geq N$, $z_n \in \Delta(a, R)$. It follows from the power series uniqueness theorem, Theorem 4.3.2, that $f(z) = g(z)$ on $\Delta(a, R)$, and $a_k = f^{(k)}(a)/k! = b_k = g^{(k)}(a)/k!$, $k = 0, 1, 2, \ldots$. Thus, Corollary 1 is applicable, and $f(z) = g(z)$ for every $z \in B$. Q.E.D.

A point a in the domain of a function f at which $f(a) = 0$ is called a *zero* of the function.

Corollary 3 Let B be a region in **C** and $f: B \to \mathbf{C}$ have the LPS property on B. If there is a point $z \in B$ such that $f(z) \neq 0$, then the set of zeros of f either is empty or consists only of isolated points.

Proof. The conclusion is equivalent to saying that if a is a zero of f, then there exists a disk $\Delta(a, R)$, $R > 0$, such that no other zero of f lies in the disk. If this were not true, then in any disk $\Delta(a, 1/n)$ there would lie a zero $z_n \neq a$, $n = 1, 2, 3, \ldots$. The sequence $\langle z_n \rangle$ converges to a, and the values of f on this sequence would be equal to the value of the constant function $g(z) = 0$. Thus, by Corollary 2, f would be the constant function with value 0 on B, contrary to hypothesis. Q.E.D.

Exercises 6.5

1. Show that $f(z) = z^n$, n a positive integer, has the LPS property on **C**. (*Hint*: The binomial theorem, Exercise 1.2.10, is useful; let $z^n = [(z - a) + a]^n$.)

2. Show that $f(z) = 1/z$ has the LPS property on $\mathbf{C} \sim \{0\}$.
3. The so-called method of undetermined coefficients in the theory of ordinary differential equations is exemplified by the following prescription for solving the differential equation $f''(z) + af'(z) + bf(z) = 0$, $a, b \in \mathbf{C}$. Assume that there is a power series $f(z) = \sum_0^\infty a_k z^k$ which has radius of convergence $R > 0$ and which satisfies the equation in $\Delta(0, R)$. Find the power series for $f''(z) + af'(z) + bf(z)$, which is

$$\sum_{k=0}^{\infty} [(k+2)(k+1)a_{k+2} + a(k+1)a_{k+1} + ba_k]z^k. \tag{6.5.1}$$

Then set the coefficients in (6.5.1) equal to zero in sequence. This provides a recursion relation from which $a_3, a_4, a_5, \ldots$ can be determined in terms of ba_0 and aa_1. Accepting the assumption as to the existence of a power series solution in a neighborhood of $z = 0$, cite authorities for the validity of (6.5.1) as the power series for $f'' + af' + bf$, and for finding the coefficients $\langle a_k \rangle$ by setting the coefficients in (6.5.1) equal to zero.
4. Recall the definition in Exercise 6.4.3 of analyticity for a real-valued function ϕ on an interval (α, β). Prove that if such a ϕ and all its derivatives are equal zero at a point $x_0 \in (\alpha, \beta)$, then $\phi(x)$ is identically zero on (α, β). (*Hint*: The interval is connected, so the method of proof of Theorem 6.5.1 is applicable.)
5. Let ϕ be real valued and analytic on the interval (α, β). Let $\langle x_n \rangle_0^\infty$ be a sequence of points in (α, β) converging to $y \in (\alpha, \beta)$, with $x_n \neq y$, $\phi(x_n) = 0$, $n = 0,1,2, \ldots$. Then $\phi(x) = 0$ everywhere on (α, β). (*Hint*: Exercise 6.4.3; Proposition 4.3.1; Exercise 6.4.3 again to show that the condition in Exercise 6.5.4 is satisfied at y.)

7

The Exponential and Related Functions

7.1 The Exponential Function

In this chapter, we investigate a special family of analytic functions given by power series which contains the complex analogs of the exponential, trigonometric, and logarithmic functions of elementary calculus.

First, we define the real number e to be the sum of the series $1 + 1 + (1/2!) + (1/3!) + \cdots$. Then we introduce the complex-valued function E with values given by $E(z) = 1 + z + (z^2/2!) + (z^3/3!) + \cdots$ for z on the disk of convergence of this power series. However, the radius of convergence of this power series is infinite [Exercise 4.1.2(a)], so E is continuous and analytic on **C** [Theorem 6.4.1].

Proposition 7.1.1 $E(0) = 1$, $E(1) = e$, $E^{(n)}(z) = E(z)$, $E^{(n)}(0) = 1$, $E^{(n)}(1) = e$, $n = 1, 2, \ldots$.

These equations can be easily verified by direct computation, using Theorem 6.4.1.

Theorem 7.1.1 (*Addition Law*). *For every* z_1, $z_2 \in \mathbf{C}$, $E(z_1 + z_2) = E(z_1)E(z_2)$.

Proof. Let $f_a(z) = E(z)E(a - z)$, $a \in \mathbf{C}$. Then by the chain law and rule for differentiation of a product, $f_a'(z) = E(z)E'(a - z)(-1) + E(z)E(a - z) = -E(z)E(a - z) + E(z)E(a - z) = 0$. This is true for every $z \in \mathbf{C}$. Therefore, by Theorem 6.2.1, f_a is a constant function. The constant can be found by letting $z = 0$, and we find that $f_a(0) = 1 \cdot E(a)$.

Now let $z = z_1$ and $a = z_1 + z_2$. Substituting into $E(a) = E(z)E(a - z)$, we obtain $E(z_1 + z_2) = E(z_1)E(z_2)$. Q.E.D.

Corollary 1 $E(z) \neq 0$ for every $z \in \mathbf{C}$.

Proof. If for some $z_1 \in \mathbf{C}$, $E(z_1) = 0$, then $E(z_1 + z_2) = 0 \cdot E(z_2) = 0$ for every $z_2 \in \mathbf{C}$. Set $z = z_1 + z_2$. The implication is that $E(z) = 0$ for every $z \in \mathbf{C}$; but actually $E(0) = 1$. Q.E.D.

Corollary 2 $E(-z) = 1/E(z)$, $E(z_1 - z_2) = E(z_1)/E(z_2)$ for every z, z_1, $z_2 \in \mathbf{C}$.

The proof is an exercise.

Corollary 3 Let $z = x + i \cdot 0$, $x \in \mathbf{R}$.

(i) The function given by $u = E(x)$ is real valued, continuous, and strictly increasing on its domain $\mathbf{R}$.
(ii) It is a homeomorphism from $\mathbf{R}$ onto $\{u: u > 0\}$ in which $\{x: x \geq 0\}$ is mapped onto $\{u: u \geq 1\}$ and $\{x: x \leq 0\}$ is mapped onto $\{u: 0 < u \leq 1\}$.
(iii) The only solution of the equation $E(x) = 1$ is $x = 0$.

Proof. It is evident from the power series representation of $E(x)$ that $E(x)$ is real valued, continuous on $\mathbf{R}$ [Theorem 4.2.1(i)], strictly increasing at least for $x \geq 0$; and $E(x) > 1$ for $x > 0$. Furthermore, for any $x \in \mathbf{R}$, and in particular for any $x \in \{x: x \leq 0\}$, we have $E(x) = 1/E(-x)$. Thus, if $x_1 < x_2 \leq 0$, we find that $E(x_2) - E(x_1) = [1/E(-x_2)] - [1/E(-x_1)] = [E(-x_1) - E(-x_2)]/E(-x_1)E(-x_2) > 0$. This proves (i). A further consequence of $E(x) = 1/E(-x)$ for $x < 0$ is that $0 < E(x) < 1$ for $x < 0$. Of course, $E(0) = 1$.

From the strictly increasing character of $E(x)$ on $\mathbf{R}$, it follows that $E(x)$ is univalent [Exercise 5.3.4]. Since $E(x)$ is continuous, $E(\mathbf{R})$ is an open interval [Exercise 5.3.5], and the inverse function $x = E^{-1}(u)$ is continuous on $E(\mathbf{R})$ [Exercise 5.3.6]. In fact, by reference to the preceding paragraph we see that $u = E(x)$ gives a homeomorphism from $\{x: x \geq 0\}$ onto some interval $\{u: 1 \leq u < \beta\}$ and also a homeomorphism from $\{x: x \leq 0\}$ onto $\{u: (1/\beta) < u \leq 1\}$ [Exercise 5.3.5]. It remains to show that $\beta = \infty$.

But if it were to be assumed that $\beta < \infty$, it would be easy to see from the power series representation that a positive x could be found such that $E(x) > 1 + x > \beta$, so it must be true that $\beta = \infty$. Q.E.D.

Corollary 4 With $z = x + iy$, x, $y \in \mathbf{R}$, $\overline{E(z)} = E(\bar{z})$, $\overline{E(iy)} = E(-iy)$, $|E(iy)| = 1$, $|E(z)| = E(x)$, for every $z \in \mathbf{C}$.

The proof is an exercise.

Corollary 5 $E(p/q) = e^{p/q}$, where p and q are real integers, $q > 0$, and $e^{1/q}$ means the positive solution of $x^q = e$.

Proof.

$$\left[E\left(\frac{1}{q}\right)\right]^q = E\left(\frac{1}{q} + \frac{1}{q} + \cdots + \frac{1}{q}\right) = E(1) = e,$$

where in the second member the fraction $1/q$ is repeated q times. Thus, $E(1/q) = e^{1/q}$. Then $E(p/q) = [E(1/q)]^p = e^{p/q}$. Q.E.D.

Theorem 7.1.1 and the corollaries, particularly Corollaries 3 and 5, show that the restriction of $E(z)$ to **R** has the properties which are developed in one way or another for the exponential function e^x in elementary calculus. We therefore use the notation e^z instead of $E(z)$. It should be noticed that the discussion here is independent of any of the various equivalent definitions of e^x in elementary real analysis and thus furnishes an alternative rigorous treatment of this real-valued function.

Exercises 7.1

1. Prove Corollary 2. (*Hint*: What is the value of $E(-z)E(z)$?)
2. Prove Corollary 4.
3. Suppose that the domain of the complex-valued function f is **C**; that f has a power series representation in a neighborhood of $z = 0$ in powers of z; that $f(0) = 1 = f'(0)$; and finally that for every $z_1, z_2 \in \mathbf{C}$, f satisfies $f(z_1 + z_2) = f(z_1)f(z_2)$. Prove that $f(z)$ is e^z. (*Hint*: Show that f' exists and $f'(z) = f(z)$, everywhere.)
4. Prove the identity

$$\frac{2^n}{n!} = \frac{1}{n!} + \frac{1}{(n-1)!} + \frac{1}{2!(n-2)!} + \cdots + \frac{1}{k!(n-k)!} + \cdots + \frac{1}{(n-1)!} + \frac{1}{n!}.$$

(*Hint*: Look at the $(n + 1)$th term of the power series for e^{2z}, and the Cauchy product series for $e^z e^z$.)

7.2 The Trigonometric Functions

We make these definitions:

$$\cos z = \frac{e^{iz} + e^{-iz}}{2}, \qquad \sin z = \frac{e^{iz} - e^{-iz}}{2i},$$

and call these analytic functions, respectively, the cosine function (cos) and the sine function (sin). These are the basic complex trigonometric functions.

In the elementary textbook treatments of the trigonometric functions as functions of a real variable, their analytical properties (continuity, differentiability, power series representation) are usually derived by methods which involve a certain amount of rather loose geometric reasoning. For example, a standard first approach to the continuity and differentiability of the sine functions is based on a geometric picture and an assumed theory of arc length. The program here presents a rigorous development of the properties of the complex trigonometric functions as already defined, which proceeds entirely within the framework of analysis. It is shown in the course of the program that the complex trigonometric functions as already defined, when restricted to **R**, have the familiar algebraic and analytic properties of the cosine and sine as presented in the elementary treatment.

By using the rules for adding convergent power series [Exercise 4.1.4], we easily obtain the following MacLaurin series from those for e^{iz} and e^{-iz}:

$$\left.\begin{aligned} \sin z &= z - \frac{z^3}{3!} + \frac{z^5}{5!} - \cdots, \\ \cos z &= 1 - \frac{z^2}{2!} + \frac{z^4}{4!} - \cdots, \end{aligned}\right\} \tag{7.2.1}$$

with radii of convergence ∞. When z is real, these series are identical with the MacLaurin series derived in calculus textbooks for the cosine and sine functions. Thus, by the uniqueness theorem for power series [Theorem 4.3.2], our functions cosine and sine, as defined at the beginning of this section, are the unique complex-valued functions representable everywhere on **C** by series of powers of z, which have values which coincide on **R** with those of the real-valued cosine and sine functions of elementary calculus and trigonometry. In making this statement, we overlook any possible fuzziness in the derivation of the MacLaurin series in the elementary treatments, or else we assume that the real-valued cosine and sine are simply defined outright by the series in (7.2.1) with z real.

The following identities are easily proved consequences of our definition of cosine and sine:

$$\cos' z\left(=\frac{d}{dz}\cos z\right) = -\sin z, \qquad \sin' z = \cos z. \tag{7.2.2}$$

$$\begin{aligned} e^{iz} &= \cos z + i \sin z, \\ e^{x+iy} &= e^x e^{iy} \\ &= e^x(\cos y + \sin y). \end{aligned} \tag{7.2.3}$$

$$(\cos z + i \sin z)^n = \cos nz + i \sin nz, \qquad n \text{ a real integer.} \tag{7.2.4}$$

This is known as De Moivre's law. The proof is immediate from (7.2.3) and $(e^{iz})^n = e^{inz}$.

$$\cos^2 z + \sin^2 z = 1. \tag{7.2.5}$$

This identity implies that $-1 \le \cos x, \sin x \le 1$, $x \in \mathbf{R}$. The situation is strikingly different for $\cos iy = (e^{-y} + e^y)/2$, $i \sin iy = (e^{-y} - e^y)/2$.

By Corollary 3 of Theorem 7.1.1, each of these functions increases in absolute value without bound as $y \to \pm\infty$.

$$\begin{aligned} \cos(z_1 + z_2) &= \cos z_1 \cos z_2 - \sin z_1 \sin z_2, \\ \sin(z_1 + z_2) &= \sin z_1 \cos z_2 + \sin z_2 \cos z_1. \end{aligned} \tag{7.2.6}$$

The proof is an exercise.

$$\cos(-z) = \cos z, \qquad \sin(-z) = -\sin z. \tag{7.2.7}$$

We define $\tan z = \sin z/\cos z$, $\cot z = 1/\tan z$, $\sec z = 1/\cos z$, $\operatorname{cosec} z = 1/\sin z$, as in elementary trigonometry. Notice that all of our extensions of the elementary trigonometric functions into the complex field are simple rational functions of e^{iz}.

Exercises 7.2

1. Prove the following identities for complex z_1, z_2:

$$\cos(z_1 \pm z_2) = \cos z_1 \cos z_2 \mp \sin z_1 \sin z_2,$$
$$\sin(z_1 \pm z_2) = \sin z_1 \cos z_2 \pm \sin z_2 \cos z_1,$$

$$\cos z_2 - \cos z_1 = -2 \sin\frac{z_2 + z_1}{2} \sin\frac{z_2 - z_1}{2},$$

$$\sin z_2 - \sin z_1 = 2 \cos\frac{z_2 + z_1}{2} \sin\frac{z_2 - z_1}{2},$$

$$\tan(z_1 + z_2) = \frac{\tan z_1 + \tan z_2}{1 - \tan z_1 \tan z}.$$

2. Prove the half-angle formulas:

$$\sin^2\frac{z}{2} = \frac{1 - \cos z}{2},$$

$$\cos^2\frac{z}{2} = \frac{1 + \cos z}{2}.$$

3. Prove this identity for complex z_1, z_2:

$$e^{iz_1} - e^{iz_2} = 2ie^{i(z_1 + z_2)/2} \sin\frac{z_1 - z_2}{2}.$$

Remark. Notice that for $z_1 = \alpha$, $z_2 = \beta$, α, β real, the identity shows that the distance between the two points $e^{i\alpha}, e^{i\beta} \in C(0, 1)$ is given by the formula $|e^{i\alpha} - e^{i\beta}| = 2|\sin(\alpha - \beta)/2|$.

4. The functions $\sinh z = (e^z - e^{-z})/2$, $\cosh z = (e^z + e^{-z})/2$ are called the hyperbolic sine and hyperbolic cosine, respectively. Prove these identities ($z = x$

$+ iy)$:

$$\cosh^2 z - \sinh^2 z = 1,$$
$$\sin z = \sin x \cosh y + i \cos x \sinh y,$$
$$\cos z = \cos x \cosh y - i \sin x \sinh y,$$
$$|\sin z|^2 = \sin^2 x + \sinh^2 y,$$
$$|\cos z|^2 = \cos^2 x + \sinh^2 y.$$

5. An expression of the form $\sum_{k=0}^{n} (a_k \cos k\theta + b_k \sin k\theta)$, $a_k, b_k \in \mathbf{R}$, $k = 0, \ldots, n$, and $\theta \in \mathbf{R}$, is called a trigonometric polynomial. Show that it can be written in the complex form $\sum_{-n}^{n} c_k e^{ik\theta}$, where $c_{-k} = \overline{c_k}$.

7.3 The Mapping Given by $w = e^{iy}$

A function $f(z)$ with domain $\mathbf{C}$ is said to be *periodic* with period $c \in \mathbf{C}$ if the equation $f(z + c) = f(z)$ is valid for every $z \in \mathbf{C}$. The following theorem is basic to proving that $\cos z$, $\sin z$, and e^z are periodic, and also to the possibility of interpreting y in (7.2.3) geometrically as an angle.

Theorem 7.3.1 *The function ϕ with values given by $w = e^{iy}$ is continuous for every $y \in \mathbf{R}$. There exists a positive real number p such that the restrictions of ϕ to the intervals $[0, 4p)$ and $(0, 4p]$ are, respectively, univalent on each of these intervals and map each of them onto the unit circle $C(0, 1)$. Furthermore, $e^{i4p} = e^{i0} = 1$.*

Proof. The functions $\phi = \cos + i \sin$, and also cosine and sine, are each the sum of a power series convergent on $\mathbf{C}$, so they are continuous on $\mathbf{C}$ [Theorem 4.2.1] and, therefore, on $\mathbf{R}$. By Corollary 4 of Theorem 7.1.1, $|e^{iy}| = 1$, so the range of ϕ as a function $\mathbf{R} \to \mathbf{C}$ is contained in $C(0, 1)$.

Lemma 1 There is a root of the equation $\cos y = 0$ on the interval $(0, 2)$.

Proof. The example which follows the corollary of Theorem 2.6.3 establishes that $\cos 2 < -1/3$. Now $\cos 0 = 1$, and $\cos x$ is continuous on $[0, 2]$ (in fact, on $\mathbf{R}$), so the lemma follows from Proposition 5.3.2 (the intermediate value theorem).

According to Exercise 3.2.4, since $\cos 0 \neq 0$, there is a least root of $\cos y = 0$ on $(0, 2)$. Let this least root be denoted by p. Clearly, $\cos y > 0$ on $[0, p)$. Also, from $\sin^2 p + \cos^2 p = 1$, we know that $\sin p$ is $+1$ or -1.

The plan of this proof now is to show first that $\phi = \cos + i \sin$ gives a univalent mapping of $\{y: 0 \leq y \leq p\}$ into C_1 $\{w: |w| = 1, \operatorname{Re} w \geq 0, \operatorname{Im} w \geq 0\}$ (the subset of $C(0, 1)$ in the closed first quadrant). The mapping then is extended by the identities for the trigonometric functions into the remainder of $C(0, 1)$. It is finally proved that the mapping is onto $C(0, 1)$.

Lemma 2 ϕ maps $[0, p]$ into C_1, and the mapping is univalent.

Proof. Since $\sin' y = \cos y > 0$, $0 < y < p$, and the function sine is continuous on $[0, p]$, it follows from the mean value theorem (see the remarks after Exercise 6.1.7) that sine is strictly increasing on $[0, p]$. Also, since $\sin 0 = 0$, it must follow that $\sin y > 0$, $y \in (0, p]$. Then $\cos' y = -\sin y < 0$, $y \in (0, p]$ so the function cosine is strictly decreasing on $[0, p]$. By Exercise 5.3.4, both cosine and sine are univalent on $[0, p]$, so ϕ is univalent on $[0, p]$. (According to Exercise 1.11.13, it would have sufficed for the univalence of ϕ if only sine or only cosine were known to be univalent.) The inequalities for cosine and sine clearly show that the mapping $\phi = \cos + i \sin$ sends $[0, p]$ *into* C_1.

We observe in passing that $e^{ip} = \cos p + i \sin p = i$, $\sin p = 1$, $e^{i2p} = i^2 = -1$, $\sin 2p = 0$, $\cos 2p = -1$, $e^{i4p} = (-1)^2 = 1$, $\cos 4p = 1$.

Lemma 3 ϕ maps $[p, 2p]$ into $C_2 = \{w: |w| = 1, \operatorname{Re} w \le 0, \operatorname{Im} w \ge 0\}$ and the mapping is univalent.

Proof. For any real y,

$$\begin{aligned} e^{iy} &= \cos y + i \sin y \\ &= e^{i(y-p)} e^{ip} \\ &= i\,[\cos(y - p) + i \sin(y - p)] \\ &= i \cos(y - p) - \sin(y - p), \end{aligned}$$

so $\cos y = -\sin(y - p)$. By (a), $\sin(y - p)$ is strictly increasing on $[p, 2p]$; therefore, $\cos y$ is strictly decreasing on $[p, 2p]$. Since $\cos p = 0$, it follows that $\cos y < 0$ for $y \in (p, 2p]$. Now $\sin' y = \cos y < 0$ on $(p, 2p)$, so $\sin y$ is strictly decreasing on $[p, 2p]$. Since $\sin 2p = 0$, it follows that $\sin y > 0$ for $y \in [p, 2p)$. These facts prove that the range of $\phi|[p, 2p]$ lies in C_2, and that ϕ is univalent on $[p, 2p]$ [Exercise 1.11.13].

Lemma 4 ϕ maps $[2p, 4p]$ into $C_3 = \{w: |w| = 1, \operatorname{Im} w \le 0\}$ and the mapping is univalent.

Proof. This time we use the identities $e^{iy} = e^{i(y-2p+2p)} + (-1)e^{i(y-2p)}$, $\cos y = -\cos(y - 2p)$, $\sin y = -\sin(y - 2p)$. In Lemmas 2 and 3, we have established that $\cos(y - 2p)$ is strictly decreasing on $[2p, 4p]$. Therefore, $\cos y$ is strictly increasing (and so univalent) on this interval. Also according to Lemmas 2 and 3, $\sin(y - 2p) \ge 0$ on $[2p, 4p]$ so $\sin y \le 0$ on this interval. These facts prove that the range of $\phi|[2p, 4p]$ lies in C_3 and that ϕ is univalent on $[2p, 4p]$.

To summarize, we have now shown that ϕ gives a continuous mapping of $\mathbf{R}$ into $\mathbf{C}$ (and of $[0, 4p]$ into $C(0, 1)$); and gives univalent mappings of $[0, p]$ into $C_1 = C(0, 1) \cap \{w: \operatorname{Re} w \ge 0, \operatorname{Im} w \ge 0\}$; of $[p, 2p]$ into $C_2 =$

$C(0, 1) \cap \{w: \text{Re } w \leq 0, \text{ Im } w \geq 0\}$; and of $[2p, 4p]$ into $C_3 = C(0, 1) \cap \{w: \text{Im } w \leq 0\}$. The only point that C_1 and C_2 have in common is $w = i$, which by the univalence has the unique preimage p in $[0, 2p]$, so ϕ is univalent on $[0, 2p]$. The only points that $C_1 \cup C_2$ and C_3 have in common are $w = -1$ and $w = +1$. The unique preimage of $w = -1$ in $[0, 4p]$ is $2p$, but $1 = \phi(0) = \phi(4p)$. The restriction of ϕ to $[0, 4p)$ is univalent and the range lies in $C(0, 1)$. The same is true of the restriction of ϕ to $(0, 4p]$.

Lemma 5 ϕ maps $[0, 4p)$ and $(0, 4p]$ onto $C(0, 1)$.

Proof. We show that any point $w_0 \in C(0, 1)$ has a preimage y_0 on $[0, 4p)$. Let $w_0 = u_0 + iv_0$. Then $u_0^2 + v_0^2 = 1$, so $-1 \leq u_0 \leq 1$ and $v_0 = \pm\sqrt{1 - u_0^2}$. To be specific, let $v_0 = +\sqrt{1 - u_0^2}$. (This case covers that of any $w_0 \in C(0, 1) \cap \{w: \text{Im } w \geq 0\}$.) The function cosine is continuous on $[0, 4p]$ and $\cos 0 = 1$, $\cos 2p = -1$, $\cos 4p = 1$. If $-1 < u_0 < 1$, by the intermediate value theorem [Proposition 5.3.2], there exists $y_0 \in [0, 2p]$ such that $\cos y_0 = u_0$. (The statement is obviously true if $u_0 = \pm 1$.) We have shown that $\sin y \geq 0$ on $[0, 2p]$, so the identity (7.2.5), $\cos^2 y + \sin^2 y = 1$, here implies that $\sin y_0 = +\sqrt{1 - \cos^2 y_0} = \sqrt{1 - u_0^2} = v_0$. Thus, $\phi(y_0) = \cos y_0 + i \sin y_0 = u_0 + iv_0 = w_0$.

Now we take the case in which $-1 < u_0 < 1$ and $v_0 = -\sqrt{1 - u_0^2}$. There exists $y_0 \in [2p, 4p]$ such that $\cos y_0 = u_0$, and in this case $y_0 \in (2p, 4p)$ because $\cos 2p = -1$, $\cos 4p = 1$. Now $\sin y \leq 0$ on $[2p, 4p]$, so $\sin y_0 = -\sqrt{1 - \cos^2 y_0} = -\sqrt{1 - u_0^2} = v_0$.

Thus, ϕ maps $[0, 4p)$ onto $C(0, 1)$. A slight change in the argument shows that ϕ maps $(0, 4p]$ onto $C(0, 1)$ also. Q.E.D.

Corollary Let y_0 be any real number. The function ϕ with values $w = e^{iy}$, $y \in \mathbf{R}$, when restricted to the interval $(y_0, y_0 + 4p]$, gives a one-to-one mapping of this interval onto $C(0, 1)$. The function is continuous on $(y_0, y_0 + 4p]$.

Proof. According to the theorem, the function $w = e^{i(y-y_0)}$ gives a one-to-one continuous mapping from $(y_0, y_0 + 4p]$ onto $C(0, 1)$. The equation $\zeta = we^{iy_0}$ defines a homeomorphism of $\mathbf{C}$ onto $\mathbf{C}$, and the image of $C(0, 1)$ is $C(0, 1)$. Therefore, the composition function represented by $\zeta = e^{i(y-y_0)}e^{iy_0} = e^{iy}$ gives a one-to-one mapping of $(y_0, y_0 + 4p]$ onto $C(0, 1)$. The continuity statement is established in the theorem. Q.E.D.

The interval $[y_0, y_0 + 4p)$ can be substituted for $(y_0, y_0 + 4p]$ in the corollary.

There are a number of ways in which the number $2p$ can be identified with the transcendental number $\pi = 3.1415926535\ldots$. The simplest approach is to define the number $\pi/2$ as the first zero of $\cos \theta$ on $\{\theta: \theta \geq 0\}$.

This definition causes no conflict with other geometric and analytical definitions of π.

But perhaps a more intuitively satisfying identification is as follows: Historically π was in effect defined geometrically as the area of a closed circular a disk with radius 1. In elementary calculus, it is shown that this area, when represented analytically as the limit of the areas of inscribed rectangles, is given by the integral

$$\pi = 2\int_{-1}^{1} \sqrt{1 - t^2}\, dt = 4\int_{0}^{1} \sqrt{1 - t^2}\, dt.$$

We show that the second integral is equal to the number $p/2$.

The mapping given by $t = \sin y$ is strictly increasing on $[0, p]$ and has a continuous derivative $\cos y$; also $0 = \sin 0$ and $1 = \sin p$. By the change of variables rule in integral calculus and the identity $\cos^2 z = (1 + \cos 2z)/2$ (see Exercise 7.2.2), we have

$$\begin{aligned} \frac{\pi}{4} &= \int_0^1 \sqrt{1 - t^2}\, dt \\ &= \int_0^p \sqrt{1 - \sin^2 y}\, \cos y\, dy \\ &= \int_0^p \cos^2 y\, dy \\ &= \int_0^p \frac{1 + \cos 2y}{2}\, dy \\ &= \frac{y}{2} + \frac{1}{4}\sin 2y \Big]_0^p \\ &= \frac{p}{2} + \frac{1}{4}\sin 2p - 0 = \frac{p}{2} + 0. \end{aligned}$$

We henceforth use π instead of $2p$. The various special values of e^{iy} appearing in the proof of Theorem 7.2.1 now look like this:

$$\begin{array}{lll} \sin\frac{\pi}{2} = 1, & \cos\frac{\pi}{2} = 0, & e^{i\pi/2} = i, \\ \sin \pi = 0, & \cos \pi = -1, & e^{i\pi} = -1, \\ \sin 0 = \sin 2\pi = 0, & \cos 2\pi = \cos 0 = 1, & e^{2\pi i} = 1 = e^0. \end{array}$$

For any integer k, $e^{2\pi ik} = 1$, $e^{i\pi k} = (-1)^k$.

We conclude this section by establishing a useful inequality for the sine function which is proved analytically by methods reminiscent of those used to establish Theorem 7.3.1.

Proposition 7.3.1 *For every* $y \in [0, \pi/2]$, *we have* $2y/\pi \le \sin y \le y$, *with* *equality on the right-hand side only at $y = 0$ and equality on the left-hand side only at $y = \pi/2$.*

Proof. Geometrically, the inequality says simply that the graph of $\sin y$ lies above the chord joining $(0, 0)$ to $(\pi/2, 1)$, and lies below the tangent line at $(0, 0)$. The problem here is to prove the result analytically.

The right-hand side follows from the power series for $\sin y$ and from the alternating series test [Theorem 2.6.3]. It is readily verified that the successive terms of the series for $\sin y$ are alternating in sign and strictly decreasing for $0 \le y < 4$.

For the left-hand side, let $\phi(y) = \sin y - (2y/\pi)$. We have $\phi(0) = 0$, $\phi(\pi/2) = 0$. The problem is to show that $\phi(y) > 0$, $y \in (0, \pi/2)$.

Now $\phi'(y) = \cos y - (2/\pi)$. It was shown in the proof of Lemma 2 for Theorem 7.3.1 that this function is strictly decreasing on $[0, \pi/2]$. It is continuous, and $\phi'(0) = 1 - (2/\pi) > 0$, $\phi'(\pi/2) = -(2/\pi) < 0$. By the intermediate value theorem [Proposition 5.3.2], there exists y_0 such that $\phi'(y_0) = 0$, and by univalence, this root is unique. Thus, $\phi'(y) > 0$ for $y \in [0, y_0)$ and $\phi'(y) < 0$ for $y \in (y_0, \pi/2]$. By the mean value theorem for differential calculus (see Remarks (ii) and (iii) after Exercise 6.1.7), it must follow that $\phi(y)$ is strictly increasing in $[0, y_0)$ and strictly decreasing in $(y_0, \pi/2]$. But then in the light of $\phi(0) = \phi(\pi/2) = 0$, the conclusion to be drawn is that $\phi(y) > 0$, $\sin y > 2y/\pi$, for every $y \in (0, \pi/2)$. Q.E.D.

Exercises 7.3

1. Find the values of $\cos k\pi$, $\sin k\pi$, $\cos(k\pi/2)$, and $\sin(k\pi/2)$, $k = 0, \pm 1, \pm 2, \ldots$.
2. Evaluate $\cos(\pi/4)$, $\sin(\pi/4)$, $\cos(\pi/3)$, and $\sin(\pi/3)$ by using the identities in Sec. 7.2 and the values given in Sec. 7.3.
3. Show that the equation $z = a + re^{i\theta}$ gives a one-to-one mapping of $\{\theta : 0 \le \theta < 2\pi\}$ onto the circle $C(a, r)$.
4. Let $w = e^{2\pi i/n}$, where n is a nonnegative integer. The numbers $1, w, w^2, \ldots, w^{n-1}$ are called the nth roots of unity, because $(w^k)^n = e^{2\pi ikn/n} = e^{2\pi ik} = 1$, $k = 0, 1, \ldots, n-1$. Prove the following:
$$\sum_{j=0}^{n-1} (w^k)^j = \begin{cases} 0, & k \not\equiv 0 \pmod{n}. \\ n, & k \equiv 0 \pmod{n}. \end{cases}$$
5. Prove that $|\sin y| = \sin |y|$ for every $y \in [-\pi, \pi]$.
6. Prove that $\lim_{R \to \infty} R \int_0^{\pi/2} e^{-R^2 \sin y} dy = 0$. (*Hint*: Proposition 7.3.1.)

7.4 The Periods of e^z and of the Trigonometric Functions

First, we deal with the matter of the roots of the equation $e^z = 1$.

Theorem 7.4.1 *The equation $e^z = 1$ is satisfied if and only if $z = 2\pi ik$, $k = 0, \pm 1, \pm 2, \ldots$.*

Proof. If $z = 2\pi ik$, we have already observed that $e^z = 1$. We show that there are no other roots.

Suppose $e^z = e^{x+iy} = 1$, $x, y \in \mathbf{R}$. Then $|e^z| = 1 = e^x$ [Theorem 7.1.1, Corollary 4], and this implies that $x = 0$ [Theorem 7.1.1, Corollary 3]. Therefore, the equation $e^{iy} = 1$ must be true for every z such that $e^z = 1$.

By Theorem 7.3.1 the only solution on $\{y: 0 \le y < 2\pi\}$ is $y = 0$. Now take $y > 2\pi$, but $y \ne 2\pi k$, k any integer. Choose an integer n so that $n < y/2\pi < (n + 1)$, which is equivalent to $2\pi n < y < 2\pi(n + 1)$. Then $e^{iy} = e^{i(y-2\pi n+2\pi n)} = e^{i(y-2\pi n)}e^{i2\pi n} = e^{i(y-2\pi n)}$. Since $0 < y - 2\pi n < 2\pi$, it is impossible for e^{iy} to equal 1, again by Theorem 7.3.1.

To summarize, we have shown that the only nonnegative solutions of $e^{iy} = 1$ are of the form $y = 2\pi k$, $k = 0, 1, 2, \ldots$, and this implies that the only solutions of $e^z = 1$ with nonnegative imaginary parts are $z = 0 + 2\pi ik$, $k = 0, 1, 2, \ldots$.

Now let y be negative; say $y = -t$, $t > 0$. The equation $e^{iy} = 1$ is equivalent to $e^{-it} = 1$, or $1 = e^{it}$. The only solutions with $t > 0$ are $t = 2\pi k$, $k = 1, 2, \ldots$, which correspond to $y = -2\pi k$, $k = 1, 2, \ldots$. We have now shown that the only solutions of $e^z = 1$ are $z = 0 + 2\pi ik$, $k = 0, \pm 1, \pm 2, \ldots$. Q.E.D.

Corollary 1 The function $f(z) = e^z$, $z \in \mathbf{C}$, is periodic, with period $2\pi ik$, where k is any positive or negative integer. There are no other periods. The function $\phi(y) = e^{iy}$, $y \in \mathbf{R}$, is periodic, and the periods are $2\pi k$, $k = \pm 1, \pm 2, \ldots$.

Proof. By Corollary 1 of Theorem 7.1.1, $e^z \ne 0$ for every $z \in \mathbf{C}$. Therefore, the equation $e^{z+c} = e^z$, $e^z e^c = e^z$, is equivalent to $e^c = 1$, and this holds if and only if $c = 2\pi ik$, $k = 0, \pm 1, \pm 2, \ldots$, by the theorem. Q.E.D.

Corollary 2 If $e^{z_1} = e^{z_2}$, then $z_2 = z_1 + 2\pi ik$ for some integer k.

Proof. $e^{z_1} = e^{z_2}$ is equivalent to $e^{z_1 - z_2} = 1$, and Theorem 7.4.1 indicates that $z_1 - z_2 = 2\pi ik$ for some integer k. Q.E.D.

Corollary 3 The only periods of $\cos z$ and $\sin z$ are $2\pi k$, and the only periods of $\tan z$ are πk, $k = 0, \pm 1, \pm 2, \pm 3$.

Proof. It can readily be verified from the definitions of the trigonometric functions that the indicated numbers are indeed periods. There remains a question as to whether they are the only periods. If, for example, the complex number c is a period of $\cos z$, then $\cos(z + c) = \cos z$ for every $z \in \mathbf{C}$, and in particular for $z = 0$. Thus $\cos c = \cos 0 = 1$, which is the same as

$e^{ic} + e^{-ic} = 2$, or $e^{2ic} - 2e^{ic} + 1 = 0$, or $(e^{ic} - 1)^2 = 0$. Thus, $e^{ic} = 1$, and by Theorem 7.4.1, it follows that $ic = 2\pi ik$, $c = 2\pi k$, $k = 0, \pm 1, \pm 2, \ldots$. The proof for $\sin z$ is similar. The proof for $\tan z$ is left as an exercise.

Exercises 7.4

1. Prove Corollary 3 for $\tan z$.
2. Show that the only zeros of $\sin z$ and $\cos z$ in **C** are real ones, and find all of them.
3. Find all the zeros of $1 - \cos z$.

7.5 Logarithm and Argument

Theorem 7.5.1 *The restriction of the function $w = e^z$ to the strip $S = \{z = x + iy: -\infty < x < \infty, y_0 < y \leq y_0 + 2\pi\}$, where $y_0 \in \mathbf{R}$ is arbitrary, gives a one-to-one mapping of S onto $\mathbf{C} \sim \{0\}$.*

Proof. We begin by showing that the function $w = e^z$ so restricted is univalent. Let w_1 be the image of some point z_1 in S, and suppose $e^{z_1} = w_1 = e^{z_2}$. Then by Corollary 2 of Theorem 7.4.1, there is an integer k such that $z_2 = z_1 + 2\pi ik$. Letting $z_1 = x_1 + iy_1$, $z_2 = x_2 + iy_2$, we have the equation $y_2 = y_1 + 2\pi k$. But if $k \neq 0$, clearly y_2 cannot satisfy $y_0 < y_2 \leq y_0 + 2\pi$ at the same time that y_1 satisfies $y_0 < y_1 \leq y_0 + 2\pi$; so k must be zero and $z_1 = z_2$.

So there is a uniquely defined inverse function for $w = e^z$, with z so restricted. The problem now is to show that the domain of the inverse function is all of $C \sim \{0\}$. To do this, we solve the equation $e^z = w$ in steps.

Let $z = x + iy$. With $w \neq 0$, we write $e^x e^{iy} = |w|(w/|w|)$. According to Corollary 3(ii) of Theorem 7.1.1, for each w, $|w| > 0$, there is a (unique) solution $x = x(w)$ of the equation $e^x = |w|$. Now let $\zeta = w/|w|$, $w \neq 0$, and notice that $|\zeta| = 1$. By the corollary of Theorem 7.3.1, there is a (unique) solution $y = y(\zeta)$ of the equation $e^{iy} = \zeta$ in the interval $(y_0, y_0 + 2\pi]$. Thus, for any $w \neq 0$, there exists $z(w) = x(w) + iy(w/|w|)$ in S such that $e^{z(w)} = w$ and $y(w/|w|) \in (y_0, y_0 + 2\pi]$. Thus, the mapping of S into $C \sim \{0\}$ given by $w = e^z$ is not only univalent but also onto. Q.E.D.

Remark (i). *Terminology and notation.* With $w \neq 0$, the solution $x(w)$ of $e^x = |w|$ is written $\ln|w|$ or $\log_e|w|$. Any solution $z(w)$ of the equation $e^z = w$ is called a *logarithm* of w, written $\log w$. Any real solution y of $e^{iy} = w/|w|$ is called an argument of w, written $\arg w$. In this notation, the inverse image of w under the function $w = e^z$, as found in the proof of the theorem, is written $z(w) = x(w) + iy(w/|w|) = \log w = \ln|w| + i \arg w$. The ambiguity of this notation in the absence of a restriction of z to a strip like that indicated in the theorem is discussed in the next two remarks.

Remark (ii). By Corollary 1 of Theorem 7.4.1, for any $w \neq 0$, if $z(w)$ satisfies $e^{z(w)}$

$= w$, then so does $z(w) + 2\pi ik$, $k = \pm 1, \pm 2, \ldots$, and by Corollary 2 of Theorem 7.4.1, these are the only roots of the equation. They constitute the complete set of logarithms of w.

Remark (iii). Again by Corollary 1 of Theorem 7.4.1, for any $w \neq 0$, if $y(w)$ satisfies $e^{iy} = w/|w|$, then so does $y(w) + 2\pi k$, $k = \pm 1, \pm 2, \ldots$, and by Corollary 2 these are the only roots of the equation. They constitute the complete set of arguments of w. The ambiguity in the meaning of log w resides in the imaginary part of log w.

Remark (iv). Our definition of function has required that with each element in the domain there is associated one and only one element in the range. It is inconvenient to insist on single-valuedness of functions in more advanced parts of complex analysis, but for the present we do so. It is then clear from Theorem 7.5.1 and Remarks (i), (ii), and (iii) that for arg w to be regarded as a function, its range must be restricted to a set $E \in \mathbf{R}$ such that if $y \in E$, then $y + 2\pi k \notin E$ for any integer k. With such a restriction on its imaginary part, log w becomes a function.

Remark (v). The *principal argument* of w, written Arg w, is the restriction of the range of arg w to $(-\pi, \pi)$. The correspondingly restricted logarithm of w, Log w $= \ln|w| + i$ Arg w, is called the *principal logarithm* of w.

Remark (vi). Let u be a positive real number. The arguments of u are the solutions y of $e^{iy} = u/|u| = 1$, which are $y = 2\pi k$, $k = 0, \pm 1, \pm 2, \ldots$. It is customary to speak of "the" logarithm of such a number u, and the complex-field interpretation of this is that it means Log $u = \ln u + 0$.

Remark (vii). *Polar representation of complex numbers.* It is apparent from the preceding remarks that any complex number $w \neq 0$ can be represented in the form $w = |w|(w/|w|) = |w|e^{i \operatorname{Arg} w} = |w|\,[\cos(\operatorname{Arg} w) + i \sin(\operatorname{Arg} w)]$. With $w = u + iv$, $u, v \in \mathbf{R}$, we see that $\cos(\operatorname{Arg} w) = u/|w| = u/\sqrt{u^2 + v^2}$ and $\sin(\operatorname{Arg} w) = v/|w| = v/\sqrt{u^2 + v^2}$. Now let $w = u + iv = (u, v)$ be interpreted as a point in $\mathbf{R}^2$. According to the definitions used in elementary trigonometry and analytic geometry, the cosine of the angle θ measured in radians, from the positive u-axis to the half-line joining 0 to w, is given by $u/\sqrt{u^2 + v^2}$; the sine of this angle is $v/\sqrt{u^2 + v^2}$. A pair of "polar coordinates" for w is (r, θ) with $r = \sqrt{u^2 + v^2}$. Thus, we see that if we interpret any w as the angle θ so described, our definition of cosine and sine are compatible with the geometric definitions of the sine and cosine of Arg w, and a pair of polar coordinates for w is $(|w|, \operatorname{Arg} w)$. The representation $w = |w|e^{i \operatorname{Arg} w} = |w| \cos(\operatorname{Arg} w) + i|w| \sin(\operatorname{Arg} w)$ is called the *polar representation* of w.

Since any two arguments of w differ by only an integer multiple of 2π, any determination of the argument can be used in the polar representation. Often when no confusion can result, we use a notation such as $w = re^{i\theta}$, where r means $|w|$, and θ is any argument of w.

Proposition 7.5.1 *With $w_1 \neq 0$, $w_2 \neq 0$, the following congruences are true*:

(i) $\arg w_1 w_2 \equiv \arg w_1 + \arg w_2 \pmod{2\pi}$.
(ii) $\log w_1 w_2 \equiv \log w_1 + \log w_2 \pmod{2\pi i}$.
(iii) $\arg 1/w_1 \equiv -\arg w_1 \pmod{2\pi}$.

(iv) $\log 1/w_1 \equiv -\log w_1 \pmod{2\pi i}$.
(v) $\arg(-w_1) \equiv \pi + \arg w_1 \pmod{2\pi}$.
(vi) $\log(-w_1) \equiv \pi + \log w_1 \pmod{2\pi i}$.

In each case, on both the left- and right-hand sides, arg or log denotes any particular determination.

Proof of (i). The number $\arg w_1w_2$ satisfied the equation $e^{i \arg w_1w_2} = w_1w_2/|w_1w_2| = (w_1/|w_1|)(w_2/|w_2|)$. The numbers $\arg w_1$ and $\arg w_2$ satisfy, respectively, the equations $e^{i \arg w_1} = w_1/|w_1|$, $e^{i \arg w_2} = w_2/|w_2|$. Thus, we have

$$e^{i \arg w_1w_2} = e^{i \arg w_1}e^{i \arg w_2} = e^{i(\arg w_1 + \arg w_2)}$$

and

$$e^{i[\arg w_1w_2 - \arg w_1 - \arg w_2]} = 1.$$

But then by Theorem 7.4.1, for some integer k, $\arg w_1w_2 - \arg w_1 - \arg w_2 = 2\pi k$. Q.E.D.

The remaining proofs are similar and are left as exercises.

Corollary For w_1, $w_2 \neq 0$, with proper choice of the integer m from the possibilities indicated, the following are valid:

(i) $\operatorname{Arg} w_1w_2 = \operatorname{Arg} w_1 + \operatorname{Arg} w_2 + 2m\pi$, $m = 0, \pm 1$.
(ii) $\operatorname{Log} w_1w_2 = \operatorname{Log} w_1 + \operatorname{Log} w_2 + 2m\pi i$, $m = 0, \pm 1$.
(iii) If $\alpha > 0$, $\operatorname{Arg} \alpha w_1 = \operatorname{Arg} w_1$.
(iv) If $\alpha > 0$, $\operatorname{Log} \alpha w_1 = \ln \alpha + \operatorname{Log} w_1$.

The proof again is an exercise.

Examples

(i) Let $w_1 = r_1e^{i3\pi/4}$, $w_2 = r_2e^{i3\pi/4}$, $r_1, r_2 \neq 0$. $\operatorname{Arg} w_1w_2 = \operatorname{Arg}(r_2r_1e^{i6\pi/4}) = -2\pi/4 = \operatorname{Arg} w_1 + \operatorname{Arg} w_2 - 2\pi$. Here m in the corollary (part (i)) to Proposition 7.5.1 is -1.

(ii) The complete set of arguments for $i^2 = -1$ is $\{\pi, \pm 3\pi, \pm 5\pi, \ldots\}$. Look at the values given by Proposition 7.5.1(i), treated as an equation, not a congruence. We would have $\arg i^2 = \arg i \cdot i = \arg i + \arg i = 2 \arg i$. A complete set of arguments for $\arg i$ is $\{\pi/2, 5\pi/2, 9\pi/2, \ldots, -3\pi/2, -7\pi/2, -11\pi/2, \ldots\}$. The corresponding set for $2 \arg i$ is $\{\pi, 5\pi, 9\pi, \ldots, -3\pi, -7\pi, \ldots\}$, and this omits an infinite number of arguments of i^2, for example $3\pi, 7\pi, -5\pi$. The congruences in Proposition 7.5.1 cannot in general be replaced by equations.

Exercises 7.5

1. Evaluate the following in $u + iv$ form, $u, v \in \mathbf{R}$: $\operatorname{Log}(1 + i)$, $\log i$, $\log(-1)$. (*Answer*: $\operatorname{Log}(1 + i) = \ln 2^{1/2} + (\pi/4)$.)

2. Prove Proposition 7.5.1(ii)–(iv).
3. Prove the corollary of Proposition 7.5.1.
4. Show that if $w \neq 0$ and w is not a negative real number, then Arg $\bar{w} = -$Arg w.
5. Find Arg w^3 in terms of $\theta =$ Arg w. (*Hint*: The interval $-\pi < \theta \leq \pi$ must be divided into three subintervals.)
6. Sketch the set $\{w: |w|$ Arg $w = \pi\}$. (*Hint*: Write the equation in polar coordinates.)
7. Let $f(z)$ be analytic in a region B, $f(z) \neq 0$ in B, and suppose that it is known that Arg $f(z) = \alpha$, a constant, for all $z \in B$. Prove that $f(z)$ is a constant function on B. (*Hint*: $f(z) = |f(z)|e^{i\alpha}$, $e^{-i\alpha} f(z) = |f(z)|$, real. But $e^{-i\alpha} f(z)$ is analytic on B if α is a constant.)
8. Use the definition of the natural logarithm given in this section to prove that if α, x are real, $x > 0$, then $\ln x^\alpha = \alpha \ln x$. (Assume the laws of exponents in **R**.)
9. Show that the function $z = i(y_0 + \pi) + \text{Log } w$ gives a one-to-one mapping of $\mathbf{C} \sim \{0\}$, onto the strip S in Theorem 7.5.1.
10. In the integration theory to be developed in succeeding chapters, a point set of the following type appears from time to time: $B = \{z: z = a + sr(t)e^{it},\ 0 \leq t \leq 2\pi,\ 0 \leq s \leq 1\}$, where $r(t)$ is continuous, $r(t) > 0$ for every $t \in [0, 2\pi]$, and $r(0) = r(2\pi)$. Given $z = a + sr(t)e^{it} \in B$, the line segment joining a to z has the equation $w = (1 - \tau)a + \tau[a + sr(t)e^{it}] = a + \tau sr(t)e^{it}$, $0 \leq \tau \leq 1$, so $w \in B$. Thus, B is starlike with respect to a and so is connected. The purpose of this exercise is to establish further that B is open and, therefore, is a region. Without loss of generality, it can be assumed that $a = 0$, because if $B_0 = \{w: w = sr(t)e^{it},\ 0 \leq t \leq 2\pi,\ 0 \leq s \leq 1\}$ is shown to be open, then the transformation $w = z - a$, $z = a + w$, gives a homeomorphism of **C** onto **C** which maps B_0 onto B.
    (a) Extend the definition of $r(t)$ to **R** by periodicity; that is, for every $t \in [0, 2\pi]$ define $r(t + 2\pi k) = r(t)$, $k = \pm 1, \pm 2, \ldots$. Let $B_\infty = \{z: z = sr(t)e^{it}, -\infty < t < \infty, 0 \leq s < 1\}$. Prove that $B_\infty = B$ (now with $a = 0$).
    (b) Take $z_0 \in B_\infty$, $z_0 = s_0 r(t)e^{it_0}$, $0 < s_0 < 1$. Prove by using the continuity of $r(t)$ at t_0 that there exists δ, $0 < \delta < \pi/2$, such that $r(t) > s_0 r(t_0)$ for every t, $t_0 - \delta \leq t \leq t_0 + \delta$. (*Hint*: In the "$\varepsilon$, δ" formulation of continuity, take $\varepsilon = (1 - s_0)r(t_0)$.)
    (c) Let $m = \min\{r(t): t_0 - \delta \leq t \leq t_0 + \delta\}$, and let $S = \{z: z = \sigma m e^{it},\ t_0 - \delta < t < t_0 + \delta,\ 0 < \sigma < 1\}$. Show that $z_0 \in S$ and $S \subset B_\infty \sim \{0\}$. (*Hint*: For $S \subset B_\infty \sim \{0\}$, let $z_1 = \sigma_1 m e^{it_1}$, $t_0 - \delta < t_1 < t_0 + \delta$, $0 < \sigma_1 < 1$; it is necessary only to show that there exists an admissible value of s in the definition of B_∞, $s > 0$, such that $z_1 = sr(t_1)e^{it_1}$.)
    (d) Show that S is an open set in **C**. (*Hint*: This has been anticipated in Exercise 5.2.8, since S is a sector of a circular disk. First, show that the homeomorphism of **C** onto **C** given by $w = ze^{-it_0}$ takes S onto $S' = \{w: w = \sigma m e^{i\tau},\ -\delta < \tau < \delta, 0 < \sigma < 1\}$. Then show that an equivalent definition of S' is $S' = \{w = u + iv: -u \tan \delta < v < u \tan \delta, 0 < u^2 + v^2 < m^2\}$. Thus, Exercise 5.2.8 applies to S'.)

(e) Again using the continuity of $r(t)$, show that there is a disk $\Delta(0, r) \subset B_\infty$, $r > 0$. The proof that B_∞ (and, therefore, B) is open is then complete, because it has been shown that every $z \in B_\infty$ lies in a subset of B_∞ open in **C**.

11. Let the set B be defined as in Exercise 7.5.10, and let Γ be the graph of the arc given by $z = a + r(t)e^{it}, 0 \le t \le 2\pi$. Prove that $B \cup \Gamma$ is bounded and closed and in fact $B \cup \Gamma = \bar{B}$. (*Hint*: The boundedness is trivial, since if $z \in B \cup \Gamma$, then $|z - a| \le r(t)$ and $r(t)$ is continuous. To show that $B \cup \Gamma$ is closed, examine $(B \cup \Gamma)^c = \{z : z = a + sr(t)e^{it}, 0 \le t \le 2\pi, s > 0\}$. The transformation given by $w = 1/(\bar{z} - \bar{a})$ is a homeomorphism of $\mathbf{C} \sim \{a\}$ onto $\mathbf{C} \sim \{0\}$ in which any open set in one of these spaces is mapped onto an open set in the other space; such open sets are also open in **C**. Show that this mapping transforms $(B \cup \Gamma)^c$ into a set $B' \sim \{0\}$ in the w-plane in which B' is of the same general type as B, and so is open by Exercise 7.5.10.)

12. Let S be the circular sector $\{z : z = a + \sigma Re^{it}, t_0 - \delta < t < t_0 + \delta, 0 < \sigma < 1\}$, where $a \in \mathbf{C}$, $R > 0$, $0 < \delta \le \pi/2$. Show that S is a convex region. (*Hint*: The problem can be solved by relating it to Exercise 5.3.13 by a linear transformation, as in the hint for Exercise 7.5.10(d). Try $w = (z - a)e^{-it_0}$.)

7.6 The General Exponential Function

Given $w, z \in \mathbf{C}$, $w \ne 0$, we define w^z by $w^z = e^{z \log w} = e^{z(\ln|w| + i \arg w)}$. It follows that w^z is multivalued because $\arg w$ is multivalued. Some authors (for example, Pennisi [15]) use the convention that w^z is taken to mean $e^{z \operatorname{Log} w}$. A more usual convention, which we adopt here, is that w^z means $e^{z \log w}$ only when w and z are both real, with $w > 0$.

Exercises 7.6

1. Find the value(s) of the following:

(a) i^i. *Answer*: $e^{(-\pi/2) + 2k\pi}$, $k = 0, \pm 1, \pm 2, \ldots$.

(b) $|z^\alpha|$, α real. *Answer*: $|z|^\alpha$.

2. Show that there are only a finite number of different determinations of $w^{1/n}$, where n is a positive integer, $w \in \mathbf{C}$, $w \ne 0$, and that these are $|w|^{1/n}e^{i[(\operatorname{Arg} w) + 2\pi k]/n}$, $k = 0, 1, 2, \ldots, n - 1$.

3. Show that $w^{z_1}w^{z_2} = w^{z_1 + z_2}$ if the same determination of the logarithm of w is used for all three powers.

4. Show that $(w_1 w_2)^z = w_1^z w_2^z$ implies a special equation between the determinations of the logarithms of w_1, w_2, and $w_1 w_2$.

5. Fing $\sup|(1 + z)^i|$, $z \in \Delta(0, 1)$. *Answer*: $e^{\pi/2}$.

6. Show that if principal logarithms are used on each side, then $\overline{z^w} = \bar{z}^{\bar{w}}$, $z \notin (-\infty, 0)$. (*Hint*: Remember that $\overline{e^\zeta} = e^{\bar{\zeta}}$, $\zeta \in \mathbf{C}$.)

7. Prove that the series $\sum_{k=1}^{\infty} 1/k^z$ (using principal logarithms) converges absolutely

for every z such that Re $z > 1$, and converges uniformly in any half-plane $\{z = x + iy : x \geq p > 1\}$.

8. Criticize this apparent paradox: $e^{2\pi i} = 1$ so $(e^{2\pi i})^{1/2\pi i} = 1^{1/2\pi i} = 1$. But also $(e^{2\pi i})^{1/2\pi i} = e$; so $e = 1$.
9. Find Re z^z, Im z^z, $z \in \mathbf{C}$, $z \neq 0$.
10. If α is a positive real number, show that at least one and at most two values of α^α are real when the convention in the last sentence of this section is ignored.
11. Show that the function $w = z^\alpha$, α real, $0 < \alpha < 1$ (principal logarithm), maps $\mathbf{C} \sim (-\infty, 0]$ in a one-to-one manner onto a sector containing the positive real axis and bounded by rays $w = te^{i\alpha\pi}$, $0 < t < \infty$, and $w = te^{-i\alpha\pi}$, $0 < t < \infty$.

7.7 Continuity of Arg and Log; Analyticity of Log

Theorem 7.7.1 Arg w *is continuous at every* $w \in \mathbf{C} \sim (-\infty, 0]$.

Proof. We first show that Arg w is continuous on $C(0, 1) \sim \{-1\}$. First, consider the restriction of Arg w to the "upper half" of this point set, that is, to $C^+ = C(0, 1) \cap \{w: \text{Im } w \geq 0\} \sim \{-1\}$. It is apparent from Lemmas 2 and 3 for Theorem 7.3.1 that for any $\zeta \in C^+$, Arg $\zeta \in [0, \pi)$. By the definition of Arg, and Exercise 7.2.3 and the remark after it, we have for any ζ_0, ζ on $C(0, 1)$,

$$|\zeta - \zeta_0| = |e^{i \text{Arg} \zeta} - e^{i \text{Arg} \zeta_0}| = 2\left|\sin\left(\frac{\text{Arg } \zeta - \text{Arg } \zeta_0}{2}\right)\right|. \tag{7.3.1}$$

Now take ζ_0, $\zeta \in C^+$. Then $|\text{Arg } \zeta - \text{Arg } \zeta_0|/2 \in [0, \pi/2)$, and by using Proposition 7.3.1 and Exercise 7.3.5, we obtain

$$|\zeta - \zeta_0| = 2 \sin\left|\frac{\text{Arg } \zeta - \text{Arg } \zeta_0}{2}\right| \geq 2\left[\frac{2}{\pi}\left|\frac{\text{Arg } \zeta - \text{Arg } \zeta_0}{2}\right|\right]. \tag{7.3.2}$$

Thus,

$$|\text{Arg } \zeta - \text{Arg } \zeta_0| \leq \frac{\pi}{2}|\zeta - \zeta_0|. \tag{7.3.3}$$

From this, it readily follows that $\lim_{\zeta \to \zeta_0} \text{Arg } \zeta = \text{Arg } \zeta_0$, $\zeta \in C^+$.

If now we restrict ζ, ζ_0 to $C^- = C(0, 1) \cap \{w: \text{Im } w \leq 0\} \sim \{-1\}$, then it follows from the corollary of Theorem 7.3.1 (with $y_0 = -\pi$) that Arg ζ, Arg $\zeta_0 \in (-\pi, 0]$. Equations (7.3.1) and inequalities (7.3.2) and (7.3.3), remain valid for this case, and again we find that Arg $\zeta | C^-$ is continuous at $\zeta_0 \in C^-$. The values of Arg $\zeta | C^+$ and Arg $\zeta | C^-$ of course agree at $\zeta = 1$, so Arg ζ is seen to be continuous on $C(0, 1) \sim 1$.

Now for any $w \in \mathbf{C}$, $w \neq 0$, by the corollary (part (iii)) to Proposition

7.5.1, Arg w = Arg $(w/|w|)$. The function $\zeta = w/|w|$ is continuous on $\mathbf{C} \sim (-\infty, 0]$ (indeed, on $\mathbf{C} \sim 0$) [Exercise 3.2.8] and maps this set onto $C(0, 1) \sim \{-1\}$. The function with values Arg ζ has been shown to be continuous on $C(0, 1) \sim \{-1\}$. Thus, the composition of the function Arg ζ with the function $\zeta = w/|w|$ must be continuous on $\mathbf{C} \sim (-\infty, 0]$ [Exercises 5.2.2 and 3.2.10], and this composition function has values Arg $(w/|w|)$ = Arg w. Q.E.D.

Remark. If w_0 lies on $(-\infty, 0)$, the negative real axis, a slight modification of the proof shows that $\lim_{w \to w_0} \text{Arg } w = Arg\ w_0 = \pi$, Im $w > 0$.

Theorem 7.7.2 Log w *is continuous at every* $w \in \mathbf{C} \sim (-\infty, 0]$.

Lemma The function $x = \ln |w|$ is continuous on $C \sim \{0\}$.

Proof. The number $\ln |w|$ is the unique solution $x(w)$ of $e^{x(w)} = |w|$. (See Remark (i) in Section 7.5.) By Corollary 3(ii) of Theorem 7.1.1, the function given by $u = e^x$ is a homeomorphism from $\mathbf{R}$ onto $\{u: u > 0\}$, so the inverse function $x = \ln u$ is continuous on $\{u: u > 0\}$. The function of w, $u = |w|$, is continuous on $\mathbf{C}$. The function from $\mathbf{C} \sim 0$ to $\mathbf{R}$ with values $\ln |w|$ is the composition of $x = \ln u$ with $u = |w|$ and so is continuous on $\mathbf{C} \sim 0$ [Exercises 5.2.2 and 3.2.10].

Theorem 7.7.2 now follows immediately from Theorem 7.7.1 and the lemma, since the real and imaginary parts of Log w so restricted are continuous [Theorem 3.2.1]. Q.E.D.

Corollary The equation $w = e^z$ gives a homeomorphism from $S = \{z = x + iy: -\infty < x < \infty, -\pi < y < \pi\}$ onto $\mathbf{C} \sim (-\infty, 0]$.

Proof. The result follows at once from the continuity of e^z and Theorems 7.5.1 and 7.7.2. Q.E.D.

Theorem 7.7.3 Log w *is analytic on* $\mathbf{C} \sim (-\infty, 0]$, *and at any point* w *in this set*, $(\text{Log } w)' = 1/w$.

Proof. The set S is open in $\mathbf{C}$, so any point in it is an accumulation point. Hence, in the corollary of Theorem 7.7.2, we can apply Theorem 6.1.3 (the inverse function law for differentiation) with $w = f(z) = e^z$, $E = S$, $f(E) = C \sim (-\infty, 0]$ and $f^{-1}(w) = \text{Log } w$. Theorem 6.1.3 assures the existence of $(\text{Log } w)'$ for each $w \in f(E)$ and gives its value as $1/(e^z)'$. However, $(e^z)' = e^z = w$. Q.E.D.

Corollary For a fixed real θ, let $\log_\theta w = i(\theta + \pi) + \text{Log } w$. Then $\log_\theta w$ is analytic on $\mathbf{C} \sim (-\infty, 0]$ and $(\log_\theta w)' = 1/w$.

This follows at once from the laws of differentiation. The mapping given by $z = \log_\theta w$ is described in Exercise 7.5.9.

Exercises 7.7

1. Let $w = z^{\zeta}$, $z, \zeta \in \mathbf{C}$, $z \neq 0$. Use the principal logarithm in defining this function of z and ζ. Show that $\partial w/\partial\zeta = z^{\zeta}$ Log z, and $\partial w/\partial z = \zeta z^{\zeta-1}$.
2. Show that the power series $\sum_1^\infty (-1)^{k+1}(w-1)^k/k$ converges to the function Log w in $\Delta(1, 1)$ by carrying out the following steps.
   (a) Show that the radius of convergence is unity, so the sum is analytic. Call it $L(w)$.
   (b) Show that $L'(w) = 1/w$ by summing the series for $L'(w)$.
   (c) Invoke the corollary of Theorem 6.2.1 (the constant can be determined by letting $w = 1$).
3. Prove that the "binomial series" $f(z) = 1 + \sum_{k=1}^\infty [a(a-1)(a-2)\cdots(a-k+1)/k!]z^k$, $a, z \in \mathbf{C}$, converges on $\Delta(0, 1)$ to the function $(1+z)^a$, with the principal logarithm used in the determination. The suggested steps are as follows:
   (a) According to Exercise 2.6.6, the radius of convergence is at least unity. The ratio test can be used to show that it is exactly unity.
   (b) By differentiating the power series show that $(1+z)f'(z) = af(z)$, so $f'(z) = af(z)/(1+z)$.
   (c) Let $g(z) = f(z)/(1+z)^a$. Use (b) to show that $g'(z) \equiv 0$ in $\Delta(0, 1)$.
   (d) Invoke Theorem 6.2.1, and find the constant by substitution.

8

Complex Line Integrals

8.1 The Riemann Integral of a Complex-valued Function

Let f be a real-valued function continuous on the interval $[\alpha, \beta]$ in **R**. According to Corollary 1 of Theorem 5.5.2, f is bounded on $[\alpha, \beta]$. It is assumed that the reader is familiar with the properties of the Riemann integral $\int_\alpha^\beta f(t)\,dt$ of f over the interval $[\alpha, \beta]$ as developed in the elementary calculus textbooks. Recall in particular that if $F(\tau) = \int_\alpha^\tau f(t)\,dt$, $\alpha \le \tau \le \beta$, then $F'(\tau)$ exists on $[\alpha, \beta]$ and $F'(\tau) \equiv f(\tau)$. (Of course, the derivative is a one-sided derivative at $\tau = \alpha$ and $\tau = \beta$.) It follows that if the function $\phi : [\alpha, \beta] \to R$ is known to have the continuous derivative $\phi' = f$ on $[\alpha, \beta]$, then $\phi(\tau) - \phi(a)$ differs from $F(\tau)$ by at most a constant. (See the lemma for Theorem 6.2.1.) The constant turns out to be 0, so we have the important equation

$$\phi(\tau) - \phi(\alpha) = \int_\alpha^\tau \phi'(t)\,dt, \qquad \alpha \le \tau \le \beta. \tag{8.1.1}$$

The function ϕ is called a *primitive* or *antiderivative* or *indefinite integral* of f. Another fact which is worth reviewing is that if f and g are each continuous on $[\alpha, \beta]$, and if $f(t) \le g(t)$ for every $t \in [\alpha, \beta]$, then

$$\int_\alpha^\beta f(t)\,dt \le \int_\alpha^\beta g(t)\,dt. \tag{8.1.2}$$

Now let f be a complex-valued function continuous on the real interval $[\alpha, \beta]$, and let $f(t) = u(t) + iv(t)$ where $u(t)$ and $v(t)$ are real. Both u and v are continuous on $[\alpha, \beta]$ according to Theorem 3.2.1. We *define* the Riemann integral of f over the interval $[\alpha, \beta]$ by the equation

$$\int_\alpha^\beta f(t)\,dt = \int_\alpha^\beta u(t)\,dt + i\int_\alpha^\beta v(t)\,dt. \tag{8.1.3}$$

Notice that Re $\int_\alpha^\beta f(t)\,dt = \int_\alpha^\beta \operatorname{Re} f(t)\,dt$ and Im $\int_\alpha^\beta f(t)\,dt = \int_\alpha^\beta \operatorname{Im} f(t)\,dt$. This integral has most of the properties of the Riemann integral of a real-valued function. A few of them follow.

Proposition 8.1.1 *Let f and g be complex-valued functions continuous on $[\alpha, \beta] \subset R$.*

(a) $\int_\alpha^\beta cf(t)\,dt = c\int_\alpha^\beta f(t)\,dt$, *for any complex number c.*

(b) $\int_\alpha^\beta [f(t) + g(t)]\,dt = \int_\alpha^\beta f(t)\,dt + \int_\alpha^\beta g(t)\,dt$.

(c) *If $\alpha = t_0 < t_1 < \cdots < t_n = \beta$, then $\int_\alpha^\beta f(t)\,dt = \sum_{n=1}^k \int_{t_{k-1}}^{t_k} f(t)\,dt$.*

(d) *If f is continuous on $[\alpha, \beta]$, then the function $F(\tau) = \int_\alpha^\tau f(t)\,dt$, $\alpha \le \tau \le \beta$, has a continous derivative $F'(\tau) = f(\tau)$ on $[\alpha, \beta]$.*

(e) *If the derivative f' of f exists and is a continuous function on $[\alpha, \beta]$, then $f(\tau) - f(\alpha) = \int_\alpha^\tau f'(t)\,dt$, $\alpha \le \tau \le \beta$.*

(f) $\left|\int_\alpha^\beta f(t)\,dt\right| \le \int_\alpha^\beta |f(t)|\,dt$.

We defer the proofs of (a) through (e) to the exercises. There is a little more to (f) than meets the eye, because the result is not obtainable by simply applying the triangle inequality and (8.1.2) to the definition (8.1.3). That technique yields the looser inequality $\left|\int_\alpha^\beta f(t)\,dt\right| \le \int_\alpha^\beta (|u(t)| + |v(t)|)\,dt$.

If the integral on the left in (f) is zero, then there is nothing to prove. Notice that for any complex $z \ne 0$, $|z| = ze^{-i\operatorname{Arg} z}$, a real number. Now suppose that the left-hand side in (f) is not zero. Let $\theta = \operatorname{Arg}\int_\alpha^\beta f(t)\,dt$. Then by part (a),

$$\begin{aligned}\left|\int_\alpha^\beta f(t)\,dt\right| &= e^{-i\theta}\int_\alpha^\beta f(t)\,dt\\ &= \int_\alpha^\beta e^{-i\theta} f(t)\,dt\\ &= \int_\alpha^\beta \operatorname{Re}\,[e^{-i\theta}f(t)]\,dt + i\int_\alpha^\beta \operatorname{Im}\,[e^{-i\theta}f(t)]\,dt.\end{aligned}$$

But the first member of this continued equality is a real number, so all members are real numbers. By using (8.2.1) in conjunction with the fact that for any complex z, Re $z \le |z|$, we obtain

$$\left|\int_\alpha^\beta f(t)\,dt\right| = \int_\alpha^\beta \operatorname{Re}\,[e^{-i\theta}f(t)]\,dt + i\cdot 0 \le \int_\alpha^\beta |e^{-i\theta}|\,|f(t)|\,dt = \int_\alpha^\beta |f(t)|\,dt.$$

Q.E.D.

Exercises 8.1

1. Show that $\int_0^\pi e^{it}\,dt = 2i$, $\int_0^\pi |e^{it}|\,dt = \pi$, and $\int_0^\pi (|\operatorname{Re} e^{it}| + |\operatorname{Im} e^{it}|)\,dt = 4$.

2. Prove Proposition 8.1.1(a) through (e) using the corresponding results for integrals of real-valued functions.

8.2 Complex Line Integrals

First, recall from Sec. 5.3 that an *arc* in a metric space X is a continuous mapping γ of a closed interval $[\alpha, \beta]$ in $\mathbf{R}$ into X. The range of γ is the *graph* of the arc. The graph is a compact set in X [Sec. 5.5, Example (ii)]. The point $\gamma(\alpha)$ is the *initial point* and $\gamma(\beta)$ is the *terminal point* or *endpoint*. The arc is said to join $\gamma(\alpha)$ to $\gamma(\beta)$.

The ordering of reals on $[\alpha, \beta]$ induces a kind of ordering of points on the graph Γ of the arc $\gamma: [\alpha, \beta] \to X$, which can be precisely described by the fact that for every $t_1, t_2, \alpha \le t_1 \le t_2 \le \beta$, the graph of the subarc $\gamma|[\alpha, t_1]$ is a subset of the graph of the subarc $\gamma|[\alpha, t_2]$. The ordering on Γ is called the orientation of the arc; this concept is important for complex integration theory.

The case of a constant arc or point arc, $\gamma: [\alpha, \beta] \to a$, where a is a point in X, is not excluded.

Traditionally, the graph of an arc has been verbally identified with the arc itself. We use this loose terminology in later sections when no confusion can result.

The equation $x = \gamma(t)$, $\alpha \le t \le \beta$, is called the parametric equation of the arc γ. Given an arc γ, let g be a one-to-one continuous mapping of an interval $[\alpha_1, \beta_1]$ onto $[\alpha, \beta]$ such that $g(\alpha_1) = \alpha$, $g(\beta_1) = \beta$. (The implication is that g is strictly increasing.) The mapping given by the function $\gamma_1 = \gamma \circ g$ with values $\gamma_1(\tau) = \gamma[g(\tau)]$, $\alpha_1 \le \tau \le \beta_1$, is again an arc. It has the same graph as that of γ; its initial and terminal points are, respectively, the same points in X as those of γ, and the ordering of points on the graph remains the same. The new arc is said to be obtained from the old one by a *change of parameter*. A set of arcs, each of which is obtainable from any other by a change of parameter, is viewed as an equivalence class of arcs. When the word "arc" is used loosely in the geometric sense identified in the preceding paragraph, the various parametric equations of the members of the equivalence class are each said to "represent the same arc."

If the arc $\gamma: [\alpha, \beta] \to X$ is such that $\gamma(\alpha) = \gamma(\beta)$, the arc is called a *closed curve*.

Henceforth we let $X = \mathbf{C}$.

We now define the line integral of a complex-valued function along an arc $\gamma: [\alpha, \beta] \to \mathbf{C}$ with graph Γ. Let f be a complex-valued function with domain Γ. Let $\pi = \{\alpha = t_0 < t_1 < \cdots < t_n = \beta\}$ be a partition of $[\alpha, \beta]$ and let $U = U_\pi = \langle u_k \rangle_{n=1}^{k}$ be a sequence of real numbers such that $t_0 \le u_1 \le t_1 \le u_2 \le t_2 \le \cdots \le t_{n-1} \le u_n \le t_n$. Such a sequence U for a given

π is called π-admitted. Let $\|\pi\| = \max(t_k - t_{k-1})$, $k = 1, \ldots, n$. Finally, let

$$S(f; \gamma; \pi; U) = \sum_{k=1}^{\infty} f[\gamma(u_k)][\gamma(t_k) - \gamma(t_{k-1})].$$

If there exists a complex number $I(f; \gamma)$ with the property that for any $\varepsilon > 0$ there exists $\delta = \delta_\varepsilon > 0$ such that for every π and U_π with $\|\pi\| < \delta$, the inequality $|S(f; \gamma; \pi; U) - I(f; \gamma)| < \varepsilon$ is satisfied, then $I(f; \gamma)$ is called the *complex line integral of f along* γ.

We often shorten this nomenclature when the meaning is clear; for example we might call I the integral of f over Γ. The classical notation for I is $\int_\Gamma f(z)\,dz$, but this notation is ambiguous because it does not indicate the orientation of Γ. We remedy this by using a notation such as $\int_\gamma f(z)\,dz$ when the function γ is adequately specified.

The integral which we define is a complex version of what is known in real analysis as the Riemann–Stieltjes integral. A fairly complete treatment of integrals of this type is found in the appendix of [7]. The present discussion is confined to features essential for complex analysis.

Notice that if γ is a point arc, $I(f; \gamma)$ exists for any f and equals zero.

Proposition 8.2.1 *Let the complex line integral $I(f; \gamma)$ exist. Let γ_1 be any arc obtained from γ by a change of parameter. Then the complex line integral $I(f; \gamma_1)$ of f along γ_1 exists and equals $I(f; \gamma)$.*

This is a version of the change-of-variable formula for Riemann–Stieltjes integration.

Proof. Let the domain of γ be $[\alpha, \beta]$; let the change of parameter be given by the strictly increasing continuous function g mapping $[\alpha_1, \beta_1]$ onto $[\alpha, \beta]$, with $g(\alpha_1) = \alpha$, $g(\beta_1) = \beta$. Then $\gamma_1 = \gamma \circ g$. Let $\pi_1 = \{\alpha_1 = \tau_0 < \tau_1 < \cdots < \tau_n = \beta\}$ be an arbitrary partition of $[\alpha_1, \beta_1]$ and let $U_1 = \langle v_k \rangle_{k=1}^n$ be π_1-admitted. Let $t_k = g(\tau_k)$, $k = 0, 1, \ldots, n$ and $u_k = g(v_k)$, $k = 1, \ldots, n$. The monotonicity of g implies that $\alpha = t_0 < t_1 < t_2 < \cdots < t_n = \beta$ and $t_0 \le u_1 \le t_1 \le u_2 \le t_2 \le \cdots \le t_{n-1} \le u_n \le t_n$, so $\pi = \{\alpha = t_0 < t_1 < \cdots < t_n = \beta\}$ is a partition of $[\alpha, \beta]$ and $U = \langle u_k \rangle_{k=1}^n$ is π-admitted. We have the equation

$$\begin{aligned} S(f; \gamma_1; \pi_1; U_1) &= \sum_{k=1}^{n} f[\gamma_1(v_k)][\gamma_1(\tau_k) - \gamma_1(\tau_{k-1})] \\ &= \sum_{k=1}^{n} f[\gamma(g(v_k))][\gamma[g(\tau_k)] - \gamma[g(\tau_{k-1})]] \\ &= \sum_{k=1}^{n} f[\gamma(u_k)][\gamma(t_k) - \gamma(t_{k-1})] \\ &= S(f; \gamma; \pi; U). \end{aligned} \tag{8.2.1}$$

Now g is uniformly continuous [Theorem 5.5.3], so given any $\delta > 0$, there exists η_δ such that if $\|\pi_1\| < \eta_\delta$, then $\|\pi\| < \delta$. The integral $I(f; \gamma)$ is a number such that for any $\varepsilon > 0$, there exists $\delta > 0$ with the property that

$$|S(f; \gamma; \pi; V) - I(f; \gamma)| < \varepsilon \tag{8.2.2}$$

whenever $\|\pi\| < \delta$. Now take $\varepsilon > 0$ and let η_δ be the number described in the first sentence of this paragraph. For any partition π_1 of $[\alpha_1, \beta_1]$ with $\|\pi_1\| < \eta_\delta$, π of $[\alpha, \beta]$ satisfies $\|\pi\| < \delta$, and then (8.2.2) is satisfied. By using (8.2.1), we can replace $S(f; \gamma; \pi; U)$ in (8.2.2) by $S(f; \gamma_1; \pi_1; U_1)$, and so $I(f; \gamma)$ is at the same time the complex line integral of f along γ_1. Q.E.D.

Proposition 8.2.2 *Let γ be an arc with graph Γ, and let f, f_1, and f_2 be functions from Γ into* **C**.

(a) *If $\int_\gamma f_1(z)\,dz$ and $\int_\gamma f_2(z)\,dz$ both exist, then $\int_\gamma (f_1(z)+f_2(z))\,dz$ exists and equals $\int_\gamma f_1(z)\,dz + \int_\gamma f_2(z)\,dz$.*
(b) *If $\int_\gamma f(z)\,dz$ exists and c is any complex constant, then $\int_\gamma cf(z)\,dz$ exists and equals $c\int_\gamma f(z)\,dz$.*

The proof is an exercise.

Given arcs $\gamma_1: [\alpha, \beta] \to \mathbf{C}$ and $\gamma_2: [\beta, \delta] \to \mathbf{C}$, with $\gamma_1(\beta) = \gamma_2(\beta)$ (so the endpoint of γ_1 coincides with the initial point of γ_2), the parametric equation

$$z = \gamma(t) = \begin{cases} \gamma_1(t), & \alpha \le t \le \beta, \\ \gamma_2(t), & \beta \le t \le \delta, \end{cases}$$

again represents an arc of which the graph Γ is the union of the graphs of γ_1 and of γ_2. We call γ the *sum* of γ_1 and γ_2, and write symbolically $\gamma = \gamma_1 + \gamma_2$. (Some authors call γ the *product arc* of γ_1 and γ_2; see, e.g., [8, p. 49].) The extension of the concept to the sum of more than two arcs is obvious. The polygonal type of arc introduced in Section 5.3 is an example of such a sum, in which the graphs of the subarcs are line segments.

Proposition 8.2.3 *Let Γ_1 and Γ_2 be the graphs, respectively, of arcs $\gamma_1: [\alpha, \beta] \to$ **C** and $\gamma_2: [\beta, \delta] \to$ **C**, with $\gamma_1(\beta) = \gamma_2(\beta)$. Let $\gamma = \gamma_1 + \gamma_2$. Let $\Gamma = \Gamma_1 \cup \Gamma_2$ and let $f: \Gamma \to$ **C** be bounded. If $\int_{\gamma_1} f(z)\,dz$ and $\int_{\gamma_2} f(z)\,dz$ both exist, then $\int_{\gamma_1+\gamma_2} f(z)\,dz$ exists and equals $\int_{\gamma_1} f(z)\,dz + \int_{\gamma_2} f(z)\,dz$.*

The proof is an exercise.

Given the arc $\gamma: [\alpha, \beta] \to \mathbf{C}$, the *opposite arc* $-\gamma$ is the arc represented by the equation $z = \gamma(2\beta - t)$, $\beta \le t \le 2\beta - \alpha$. It has the same graph as γ, but the orientation is reversed, and the initial and terminal points are interchanged.

Proposition 8.2.4 *If the arc γ, with graph Γ, and the function $f: \Gamma \to$ **C**, are such that $\int_\gamma f(z)\,dz$ exists, then so does $\int_{-\gamma} f(z)\,dz$ and $\int_\gamma f(z)\,dz = -\int_{-\gamma} f(z)\,dz$.*

The proof is an exercise. Notice that $\int_{\gamma-\gamma} f(z)\,dz$ (defined as $\int_{\gamma+(-\gamma)} f(z)\,dz$) has the value 0.

Exercises 8.2

1. Let $\gamma:[\alpha, \beta] \to \mathbf{C}$ be any arc. Show that for any complex number $c, \int_\gamma c\,dz = [\gamma(\beta) - \gamma(a)]c$.
2. Let $\gamma:[\alpha, \beta] \to \mathbf{C}$ be an arc such that $I = \int_\gamma z\,dz$ exists. Show that $\int_\gamma z\,dz = ([\gamma(\beta)]^2 - [\gamma(\alpha)^2)/2$. (*Hint*: Let $\pi = \{\alpha = t_0 < t_1 < \cdots < t_n = \beta\}$ be an arbitrary partition of $[\alpha, \beta]$, and let $U_1 = \langle t_{k-1}\rangle_{k=1}^n$ and $U_2 = \langle t_k\rangle_{k=1}^n$. Verify that with $f(z) = z$, $S(f;\gamma;\pi;U_1) + S(f;\gamma;\pi;U_2) = [\gamma(\beta)]^2 - [\gamma(\alpha)]^2$. Then use the definition of I.)
3. Prove Proposition 8.2.2.
4. Prove Proposition 8.2.3.
5. Prove Proposition 8.2.4. (*Hint*: The proof is something like that of Proposition 8.2.1, but easier. Let the equation of γ be $z = \gamma(t)$, $\alpha \le t \le \beta$, and that of $-\gamma$ be $z = \gamma(2\beta - \tau)$, $\beta \le \tau \le 2\beta - \alpha$. Let $\pi_1 = \{\beta = \tau_0 < \tau_1 < \cdots < \tau_n = 2\beta - \alpha\}$, and let $U_1 = \langle v_k\rangle_1^n$ be π_1-admitted. Set $u_k = 2\beta - v_{n-k+1}$, $k = 1,2,\ldots,n$ and $t_k = 2\beta - \tau_{n-k}$, $k = 0,1,2,\ldots,n$. Then $\pi = \{\alpha = t_0 < t_1 < \cdots < t_n = \beta\}$, a partition of $[\alpha, \beta]$, and $U = \langle u_k\rangle_1^n$ is π-admitted.)
6. Given an arc $\gamma:[\alpha, \beta] \to \mathbf{C}$ and a partition $\pi = \{\alpha = t_0 < t_1 < \cdots < t_n = \beta\}$, the variation of γ over π is defined to be $V(\gamma;\pi) = \sum_{k=1}^n |\gamma(t_k) - \gamma(t_{k-1})|$. The *total variation* of γ, finite or infinite, is defined by $T(\gamma) = \sup\{V(\gamma;\pi)\}$ taken over all partitions π of $[\alpha, \beta]$. If $T(\gamma) < \infty$, the arc is said to be *rectifiable* and $T(\gamma)$ is called its *length*. Let Γ be the graph of the retifiable arc γ, let $f:\Gamma \to \mathbf{C}$ be bounded, and let $M > 0$ be such that $|f(z)| \le M$ on Γ. Show that if $I = \int_\gamma f(z)\,dz$ exists, then $|I| \le MT(\gamma)$. (*Hint*: In the notation of this section, $|I| \le |I - S(f;\gamma;\pi;U)| + |S(f;\gamma;\pi;U)|$.)
7. Given a rectifiable arc $\gamma:[\alpha, \beta] \to \mathbf{C}$ with length $T(\gamma)$; let $\gamma_1[\alpha_1, \beta_1] \to \mathbf{C}$ be obtained from γ by a change of parameter. Then γ_1 is rectifiable and $T(\gamma_1) = T(\gamma)$. (*Hint*: Proceeding as in the proof of Proposition 8.2.1, establish a one-to-one correspondence between partitions of $[\alpha_1,\beta_1]$ and of $[\alpha,\beta]$. With $\pi_1 \leftrightarrow \pi$, for any π_1, $V(\gamma_1, \pi_1) = V(\gamma, \pi) \le T(\gamma)$. Take $\varepsilon > 0$, and choose π so that $V(\gamma, \pi) > T(\gamma) - \varepsilon$. Use these inequalities to locate $T(\gamma_1)$ with respect to $T(\gamma)$.)

8.3 Integration over Paths

To insure the existence of the complex line integral defined in the preceding section, certain regularity conditions must be imposed on the functions involved. Perhaps the most aesthetically satisfying of the various existence theorems is the following complex counterpart of the basic existence theorem for Riemann–Stieltjes integrals.

Theorem 8.3.1 *Let the arc $\gamma: [\alpha, \beta] \to \mathbf{C}$, with graph Γ, be rectifiable, and let $f: \Gamma \to \mathbf{C}$ be continuous. Then $\int_\gamma f(z)\,dz$ exists.*

The definition of rectifiable appears in Exercise 8.2.6. The theorem plays a relatively minor role in complex function theory and no role at all in the elementary parts of the theory and in the applications. Therefore, we shall not give a proof here; proofs are found, for instance, in the appendix of [7] and in [8, p. 98].

For the purposes at hand, the most useful regularity conditions retain the continuity of f on γ but replace rectifiability of γ by a differentiability property which is specified in the following definition. An arc $\gamma: [\alpha, \beta] \to \mathbf{C}$ is called a *path* if there is a partition $\{\alpha = s_0 < s_1 < \cdots < s_m = \beta\}$ such that the restriction of γ to each subinterval $[s_{j-1}, s_j]$ has a continuous derivative. At the points $s_1, s_2, \ldots, s_{m-1}$ the right- and left-hand derivatives of γ may differ. If $\gamma(\alpha) = \gamma(\beta)$, the arc is called a closed path.

A *change of parameter* for such a path is a function g which gives a one-to-one continuous mapping of an interval $[\alpha_1, \beta_1]$ onto $[\alpha, \beta]$, with $g(\alpha_1) = \alpha$, $g(\beta_1) = \beta$, and with the further property that the derivative function $g'(r) > 0$ exists on $[\alpha_1, \beta_1]$ and is continuous. The "reparametrized" arc $\gamma_1: [\alpha_1, \beta_1] \to \mathbf{C}$ has a parametric equation $z = \gamma_1(\tau) = \gamma[g(\tau)]$, $\alpha_1 \leq \tau \leq \beta_1$. By the chain law for differentiation, this arc is again a path.

Our use of the word "path" to indicate a piecewise continuously differentiable arc is not a standard practice in the literature of complex analysis. The fact is that there are no universally accepted terms to indicate degrees of regularity for arcs and closed curves. Some of the various terms which have been used to indicate that an arc or closed curve is what we here call a path or closed path are "contour," piecewise differentiable arc," "smooth arc," "smooth curve," "piecewise smooth arc," "regular curve"; but these terms have also been used to indicate other types of regularity. The only way to be sure of what an author means by such an expression is to consult his definition.

Our basic existence theorem follows.

Theorem 8.3.2 *Let $\gamma: [\alpha, \beta] \to \mathbf{C}$ be a path with graph Γ, and let the function $f: \Gamma \to \mathbf{C}$ be continuous. Then the complex line integral of f along γ exists and is given by the formula*

$$\int_\gamma f(z)\, dz = \int_\alpha^\beta f[\gamma(t)]\gamma'(t)\, dt. \tag{8.3.1}$$

Remark. Since γ' may have two different values at a point of discontinuity, the second integral in (8.3.1) requires an explanation. Let the discontinuities of γ' be at points $s_1 < s_2 < \cdots < s_{m-1}$ on $[\alpha, \beta]$, and let $s_0 = \alpha$, $s_m = \beta$. The integral is to be interpreted as follows:

$$\int_\alpha^\beta f[\gamma(t)]\gamma'(t)dt = \sum_{j=1}^{m} \int_{s_{j-1}}^{s_j} f[\gamma(t)]\gamma'(t)dt. \tag{8.3.2}$$

Proof. We first assume that γ' is continuous on the entire interval $[\alpha, \beta]$.

The function with values $f[\gamma(t)]$ is the composition of a continuous function with a continuous function and, therefore, is continuous on $[\alpha, \beta]$. Let $J = \int_\alpha^\beta f[\gamma(t)]\gamma'(t)\,dt$ and $L = \int_\alpha^\beta |\gamma'(t)|\,dt$; the integrands are continuous, so these Riemann integrals surely exist. Let $\pi = \{\alpha = t_0 < t_1 < \cdots < t_n = \beta\}$, and let $U = \langle u_k \rangle_{k=1}^n$ be π-admitted. By Proposition 8.1.1(c), $J = \sum_{k=1}^n \int_{t_{k-1}}^{t_k} f[\gamma(t)]\gamma'(t)\,dt$. By Proposition 8.1.1(d) and (a),

$$f[\gamma(u_k)][\gamma(t_k) - \gamma(t_{k-1})] = f[\gamma(u_k)] \int_{t_{k-1}}^{t_k} \gamma'(t)\,dt = \int_{t_{k-1}}^{t_k} f[\gamma(u_k)]\gamma'(t)\,dt.$$

Then

$$S = S(f; \gamma; \pi; U) = \sum_{k=1}^n f[\gamma(u_k)][\gamma(t_k) - \gamma(t_{k-1})] = \sum_{k=1}^n \int_{t_{k-1}}^{t_k} f[\gamma(u_k)]\gamma'(t)\,dt;$$

$$|S - J| = \left| \sum_{k=1}^n \left[\int_{t_{k-1}}^{t_k} f[\gamma(u_k)]\gamma'(t)\,dt - \int_{t_{k-1}}^{t_k} f[\gamma(t)]\gamma'(t)\,dt \right] \right| = \left| \sum_{k=1}^n \int_{t_{k-1}}^{t_k} [f[\gamma(u_k)] - f[\gamma(t)]]\,\gamma'(t)\,dt \right| \le \sum_{k=1}^n \int_{t_{k-1}}^{t_k} |f[\gamma(u_k)] - f[\gamma(t)]|\,|\gamma'(t)|\,dt.$$

The inequality follows from the triangle inequality and Proposition 8.1.1(f).

Now $f \circ \gamma$ is uniformly continuous on $[\alpha, \beta]$ [Exercise 5.5.2], so for any $\varepsilon > 0$, there exists $\delta_\varepsilon > 0$ such that if $|t'' - t'| < \delta_\varepsilon$, then $|f[\gamma(t'')] - f[\gamma(t')]| < \varepsilon/L$, where $\alpha \le t', t'' \le \beta$. If $\|\pi\| < \delta_\varepsilon$, then $|u_k - t| < \delta_\varepsilon$ for every $t \in [t_{k-1}, t_k]$. Thus, if $\|\pi\| < \delta_\varepsilon$,

$$|S - J| \le \sum_{k=1}^n \int_{t_{k-1}}^{t_k} \frac{\varepsilon}{L} |\gamma'(t)|\,dt = \frac{\varepsilon}{L} \sum_{k=1}^n \int_{t_{k-1}}^{t_k} |\gamma'(t)|\,dt = \frac{\varepsilon}{L} \int_\alpha^\beta |\gamma'(t)|\,dt = \frac{\varepsilon}{L} \cdot L = \varepsilon.$$

Thus, $\int_\gamma f(z)\,dz$ exists and equals J, by our definition of the complex line integral.

To establish the theorem for a general path $\gamma: [\alpha, \beta] \to \mathbf{C}$, as usual let the discontinuities of γ' be at the points $s_1 < s_2 < \cdots < s_{m-1}$, and let $\alpha = s_0 < s_1$, $s_{m-1} < s_m = \beta$. Let γ_j be the restriction of γ to $[s_{j-1}, s_j]$. Then the foregoing proof shows that $\int_{\gamma_j} f(z)\, dz = \int_{s_{j-1}}^{s_j} f[\gamma(t)]\gamma'(t)\, dt$. Since $\gamma = \gamma = \gamma_1 + \gamma_2 + \cdots + \gamma_m$, by Proposition 8.2.1 we have $\int_\gamma f(z)\, dz = \sum_1^n \int_{s_{j-1}}^{s_j} f[\gamma(t)]\gamma'(t)\, dt$, and reference to (8.3.2) completes the proof. Q.E.D.

Our first corollary contains an estimate which we use repeatedly in the sequel. First, notice that if Γ is the graph of an arc and if the function $f\colon \Gamma \to C$ is continuous, then Γ is compact [Sec. 5.5, Example (ii)] and $|f(z)|$ has a maximum value on Γ [Theorem 5.5.2, Corollary 1.]

Corollary 1 Let $\gamma\colon [\alpha, \beta] \to C$ be a path with graph Γ, and let $f\colon \Gamma \to C$ be continuous. Then

$$\left|\int_\gamma f(z)\, dz\right| \le \int_\alpha^\beta |f[\gamma(t)]|\, |\gamma'(t)|\, dt$$
$$\le \max_{z\in\Gamma} |f(z)|L, \tag{8.3.3}$$

when $L = L(\gamma) = \int_\alpha^\beta |\gamma'(t)|\, dt$.

Proof. The integrals involving γ' in this corollary are to be interpreted, as in (8.3.2), as the sum of m integrals over the respective subintervals $[s_{j-1}, s_j]$, $j = 1, \ldots, m$. We have, using Proposition 8.1.1(f) and (8.1.2):

$$\begin{aligned}\left|\int_\gamma f(z)\, dz\right| &= \left|\sum_{j=1}^m \int_{s_{j-1}}^{s_j} f[\gamma(t)]\gamma'(t)\, dt\right| \\ &\le \sum_{j=1}^m \int_{s_{j-1}}^{s_j} |f[\gamma(t)]|\, |\gamma'(t)|\, dt \\ &\le \sum_{j=1}^m \int_{s_{j-1}}^{s_j} \max_{z\in\Gamma} |f(z)|\, |\gamma'(t)|\, dt \\ &= \max_{z\in\Gamma} |f(z)| \sum_{j=1}^m \int_{s_{j-1}}^{s_j} |\gamma'(t)|\, dt \\ &= \max_{z\in\Gamma} |f(z)|L(\gamma).\end{aligned}$$

Q.E.D.

Corollary 2 Let γ be a path with graph Γ and let $f\colon \Gamma \to \mathbf{C}$ be continuous on Γ. Let $\gamma_1\colon [\alpha_1, \beta_1] \to \mathbf{C}$ be a path obtained from γ by a change of parameter. Then

$$\int_\gamma f(z)\, dz = \int_{\gamma_1} f(z)\, dz = \int_{\alpha_1}^{\beta_1} f[\gamma_1(\tau)]\gamma_1'(\tau)\, d\tau. \tag{8.3.4}$$

Proof. Proposition 8.2.1 and Theorem 8.3.2 establish the existence and equality of the first two integrals in (8.3.4). As noted previously, a change of parameter for a path produces another path with the same graph, so the second equality in (8.3.4) follows from the theorem. Q.E.D.

Remark (i). Under the hypotheses of the theorem, the first integral in (8.3.4) exists. Then by Proposition 8.2.1, the second integral would exist and equal the first one if the arc γ_1 were obtained from γ by *any* type of change of parameter as described in Sec. 8.2, whether or not the change function satisfied the condition on its derivative specified above for a change of parameter for a path. The reason for imposing this regularity condition on the change function in Corollary 2 is simply to insure that γ_1 is itself a path.

Remark (ii). It is possible to show that a path γ is rectifiable and that the number $L(\gamma)$, which appears in Corollary 1, is its length. (See Exercise 8.2.6 for this terminology. For a proof of these statements, see for instance [7, p. 299].) Thus, Corollary 1 was anticipated in Exercise 8.2.6. According to Exercise 8.2.7, $L(\gamma)$ is invariant under a change of parameter.

Examples

(i) The graph of the path γ given by the parametric equation $z = \gamma(t) = a + t(b - a)$, $0 \le t \le 1$, $a, b \in \mathbf{C}$, is a line segment joining a to b, as was previously noted in Sec. 5.3. The change of parameter given by $t = (\tau - \alpha)/(\beta - \alpha)$, $\alpha \le \tau \le \beta$, produces an equivalent arc with parametric equation $z = a + [(\tau - \alpha)/(\beta - \alpha)](b - a)$, $\alpha \le \tau \le \beta$. We denote such an arc and all its equivalent arcs and also its graph by the one symbol $[a, b]$. The opposite arc is denoted by $[b, a]$ (Exercise 8.3.5). If $f:[a, b] \to \mathbf{C}$ is continuous then

$$\int_{[a,b]} f(z)\,dz = \int_0^1 f[a + t(b-a)](b-a)\,dt$$

$$= \int_\alpha^\beta f\left[a + \frac{\tau - \alpha}{\beta - \alpha}(b-a)\right]\frac{b-a}{\beta-\alpha}\,d\tau,$$

and

$$L([a, b]) = \int_0^1 |b - a|\,dt = \int_\alpha^\beta \left|\frac{b-a}{\beta-\alpha}\right| d\tau = |b - a|,$$

as expected.

(ii) The parametric equation $z = \gamma(t) = a + Re^{it}$, $\theta \le t \le 2\pi + \theta$, represents a closed path which we call a positively oriented circle. The graph is $C(a, R)$, which we also call a circle. The function $f(z) = 1/(z - a)$ is continuous on $C(a, R)$. We have the important integral:

$$\int_\gamma \frac{1}{z-a}\,dz = \int_\theta^{2\pi+\theta} \frac{Rie^{it}\,dt}{a + Re^{it} - a} = \int_\theta^{2\pi+\theta} i\,dt = 2\pi i.$$

The opposite arc has a parametric equation $z = \gamma_1(\tau) = a + Re^{i(4\pi + 2\theta - t)} = a + Re^{i(2\theta - t)}$, $2\pi + \theta \le 4\pi + \theta$, and

$$\int_{\gamma_1} \frac{1}{z-a}\,dz = \int_{2\pi+\theta}^{4\pi+\theta} \frac{(-Rie^{i(2\theta-t)}\,dt}{a + Re^{i(2\theta-t)} - a}$$

$$= \int_{2\pi+\theta}^{4\pi+\theta} (-i)\,dt = -2\pi i.$$

(iii) Let $T = (a, b, c)$ be an ordered triple of complex numbers, and let T^* be the convex hull of T (see Exercises 5.3.10 and 5.3.11). We call T^* the triangle determined by T and denote by ∂T the polygonal closed path $[a, b] + [b, c] + [c, a]$. (To justify the + signs, the line segments are assumed to be suitably parametrized.) Then if the function $f: \partial T \to \mathbf{C}$ is continuous, by Proposition 8.2.3 we have

$$\int_{\partial T} f(z)\,dz = \int_{[a,b]} f(z)\,dz + \int_{[b,c]} f(z)\,dz + \int_{[c,a]} f(z)\,dz.$$

Recall that T^* is the locus of all weighted averages of the points a, b, c [Exercise 5.3.10]. From the parametric equation for $[a, b]$: $z = a + t(b - a) = (1 - t)a + tb, 0 \le t \le 1$, it is seen that each point of $[a, b]$ is a weighted average of a and b; similarly for $[b, c]$ and $[c, d]$; so the graph of ∂T lies in T^*. If T is replaced by $T_1 = (a, c, b)$, then $\partial T_1 = [a, c] + [c, b] + [b, a] = -\partial T$.

We conclude this section by considering termwise integration of sequences and series.

Proposition 8.3.1 *Let $\gamma: [\alpha, \beta] \to \mathbf{C}$ be a path with graph Γ. Let $f_k: \Gamma \to \mathbf{C}$ be continuous on Γ, $k = 0, 1, 2, \ldots$, and let the sequence $\langle f_k \rangle_0^\infty$ converge uniformly of Γ to the function f. Then*

$$\int_\gamma f(z)\,dz = \lim_{k \to \infty} \int_\gamma f_k(z)\,dz. \tag{8.3.5}$$

In other words, under these hypotheses the sequence can be integrated term by term over γ and the limit of the integrals is the integral of the limit function. According to Theorem 3.3.3, the limit function is continuous on Γ, so there is no problem concerning the existence of the integrals in (8.3.5).

Proof. Given any $\varepsilon > 0$, by hypothesis there exists K_ε such that $|f_k(z) - f(z)| < \varepsilon / L(\gamma)$ for every $k \ge K_\varepsilon$ and every $z \in \Gamma$. For such values of k, by Corollary 1 of Theorem 8.3.2,

$$\left| \int_\gamma f_k(z)\,dz - \int_\gamma f(z)\,dz \right| = \left| \int_\gamma [f_k(z) - f(z)]\,dz \right|$$
$$\le \frac{\varepsilon}{L(\gamma)} L(\gamma) = \varepsilon.$$

Q.E.D.

Corollary Let $\gamma: [\alpha, \beta] \to \mathbf{C}$ be a path graph Γ. Let $f_k: \Gamma \to \mathbf{C}$ be continuous on Γ, $k = 0, 1, 2, \ldots$, and let $\sum_{k=0}^\infty f_k$ converge uniformly on Γ to the function f. Then

$$\int_\gamma f(z)\,dz = \sum_{k=0}^\infty \int_\gamma f_k(z)\,dz.$$

Exercises 8.3

1. Prove the corollary of Proposition 8.3.1.
2. Let γ be a positively oriented circle with center a and radius R. Show that $\int_\gamma (z-a)^m\,dz = 0$, m an integer, $m \neq 1$.
3. Let γ be the closed rectangular path joining $\alpha + i\beta$ to $\alpha + h + i\beta$ to $\alpha + h + i(\beta + k)$ to $\alpha + i(\beta + k)$ to $\alpha + i\beta$, where α, β, h, k are real, $h > 0$, $k > 0$. Show that $\int_\gamma [z - \alpha - (h/2) - i\beta - i(k/2)]^{-1}\,dz = 2\pi i$. A path γ of this type is called a *positively oriented rectangular path*.
4. Show that $|\int_{[i,2+i]} (1/z^2)\,dz| \leq 1.445$. (Use tables if necessary.)
5. Let $[a, b]$ have the parametric equation $z = a + t(b-a)$, $0 \leq t \leq 1$. Prove that $-[a, b] = [b, a]$. (*Hint*: Refer to the parametric equation for the opposite arc in Sec. 8.2. Here it gives for $-[a, b]$ the parameter interval $1 \leq t \leq 2$. Set up the parametric equations for $-[a, b]$ and $[b, a]$ over this parameter interval and compare.)
6. Let $\gamma: [\alpha, \beta] \to \mathbf{C}$ be a path with graph Γ, and let $f: \Gamma \to \mathbf{C}$ be continuous on Γ. Prove that the function $\phi(\tau) = \int_\alpha^\tau f[\gamma(t)]\gamma'(t)\,dt$, $\alpha \leq \tau \leq \beta$, is continuous on $[\alpha, \beta]$.
7. Let $\gamma: [\alpha, \beta] \to \mathbf{C}$ be a path, and let $f(t, \sigma)$ be a complex-valued function of the real variables t and σ which is continuous on the rectangle $R = \{(t, s): \alpha \leq t \leq \beta, \alpha' \leq \sigma \leq \beta'\}$. Prove that the function ϕ given by $\phi(\sigma) = \int_\alpha^\beta f(t, \sigma)\gamma'(t)\,dt$ is continuous on $[\alpha', \beta']$. The integral is to be interpreted as indicated in (8.3.2). (*Hint*: First prove the statement for $\gamma'(t)$ continuous on $[\alpha, \beta]$ and then generalize as in the proof of Theorem 8.3.2 to the piecewise continuous case. The rectangle R is a compact set in $\mathbf{R}^2$, so by Theorem 5.5.3, $f(t, \sigma)$ is uniformly continuous on R. Use the arithmetical formulation of uniform continuity together with Proposition 8.1.1(b) and (f) and the inequality (8.1.2) to show that for any $\varepsilon > 0$, there exists $\delta > 0$ such that if $\sigma, \sigma_0 \in [\alpha', \beta']$ satisfy $|\sigma - \sigma_0| < \delta$, then $|\phi(\sigma) - \phi(\sigma_0)| = |\int_\alpha^\beta [f(t, \sigma) - f(t, \sigma_0)]\gamma'(t)dt| < \varepsilon$.)
8. Let γ be a closed path with graph Γ, such that Γ^c has only two components, and with parametric equation $z = \gamma(t) = \gamma_1(t) + i\gamma_2(t)$, $\alpha \leq t \leq \beta$, γ_1 and γ_2 real valued. Under certain restrictions on γ it can be shown that the area of the bounded component of Γ^c [Proposition 5.4.2] is given by $A = (1/2)\int_\alpha^\beta [\gamma_2(t)\gamma_1'(t) - \gamma_1(t)\gamma_2'(t)]\,dt$. (This is a special case of Green's theorem; see [6, pp. 407–414].) Show that in complex variable notation the formula becomes $A = -\mathrm{Im}[(1/2)\int_\gamma \bar{z}\,dz]$.

8.4 A Special Type of Complex Line Integral

A line integral which is of particular importance in complex function theory is that which appears in the following theorem.

Theorem 8.4.1 *Let $\gamma: [\alpha, \beta] \to \mathbf{C}$ be a path with graph Γ, let the functions $g: \Gamma \to \mathbf{C}$ and $h: \Gamma \to \mathbf{C}$ be continuous on Γ, and let O be the complement of $h(\Gamma)$ with respect to $\mathbf{C}$. The function $f: O \to \mathbf{C}$ given by*

$$f(z) = \int_\gamma \frac{g(\zeta)\, d\zeta}{h(\zeta) - z} \tag{8.4.1}$$

has the following properties:

(i) *For each point $a \in O$, there exists a power series $\sum_0^\infty a_k(a)(z - a)^k$, which converges to $f(z)$ for every $z \in \Delta[a, d(a)]$, where $d(a) = \text{dist}\,[a, h(\Gamma)]$. (Thus, f has the LPS property on O.)*

(ii) *The coefficients $a_k(a)$ and the derivatives $f^{(k)}(a)$ ($f^0 = f$) are given by*

$$a_k(a) = \frac{f^{(k)}(a)}{k!} = \int_\gamma \frac{g(\zeta)\, d\zeta}{[h(\zeta) - a]^{k+1}}, \qquad k = 0, 1, 2, \ldots. \tag{8.4.2}$$

Proof. According to Example (iii) of Sec. 5.5, the set $h(\Gamma)$ is compact. Then by Theorem 5.6.1, $d(a) > 0$. For each point $a \in O$, the open disk $\Delta[a, d(a)]$ is contained in O. We take $a \in O$ and $z \in \Delta[a, d(a)]$ and hold a and z fixed in the remainder of the proof. In intuitive terms, the "independent variable" is ζ.

We write

$$\begin{aligned} \frac{g(\zeta)}{h(\zeta) - z} &= \frac{g(\zeta)}{h(\zeta) - a - (z - a)} \\ &= \frac{1}{h(\zeta) - a}\left[\frac{g(\zeta)}{1 - (z - a)/(h(\zeta) - a)}\right]. \end{aligned} \tag{8.4.3}$$

Let $d = d(a)$. Now $|h(\zeta) - a| \geq d$ for every $\zeta \in \Gamma$, but $|z - a| < d$. It follows that $|z - a|/|h(\zeta) - a| < 1$. The geometric series $1/(1 - w) = 1 + w + w^2 + \cdots$, $|w| < 1$ can be used to expand the right-hand member of (8.4.3) as follows:

$$\frac{g(\zeta)}{h(\zeta) - z} = \sum_{k=0}^{\infty} \frac{g(\zeta)}{h(\zeta) - a}\left(\frac{z - a}{h(\zeta) - a}\right)^k. \tag{8.4.4}$$

We now set up an M-test for the uniform convergence of this series, with the terms of the series regarded as functions only of ζ. Since Γ is a compact subset of C and $g: \Gamma \to C$ is continuous, according to Corollary 1 of Theorem 5.5.2, $|g(\zeta)|$ attains a maximum value G on Γ. Let

$$M_k = \frac{(z - a)^k \cdot G}{d^{k+1}} = \frac{G}{d} \cdot \rho^k,$$

where $\rho = (z - a)/d$, so $|\rho| < 1$. Then for every $\zeta \in \Gamma$ and $k = 0, 1, 2, \ldots,$

$$\left|\frac{g(\zeta)}{h(\zeta) - a}\left(\frac{(z - a)}{h(\zeta) - a}\right)^k\right| < M_k.$$

The series $\sum_{k=0}^\infty M_k$ obviously converges. Thus the series in (8.4.4) converges uniformly on Γ, and by the corollary of Proposition 8.3.1 it can be integrated term by term. We take the "constants" $(z - a)^k$, $k = 0, 1, 2, \ldots$, out of the integrals and obtain

$$f(z) = \sum_{k=0}^{\infty} (z - a)^k \int_\gamma \frac{g(\zeta)\, d\zeta}{[h(\zeta) - a]^{k+1}}.$$

But this is a power series in z with terms $a_k(a)(z - a)^k$, $k = 0, 1, 2, \ldots$, where $a_k(a)$ is given by (8.4.2). The argument shows that the series converges for every $z \in \Delta(a, d)$. The relationships with the derivatives of f at a indicated in (8.4.2) were established in the corollary of Theorem 6.4.1. Q.E.D.

The specialization $g(\zeta) = 1$, $h(\zeta) = \zeta$ and multiplication by $1/2\pi i$ reduces (8.4.1) to what is known as the *topological index of z with respect to γ*, or *winding number of γ with respect to z*. The notation we use is $\text{Ind}_\gamma(z)$. Then

$$\text{Ind}_\gamma(z) = \frac{1}{2\pi i} \int_\gamma \frac{d\zeta}{\zeta - z}, \qquad z \in \Gamma^c.$$

Theorem 8.4.2 *Let $\gamma: [\alpha, \beta] \to \mathbf{C}$ be a closed path with graph Γ. Then $\text{Ind}_\gamma(z)$ is constant in each component of Γ^c. The constant value for a given component is a real integer, and the constant value for the unbounded component is* 0.

Remark. The components of Γ^c were introduced in Sec. 5.4. Since Γ is compact, there is one and only one unbounded component of Γ^c [Proposition 5.4.2].

Proof of the Theorem. We first show that for any fixed $z \in \Gamma^c$, $\text{Ind}_\gamma(z)$ is a real integer. Let

$$\phi(\tau) = \int_\alpha^\tau \frac{\gamma'(t)\, dt}{\gamma(t) - z}, \qquad z \in \Gamma^c, \qquad \alpha \le t \le \beta.$$

Then $\phi(\alpha) = 0$, $\phi(\beta) = 2\pi i \,\text{Ind}_\gamma(z)$ [Theorem 8.3.2], and according to Exercise 8.3.6, $\phi(\tau)$ is continuous on $[\alpha, \beta]$. Also, it follows from Proposition 8.1.1(d) and the interpretation (8.3.2) that the derivative ϕ' exists at every point of (α, β) except possibly at the points $s_1, s_2, \ldots, s_{m-1}$ identified in connection with (8.3.2).

Now let $\psi(\tau) = e^{-\phi(\tau)}[\gamma(\tau) - z]$. This composite function is continuous on $[\alpha, \beta]$. Moreover, by using the chain and product laws for differentiation, it can be seen that ψ' exists for every τ for which ϕ' exists, and in fact has the value zero wherever it exists on (α, β). According to the lemma for Theorem 6.2.1, ψ must have a constant value on each of the intervals $[\alpha, s_1]$, $[s_1, s_2], \ldots, [s_{m-1}, \beta]$. By continuity, these constant values must all be equal to $\psi(\alpha) = \gamma(\alpha) - z$. We have established the identity $e^{-\phi(\tau)}[\gamma(\tau) - z] \equiv \gamma(\alpha) - z$, valid for every $\tau \in [\alpha, \beta]$. Letting $\tau = \beta$, we obtain $e^{-2\pi i \,\text{Ind}_\gamma(z)}[\gamma(\beta) - z] = \gamma(\alpha) - z$. But γ is a *closed* path, so $\gamma(\alpha) = \gamma(\beta)$. Thus, $e^{-2\pi i \,\text{Ind}_\gamma(z)} = 1$, so by Theorem 7.4.1, $\text{Ind}_\gamma(z)$ must be a real integer.

Now by Theorems 8.4.1 and 4.2.1, $\text{Ind}_\gamma(z)$ is continuous on Γ^c. Let K be a component of Γ^c; K by definition is a connected set. Suppose that for a certain pair of points $z_1, z_2 \in K$, $z_1 \neq z_2$, $\text{Ind}_\gamma(z_1) = k_1$ and $\text{Ind}_\gamma(z_2) = k_2$, where k_1 and k_2 are unequal integers. Then by the intermediate value theorem [Proposition 5.3.2], if c is a noninteger lying between k_1 and k_2, there exists $z_c \in K$ such that $\text{Ind}_\gamma(z_c) = c$. But this contradicts the fact that $\text{Ind}_\gamma(z)$ is an integer for every $z \in K$. Therefore, $\text{Ind}_\gamma(z)$ is constant on K.

We now show that the constant value of $\text{Ind}_\gamma(z)$ on the unbounded component of Γ^c is 0. For any z on Γ^c, Corollary 1 of Theorem 8.3.2 yields the following inequality:

$$|\text{Ind}_\gamma(z)| \leq \frac{1}{2\pi} \max_{\zeta \in \Gamma} \frac{1}{|\zeta - z|} L(\gamma). \tag{8.4.5}$$

Since Γ is bounded, there exists $R > 0$ such that $|\zeta| < R$ for every $\zeta \in \Gamma$. The unbounded component of Γ^c contains $\Delta'(\infty, R) = \{z: |z| > R\}$ [Proposition 5.4.2]. If we place z in $\Delta'(\infty, R)$, we have $|\zeta - z| \geq \big||\zeta| - |z|\big| = |z| - |\zeta| > |z| - R$ for every $\zeta \in \Gamma$. Under these circumstances, (8.4.5) becomes

$$|\text{Ind}_\gamma(z)| < \frac{L(\gamma)}{2\pi} \cdot \frac{1}{|z| - R}. \tag{8.4.6}$$

If now $|z| > [L(\gamma)/2\pi] + R$, the right-hand member of (8.4.6) is less than 1. But $\text{Ind}_\gamma(z)$ has a single integer value throughout the unbounded component of Γ^c, so this value must be 0. Q.E.D.

Examples

(i) Let γ be a positively oriented circle with center a and radius R. In Example (ii) of Sec. 8.3, it was shown that

$$\text{Ind}_\gamma(a) = \frac{1}{2\pi i} \int_\gamma \frac{d\zeta}{\zeta - a} = 1.$$

Now $\Delta = \Delta(a, R)$ is a convex set [Proposition 5.3.6] and so is connected. Therefore, by Theorem 5.4.1, it lies in a component of $[C(a, R)]^c$, and so according to Theorem 8.4.2, $\text{Ind}_\gamma(z) = 1$ for every $z \in \Delta(a, R)$. Now $C(a, R) \subset \bar{\Delta}(a, R)$, so by Proposition 5.4.2, $\bar{\Delta}(a, R)^c$ is contained in the unbounded component of $[C(a, R)]^c$. Therefore, $\text{Ind}_\gamma(z) = 0$ for every $z \in \bar{\Delta}(a, R)^c$. (Incidentally, it follows from $[C(a, R)]^c = \Delta(a, R) \cup \bar{\Delta}(a, R)^c$ that $\Delta(a, R)$ and $\bar{\Delta}(a, R)^c$ actually *are* the components of $[C(a, R)]^c$ in $\mathbf{C}$.)

(ii) Let γ be the positively oriented rectangular path described in Exercise 8.3.3, and denote the graph of this path by Γ. Let $z = x + iy$, x and y real, let $S = \{z: \alpha < x < \alpha + h,\ \beta < y < \beta + k\}$, $U = \{z: x < \alpha\} \cup \{z: x > \alpha + h\} \cup \{z: y < \beta\} \cup \{z: y > \beta + k\}$. It is readily verified that S, Γ, U are pairwise disjoint and $S \cup \Gamma \cup U = \mathbf{C}$. The set S is the intersection of four open half-planes, so by Exercises 5.3.8 and 5.3.12, S is a convex region. This region is called the interior region of Γ (or of γ). The set U is the union of four open half-planes which taken in sequence have pairwise nonempty intersections, so by Exercise 5.3.12 and the corollary to Proposition 5.4.1, U is a region. Clearly, U is unbounded and, therefore, must be the unbounded component of Γ^c. The point $z_0 = \alpha + (h/z) + i(\beta + (k/2))$ lies in S, and by Exercise 8.3.3, $\text{Ind}_\gamma(z_0) = 1$. Therefore, $\text{Ind}_\gamma(z) = 1$ for every $z \in S$. Also, $\text{Ind}_\gamma(z) = 0$ for every $z \in U$.

(iii) Let γ be a closed path with graph Γ and with parametric equation $\zeta = a + r(t)e^{it}$, $a \in \mathbf{C}$, $0 \le t \le 2\pi$, where $r(0) = r(2\pi)$, $r(t) > 0$, and $r(t)$ is piecewise continuously differentiable. (This term is defined as in the definition of path.) The *interior region* of this path is defined to be the set $B = \{z : z = a + sr(t)e^{it}, 0 \le t \le 2\pi, 0 \le s < 1\}$, which is introduced in Exercise 7.5.10. It is shown in that exercise that B is open and connected. Because of the latter property, it lies in a component of Γ^c [Theorem 5.4.1]. (Actually, B *is* a component of Γ^c, but it is not necessary for present purposes to establish this.) We have

$$\text{Ind}_\gamma(a) = \frac{1}{2\pi i}\int_\gamma \frac{d\zeta}{\zeta - a} = \frac{1}{2\pi i}\int_0^{2\pi} \frac{[r'(t)e^{it} + ie^{it}r(t)]\,dt}{r(t)e^{it}}$$

$$= \frac{1}{2\pi i}\int_0^{2\pi} \frac{r'(t)}{r(t)}\,dt + \frac{1}{2\pi i}\int_0^{2\pi} i\,dt$$

$$= \left[\frac{1}{2\pi i}\ln r(t)\right]_0^{2\pi} + \frac{1}{2\pi}\cdot 2\pi = 0 + 1.$$

Thus, $\text{Ind}_\gamma(z) = 1$ for every $z \in B$.

Exercises 8.4

1. Let γ be a path with graph Γ. Show that for $z \notin \Gamma$, $\mathit{Ind}_{-\gamma}(z) = -\text{Ind}_\gamma(z)$.
2. Let γ_1 and γ_2 be paths with the terminal point of γ_1 coinciding with the initial point of γ_2. Show that for z in the complement of the graph of $\gamma_1 + \gamma_2$, $\text{Ind}_{\gamma_1+\gamma_2}(z) = \text{Ind}_{\gamma_1}(z) + \text{Ind}_{\gamma_2}(z)$.

Remark. Exercise 8.4.2 can be used to lend credibility to an intuitive interpretation of $\text{Ind}_\gamma(z)$ as the number of times γ "winds around" z (hence the term "winding number"). Let γ be a closed path with graph Γ, and take $z_0 \in \Gamma^c$. Now $\text{Ind}_\gamma(z)$ is an integer, and it is the same for every z in the component of Γ^c in which z_0 lies. These facts are at least consistent with the "winding number" interpretation. Let a parametric equation of γ be $z = \gamma(t)$, $\alpha \le t \le \beta$. The path with parametric equation

$$z = \gamma^*(t) = \begin{cases} \gamma(t), & \alpha \le t \le \beta \\ \gamma_1(t) = \gamma(t - \beta + \alpha), & \beta \le t \le 2\beta - \alpha \\ \gamma_2(t) = \gamma(t - 2\beta + 2\alpha), & 2\beta - \alpha \le t \le 3\beta - 2\alpha \\ \vdots & \vdots \\ \gamma_m(t) = \gamma(t - m\beta + m\alpha), & m\beta - (m-1)\alpha \le t \le (m+1)\beta \\ & - m\alpha \end{cases}$$

has the graph of γ repeated $m + 1$ times, in the sense that as t moves along the real axis from α to $(m + 1)\beta - m\alpha$, the point $z = \gamma^*(t)$ describes the graph of γ $m + 1$ times. For each k, the arc γ_k is obtained from γ by the change of parameter $t = \tau - k\beta + k\alpha$, $k\beta - (k - 1)\alpha \le \tau \le (k + 1)\beta - k\alpha$, so $\text{Ind}_{\gamma_k}(z) = \text{Ind}_\gamma(z)$. Since $\gamma^* = \gamma + \gamma_1 + \gamma_2 + \cdots + \gamma_m$, it follows from the exercise that $\text{Ind}_{\gamma^*}(z_0) = (m + 1)$ $\text{Ind}_\gamma(z_0)$. This seems to say that γ^* is "winding around" z_0 $m + 1$ times as often as does γ, which accords with an intuitive interpretation of "winding number." The argument is particularly appealing if $\text{Ind}_\gamma(z) = 1$.

8.5 Primitives and Integration over Paths

Recall from Sec. 6.2 that a primitive of a complex-valued function f, of which the domain is an open set $O \subset \mathbf{C}$, is a function $F: O \to \mathbf{C}$ analytic on O and such that $F'(z) \equiv f(z)$ on O. The existence of a primitive of a complex-valued function f on an open set in $\mathbf{C}$ turns out to be a very strong condition on f, in contrast to the analogous situation for a real-valued function with domain in $\mathbf{R}$.

More specifically, let $f: [\alpha, \beta] \to R$, $[\alpha, \beta] \subset \mathbf{R}$, be continuous. A primitive of f always exists and is given by the formula $F(t) = \int_\alpha^t f(\tau)\, d\tau$, $\alpha \le t \le \beta$. But a continuous function $f: O \subset \mathbf{C} \to \mathbf{C}$, with O an open set, does not necessarily have a primitive. For example, if $f(z) = \bar{z}$, $O = \{z: \text{Im } z \ne 0\}$, then there cannot exist a function F analytic on O such that $F'(z) \equiv \bar{z}$ on O, for if there did, the Cauchy–Riemann equations would be violated. We defer a detailed proof to an exercise.

But when a primitive does exist for a complex-valued function, it can be used to compute the integral of the function over paths.

Theorem 8.5.1 *Let $f: O \to \mathbf{C}$ be continuous on the open set $O \subset \mathbf{C}$, and let F be a primitive of f in O. For any path γ with initial point a and endpoint b, and with graph lying in O, $\int_\gamma f(z)\, dz = F(b) - F(a)$. In particular, if γ is a closed path, then $\int_\gamma f(z)\, dz = 0$.*

Proof. Let a parametric equation of γ be $z = \gamma(t)$, $\alpha \le t \le \beta$, so $a = \gamma(\alpha)$, $b = \gamma(\beta)$. Let the discontinuities of γ' on (α, β) be at $s_1 < s_2 < \cdots < s_{m-1}$, and let $s_1 > s_0 = \alpha$ and $s_{m-1} < s_m = \beta$. By Theorem 8.3.2,

$$\int_\gamma f(z)\, dz = \sum_{j=1}^{m} \int_{s_{j-1}}^{s_j} f[\gamma(\tau)]\gamma'(\tau)\, d\tau$$
$$= \sum_{j=1}^{m} \int_{s_{j-1}}^{s_j} F'[\gamma(\tau)]\gamma'(\tau)\, d\tau.$$

The symbol F' here means dF/dz, not $dF/d\tau$. By the chain law for differentiation, $F'[\gamma_1(\tau)]\gamma_1'(\tau) = dF[\gamma_1(\tau)]/d\tau$, so we have

$$\int_\gamma f(z)\,dz = \sum_{j=1}^{m} \int_{s_{j-1}}^{s_j} \frac{dF[\gamma_1(\tau)]}{d\tau}\,d\tau$$

$$= \sum_{i=1}^{m} \{F[\gamma(s_j)] - F[\gamma(s_{j-1})]\}$$

$$= F[\gamma(\beta)] - F[\gamma(\alpha)].$$

Proposition 8.1.1(e) was used to evaluate the integrals in the second member of this equation. Q.E.D.

Corollary Let B be a region in **C**, let $f\colon B \to \mathbf{C}$ be continuous, and let f possess a primitive in B. For any two points a and b in B, the value of $\int_\gamma f(z)\,dz$ is the same for all paths γ with graphs in B and with initial point a and terminal point b.

Remark. The line integral $\int_\gamma f(z)\,dz$ is said to be "independent of the path" if it satisfies the condition in the conclusion of the theorem.

The theorem and its corollary open up a wide variety of complex line integrals to direct computation, but in using it care must be taken to be sure that the alleged primitive is a single-valued analytic function on the relevant open set O.

Examples

(i) Let m be a nonnegative integer. The function $(z-a)^{m+1}/(m+1)$ is a primitive in **C** for $(z-a)^m$, where a is a complex constant. Thus, given any path γ in **C** with initial point z_1 and terminal point z_2, $\int_\gamma (z-a)^m\,dz = [(z_2-a)^{m+1} - (z_1-a)^{m+1}]/(m+1)$; and if γ is a closed path, the integral has the value 0.

(ii) For $m = -2, -3, -4, \ldots$, the function $(z-a)^{m+1}/(m+1)$ is a primitive of $(z-a)^m$ in $\mathbf{C} \sim \{a\}$. Thus, given any path γ in $\mathbf{C} \sim \{a\}$ with initial point z_1 and terminal point z_2, $\int_\gamma (z-a)^m\,dz = [(z_2-a)^{m+1} - (z_1-a)^{m+1}]/(m+1)$; and if γ is a closed path in $\mathbf{C} \sim \{a\}$, the integral has the value 0.

(iii) Let f be the sum function of a power series $\sum_0^\infty a_k(z-a)^k$ with disk of convergence $\Delta(a, R)$, $R > 0$. According to Exercise 6.4.1, f has a primitive in $\Delta(a, R)$ wich is given by $\sum_{k=0}^\infty [a_k(z-a)^{k+1}/(k+1)]$. Therefore, if γ is any closed path in $\Delta(a, R)$, $\int_\gamma f(z)\,dz = 0$. (This is a special case of the Cauchy integral theorem.)

(iv) Let $f(z) = 1/z$, and let γ be a positively oriented circle with center 0 and radius 1. We know from Example (i) of Sec. 8.3 that $\int_\gamma (1/z)\,dz = 2\pi i$. Let the equation of γ be $z = e^{it}, -\pi \le t \le \pi$. The function Log z is analytic on $\mathbf{C} \sim (-\infty, 0)$, and its derivative there is $1/z$ [Theorem 7.7.3]; and Log z is well defined at each point of $C(0, 1)$. But the computation $\int_\gamma (1/z)\,dz = \operatorname{Log}(e^{i\pi}) - \operatorname{Log}(e^{-i\pi}) = \operatorname{Log}(-1) - \operatorname{Log}(-1) = i\pi - i\pi = 0$ is invalid because the graph of γ does not lie in a region in which Log z is analytic. Notice that if we replace γ by γ_ε with equation

$z = e^{it}, -\pi + \varepsilon \le t \le \pi - \varepsilon$, $0 < \varepsilon < \pi$, we obtain the valid equation

$$\begin{aligned}\int_{\gamma_\varepsilon} \frac{1}{z}\,dz &= \operatorname{Log} e^{i(\pi-\varepsilon)} - \operatorname{Log} e^{i(-\pi+\varepsilon)} \\ &= 0 + i(\pi - \varepsilon) - [0 + i(-\pi + \varepsilon)] \\ &= 2\pi i - 2i\varepsilon.\end{aligned}$$

It is easy to show that as $\varepsilon \to 0$, the integral of $1/z$ over γ_ε approaches the integral over γ, which provides a proof via primitives that $\int_\gamma (1/z)dz = 2\pi i$.

Although a continuous function from an open set in **C** into **C** does not necessarily have a primitive, there is a partial converse to the corollary of Theorem 8.5.1 which does assure the existence of a primitive. We develop this converse in a form which anticipates the results in the next chapter. By imposing a restriction on the region B, we are able to weaken the hypothesis of independence of the path dramatically.

Recall the definition of a starlike set in Chapter 5: It has the property that there is a point a in it such that if z is any other point in it, then $[a, z]$ lies in the set.

Theorem 8.5.2 *Let O be an open set which is starlike with respect to $a \in O$. Let the function $f\colon O \to \mathbf{C}$ be continuous on O and have the property that if z_0 and z_1 are any two points in O such that the graph of the polygonal path ∂T determined by the ordered triple $T = (a, z_0, z_1)$ lies in O, then $\int_{\partial T} f(z) = 0$. Then f has a primitive F in O with values $F(z) = \int_{[a,z]} f(\zeta)\,d\zeta$.*

(See Example (iii), Sec. 8.3, for the definition of ∂T.)

Proof. The starlike character of O with respect to a assures that the function F specified in the theorem is well defined for each $z \in O$. The problem is to prove that $F'(z_0)$ exists for every $z_0 \in O$ and equals $f(z_0)$.

Since O is open, for each $z_0 \in O$ there exists a closed disk $\bar{\Delta}(z_0, R)$,. $R = R_{z_0} > 0$, such that $\bar{\Delta}(z_0, R) \subset O$. We hold z_0 fixed in what follows. Given any $z_1 \in \bar{\Delta}(z_0, R)$ with $z_1 \ne z_0$, both $[a, z_1] \subset O$ and $[z_0, z_1] \subset \bar{\Delta}(z_0, R) \subset O$. Then our hypothesis on f implies that for every $z_1 \in \bar{\Delta}(z_0, R)$,

$$0 = \int_{[a,z_0]} f(z)\,dz + \int_{[z_0,z_1]} f(z)\,dz + \int_{[z_1,a]} f(z)\,dz. \tag{8.5.1}$$

Now $[z_1, a] = -[a, z_1]$, so by Proposition 8.2.4, $\int_{[z_1,a]} f(z)\,dz = -\int_{[a,z_1]} f(z)\,dz$. With the definition of $F(z)$ given in the theorem, (8.5.1) can be rewritten in the form $F(z_1) - F(z_0) = \int_{[z_0,z_1]} f(z)\,dz$. Notice that $\int_{[z_0,z_1]} f(z_0)\,dz = (z_1 - z_0)f(z_0)$ [Exercise 8.2.1]. By using the elementary properties of line integrals given in Sec. 8.2, we obtain

$$D = \left|\frac{F(z_1) - F(z_0)}{z_1 - z_0} - f(z_0)\right|$$
$$= \left|\frac{1}{z_1 - z_0}\int_{[z_0,z_1]} f(z)\,dz - \frac{1}{z_1 - z_0}\int_{[z_0,z_1]} f(z_0)\,dz\right|$$
$$= \left|\frac{1}{z_1 - z_0}\right|\left|\int_{[z_0,z_1]} [f(z) - f(z_0)]\,dz\right|. \tag{8.5.2}$$

According to Theorem 5.5.2, Corollary 1, the continuous function $|f((z) - fz_0)|$ attains a maximum value on the compact set $[z_0, z_1]$, so the corollary of Theorem 8.3.2 yields the inequality

$$D \le \frac{1}{|z_1 - z_0|}\max_{z\in[z_0,z_1]} |f(z) - f(z_0)|\,|z_1 - z_0|$$
$$= \max_{z\in[z_0,z_1]} |f(z) - f(z_0)|. \tag{8.5.3}$$

The continuity of f at z_0 implies that for any $\varepsilon > 0$, there exists δ_ε, $0 < \delta_\varepsilon < R$ such that for all $\zeta \in \Delta(z_0, \delta_\varepsilon)$, $|f(\zeta) - f(z_0)| < \varepsilon$. Then for every $z_1 \in \Delta(z_0, \delta_\varepsilon)$, we have (i) $z_1 \in \Delta(z_0, R)$ (so (8.5.2) and (8.5.3) are valid); (ii) $[z_0, z_1] \in \Delta(z_0, \delta_\varepsilon)$, and, therefore, $\max_{z\in[z_0,z_1]} |f(z) - f(z_0)| < \varepsilon$; and so (iii) $D < \varepsilon$. Thus by the definition of derivative, $F'(z_0)$ exists and equals $f(z_0)$. Q.E.D.

Exercises 8.5

1. Let γ be any closed path in **C**. Show that $\int_\gamma e^{-z^2}\,dz = 0$ and $\int_\gamma[(\sin z)/z]\,dz = 0$. (*Hint*: See Example (iii).)
2. Let γ be a path in **C** with terminal point z_1 and initial point z_0. Let $a \neq 0$ and b be complex numbers. Evaluate $\int_\gamma e^{az+b}\,dz$. *Answer*. $(e^b/a)[e^{az_1} - e^{az_0}]$.
3. If γ is any closed path in **C** with graph Γ, then for any complex a and any complex $z \in \Gamma^c$, show that
$$(z - a)^k \operatorname{Ind}_\gamma(z) = \frac{1}{2\pi i}\int_\gamma \frac{(\zeta - a)^k}{(\zeta - z)}\,d\zeta, \qquad k = 0, 1, 2, \ldots.$$
(*Hint*: Write $(\zeta - z)^k = [(\zeta - z) + (z - a)]^k$ and expand by the binomial theorem. Use Example (i) to evaluate the integrals.)
4. Let f be the sum function of a power series $\sum_0^\infty a_k(z - a)^k$ with radius of convergence $R > 0$. Prove that if γ is any closed path with graph $\Gamma \subset \Delta(a, R)$, then the Cauchy integral formula
$$f(z) \operatorname{Ind}_\gamma(z) = \frac{1}{2\pi i}\int_\gamma \frac{f(\zeta)}{\zeta - z} \tag{8.5.3}$$
is valid for every $z \in \Gamma^c \cap \Delta(a, R)$. (*Hint*: From Exercise 8.5.3 it can be seen that the problem boils down to justifying the integration of $\sum_0^\infty a_k(\zeta - z)^{-1}(\zeta - a)^k$ term-by-term along γ. Use Exercise 5.6.1.)

5. Let γ be any closed path with graph not containing 0. Evaluate $\int_\gamma (\sin z/z^2) dz$.
Answer. $2\pi i \operatorname{Ind}_\gamma(0)$.

6. Let f be continuous on a region B. Show that if for any two points $a \in B$, $b \in B$, the integral $\int_\gamma f(z)\, dz$ is independent of the path γ joining a to b, then $\int_{\gamma_0} f(z)\, dz = 0$ for every closed path γ_0 in B.

7. Show that the function $f(z) = \bar{z}$ does not have a primitive on $O = \{z : \operatorname{Im} z \neq 0\}$.

9

Introduction to the Cauchy Theory

9.1 General Remarks

In Secs. 6.4 and 6.5 it was brought out that if a complex-valued function f defined on a open set O in $\mathbf{C}$ possesses a local power series representation at every point of O, then f is analytic on O and has analytic derivatives of all orders on O. In this chapter, we develop a converse to this proposition: If a function is analytic on O, it has the LPS property on O. The tool which we use for achieving this result is a striking theorem on complex integration associated with the name of Cauchy. In loose terms, this theorem says that a necessary condition for a complex-valued function f, continuous on O, to be analytic on O is that the integral of the functions along any closed path γ with graph Γ in O is equal to zero. The curve γ must satisfy certain restrictions, of which the most notable is that O must contain the bounded components of Γ^c.

The rigorous proofs of this theorem in its more general formulations encounter certain topological difficulties. For the LPS application, and also for many other famous classical applications, only a rather limited formulation is required in which Γ lies in a special type of region, which here we take to be starlike. Accordingly, this chapter is devoted to establishing the Cauchy integral theorem for functions analytic on starlike regions, and to presenting a number of the more immediate applications to complex analysis. In many of these applications, the starlike condition on the domain can be dropped.

9.2 The Cauchy-Goursat Theorem for a Triangle

Theorem 9.2.1 *Let O be an open set in* **C**, *let $T = (z_1, z_2, z_3)$ be an ordered triple of distinct points such that the triangle T^* determined by T (the convex hull of T) lies in O. Let $f: O \to$* **C** *be analytic on O. Then $\int_{\partial T} f(z)\,dz = 0$, where ∂T is the path $[z_1, z_2] + [z_2, z_3] + [z_3, z_1]$.*

Proof. We begin by reviewing briefly some facts about triangles established earlier. In the first place, T^* is compact [Sec. 5.5, Example (iv)]. The diameter $\operatorname{diam}(T^*) = \max\{|z_1 - z_2|, |z_2 - z_3|, |z_3 - z_1|\}$ [Proposition 5.6.2]. The length of ∂T, denoted by $L(\partial T)$, is obviously $|z_1 - z_2| + |z_2 - z_3| + |z_3 - z_1|$.

The *set of quadrisections* of $T = (z_1, z_2, z_3)$ is defined to be the set of ordered triples $T_1 = (z_1, m_1, m_3)$, $T_2 = (m_1, z_2, m_2)$, $T_3 = (m_2, z_3, m_3)$, $T_4 = (m_3, m_1, m_2)$, where $m_1 = (z_1 + z_2)/2$, $m_2 = (z_2 + z_3)/2$, $m_3 = (z_3 + z_1)/2$. The points m_1, m_2, m_3 are the midpoints of the line segments determined by T. (See Figure 1.)

The convex hulls $T_1^*, T_2^*, T_3^*, T_4^*$ all lie in T^* because they are the minimal convex sets respectively containing T_1, T_2, T_3, T_4 [Exercise 5.3.11], and each of the latter sets lies in the convex set T^*. Notice that

$$|m_1 - m_3| = \left|\frac{z_1 + z_2}{2} - \frac{z_3 + z_1}{2}\right| = \frac{1}{2}|z_2 - z_3|.$$

(This can also be deduced by elementary plane geometry.) Similarly, $|m_2 - m_3| = (1/2)|z_1 - z_2|$ and $|m_1 - m_2| = (1/2)|z_1 - z_3|$. Then since m_1, m_2, m_3 are the midpoints of the respective segments determined by T on which they lie, we have the equations $L(\partial T_j) = (1/2)L(\partial T)$ and $\operatorname{diam}(T_j^*) = \max((1/2)|z_1 - z_2|, (1/2)|z_2 - z_3|, (1/2)|z_3 - z_1|) = (1/2)\ \operatorname{diam}(T)$, $j = 1, 2, 3, 4$.

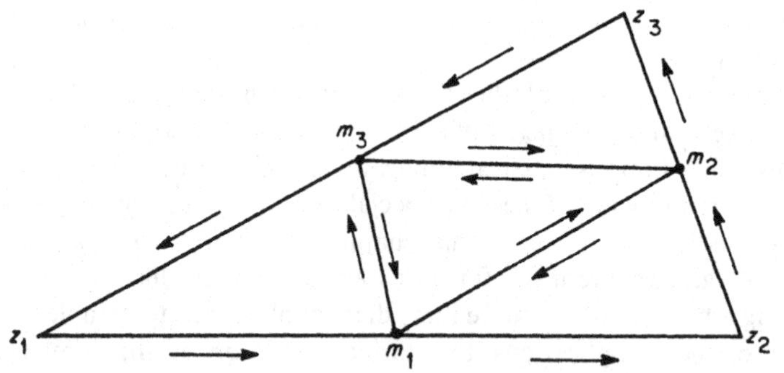

Fig. 1

We henceforth assume that $T^* \subset O$, where O is the open set identified in the statement of the theorem.

Since the function f has a derivative at each point of O, it is continuous on T^*. Recall that if a function g is continuous on the line segment $[a, b]$, then $\int_{[a,b]} g(z)\,dz = -\int_{[b,a]} g(z)\,dz$ [Proposition 8.2.4; Exercise 8.3.5]. It is easily verified that

$$\int_{\partial T} f(z)\,dz = \int_{\partial T_1} f(z)\,dz + \int_{\partial T_2} f(z)\,dz + \int_{\partial T_3} f(z)\,dz + \int_{\partial T_4} f(z)\,dz. \quad (9.2.1)$$

(Figure 1 suggests the required algebra.)

We select a triple T_j for which the integral over ∂T_j on the right-hand side of (9.2.1) has the maximum absolute value of the four integrals. This T_j is relabeled T^1. We have the following relations:

$$\left|\int_{\partial T} f(z)\,dz\right| \le 4\left|\int_{\partial T^1} f(z)\,dz\right|,$$

$$L(\partial T^1) = \frac{1}{2} L(\partial T),$$

$$\operatorname{diam}(T^{1*}) = \frac{1}{2}\operatorname{diam}(T^*).$$

We now set up the equation (9.2.1) with T^1 replacing T and the quadrisections of T^1 replacing those of T. Again we select the integral $\int_{\partial T_j{}^1} f(z)\,dz$ of maximum absolute value and relabel the corresponding triple T_j^1 as T^2. We have

$$\left|\int_{\partial T} f(z)\,dz\right| \le 4\left|\int_{\partial T^1} f(z)\,dz\right| \le 4^2\left|\int_{\partial T^2} f(z)\,dz\right|.$$

$$L(\partial T^2) = \frac{1}{2} L(\partial T^1) = \left(\frac{1}{2}\right)^2 L(\partial T),$$

$$\operatorname{diam}(T^{2*}) = \frac{1}{2}\operatorname{diam}(T^{1*}) = \left(\frac{1}{2}\right)^2 \operatorname{diam}(T^*).$$

We continue this process, obtaining triangles $T^* \supset T^{1*} \supset T^{2*} \supset \cdots \supset T^{n*} \supset \cdots$. At the nth stage, we have the relations

$$\left|\int_{\partial T} f(z)\,dz\right| \le 4^n\left|\int_{\partial T^n} f(z)\,dz\right|, \quad (9.2.2\text{a})$$

$$L(\partial T^n) = \left(\frac{1}{2}\right)^n L(\partial T), \quad (9.2.2\text{b})$$

$$\operatorname{diam}(T^{n*}) = \left(\frac{1}{2}\right)^n \operatorname{diam}(T^*). \quad (9.2.2\text{c})$$

Equation (9.2.2c) shows that $\lim_{n\to\infty} \operatorname{diam}(T^{n*}) = 0$. Therefore, by

Proposition 5.6.1, the intersection $\bigcap_{n=1}^{\infty} T^{n*}$ consists of a single point, say w. Now we use the differentiability hypothesis on the function f. According to Proposition 6.1.1, there exists a function $\delta(z)$ with domain O, continuous at w, such that $f(z) - f(w) = \delta(z)(z - w)$ for every $z \in O$. Since f is continuous on O, we can further assert that $\delta(z)$ is continuous on O. Recall that if γ is any closed path, $\int_\gamma (z - w)\, dz = 0$ and $\int_\gamma c\, dz = 0$, where c is a constant. Thus, we have

$$\begin{aligned}\int_{\partial T^n} f(z)\, dz &= \int_{\partial T^n} [f(w) + \delta(z)(z - w)]\, dz \\ &= \int_{\partial T^n} [f(w) + (\delta(z) - \delta(w))(z - w) + \delta(w)(z - w)]\, dw \\ &= 0 + \int_{\partial T^n} [\delta(z) - \delta(w)](z - w)\, dz + 0. \qquad (9.2.3)\end{aligned}$$

The existence of the last integral is assured by the continuity of $\delta(z)$. Take any $\varepsilon > 0$. By the continuity of $\delta(z)$ at w, there exists $\eta = \eta_\varepsilon > 0$ such that if $|z - w| < \eta_\varepsilon$, $z \in O$, then

$$|\delta(z) - \delta(w)| < \varepsilon/[L(\partial T) \operatorname{diam}(T^*)].$$

Again by Proposition 5.6.1, there exists N_η such that for every $n \geq N_\eta$, $T^{n*} \subset \Delta(w, \eta_\varepsilon)$. With such a value of n, we have, using (9.2.2a), (9.2.2b), and (9.2.3),

$$\begin{aligned}\left|\int_{\partial T} f(z)\, dz\right| &\leq 4^n \left|\int_{\partial T^n} f(z)\, dz\right| \\ &= 4^n \left|\int_{\partial T^n} [\delta(z) - \delta(w)](z - w)\, dz\right| \\ &\leq 4^n \max_{z \in \partial T^n} |[\delta(z) - \delta(w)](z - w)| L(\partial T^n) \\ &\leq 4^n \left(\frac{\varepsilon}{L(\partial T) \operatorname{diam}(T^*)}\right) \cdot \operatorname{diam}(\partial T^{n*}) L(\partial T^n) \\ &\leq 4^n \frac{\varepsilon}{L(\partial T) \operatorname{diam}(T^*)} \cdot \left(\frac{1}{2}\right)^n \cdot \operatorname{diam}(T^*) \cdot \left(\frac{1}{2}\right)^n L(\partial T) \\ &= \varepsilon.\end{aligned}$$

But, of course, ε is arbitrary, so it follows that $\int_{\partial T} f(z)\, dz = 0$. Q.E.D.

Theorem 9.2.2 (*Cauchy Integral Theorem for Starlike Regions*). *Let B be a starlike region in* **C**, *and let f: $B \to$* **C** *be analytic on B. Then there exists a primitive for f in B. Consequently, for any two points $z_1, z_2 \in B$, the integral $\int_\gamma f(z)\, dz$ has the same value for every path γ with graph in B joining z_1 to z_2, and if γ is a closed path, $\int_\gamma f(z)\, dz = 0$.*

Proof. According to Theorem 9.2.1, if a is the star center of B and if $T = (a, z_1, z_2)$ is such that T^* lies in B, then the integral of f over ∂T is 0. Theorem 8.5.2 guarantees the existence of a primitive under these conditions, and the consequences of its existence, as stated in the present theorem, are established in Theorem 8.5.1 and its corollary. Q.E.D.

9.3 Some Extensions of the Cauchy Integral Theorem

For some of the applications of the Cauchy integral theorem as it appears in Theorem 9.2.2, a slight weakening of the hypotheses is required. This is embodied in the following:

Proposition 9.3.1 *Let O be an open set in* **C**, *let $f\colon O \to \mathbf{C}$ be analytic on O except possibly at one point $a \in O$, and let f be continuous on O. Then if $T = (z_1, z_2, z_3)$ is an ordered triple of distinct points such that $T^* \subset O$, again $\int_{\partial T} f(z)\, dz = 0$.*

Proof. Notice that $0 \sim \{a\}$ is an open set. Therefore, if $a \notin T^*$, Theorem 9.2.1 yields the conclusion of this proposition immediately for such a triangle T^*. The important case to consider turns out to be the one in which a is at one of the vertices of T^*, say $a = z_1$. If a, z_2, z_3 are collinear, the conclusion of the proposition follows at once merely by cancelation of integrals along opposite line segements. We henceforth assume that a, z_2, z_3 are not collinear.

Choose $\zeta_1 \in [a, z_2]$, $\zeta_1 \neq a$, $\zeta_1 \neq z_2$, and $\zeta_2 \in [a, z_3]$, $\zeta_2 \neq a$, $\zeta_2 \neq z_3$. (See Figure 2.) Let T_1^*, T_2^*, T_3^* denote, respectively, the triangles determined by the ordered triples $T_1 = (z_1, \zeta_1, \zeta_2)$, $T_2 = (\zeta_1, z_2, z_3)$, and $T_3 = (\zeta_1, z_3, \zeta_2)$, and let $\partial T_1, \partial T_2, \partial T_3$ be the closed polygonal paths determined by these triples. Each of the triangles $T_j^*, j = 1, 2, 3$, is the convex hull of a set of points lying in the convex set T^*, so each of these triangles is contained in T^* and, therefore, in O. It is readily verified that

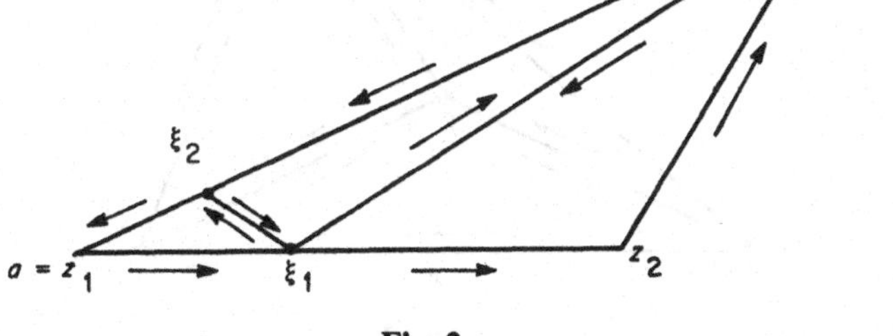

Fig. 2

$$\int_{\partial T} f(z)\, dz = \int_{\partial T_1} f(z)\, dz + \int_{\partial T_2} f(z)\, dz + \int_{\partial T_3} f(z)\, dz. \qquad (9.3.1)$$

It is intuitively apparent that $a \notin T_2^*$ and $a \notin T_3^*$; a rigorous proof is deferred to Exercise 9.3.5. It follows from the remark at the beginning of this proof that the last two integrals in (9.3.1) are both equal to zero, so we have

$$\left|\int_{\partial T} f(z)\, dz\right| = \left|\int_{\partial T_1} f(z)\, dz\right| \le \max_{z \in \partial T_1} |f(z)| L(\partial T_1). \qquad (9.3.2)$$

Since O is open, there exists $r > 0$ such that $\Delta(a, 2r) \subset O$. Let $M = \max\{|f(z)| : z \in \bar{\Delta}(a, r)\}$. Take any $\varepsilon > 0$, and choose ζ_1 and ζ_2 so that $|\zeta_1 - a| < \min(r, \varepsilon/4M)$, $|\zeta_2 - a| < \min(r, \varepsilon/4M)$. Then $\zeta_1 \in \Delta(a, r)$ and $\zeta_2 \in \Delta(a, r)$, so $[\zeta_1, \zeta_2] \subset \Delta(a, r)$ [Proposition 5.3.6]. Also, $|\zeta_1 - \zeta_2| \le |\zeta_1 - a| + |a - \zeta_2| < 2\varepsilon/4M$. We return to (9.3.2) with these inequalities and obtain

$$\left|\int_{\partial T} f(z)\, dz\right| \le M[|\zeta_1 - a| + |\zeta_2 - a| + |\zeta_1 - \zeta_2|]$$
$$< M\left(\frac{\varepsilon + 2\varepsilon + \varepsilon}{4M}\right) = \varepsilon.$$

This, of course, implies that the integral in the left-hand member is zero.

There remain some other cases to consider, but fortunately they can all be taken care of by one construction. Suppose $a \in T^*$, but $a \neq z_1, z_2, z_3$. Consider the ordered triples $T_1 = (a, z_1, z_2)$, $T_2 = (a, z_2, z_3)$, $T_3 = (a, z_3, z_1)$. As in the earlier construction, each of the triangles T_1^*, T_2^*, T_3^* is the convex hull of a set of three points lying in the convex set T^*, so each of the triangles is contained in T^* and, therefore, in O. Let $\partial T, \partial T_1, \partial T_2, \partial T_3$ have the usual meanings. It is readily verified (see Figure 3) that

$$\int_{\partial T} f(z)\, dz = \int_{\partial T_1} f(z)\, dz + \int_{\partial T_3} f(z)\, dz + \int_{\partial T_3} f(z)\, dz. \qquad (9.3.3)$$

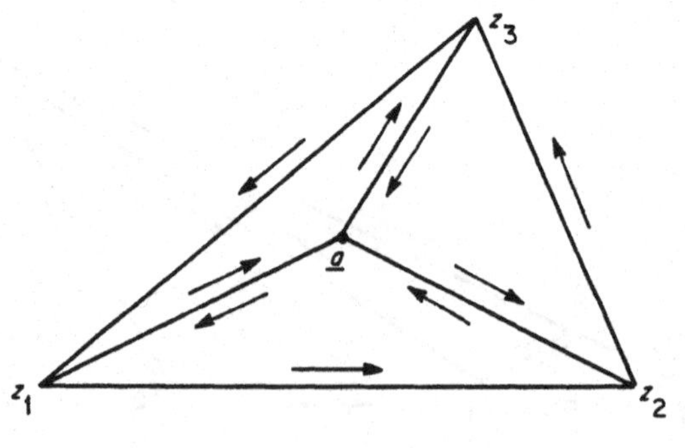

Fig. 3

The integrals on the right-hand side of the equation are all equal to zero, by the first part of this proof, so $\int_{\partial T} (f(z)\, dz = 0$ again.

Equation (9.3.3) remains valid if a lies on a segment of ∂T. In fact, it is not even necessary for the validity of the argument to assume tacitly, as in Figure 3, that if $a \in T^*$, then it lies on the "inside" the triangular path ∂T. Q.E.D.

Proposition 9.3.2 *Let B be a starlike region in* **C**, *and let* $f\colon B \to \mathbf{C}$ *be analytic on B except possibly at a single point in B, and let f be continuous on B. Then there exists a primitive for f in B. Consequently, for any two points* $z_1, z_2 \in B$, *the integral* $\int_\gamma f(z)\, dz$ *has the same value for any path* γ *with graph in B joining* z_1 *to* z_2, *and if* γ *is a closed path,* $\int_\gamma f(z)\, dz = 0$.

The proof is exactly the same as that for Theorem 9.2.2, except that Proposition 9.3.1 is substituted for Thoerem 9.2.1. Propositions 9.3.1 and 9.3.2 can be generalized to permit more than one exceptional point of non-analyticity.

There are extensions of the Cauchy integral theorem in which the integration is along the boundary of a region on which the integrand is continuous but analytic only at interior points. A general treatment is beyond the scope of this book, but the case in which the region is of the starlike type introduced in Exercise 7.5.10 can be handled with elementary analytical methods. The formal statement is as follows.

Proposition 9.3.3 *Let* γ *be the closed path given by the parametric equation* $z = \gamma(t) = a + r(t)e^{it}$, $0 \le t \le 2\pi$, *where* $r(t)$ *is continuous,* $r(t) > 0$ *for every* $t \in [0, 2\pi]$, *and* $r(0) = r(2\pi)$. *Furthermore, suppose that* $r(t)$ *is piecewise continuously differentiable. Let* $B = \{z\colon z = a + sr(t)\, e^{it}, 0 \le t \le 2\pi, 0 \le s < 1\}$, *Let* $f\colon \bar{B} \to \mathbf{C}$ *be analytic on B except possibly at one point in B, and let f be continuous on* $\bar{B}$. *Then* $\int_\gamma f(z)\, dz = 0$.

Proof. It is established in Exercises 7.5.10 and 7.5.11 that B is a starlike region and that its closure $\bar{B}$ is $\bar{B} \cup \Gamma$, where Γ is the graph of γ. For any s, $0 < s < 1$, let γ_s be the closed path with parametric equation $\zeta = a + sr(t)\, e^{it}$, $0 \le t \le 2\pi$. Clearly the graph of γ_s lies in B. Therefore, by Proposition 9.3.2, $0 = \int_{\gamma_s} f(z)\, dz = \int_0^{2\pi} sf[a + sr(t)\, e^{it}]r'(t)\, dt = I(s)$. It is easily verified by referring to Exercises 3.2.8 and 3.2.10 and Theorem 7.3.1 that $sf[a + sr(t)\, e^{it}]$ is continuous on the rectangle $\{(t, s)\colon 0 \le t \le 2\pi, 0 \le s \le 1\}$. Therefore, by Exercise 8.3.7, the integral $I(s)$ is continuous on [0, 1]. It follows that $I(1) = \int_\gamma f(z)\, dz = 0$. Q.E.D.

Exercises 9.3

1. Let $f\colon O \to C$ be analytic on the open set $O \in \mathbf{C}$ except possibly at a point $a \in O$, and let f be continuous at a. Let the rectangle $B = \{z = x + iy\colon \alpha \le x \le \alpha + h,$

$\beta \le y \le \beta + k$, $h > 0$, $k > 0$, $\alpha \in \mathbf{R}$, $\beta \in \mathbf{R}\}$ be contained in O. Show that $\int_{\partial B} f(z)\,dz = 0$, where ∂B is the perimeter of B.

2. Let f be analytic on $\Delta'(a, R) = \Delta(a, R) \sim \{a\}$, $R > 0$, except possibly at a point $z_0 \in \Delta'(a, R)$, and let f be continuous at z_0. Let χ_1 be a semicircle with equation $z = a + r_1 e^{-it}, -\pi \le t \le 0$, and χ_2 be a semicircle with equation $z = a + r_2 e^{it}$, $0 \le t \le \pi$, where $0 < r_1 < r_2 < R$. Let γ be the closed path $[a + r_1, a + r_2] + \chi_2 + [a - r_2, a - r_1] + \chi_1$. Then $\int_\gamma f(z)\,dz = 0$. (*Hint*: In Exercise 7.5.12, it is established that the circular sector $S(t_0) = \{z : z = a + \sigma R e^{it}, t_0 - \pi/2 < t < t_0 + \pi/2, 0 < \sigma < 1\}$ is a convex region and, therefore, is starlike. Let $\delta_1 = [a + r_1, a + r_2] + \chi_2|[0, \pi/2] + [a + ir_2, a + ir_1] + \chi_1|[-\pi/2, 0]$. Verify that the graph of δ_1 lies in $S(\pi/4)$, so by Proposition 9.3.2, $\int_{\delta_1} f(z)\,dz = 0$. Similarly, construct a closed path δ_2 such that $\int_{\delta_2} f(z)\,dz = 0$ and $\int_{\delta_1 + \delta_2} f(z)\,dz = \int_\gamma f(z)\,dz$. A figure might be useful.)

3. Let f be analytic on $\Delta'(a, R) = \Delta(a, R) \sim \{a\}$, $R > 0$, except possibly at a point $z_0 \in \Delta'(a, R)$, and let f be continuous at z_0. Let γ_1 be a positively oriented circle with graph $C(a, r_1)$, and let γ_2 be a positively oriented circle with graph $C(a, r_2)$, where $0 < r_1 < r_2 < R$. Then $\int_{\gamma_1} f(z)\,dz = \int_{\gamma_2} f(z)\,dz$. (*Hint*: Express $\int_{\gamma_2} f(z)\,dz - \int_{\gamma_1} f(z)\,dz$ as the sum of the integrals of f along two closed paths, one of which is the γ in Exercise 9.3.2 and the other is of a similar type composed in part of semicircles given by $z = a + r_1 e^{-it}$, $0 \le t \le \pi$, and $z = a + r_2 e^{it}$, $\pi \le t \le 2\pi$.)

   Remark. Let f be as in Exercise 9.3.3, and let γ_r be a positively oriented circle with graph $C(a, r)$, $0 < r < R$. Exercise 9.3.3 shows that $\int_{\gamma_r} f(z)\,dz$ is constant with respect to r.

4. (A theorem of Fejér and F. Riesz.) Let $f : \bar{\Delta}(0, 1) \to \mathbf{C}$ be continuous on $\bar{\Delta}(0, 1)$ and analytic on $\Delta(0, 1)$. Then with $z = x + iy$, $2\int_{-1}^{1} |f(x)|^2\,dx \le \int_0^{2\pi} |f(e^{i\theta}) f(e^{-i\theta})|\,d\theta$. (*Hint*: The reasoning for Exercise 6.3.7 shows that $\overline{f(\bar{z})}$ is analytic on $\Delta(0, 1)$. Then $f(z) \cdot \overline{f(\bar{z})}$ is analytic on $\Delta(0, 1)$ and continuous on $\bar{\Delta}(0, 1)$. Apply Proposition 9.3.3 to $\int_\gamma f(z) \cdot \overline{f(\bar{z})}\,dz$, with γ taken first to be the closed path with graph consisting of the upper half of $C(0, 1)$ and the diameter $[-1, 1]$, and then the lower half of $C(0, 1)$ and this diameter.)

5. By using the fact that the convex hull of a set of points E in $\mathbf{C}$ is the set of all weighted averages of the points in E [Exercise 5.3.10], show algebraically that the triangles T_2^* and T_3^* in the first construction in the proof of Proposition 9.3.1 (see Figure 2) do not contain the point a. [*Hint*: The proof can be given by showing that if a were contained in either triangle, it would have to lie on $[z_2, z_3]$, contrary to assumption. Suppose $a \in T_3^*$. Then (*) $a = m_1\zeta_1 + m_2\zeta_2 + m_3 z_3$, where $m_1 \ge 0$, $m_2 \ge 0$, $m_3 \ge 0$ and $m_1 + m_2 + m_3 = 1$. Now $\zeta_1 = (1 - t)z_2 + ta$, $0 < t < 1$, $\zeta_2 = (1 - s)z_3 + sa$, $0 < s < 1$. Substitute in (*).]

9.4 The Cauchy Integral Formula and the LPS Property of Analytic Functions

One version of this famous representation formula for analytic functions appears in the following theorem.

Theorem 9.4.1 *Let B be a starlike region, let f: $B \to \mathbf{C}$ be analytic on B, and let γ be a closed path with graph Γ contained in B. Then*

$$\operatorname{Ind}_\gamma (z) f(z) = \frac{1}{2\pi i} \int_\gamma \frac{f(\zeta)\, d\zeta}{\zeta - z} \tag{9.4.1}$$

for every $z \in \Gamma^c \cap B$.

Proof. Hold z fixed in $\Gamma^c \cap B$. As a function of ζ,

$$\delta(\zeta) = \begin{cases} \dfrac{f(\zeta - f(z)}{\zeta - z}, & \zeta \neq z, \qquad \zeta \in B, \\ f'(z), & \zeta = z, \end{cases}$$

is analytic on B except possibly at $\zeta = z$, where by the definition of derivative it is continuous. Then by Proposition 9.3.2, we have

$$\begin{aligned} 0 = \int_\gamma \delta(\zeta)\, d\zeta &= \int_\gamma \frac{f(\zeta) - f(z)}{\zeta - z}\, d\zeta \\ &= \int_\gamma \frac{f(\zeta)\, d\zeta}{\zeta - z} - f(z) \int_\gamma \frac{1}{\zeta - z}\, d\zeta; \\ \int_\gamma \frac{f(\zeta)\, d\zeta}{\zeta - z} &= 2\pi i \operatorname{Ind}_\gamma (z) f(z). \end{aligned}$$

Q.E.D.

Corollary Let B be a starlike region, let f: $B \to \mathbf{C}$ be analytic on B, and let γ be a positively oriented circle with graph $C(a, R)$ contained in B. Then

$$f(z) = \frac{1}{2\pi i} \int_\gamma \frac{f(\zeta)\, d\zeta}{\zeta - z}, \qquad z \in \Delta(a(R). \tag{9.4.2}$$

The integrals which appear in (9.4.1) and (9.4.2) are of the special type considered in Sec. 8.4. It follows from Theorem 8.4.1 that $\operatorname{Ind}_\gamma (z) f(z)$ (which is a constant multiple of $f(z)$ on the intersection of any component of Γ^c with B) possesses the LPS (local power series) property on $\Gamma^c \cap B$. This remark is of limited interest in the context of the general case represented by (9.4.1), because $\operatorname{Ind}_\gamma (z)$ may be zero. (It *is* zero for z on the unbounded component of Γ [Theorem 8.4.2]). But in the special case in which γ is a positively oriented circle, the remark becomes important. We formulate the result in detail as follows.

Theorem 9.4.2 *Let O be an open set, and let f: $O \to \mathbf{C}$ be analytic on O. Then f possesses the LPS property on O and, consequently, possesses analytic derivatives of all orders at each point of O, each of which itself has the LPS property* [*Proposition* 6.5.1]. *Furthermore*:

(i) *For each point $a \in O$, the local power series representation $f(z) =$*

$\sum_{k=0}^{\infty} a_k(a)(z-a)^k$ *converges to* $f(z)$ *for every* $z \in \Delta[a, \rho(a)]$, *where* $\rho(a) =$ *dist* (a, O^c). (*If* $O^c = \varnothing$, *we let* $\rho(a) = \infty$.)

(ii) *For any* r, $0 < r < \rho(a)$, *the following formulas are valid*:

$$a_k(a) = \frac{f^{(k)}(a)}{k!} = \frac{1}{2\pi i}\int_{\gamma_r} \frac{f(\zeta)\,d\zeta}{(\zeta - a)^{k+1}}, \qquad k = 0, 1, 2, \ldots, \qquad (9.4.3)$$

where γ_r *is the positively oriented circle with graph* $C(a, r)$. *The integrals in* (9.4.3) *are independent of* r, $0 < r < \rho(a)$.

Proof. The disk $\Delta(a, \rho(a))$ is a starlike region in which f is analytic, so the representation (9.4.2) is valid with $R = r$, $\gamma = \gamma_r$, where r and γ_r are described in part (ii) of the theorem. According to Theorem 8.4.1, the integral on the right-hand side of (9.4.2) can be expanded in a series of powers of $z - a$ convergent to $f(z)$ for every $z \in \Delta(a, r)$. (Here $r = \text{dist}\,(a, C(a, r))$ plays the role of $d(a)$ in Theorem 8.4.1.) The formulas for the coefficients and derivatives in (9.4.3) are simply the specialization of the formulas (8.4.2) to the case in which $g(\zeta) = f(\zeta)$, $h(\zeta) = \zeta$, $\gamma = \gamma_r$.

The fact that the integrals on the right-hand side of (9.4.3) are independent of r, $0 < r < \rho(a)$, is implied by the validity of the second equation in (9.4.3) for any such choice of r. Given any $z \in \Delta[a, \rho(a)]$, it is, of course, possible to adjust r so that $z \in \Delta(a, r)$. (For example, take $r = [\rho(a) + |z - a|]/2$.) Thus, the power series representation $\sum_{k=0}^{\infty} a_k(a)(z-a)^k = f(z)$ is valid for every $z \in \Delta[a, \rho(a)]$. Q.E.D.

Remark. In certain important applications of this theorem, the function $f: O \to \mathbf{C}$ is the restriction to O of a function analytic on a larger open set containing O. In that case, the radius of convergence of the local power series at a may exceed $\rho(a) = \text{dist}(a, O^c)$.

At this point, it is suggested that the reader review the facts in Sec. 6.5 on functions with the LPS property.

A variant of the Cauchy integral formula, which is sometime useful, can be obtained from Proposition 9.3.3.

Proposition 9.4.1 *Let* γ *be a closed path given by the parametric equation* $z = r(t)\,e^{it} + a$, $0 \le t \le 2\pi$, *where* $r(t) > 0$, $r(t)$ *is piecewise continuously differentiable, and* $r(0) = r(2\pi)$. *Let* $B = \{z: z = sr(t)\,e^{it} + a,\ 0 \le t \le 2\pi,\ 0 \le s < 1\}$. *Let* $f: \bar{B} \to \mathbf{C}$ *be continuous on* $\bar{B}$ *and analytic on* B. *Then for every* $z \in B$,

$$f(z) = \frac{1}{2\pi i}\int_{\gamma} \frac{f(\zeta)\,d\zeta}{\zeta - z}.$$

Proof. It was shown in Sec. 8.4, Example (iii), that for $z \in B$, $\text{Ind}_\gamma(z) = 1$. We proceed as in the proof of Theorem 9.4.1, setting up a function

$$\delta(\zeta) = \begin{cases} \dfrac{f(\zeta) - f(z)}{\zeta - z}, & \zeta \neq z, \zeta \in \bar{B}, z \in B. \\ f'(z), & \zeta = z. \end{cases}$$

By Proposition 9.3.3, we have $\int_\gamma \delta(\zeta)\, d\zeta = 0$, and the proof thereafter is the same as that for Theorem 9.4.1.

Corollary 1 Let γ be a positively oriented circle with equation $z = re^{it} + a$, $0 \leq t \leq 2\pi$, and let $f\colon \bar{\Delta}(a, r) \to \mathbf{C}$ be continuous on $\bar{\Delta}(a, r)$ and analytic on $\Delta(a, r)$. Then for every $z \in \Delta(a, r)$,

$$f(z) = \frac{1}{2\pi i} \int_\gamma \frac{f(\zeta)\, d\zeta}{\zeta - z} = \sum_{k=0}^{\infty} a_k (z - a)^k, \tag{9.4.4}$$

where

$$a_k = \frac{f^{(k)}(a)}{k!} = \frac{1}{2\pi i} \int_\gamma \frac{f(\zeta)}{(\zeta - a)^{k+1}}, \qquad k = 0, 1, \ldots .$$

Proof. The first equation in (9.4.4) of course follows directly from Proposition 9.4.1, specialized to the case in which B is a circular disk. The power series representation is justified by Theorem 8.4.1. Q.E.D.

Corollary 2 Let $f\colon \bar{\Delta}(0, 1) \to \mathbf{C}$ be continuous on $\bar{\Delta}(0, 1)$ and analytic on $\Delta(0, \rho)$, $\rho > 0$, and let $f(z) = U(z) + iV(z)$, $z \in \bar{\Delta}(0, 1)$, where U and V are real valued. Then in the power series representation $f(z) = \sum_{k=0}^{\infty} a_k z^k$, $z \in \Delta(0, \rho)$, we have

$$a_k = \frac{1}{\pi r^k} \int_{-\pi}^{\pi} U(re^{i\theta}) e^{-ik\theta}\, d\theta + \frac{1}{\pi r^k} \int_{-\pi}^{\pi} V(re^{i\theta}) e^{-ik\theta}\, d\theta,$$
$$0 < r \leq 1, \qquad k = 0, 1, 2, \ldots .$$

Remark. These formulas provide a link between the Cauchy theory and the theory of Fourier series. The second integral is used in Sec. 13.6 in the proof of a special case of the Bieberbach conjecture for functions analytic and univalent in the unit disk.

Proof. Let γ_r be a positively oriented circle with equation $z = re^{i\theta}$, $-\pi \leq \theta \leq \pi$, with r fixed, $0 < r \leq \rho$. Then from (9.4.4), we obtain

$$\begin{aligned} a_k &= \frac{1}{2\pi i} \int_{\gamma_r} f(\zeta) \zeta^{-k-1}\, d\zeta \\ &= \frac{1}{2\pi i} \int_{-\pi}^{\pi} f(re^{i\theta}) r^{-k-1} e^{-i(k+1)\theta} ire^{i\theta}\, d\theta \\ &= \frac{1}{2\pi r^k} \int_{-\pi}^{\pi} [U(re^{i\theta}) + iVre^{i\theta}] e^{-ik\theta}\, d\theta. \end{aligned}$$

From the version of the Cauchy integral theorem appearing in Proposition 9.3.3, we have

$$0 = \frac{1}{2\pi i}\int_{\gamma_r} f(\zeta)\zeta^{k-1}\,d\zeta$$
$$= \frac{r^k}{2}\int_{-\pi}^{\pi} [U(re^{i\theta}) + iV(re^{i\theta})]e^{ik\theta}\,d\theta.$$

The real and imaginary parts of the right-hand member here are necessarily zero, so the equation remains true when the integrand is replaced by its complex conjugate. After dividing by r^{2k}, we obtain

$$0 = \frac{1}{2\pi r^k}\int_{-\pi}^{\pi} [U(re^{i\theta}) - iV(re^{i\theta})]e^{-ik\theta}\,d\theta.$$

We first add this to the right-hand side of the forgoing formula for a_k and then subtract it from that formula. The result is the conclusion of the corollary. Q.E.D.

The fact that an analytic function possesses analytic derivatives can be used to prove a converse to the Cauchy integral theorem for a triangle, which can be formulated as follows.

Theorem 9.4.3 (*Morera's Theorem*). *Let O be an open set, and let $f\colon O \to \mathbf{C}$ be continuous on O and have the property that for every ordered triple T with the corresponding triangle T^* lying in O, $\int_{\partial T} f(z)\,dz = 0$. Then f is analytic on O.*

Proof. Let a be any point in O. There exists a disk $\Delta(a, r) \subset O$, $r > 0$; this disk is a convex region, and so is starlike with respect to a. By hypothesis, for any ordered triple $T = \{a, z_1, z_2\}$, with $z_1, z_2 \in \Delta(a, r)$, we have $\int_{\partial T} f(z)\,dz = 0$. According to Theorem 8.5.2, f has a primitive F in $\Delta(a, r)$. By definition of primitive, F is analytic in $\Delta(a, r)$ and $F' = f$. But furthermore by Theorem 9.3.2, the derivative F' is itself analytic on $\Delta(a, r)$, which, in particular, implies that $F''(a) = f'(a)$ exists. Since $a \in O$ was arbitrary, the conclusion follows at once. Q.E.D.

Remark. The proof of the theorem depends only on the existence of a local primitive, which may change as the point a is moved around O. For example, the function $f(z) = 1/z$, analytic on $\mathbf{C} \sim \{0\}$, has no global primitive in $C \sim \{0\}$ [Example (iv) of Sec. 8.5].

Another application of the fact that an analytic function possesses an analytic derivative is related to the concept of an analytic logarithm of a function. Let $f\colon O \to \mathbf{C}$ be analytic on the open set O. An *analytic logarithm* of f on O, if it exists, is a function $g\colon O \to \mathbf{C}$ analytic on O and such that $e^{g(z)} \equiv f(z)$ on O. The following proposition is in the nature of a lemma for an existence theorem which appears in the corollary that follows.

Proposition 9.4.2 *Let $f: O \to \mathbf{C}$ be analytic on the open set O and be such that $f(z) \neq 0$ on O. A necessary and sufficient condition for f to have an analytic logarithm on O is that f'/f has a primitive on O.*

Proof. If f'/f has a primitive F on O, let $\phi(z) = f(z)\, e^{-F(z)}$. The basic laws of differentiation and the analyticity of the exponential function ensure that ϕ' exists everywhere on O; in fact, $\phi'(z) = f'(z)e^{-F(z)} + f(z)e^{-F(z)}\, F'(z)$, and substitution of f'/f for F' reveals that $\phi'(z) \equiv 0$ on O. Thus, for each component of O, there exists a complex constant, say k, such that $f(z) = ke^{F(z)}$ for every z on the component [Theorem 6.2.1]. By hypothesis, $k \neq 0$. Then $f(z) = e^{F(z) + \operatorname{Log} k}$, so $F(z) + \operatorname{Log} k$ is an analytic logarithm of $f(z)$.

The proof of the converse is easy and is left as an exercise (Exercise 9.4.8). Notice that the proof of the proposition requires only the elementary rules on differentiation and properties of the exponential function. But Cauchy theory reappears in the proof of the following corollary.

Corollary Let B be a starlike region, and let $f: B \to \mathbf{C}$ be analytic on B and $f(z) \neq 0$ on B. Then f possesses an analytic logarithm on B.

Proof. By Theorem 9.4.2, f' is analytic on B, and since $f(z) \neq 0$ on B, it follows that f'/f is analytic on B. Therefore, by Theorem 9.2.2, it possesses a primitive on B. Reference to Proposition 9.4.2 completes the proof. Q.E.D.

Example. Let $p(z)$ be a polynomial, and let $\Delta(a, R)$, $R > 0$, contain no zero of $p(z)$. Then $p(z)$ has an analytic logarithm on $\Delta(a, R)$.

The analyticity of the derivative of an analytic function has an important implication for the partial derivatives of the function $f(x + iy) = F(x, y) = U(x, y) + iV(x, y)$, $(x, y) \in \mathbf{R}^2$, where U and V are real valued. Specifically, let $f: O \to \mathbf{C}$ be analytic on the open set O. According to Theorem 6.3.1, $f'(z) = F_x(x, y) = -iF_y(x, y)$ for every $z = x + iy \in O$. But the analyticity of f', taken with this equation, yields the further equation

$$\begin{aligned} f''(z) &= F_{xx}(x, y) = -iF_{yx}(x, y) = -iF_{xy}(x, y) \\ &= (-i)(-iF_{yy}(x, y)) = -F_{yy}(x, y), \end{aligned} \tag{9.4.5}$$

where $F_{xx} = (\partial/\partial x)F_x$, etc. The function f'' is itself analytic, so the partial derivatives which appear in (9.4.4) are continuous on O. From the equality of the second and last members of (9.4.5), we obtain

$$\begin{aligned} &F_{xx}(x, y) + F_{yy}(x, y) = 0, \\ &U_{xx}(x, y) + U_{yy}(x, y) = 0, \; V_{xx}(x, y) + V_{yy}(x, y) = 0, \end{aligned} \tag{9.4.6}$$

valid for every $(x, y) \in O$.

The differential equation $\Phi_{xx} + \Phi_{yy} = 0$ is called the *Laplace differential equation.* A function $\Phi: O \to \mathbf{R}$ which has continuous first and second partial derivatives on the open set O and which satisfies the Laplace equation is said to be *harmonic* on O. We formalize the message in (9.4.6) briefly:

Theorem 9.4.4 *Let O be an open set in $\mathbf{C}$ and $\mathbf{R}^2$, and let $f: O \to \mathbf{C}$ have values $f(x + iy) = U(x, y) + iV(x, y)$, where U and V are real valued. If f is analytic on O, then U and V are harmonic on O.*

Exercises 9.4

1. Let γ be a closed path with graph Γ. By using the Cauchy integral formula, evaluate the following integrals:

(a) $\displaystyle\int_\gamma \frac{e^\zeta - e^z}{\zeta - z}\, d\zeta, \quad z \notin \Gamma.$ *Answer.* 0.

(b) $\displaystyle\int_\gamma \frac{d\zeta}{\zeta - 3}, \quad 3 \notin \Gamma.$ *Answer.* $2\pi i \operatorname{Ind}_\gamma(3)$.

(c) $\displaystyle\int_\gamma \frac{\cos z}{z}\, dz, \quad 0 \notin \Gamma.$ *Answer.* $2\pi i \operatorname{Ind}_\gamma(0)$.

(d) $\displaystyle\int_\gamma \frac{1}{z^2 + 1}\, dz, \quad \pm i \notin \Gamma.$ *Answer.* $\pi[\operatorname{Ind}_\gamma(i) - \operatorname{Ind}_\gamma(-i)]$.

$\left(\textit{Hint for (d)}: \dfrac{1}{z^2+1} = \dfrac{i}{2}\left[\dfrac{1}{z+i} - \dfrac{1}{z-i}\right]\right)$

2. Find the value of the integral in Exercise 9.4.1(b) when γ is a positively oriented circle with graph (i) $C(1, 1)$, (ii) $C(1, 5)$. *Answers.* (i) 0, (ii) $2\pi i$.
3. Evaluate by Cauchy theory:

(a) $\displaystyle\int_\gamma \left(w + \frac{1}{w}\right)^3 dw,$

(b) $\displaystyle\int_\gamma \frac{z\, dz}{(9 - z^2)(z + i)},$

where in each case γ is a positively oriented circle with graph $C(0, 2)$. *Answers.* (a) $6\pi i$, (b) $\pi/5$.

4. Let f be analytic on an open set O; let $a \in O$, $\rho(a) = \operatorname{dist}(a, O^c)$; and let γ_r be any positively oriented circle with graph $C(a, r)$, $0 < r < \rho(a)$. Show that

$$k!\int_{\gamma_r} \frac{f(z)\, dz}{(z - a)^{k+1}} = \int_{\gamma_r} \frac{f^{(k)}(z)\, dz}{z - a}.$$

5. Prove that

$$\frac{1}{2\pi}\int_0^{2\pi} \frac{R^2 - r^2}{R^2 + r^2 - 2rR\cos\theta}\, d\theta = 1, \qquad 0 < r < R,$$

by carrying out the following program: Show by Cauchy theory that

$$\frac{1}{2\pi i}\int_{\gamma_r} \frac{R + z}{(R - z)z}\, dz = 1, \tag{9.4.7}$$

where γ_r has the equation $z = re^{i\theta}$, $0 \le \theta \le 2\pi$; then after substituting with this parametric equation in (9.4.7) find the real part of the integral.

6. (Cauchy–Taylor series with remainder.) Let f be analytic on an open set O; $a \in O$, $\rho(a) = \text{dist}(a, O^c)$; let γ_r be a positively oriented circle with graph $C(a, r)$, $0 < r < \rho(a)$. For every $z \in \Delta(a, r)$,

$$f(z) = f(a) + \frac{f'(a)}{1!}(z - a) + \frac{f''(a)}{2!}(z - a)^2$$
$$+ \cdots + \frac{f^{(n)}(a)}{n!}(z - a)^n + (z - a)^{n+1} R_n(z),$$

where

$$R_n(a) = \frac{1}{2\pi i}\int_{\gamma_r} \frac{f(\zeta)d\zeta}{(\zeta - a)^{n+1}(\zeta - z)}.$$

(*Hint*: $1/(1 - w) \equiv 1 + w + w^2 + \cdots + w^n + w^{n+1}/(1 - w)$, $w \neq 1$. Write the Cauchy integral formula in the form

$$f(z) = \frac{1}{2\pi i}\int_{\gamma_r} \frac{f(\zeta)d\zeta}{\zeta - z} = \frac{1}{2\pi i}\int_{\gamma_r} \frac{f(\zeta)d\zeta}{(\zeta - a)[1 - (z - a)/(\zeta - a)]}$$

and let $w = (z - a)/(\zeta - a)$.)

7. A simplified version of Theorem 8.4.1 reads as follows: Let γ be a path with graph Γ and let $g:\Gamma \to \mathbf{C}$ be continuous. The function

$$f(z) = \int_\gamma \frac{g(\zeta)d\zeta}{\zeta - z}, \qquad z \in \Gamma^c,$$

is analytic on Γ^c, and for any $a \in \Gamma^c$,

$$f^{(k)}(a) = k!\int_\gamma \frac{g(\zeta)d\zeta}{(\zeta - a)^{k+1}}, \qquad k = 1,2, \ldots. \tag{9.4.8}$$

Show that this implies that the function

$$F_n(z) = \int_\gamma \frac{g(\zeta)d\zeta}{(\zeta - z)^n}, \qquad n = 1,2, \ldots$$

is analytic on Γ^c and that $F_n'(z) = nF_{n+1}(z)$ for every $z \in \Gamma^c$.

8. Prove that if f is analytic on O, an open set, and if f has an analytic logarithm on O, then $f(z) \neq 0$ on O and f'/f has a primitive on O. (*Hint*: Merely set up the equation defining the analytic logarithm and differentiate it.)
9. Let $f: O \to \mathbf{C}$ be analytic on the open set O. An analytic square root of f on O, if it exists, is a function $\phi: O \to \mathbf{C}$ analytic on O such that $[\phi(z)]^2 \equiv f(z)$ on O. Prove that if O is starlike and $f(z) \neq 0$ on O, then f does have an analytic square root on O. (*Hint*: Let $\phi = e^{g/2}$, where g is the analytic logarithm.)
10. Let $f(z) = f(x + iy) = U(x, y) + iV(x, y)$, U, V real valued, be analytic on an open set O. Show that

$$F(z) = \left(\frac{\partial U}{\partial y} - \frac{\partial V}{\partial x}\right) + i\left(\frac{\partial U}{\partial x} + \frac{\partial V}{\partial y}\right)$$

is analytic on O.

11. Let $f(z) = f(x + iy) = U(x, y) + iV(x, y)$ be analytic and nonzero on an open set O. Show that $F(x, y) = \ln|f(x + iy)|$ is harmonic on O. (*Hint*: $U^2(x, y) + V^2(x, y) = e^{2F(x, y)}$; differentiate implicitly twice partially with respect to x and with respect to y; use the Laplace equation, the Cauchy–Riemann equations, and ingenuity. The reason for not simply treating $\ln|f(z)|$ as the real part of an analytic function $\log f(z)$ is that $\log f(z)$ might not be single-valued and analytic on O. For example, take $O = \Delta(0, 1) \sim \{0\}$ and $f(z) = z$.)
12. There is a classical theorem in advanced calculus concerning the area of the image of a set under a mapping from $\mathbf{R}^2$ into $\mathbf{R}^2$. It can be formulated as follows. Let O be an open set in $\mathbf{R}^2$, and let $T: O \to \mathbf{R}^2$ be given by a pair of functions $u = g(x, y)$, $v = h(x, y)$, which are continuous and have continuous partial derivatives on O. Let $E \subset O$ be a bounded set such that the area $|E|$ of E exists (in the sense of Jordan content; see [6, pp. 359–361]). Let T be univalent on E, and let the Jacobean $J(x, y) = g_x h_y - h_x g_y$ be positive on E. Then the area $|T(E)|$ exists and is given by $|T(E)| = \iint_E J(x, y)\, dx\, dy$. Furthermore if $F(u, v)$ is continuous on $T(E)$ then

$$\iint_{T(E)} F(u, v)\, du\, dv = \iint_E F[g(x, y), h(x, y)]J(x, y)\, dx\, dy.$$

(A rigorous proof is found in [6, pp. 390 ff.].) Now suppose that $B \subset \mathbf{C}$ is a region and $f: B \to \mathbf{C}$ is analytic and univalent on B. With $z = x + iy$, x, y real, show that the mapping given by $w = f(z)$ satisfies the above conditions on T, with $|f'(z)|^2 = J(x, y)$. (It follows that if E is any set with $E \subset B$ such that the area $|E|$ exists in the sense of Jordan content in $\mathbf{R}^2$, then $|f(E)|$ exists and is given by $\iint_E |f'(z)|\, dx\, dy$.) Show also that if E is the annular region $\{z: r_1 \leq |z| \leq r_2\}$, $0 \leq r_1 < r_2$, then $|f(E)| = \iint_R |f'(z)|\, r\, dr\, d\theta$, where $z = re^{i\theta}$ and $R = \{(r, \theta): r_1 \leq r \leq r_2,\ 0 \leq \theta < 2\pi\}$. [*Hint*: To compute J use the Cauchy–Riemann equations.]

Remark. It is known that if the boundary of E consists of a finite number of a simple closed paths, then $|E|$ exists [6, p. 365].

9.5 The Gauss Mean Value Theorem and the Cauchy Estimates

Let $f: O \to \mathbf{C}$ be analytic on the open set O. As in Theorem 9.4.2, for any point $a \in O$, let $\rho(a) = \text{dist}(a, O^c)$. Let $\sum_{k=0}^{\infty} a_k(a)(z-a)^k$ be the local power series representation of $f(z)$, valid for $z \in \Delta[a, \rho(a)]$. Let γ_r be a positively oriented circle given by $z = a + re^{it}$, $0 \le t \le 2\pi$, $0 < r < \rho(a)$. From (9.4.3) we have

$$a_k(a) = \frac{f^{(k)}(a)}{k!} = \frac{1}{2\pi i}\int_0^{2\pi} \frac{f(a+re^{it})rie^{it}\,dt}{(a+re^{it}-a)^{k+1}}$$
$$= \frac{1}{2\pi r^k}\int_0^{2\pi} f(a+re^{it})e^{-ikt}\,dt, \qquad k = 0, 1, 2, \ldots. \tag{9.5.1}$$

The case $k = 0$ is especially noteworthy:

$$f(a) = \frac{1}{2\pi}\int_0^{2\pi} f(a+re^{it})\,dt. \tag{9.5.2}$$

This equation is known as the *Gauss mean value theorem.* It states that the value of an analytic function at the center of a closed disk contained in the domain of analyticity is the mean value (in the sense of elementary integral calculus) of the values of the function on the circumference of the disk.

From (9.5.1) we obtain

$$|a_k(a)| = \left|\frac{f^{(k)}(a)}{k!}\right| \le \frac{1}{2\pi r^k}\int_0^{2\pi} |f(a+re^{it})|\,dt$$
$$\le \frac{M(r)}{r^k}, \qquad k = 0, 1, 2, \ldots, \tag{9.5.3}$$

where $M(r) = \max\{|f(z)|, z \in C(a, r)\}$. These are the *Cauchy estimates* for the coefficients of the local power series.

Proposition 9.5.1 *With f, a, and $\rho(a)$ defined as before, if there exists $M_a > 0$ such that $|f(z)| \le M_a$ for every $z \in \Delta[a, \rho(a)]$, then*

$$|a_k(a)| = \left|\frac{f^{(k)}(a)}{k!}\right| \le \frac{M_a}{[\rho(a)]^k}. \tag{9.5.4}$$

Proof. In (9.5.3), we now have $M(r) \le M_a$ for every r, $0 < r < \rho(a)$, so

$$|a_k(a)| \le \frac{M_a}{r^k}, \qquad k = 0, 1, 2, \ldots. \tag{9.5.5}$$

Now $\lim_{r\to\rho(a)} (M_a/r^k) = M_a/[\rho(a)]^k$, and the left-hand member of (9.5.5) is independent of r. Given any $\varepsilon > 0$, there exists r_ε, $0 < r_\varepsilon < \rho(a)$ such that $M_a/r_\varepsilon^k < [M_a/\rho(a)] + \varepsilon$. Thus,

$$|a_k(a)| < \frac{M_a}{[\rho(a)]^k} + \varepsilon,$$

and since ε is arbitrary, this inequality implies (9.5.4). Q.E.D.

Remark. The Cauchy estimates in a given instance may be rather generous. For example, let $f(z) = (1 - z)^{-1} = 1 + z + z^2 + \cdots$, $z \in \Delta(0, 1)$. Then for $0 < r < 1$, $M(r) = \max\{|f(z)|, z \in C(0, r)\} = (1 - r)^{-1}$, and the Cauchy estimates are

$$\frac{f^{(k)}(0)}{k!} \leq \frac{1}{r^k(1 - r)}, \qquad k = 0,1,2, \ldots .$$

But actually $f^{(k)}(0)/k! = 1$, $k = 0,1,2, \ldots$.

9.6 Liouville's Theorem and the Fundamental Theorem of Algebra

A function f: $\mathbf{C} \to \mathbf{C}$ analytic at every point of $\mathbf{C}$ is called an *entire function.* The contrapositive of the following theorem shows that the range of a nonconstant entire function is an unbounded set in $\mathbf{C}$.

Theorem 9.6.1 (*Liouville's Theorem*). *If f: $\mathbf{C} \to \mathbf{C}$ is analytic and bounded on $\mathbf{C}$, then f must be a constant function.*

Proof. Let $f(z) = \sum_0^\infty a_k z^k$ be the local power series expansion of f in $\Delta(0, r)$. Here r is any positive number because the number $\rho(a)$ in Theorem 9.4.2 is ∞. In other words, this power series representation is valid for every $z \in \mathbf{C}$. By hypothesis, there exists $M > 0$ such that $|f(z)| \leq M$ for every $z \in \mathbf{C}$. According to (9.5.3), we have $|a_k| = M/r^k$, $k = 1, 2, \ldots$. Given any ε, $0 < \varepsilon < M$, if $r > M/\varepsilon$, then $M/r^k < M/r < \varepsilon$, and $|a_k| < \varepsilon$, $k = 1, 2, \ldots$. It follows that $a_k = 0$, $k = 1, 2, \ldots$. Thus, $f(z) \equiv a_0$ for every $z \in \mathbf{C}$. Q.E.D.

A generalization of the theorem appears in Exercise 9.6.1.

Theorem 9.6.2 (*Fundamental Theorem of Algebra*). *Every polynomial in $\mathbf{C}$ of degree greater than or equal to 1 has at least one zero.*

Proof. We write the polynomial in the form $a_0 z^n + a_1 z^{n-1} + a_2 z^{n-2} + \cdots + a_{n-1} z + a_n$, with $a_0 \neq 0$ and $n \geq 1$. If $a_n = 0$, there is obviously a zero at $z = 0$. Assume that $a_n \neq 0$. If there were no zeros, then

$$F(z) = \frac{1}{a_0 z^n + \cdots + a_n}$$

would be an entire function. We show that it is bounded in **C**. We have

$$|F(z)| = \frac{1}{|a_0z^n|}\cdot\left|\frac{1}{1 + a_1/a_0z + a_2/a_0z^2 + \cdots + a_n/a_0z^n}\right|.$$

Choose $R > 2n \max\{|a_k/a_0|, k = 1, 2, \ldots, n\}$ and then adjust R, if necessary, so that $R > 1$. For $|z| > R$,

$$\alpha = \left|\frac{a_1}{a_0z} + \frac{a_2}{a_0z^2} + \cdots + \frac{a_n}{a_0z^n}\right| \le \left|\frac{a_1}{a_0z}\right| + \left|\frac{a_2}{a_0z^2}\right| + \cdots + \left|\frac{a_n}{a_0z^n}\right|$$
$$\le \left|\frac{1}{z}\right| n \max_k \left|\frac{a_k}{a_0}\right| < \frac{1}{2},$$

so

$$\left|1 + \frac{a_1}{a_0z} + \cdots + \frac{a_n}{a_0z^n}\right| \ge 1 - \alpha > 1 - \frac{1}{2} = \frac{1}{2}.$$

Thus, for $|z| > R$,

$$|F(z)| < \frac{2}{|a_0|R^n}. \tag{9.6.1}$$

Now the function $|F(z)|$ is continuous, so its restriction to the compact set $\bar{\Delta}(0, R)$ has a maximum value, say M_R [Theorem 5.5.2, Corollary 1]. The analysis shows that under the assumption of no zeros, $|F(z)|$ would be bounded on **C**.

But then by Liouville's theorem, $F(z)$ would have a constant value on **C**, say c, and since $F(0) \ne 0$, $c \ne 0$. However, there is no upper bound on the choice of R in (9.6.1), and for sufficiently large values of R, the right-hand member of (9.6.1) is less than $|c|$. This yields a contradiction with the conclusion drawn from Liouville's theorem that $F(z) \equiv c$, so F cannot be an entire function, as it would be if the polynomial had no zeros. Q.E.D.

Corollary A polynomial $p_n(z)$ in **C** of degree $n > 0$ has exactly n zeros, counted according to their multiplicities. If $a_1, a_2, \ldots, a_\mu$ are the distinct zeros, then the polynomial can be written in the form $c(z - a_1)^{m_1}(z - a_2)^{m_2}\cdots(z - a_\mu)^{m_\mu}$, where c is a constant and the positive integers m_j satisfy $\sum_{j=1}^{\mu} m_j = n$.

Proof. The result can be proved by algebra in **C**. The following analytical proof is presented as a review. The basic idea in the proof is contained in Theorem 4.3.1, but we modify it here slightly to make it apply to polynomial factorization. Let z_1 be a zero of p_n. Since p_n is analytic on **C**, there is a local power series representation $p_n(z) = \sum_0^\infty a_k(z_1)(z - z_1)^k$, valid for every $z \in \mathbf{C}$; and $a_k(z_1) = p_n^{(k)}(z_1)/k!$. Now $p_n^{(k)}(z) \equiv 0$ for $k > n$, so this power series is really a finite series $p_n(z) = \sum_{k=0}^{n} a_k(z_1)(z - z_1)^k$. Since $p_n(z_1) = 0$, we must have $a_0(z_1) = 0$. Then $p_n(z) = (z - z_1)\sum_{n=1}^{k} a_k(z_1)(z - z_1)^{k-1} =$

$(z - z_1)p_{n-1}(z)$, where p_{n-1} is a polynomial of degree $n - 1$. We apply this argument to p_{n-1} and obtain $p_n(z) = (z - z_1)(z - z_2)p_{n-2}(z)$, where p_{n-2} is of degree $n - 2$, z_2 is a root of p_{n-1}, and z_1 and z_2 may or may not be distinct. Proceeding inductively, we finally arrive at $p_n(z) = (z - z_1)(z - z_2)\cdots(z - z_n)p_0(z)$. The factorization in the statement of the corollary follows at once. Q.E.D.

The *paraconjugate polynomial* corresponding to the polynomial $p_n(z) = a_0 + a_1 z + \cdots + a_n z^n$ in $\mathbf{C}$, with $a_n \neq 0$, is the polynomial $p^*(z) = \overline{p_n(-\bar{z})} = \bar{a}_0 - \bar{a}_1 z + \bar{a}_2 z^2 - \cdots + (-1)^n \bar{a}_n z^n$. If $p_n(z) = c(z - z_1)(z - z_2)\cdots(z - z_n)$, then $p_n^*(z) = (-1)^n \bar{c}(z + \bar{z}_1)(z + \bar{z}_2)\cdots(z + \bar{z}_n)$, so the roots of p_n^* are obtained by reflecting those of p_n in the imaginary axis. If $z = x + iy$, x, y real, then $p_n^*(iy) = \overline{p_n[-(-iy)]} = \overline{p_n(iy)}$. Therefore,

$$|p_n^*(iy)| = |p_n(iy)|, \tag{9.6.2}$$

so if iy_0 is a zero of p_n it is also a zero of p^*, and conversely.

A Hurwitz polynomial is a nonconstant polynomial with all its zeros in the left half-plane $\{z\colon \operatorname{Re} z < 0\}$. Notice that if $p_n(z)$ is a Hurwitz polynomial, then $p_n(z)$ and $p_n^*(z)$ have no zero in common. Hurwitz polynomials are encountered in the problem of stability of mechanical and electrical systems. A basic result for such polynomials follows.

Proposition 9.6.1 *Let $p_n(z)$ be a Hurwitz polynomial. Define a function g from $\mathbf{C}$ into $\hat{\mathbf{C}}$ (the extended complex plane) by $g(z) = p_n^*(z)/p_n(z)$. The function g maps the half-plane $H = \{z\colon \operatorname{Re} z > 0\}$, into $\Delta(0, 1) = \{w\colon |w| < 1\}$ and maps the imaginary axis into $C(0, 1) = \{w\colon |w| = 1\}$.*

(Recall that $\hat{\mathbf{C}}$ was introduced in Sec. 1.15.)

Proof. First, notice that with $z = x + iy$, $z_0 = x_0 + iy_0$, and $x_0 < 0$, $x > 0$, we have

$$\left|\frac{z + \bar{z}_0}{z - z_0}\right|^2 = \frac{(x + x_0)^2 + (y - y_0)^2}{(x - x_0)^2 + (y - y_0)^2} < 1, \tag{9.6.3}$$

and also if $x_0 < 0$, $x = 0$, we have

$$\left|\frac{z + \bar{z}_0}{z - z_0}\right|^2 = \frac{x_0^2 + (y - y_0)^2}{x_0^2 + (y - y_0)^2} = 1. \tag{9.6.4}$$

Let $p_n(z)$ be a Hurwitz polynomial of which the roots are $z_1, z_2, \ldots, z_n$. We have, for every $z \in \mathbf{C}$,

$$g(z) = \frac{p_n^*(z)}{p_n(z)} = \left|\frac{z + \bar{z}_1}{z - z_1}\right|\left|\frac{z + \bar{z}_2}{z - z_2}\right|\cdots\left|\frac{z + \bar{z}_n}{z - z_n}\right|.$$

Since p_n is a Hurwitz polynomial, all the real parts of the z_j's are negative.

Therefore by (9.6.3) and (9.6.4), we have $|g(z)| < 1$ for $z \in H$ and $|g(z)| = 1$ for z on the imaginary axis. Q.E.D.

Conversely, we have the following proposition.

Proposition 9.6.2 *Let $p_n(z)$ be a polynomial in* **C** *of degree $n \geq 1$, and such that $p_n(z)$ and $p_n^*(z)$ have no zero in common. If the function g given by $w = g(z) = p_n^*(z)/p_n(z)$ maps the half-plane $H = \{z: \operatorname{Re} z > 0\}$ into a bounded set B in the w-plane, then $p_n(z)$ is a Hurwitz polynomial.*

Proof. Let z_0 be a zero of $p_n(z)$. By hypothesis it is not a zero of $p_n^*(z)$, so $g(z_0) = \infty \notin B$. Thus, $z_0 \notin H$. Furthermore, z_0 cannot be a pure imaginary, because if it were, then by (9.6.2) it would simultaneously be a zero of $p_n^*(z)$. Thus, $\operatorname{Re} z_0 < 0$, and then $p_n(z)$ by definition is a Hurwitz polynomial. Q.E.D.

Exercises 9.6

1. Let f be an entire function and suppose that there exist $M > 0$, $\rho \geq 0$ and $\alpha > 0$ such that $|f(z)| \leq M|z|^\alpha$ for every z, $|z| \geq \rho$. Prove that $f(z)$ is a polynomial of degree $\leq \alpha$.

   Remark. Exercise 9.6.1 shows that if the absolute value of an entire function increases more slowly than some power of $|z|$ as $|z| \to \infty$, then the function must be a polynomial.

2. Prove Liouville's theorem by showing that under its hypotheses, at each point $a \in \mathbf{C}$, $f'(a) = 0$.

3. Let $p(z)$ be a polynomial with zeros $z_1, z_2, \ldots, z_\mu$ of respective multiplicities $m_1, m_2, \ldots, m_\mu$. Prove that

$$\frac{p'(z)}{p(z)} = \frac{m_1}{z - z_1} + \frac{m_2}{z - z_2} + \cdots + \frac{m_\mu}{z - z_\mu}.$$

4. (Continuation of Exercise 9.6.3.) Let γ be a positively oriented circle with graph $C(a, R)$. Suppose in Exercise 9.6.3 that $z_j \notin C(a, R)$, $j = 1, \ldots, \mu$. Prove that

$$\frac{1}{2\pi i}\int_\gamma \frac{p'(z)}{p(z)}\,dz = N(a, R),$$

   where $N(a, R)$ is the number of zeros of $p(z)$ in $\Delta(a, R)$ counted according to their multiplicities.

   Remark. The formula is generalized in Theorem 10.2.1.

5. (Lucas' theorem.) Let $p(z)$ be a polynomial of degree $n \geq 1$. Show that the zeros of $p'(z)$ all lie in the convex hull of the set of zeros of $p(z)$. (*Hint*: Write p'/p in the form given in Exercise 9.6.3, but with $m_k/(z - z_k) = m_k(\bar{z} - \bar{z}_k)/|z - z_k|^2$. Let a be a zero of $p'(z)$ such that $a \neq z_k$, $k = 1, \ldots, \mu$. We have

$0 = \sum_{k=1}^{\mu} [m_k(\bar{a} - \bar{z}_k)/|a - z_k|^2]$. Solve for the $\bar{a}$ in the numerator, getting

$$\bar{a} = \sum_{k=1}^{\mu} \frac{m_k \bar{z}_k}{|a - z_k|^2} \left(\sum_{k=1}^{\mu} \frac{m_k}{|a - z_k|^2} \right)^{-1}.$$

Take the complex conjugate and identify the right-hand member as a weighted average of the points z_k.)

6. If $p(z) = (z - 1)(z - 2) \cdots (z - n)$, locate the roots of $p'(z)$.
7. Prove that if an entire function f has a power series representation with only real coefficients, then each of its zeros either is real or occurs in a pair of complex conjugate zeros. (*Hint*: Look at $\overline{f(z)}$).
8. Prove that the zeros of the derivative of a Hurwitz polynomial of degree $n \geq 2$ all lie in the left half-plane. (*Hint*: Lucas' theorem.)
9. Let $p(z)$ be a polynomial of degree n, and let w_0 be a complex number. Prove that the equation $p(z) = w_0$ has exactly n roots, with roots counted according to multiplicities.
10. (Eneström's Theorem.) Let $p(z) = \alpha_0 + \alpha_1 z + \alpha_2 z^2 + \cdots + \alpha_{n+1} z^{n+1}$, where α_0, $\alpha_1, \ldots, \alpha_{n+1}$ are real and $\alpha_0 \geq \alpha_1 \geq \cdots \geq \alpha_{n+1} > 0$. Prove that the zeros of $p(z)$ all lie in $\{z : |z| \geq 1\}$. (*Hint*: Apply Abel's identity [Exercise 1.2.9] to $p(z)$, letting $s_k = 1 + z + z^2 + \cdots + z^k = (1 - z^{k+1})/(1 - z)$, $k = 0,1,2,\ldots,n+1$, and $s_{-1} = 0$. Suppose $p(z_0) = 0$, $|z_0| < 1$. Obtain the equation $\alpha_0 = (\alpha_0 - \alpha_1)z_0 + (\alpha_1 - \alpha_2)z^2 + \cdots + \alpha_{n+1} z_0^{n+1}$ and arrive at a contradiction.)

9.7 The Maximum Modulus Principle

Section 1.12 mentions that the absolute value of a complex number is also called its modulus. Traditionally, if f is a function from **C** into **C**, the function $|f|$ from **C** into **R**, defined by $|f|(x) = |f(x)|$, is called the modulus function of f, or simply the modulus of f. This function is introduced in Proposition 3.1.3.

The following theorem contains one of many formulations of the so-called "maximum modulus principle," or "principle of the maximum," for analytic functions.

Theorem 9.7.1 *Let B be a region, and let f: $B \to$ **C** be analytic on B. Let $\beta = \sup \{|f(z)| : z \in B\}$. If for some point $a \in B$, $|f(a)| = \beta$, then f is a constant function on B.*

Proof. We use (9.5.2), the Gauss mean value theorem. As usual, let $\rho(a) =$ dist (a, B^c) and choose any r, $0 < r < \rho(a)$. We have

$$0 = |f(a)| - \beta = \left| \frac{1}{2\pi} \int_0^{2\pi} f(a + re^{it})\, dt \right| - \beta$$

$$\le \frac{1}{2\pi}\int_0^{2\pi} |f(a + re^{it})|\, dt - \beta$$
$$\le \frac{1}{2\pi}\int_0^{2\pi} \beta\, dt - \beta = 0.$$

So

$$0 = \beta - \frac{1}{2\pi}\int_0^{2\pi} |f(a + re^{it})|\, dt$$
$$= \frac{1}{2\pi}\int_0^{2\pi} [\beta - |f(a + re^{it})|]\, dt. \qquad (9.7.1)$$

Lemma Let $\phi: [\theta_1, \theta_2] \to \mathbf{R}$, $\theta_1 < \theta_2$, be continuous and such that $\phi(\theta) \ge 0$ for every $\theta \in [\theta_1, \theta_2]$. If $\int_{\theta_1}^{\theta_2} \phi(\theta)\, d\theta = 0$, then $\phi(\theta) = 0$ for every $\theta \in [\theta_1, \theta_2]$.

Proof of the Lemma. This basic result appears in some form in all the standard theories of integration. A quick proof for Riemann integrals: If $t \in [\theta_1, \theta_2]$, we have $0 \le F(t) = \int_{\theta_1}^{t} \phi(\theta)\, d\theta \le \int_{\theta_1}^{\theta_2} \phi(\theta)\, d\theta = 0$; $F(t) \equiv 0$ on $[\theta_1, \theta_2]$; $F'(t) = \phi(t) = 0$ for every $t \in [\theta_1, \theta_2]$.

To return to the proof of the theorem, since $\beta \ge |f(z)|$, $z \in B$, the lemma applies to (9.7.1), and we find that $|f(a + re^{it})| = \beta$, for every t, $0 \le t \le 2\pi$. This equation is valid for every r, $0 < r < \rho(a)$ and also (by hypothesis) for $r = 0$, so we have shown that $|f(z)| = \beta$, a constant, for every $z \in \Delta[a, \rho(a)]$. But then by Proposition 6.3.3(c), $f(z)$ itself is constant, say b, on $\Delta[a, \rho(a)]$. Now f has the LPS property on B, so by Corollary 2 of Theorem 6.5.1, $f(z) \equiv b$ for every $z \in B$. Q.E.D.

Corollary 1 Let B be a region, and let $f: B \to \mathbf{C}$ be analytic. Then either $f(z)$ is constant on B, or for every point $a \in B$ and every disk $\Delta(a, r) \in B$, $r > 0$, there exists $b \in \Delta(a, r)$ such that $|f(a)| < |f(b)|$.

Proof. Supose that for some $a \varepsilon B$ and some $\Delta(a, r) \subset B$, $r > 0$, $|f(z)| \le |f(a)|$ for every $z \in \Delta(a, r)$. Then $|f(a)| = \sup\{|f(z)|, z \in \Delta(a, r)\}$ and according to the theorem (with $B = \Delta(a, r)$), $f(z)$ is identically a constant on $\Delta(a, r)$. It follows from Theorem 6.5.1, Corollary 2, that f is a constant function on B. Q.E.D.

Remark. This corollary can be paraphrased by saying that the modulus of a function analytic on a region can have a local maximum at a point in the region only if the value of the function is constant on the region.

Corollary 2 Let B be a region with $\bar{B}$ bounded, and let $f: \bar{B} \to \mathbf{C}$ be analytic on B and continuous on $\bar{B}$. Then either $f(z)$ is constant on $\bar{B}$ or the maximum value of $|f(z)|$ on $\bar{B}$ is attained only on the boundary ∂B.

Proof. Since $\bar{B}$ is compact and $|f(z)|$ is continuous on $\bar{B}$, there is a point

$a \in \bar{B}$ at which $|f(z)|$ attains an overall maximum value M on $\bar{B}$ [Theorem 5.5.2, Corollary 1]. Suppose such a point a lies in B. In this case, $\sup\{|f(z)|: z \in B\} = M$, and then since $|f(a)| = M$, according to the theorem, $f(z)$ is constant on B. By continuity, $f(z)$ is constant on $\bar{B}$. The alternative is that the point or points which maximize $|f(z)|$ lie on ∂B. Q.E.D.

Let f be analytic on an open set O. The function $M(r) = M(f; a; r) = \max\{|f(z)|, z \in C(a, r)\}$, where $\Delta(a, r) \subset O$, introduced for the Cauchy estimates in Sec. 9.5, furnishes one characterization of the behavior of the modulus of an analytic function.

Proposition 9.7.1 *Let B be a region, and let $f: B \to \mathbf{C}$ be analytic and nonconstant on B. Given $a \in B$, let $\rho(a) = \text{dist}(a, B^c)$. The function $M(r) = M(f; a; r)$, $0 \le r < \rho(a)$, is a strictly increasing function of r.*

Proof. Take r_1, r_2, $0 < r_1 < r_2 < \rho(a)$. By Corollary 2 of Theorem 9.7.1, $M(r_2) = \max\{|f(z)|: z \in \bar{\Delta}(a, r_2)\}$, so $M(r_1) \le M(r_2)$. But if $M(r_1) = M(r_2)$, then the maximum value of $|f(z)|$ would be attained at an interior point of $\bar{\Delta}(a, r_2)$, which according to Corollary 2 of Theorem 9.7.1 is possible only if f is a constant function on $\Delta(a, r_2)$. In that case, by Corollary 2 of Theorem 6.5.1, f would be a constant function on B. Q.E.D.

Exercises 9.7

1. Find $\max\{|f(z)|: z \in \bar{\Delta}(0, 1)\}$ for each of these functions: (a) $f(z) = z^2 - 1$, (b) $f(z) = z(z-1)(z-2)$, (c) $f(z) = \sin z$, (d) $f(z) = (z^2 + 2)/(z - 2)$. (*Hint*: The trick is to use Corollary 2 of Theorem 9.7.1 and to choose a particular z strategically.)

   Answer. (a) 2, (b) 6, (c) $1 + 1/3! + 1/5! + \cdots$, (d) 3.

2. Let B be a bounded region, let $f, f_1, f_2, \ldots$, be functions continuous on $\bar{B}$ and analytic in B, and finally let $\lim_{n\to\infty} f_n(z) = f(z)$ uniformly on ∂B. Then this limit also exists uniformly on $\bar{B}$.

   Remark. This is a weaker form of a result which appears in Proposition 9.9.2.

3. (Minimum Modulus Principle.) Let B be a region, and let $f: B \to \mathbf{C}$ be analytic on B and $f(z) \ne 0$ on B. Then either $f(z)$ is a constant function or $|f(z)|$ cannot have a local minimum on B. (*Hint*: Look at $1/f$.)

4. Let B be a region, let $\bar{B}$ be bounded, and let $f: \bar{B} \to C$ be analytic and nonconstant on B, continuous on $\bar{B}$ and $f(z) \ne 0$ on B. Then the minimum value of $|f(z)|$ is attained only for some $z \in \partial B$. (*Hint*: If $f(z) = 0$ for some $z \in \partial B$, the statement is obviously true. If $f(z) \ne 0$ on $\bar{B}$, then Corollary 2 of Theorem 9.7.1 is applicable to $1/f$.)

5. Let B be a region, $\bar{B}$ bounded, and $f: \bar{B} \to \mathbf{C}$ analytic on B, continuous on $\bar{B}$. Then either $f(z)$ is constant on $\bar{B}$, or the maximum and minimum values of $\operatorname{Re} f(z)$ in $\bar{B}$ are attained only at points of ∂B. (*Hint*: Look at $e^{f(z)}$.)

6. Let B be a region, $\bar{B}$ bounded, and let functions $f: B \to \mathbf{C}$ and $g: B \to \mathbf{C}$ be analytic on B, continuous on $\bar{B}$. Show that if $\operatorname{Re} f(z) = \operatorname{Re} g(z)$ for every $z \in \partial B$, then $f(z)$

$= g(z) + ic$ for every $z \in \bar{B}$, where ic is an imaginary constant, and in particular that if $\operatorname{Re} f(z)$ is constant on ∂B, then $f(z)$ is constant on $\bar{B}$. (*Hint*: Exercise 9.7.5.)

9.8 Schwarz's Lemma

This is the name given to the following simple consequence of the maximum modulus principle.

Theorem 9.8.1 *Let $f\colon \Delta(0, 1) \to \mathbf{C}$ be analytic, with $f(0) = 0$; furthermore, let $|f(z)| < 1$ for every $z \in \Delta(0, 1)$. Then*

(a) *$|f(z)| \le |z|$ for every $z \in \Delta(0, 1)$, and $|f'(0)| \le 1$;*
(b) *if $|f(z_0)| = |z_0|$ for some $z_0 \in \Delta(0, 1)$, $z_0 \ne 0$, or if $|f'(0)| = 1$, then $f(z) = e^{i\alpha}z$ for every $z \in \Delta(0, 1)$, where α is some real number.*

Proof. Define

$$g(z) = \begin{cases} \dfrac{f(z)}{z}, & z \ne 0. \\ f'(0), & z = 0. \end{cases}$$

The function $f(z)$ has the power series representation $f(z) = f'(0)z + a_2z^2 + a_3z^3 + \cdots$ valid in $\Delta(0, 1)$, so $g(z) = f'(0) + a_2z + a_3z^3 + \cdots$ for every $z \in \Delta(0, 1) \sim \{0\}$. This series also represents $g(z)$ at $z = 0$. The sum function of a power series is analytic on the disk of convergence, so $g(z)$ is analytic on $\Delta(0, 1)$. Take any $z_1 \in \Delta(0, 1)$, hold it fixed, and choose r, $|z_1| < r < 1$. By Corollary 2 of Thoerem 9.7.1,

$$\begin{aligned} |g(z_1)| &\le \max\{|g(z)|: z \in C(0, r)\} \\ &= \max\left\{\frac{|f(z)|}{r}: z \in C(0, r)\right\} < \frac{1}{r}. \end{aligned} \tag{9.8.1}$$

Now $\lim_{r\to 1}(1/r) = 1$, so given $\varepsilon > 0$, there exists r_ε, $|z_1| < r_\varepsilon < 1$, such that $1/r_\varepsilon < 1 + \varepsilon$. Substitution into (9.8.1) gives $|g(z_1)| < 1 + \varepsilon$, and since ε is arbitrary, this implies that $|g(z_1)| \le 1$. The statement in part (a) of the theorem follows at once.

The hypothesis in part (b) implies that there exists $z_0 \in \Delta(0, 1)$ such that $|g(z_0)| = 1$. Now from part (a), we have $|g(z)| \le 1$ for every $z \in \Delta(0, 1)$, so the new implication is that $|g(z_0)| = \sup\{|g(z)|: z \in \Delta(0, 1)\}$. Then according to Theorem 9.7.1, g is a constant function on $\Delta(0, 1)$, say with value c. Since $|g(z_0)| = |c| = 1$, the polar representation of c is $e^{i \operatorname{Arg} c}$. With $\alpha = \operatorname{Arg} c$, we have $g(z) = f(z)/z = e^{i\alpha}$ for $z \ne 0$ and $f'(0) = e^{i\alpha}$. Q.E.D.

The are a number of variations of Schwarz's lemma which are useful, two of which we now state as corollaries. First, recall from Exercise 1.13.3

that the linear fractional transformation,

$$w = \frac{z - z_0}{1 - \bar{z}_0 z}, \qquad |z_0| < 1, \tag{9.8.2}$$

gives a univalent mapping of $\Delta(0, 1)$ onto $\Delta(0, 1)$. The inverse transformation is the analytic function of w,

$$z = \frac{w + z_0}{1 + w\bar{z}_0}. \tag{9.8.3}$$

The function $\zeta = e^{i\alpha}(z - z_0/(1 - \bar{z}_0 z)$, α real, is the composition of the rotation $\zeta = e^{i\alpha}w$ with (9.8.2), so it is a linear fractional transformation which again gives a one-to-one mapping of $\Delta(0, 1)$ onto $\Delta(0, 1)$. The derivative at z_0 is needed:

$$\left.\frac{d\zeta}{dz}\right]_{z=z_0} = \frac{e^{i\alpha}}{1 - |z_0|^2}.$$

Also, with z given by (9.8.3),

$$\left.\frac{dz}{dw}\right]_{w=0} = 1 - |z_0|^2. \tag{9.8.4}$$

Corollary 1 Let the function f: $\Delta(0, 1) \to \Delta(0, 1)$ be analytic, with $f(z_0) = 0$ for some $z_0 \in \Delta(0, 1)$. Then

$$|f(z)| \le \left|\frac{z - z_0}{1 - \bar{z}_0 z}\right| \tag{9.8.5}$$

for every $z \in \Delta(0, 1)$, and $|f'(z_0)| \le (1 - |z_0|^2)^{-1}$. Furthermore, if equality occurs in (9.8.5) for some $z_1 \in \Delta(0, 1)$, $z_1 \ne z_0$, or if $|f'(z_0)| = (1 - |z_0|^2)^{-1}$, then

$$f(z) = e^{i\alpha}\frac{z - z_0}{1 - \bar{z}_0 z} \tag{9.8.6}$$

for some real α and for every $z \in \Delta(0, 1)$. The equation (9.8.6) implies that $|f'(z_0)| = (1 - |z_0|^2)^{-1}$.

Proof. Let $z = (w + z_0)/(1 + w\bar{z}_0)$ and define $\phi(w) = f(z) = f((w + z_0)/(1 + w\bar{z}_0))$, $|w| < 1$. This function ϕ is the composition of the analytic function $f(z)$ with the function given by (9.8.3) and, therefore, is analytic on $\Delta(0, 1)$. Since $|f(z)| < 1$ for every $z \in \Delta(0, 1)$, it follows that $|\phi(w)| < 1$ for every $w \in \Delta(0, 1)$. Moreover, $\phi(0) = 0$. Therefore by Theorem 9.8.1, $|\phi(w)| \le |w|$ for every $w \in \Delta(0, 1)$, and $|\phi'(0)| \le 1$. The chain law of differentiation, together with (9.8.4), yields $\phi'(0) = f'(z_0)(1 - |z_0|^2)$. Substitution into $|\phi(w)| \le |w|$ with (9.8.2) and (9.8.3) gives (9.8.5), and the inequality $|\phi'(0)| \le 1$ is equivalent to $|f'(z_0)| \le (1 - |z_0|^2)^{-1}$.

If the last weak inequality in the preceding paragraph is in fact an equality, then $|\phi'(0)| = 1$. If there exists $z_1 \in \Delta(0, 1)$, $z_1 \neq z_0$, such that (9.8.5) is an equation when $z = z_1$, then with $w_1 = (z_1 - z_0)/(1 - \bar{z}_0 z_1)$, we have $|\phi(w_1)| = |w_1|$ and $w_1 \neq 0$. Part (b) of Theorem 9.8.1 applies in either of these situations, and there exists a real α such that $\phi(w) = e^{i\alpha}w$ for every $w \in \Delta(0, 1)$. Substitution with (9.8.2) yields (9.8.6). Q.E.D.

Remark. The corollary imposes an upper bound of $(1 - |z_0|^2)^{-1}$ on the modulus of the derivative $|f'(z_0)|$ of any analytic function f which maps $\Delta(0, 1)$ into itself with $f(z_0) = 0$. The function $f(z)$ which appears in (9.8.6) is such an analytic function. Moreover, for this function $|f'(z_0)| = (1 - |z_0|^2)^{-1}$, so the function is extremal with respect to the derivative condition on such functions. Notice that this extremal function gives a one-to-one mapping of the unit disk onto itself. This fact serves as a motivation for a proof, to be given later in this book, that there exist analytic functions which map more general domains univalently onto the unit disk. This existence theorem is known as the Riemann mapping theorem.

Corollary 2 Let f: $\Delta(0, 1) \to \Delta(0, 1)$ be analytic, with $f(0) = z_0$. Then

$$\left| \frac{f(w) - z_0}{1 - \bar{z}_0 f(w)} \right| \leq |w| \tag{9.8.7}$$

for every $w \in \Delta(0, 1)$, and $|f'(0)| \leq 1 - |z_0|^2$. Furthermore, if $|f'(0)| = 1 - |z_0|^2$, or if (9.8.7) is an equation for some $w_1 \in \Delta(0, 1)$, $w_1 \neq 0$, then

$$f(w) = \frac{e^{i\alpha}w + z_0}{1 + e^{i\alpha}w\bar{z}_0}$$

for some real α and for every $w \in \Delta(0, 1)$.

The proof is an exercise.

With the aid of these two corollaries, it is possible to prove certain uniqueness theorems concerning mappings by analytic functions.

Proposition 9.8.1 *Any analytic function* f: $\Delta(0, 1) \to \mathbf{C}$ *which gives a one-to one mapping of* $\Delta(0, 1)$ *onto itself and which has an analytic inverse must be a linear fractional transformation of the type* $f(z) = e^{i\alpha}(z - z_0)(1 - \bar{z}_0 z)^{-1}$, $|z_0| < 1$.

Remark. It is shown in Sec. 10.4 that the requirement here that the inverse function be analytic is redundant. In fact, the inverse of a function f which is analytic and univalent on a region B is necessarily analytic on $f(B)$ [Proposition 10.4.1(ii) and Remark (vi) thereafter].

Proof of the Proposition. Let f be such a function, and let $f(z_0) = 0$, $z_0 \in \Delta(0, 1)$. Then according to Corollary 1 of Theorem 9.8.1, we have

$$|f(z)| \leq \left| \frac{z - z_0}{1 - \bar{z}_0 z} \right|, \qquad z \in \Delta(0, 1). \tag{9.8.8}$$

We apply Corollary 2 to the inverse f^{-1}, for which $f^{-1}(0) = z_0$. We obtain, with $w = f(z)$, $z = f^{-1}(w)$, $|(f^{-1}(w) - z_0)/(1 - \bar{z}_0 f^{-1}(w))| \leq |w|$, $w \in \Delta(0, 1)$, or in terms of z,

$$\left|\frac{z - z_0}{1 - \bar{z}_0 z}\right| \le |f(z)|, \qquad z \in \Delta(0, 1). \tag{9.8.9}$$

The inequalities (9.8.8) and (9.8.9) imply that for every $z \in \Delta(0, 1)$,

$$|f(z)| = \left|\frac{z - z_0}{1 - \bar{z}_0 z}\right|.$$

But then by Corolllary 1, for some real α, $f(z) = e^{i\alpha}(z - z_0)(1 - \bar{z}_0 z)^{-1}$ for every $z \in \Delta(0, 1)$. Q.E.D.

Proposition 9.8.2 *Let B be a region, let f: $B \to \mathbf{C}$ be analytic and univalent, and let f map B onto $\Delta(0, 1)$ so that for a given $z_0 \in B, f(z_0) = 0$. Furthermore, suppose that the inverse function is analytic. These conditions determine f uniquely to within a rotation. That is, if g: $B \to \mathbf{C}$ is also analytic and univalent and maps B onto $\Delta(0, 1)$ so that $g(z_0) = 0$, then $g(z) \equiv e^{i\alpha}f(z)$, $z \in B$, for some real α. Also $|f'(z_0)|$ is uniquely determined.*

Note. See the Remark which follows Proposition 9.8.1.

Proof. Let $\zeta = \phi(w) = g[f^{-1}(w)]$. By the chain law, ϕ is analytic on $\Delta(0, 1)$. Also ϕ gives a one-to-one mapping of $\Delta(0, 1)$ onto itself. Therefore, by Proposition 9.8.1, for some real α and complex w_0, $\zeta = \phi(w) = e^{i\alpha}(w - w_0)(1 - \bar{w}_0 w)^{-1}$. But $0 = \phi(0)$, so $w_0 = 0$. Therefore, $g[f^{-1}(w)] \equiv e^{i\alpha}w$, $g[f^{-1}(f(z))] = g(z) \equiv e^{i\alpha}f(z)$, and $|g'(z_0)| = |f'(z_0)|$. Q.E.D.

Corollary The mapping function in the proposition is uniquely determined by either one of the following two additional conditions: (i) for a given $z_1 \in B$, $w_1 \in \Delta(0, 1)$, $z_1 \neq z_0$, $f(z_1) = w_1$; (ii) $\arg f'(z_0)$ is given; in particular, $f'(z_0)$ is real and positive.

Exercises 9.8

1. Let f: $(a, R) \to \Delta(0, M)$ be analytic, with $f(a) = 0$. Prove that $|f(z)| \le M|z - a|/R$ for every $z \in \Delta(a, R)$, and $|f'(a)| \le M/R$; and that, furthermore, if for some $z_0 \in \Delta(a, R)$, the first of these inequalities is an equation, then $f(z) \equiv Me^{i\alpha}(z - a)/R$, $z \in \Delta(a, R)$. (*Hint*: Consider $\phi(\zeta) = f(R\zeta + a)/M$, $|\zeta| < 1$; use Theorem 9.8.1.)
2. Prove Corollary 2 of Theorem 9.8.1. (*Hint*: Let $\phi(w) = [f(w) - z_0]/[1 - \bar{z}_0 f(w)]$ and show that ϕ satisfies the hypotheses of Theorem 9.8.1.)
3. Let f: $\Delta(0, R) \to \mathbf{C}$ be analytic, $|f(z)| \le M > 0$ for every $z \in \Delta(0, R)$, and let $f(z_0) = w_0$ for some $z_0 \in \Delta(0, R)$. Prove that for every $z \in \Delta(0, R)$

$$\left|\frac{M[f(z) - w_0]}{M^2 - \bar{w}_0 f(z)}\right| \le \frac{R|z - z_0|}{|R^2 - \bar{z}_0 z|}.$$

(*Hint*: Look at

$$\phi(\zeta) = \frac{f(R\zeta)/M - w_0/M}{1 - (\overline{w}_0/M)\cdot(f(R\zeta)/M)}.$$

4. Let $f:\Delta(0,1)\to \mathbf{C}$ map $\Delta(0,1)$ into $\Delta(0,1)$, and let f be analytic. Show that if there are two distinct fixed points of the mapping (that is, if there exist $z_1, z_2 \in \Delta(0,1)$, $z_1 \neq z_2$, such that $f(z_1) = z_1, f(z_2) = z_2$), then the function f must be the identity transformation $f(z) = z$. (*Hint*: Define $h(z) = [f(z) - z_1]/[1 - f(z)\bar{z}_1]$, and use Corollary 1 of Theorem 9.8.1. Substitute $z = z_2$ to obtain an equation. Solve for $f(z)$.)
5. The purpose of this exercise is to establish a fact known as Lindelöf's principle or the "principle of subordination." (The latter term is also applied to certain related results.) It can be stated as follows: Let $f:\Delta(0,1)\to \mathbf{C}$ be analytic, and let $g:\Delta(0,1)\to \mathbf{C}$ be analytic and univalent. Suppose further that $f[\Delta(0,1)] \subset g[\Delta(0,1)]$ and that $f(0) = g(0)$. Then $f[\Delta(0,r)] \subset g[\Delta(0,r)]$ for every r, $0 < r < 1$. (*Hint*: To prove this, anticipate Proposition 10.3.1(ii) by assuming that the inverse function g^{-1} is analytic. Show that $h = g^{-1}\circ f$ satisfies the conditions of Schwarz's lemma. Then look at $f(z) = g[h(z)]$.)

9.9 Almost Uniform Convergence of Sequences of Analytic Functions

Given an open set $O \subset \mathbf{C}$ and a sequence of complex-valued function $\langle f_n\rangle$, each with domain O, the sequence is said to *converge almost uniformly* on O if it converges uniformly on each compact subset of O. We abbreviate "almost uniformly" to "a.u." In terms of inequalities, $\langle f_n\rangle$ converges a.u. on O to a function f if and only if for each compact subset $E \subset O$, given $\varepsilon > 0$, there exists $N_{E,\varepsilon}$ such that for every $n \geq N_{E,\varepsilon}$ and every $z \in E$, $|f_n(z) - f(z)| < \varepsilon$. The dependence of $N_{E,\varepsilon}$ on E indicated by the notation is typical of a.u. convergence. An infinite series of functions $\sum_0^\infty f_j$, each with domain O, is said to converge almost uniformly (a.u.) on O if the sequence of partial sums $\langle \sum_{j=0}^n f_j\rangle$ converges a.u. on O. Clearly, if a sequence or series converges uniformly on O, it convergence a.u. on O. An example of a.u. convergence is given by any power series, with O the disk of convergence [Exercise 5.6.1].

Almost uniform convergence of a sequence of complex-valued analytic functions on a complex domain is a strong hypothesis, as the following theorem shows.

Theorem 9.9.1 *Let O be an open set, let $f_n: O \to \mathbf{C}$, $n = 0, 1, 2, \ldots$, be analytic on O, and let $\langle f_n\rangle$ converge a.u. on O to a function $f: O \to \mathbf{C}$. Then f is analytic on O, and $\langle f_n^{(k)}\rangle$ converges a.u. on O to $f^{(k)}$, for each k, $k = 1, 2, \ldots$.*

Proof. For any $z_0 \in O$, there exists $\Delta(z_0, r) \subset O$, $r > 0$, and then $\bar{\Delta}(z_0, r/2) \subset O$. The closed disk $\bar{\Delta}(z_0, r/2)$ is compact, so $\langle f_n \rangle$ converges uniformly on it. The functions f_n are each continuous on O, so by Theorem 3.3.3, the limit function f is continuous on $\bar{\Delta}(z_0, r/2)$ and in particular f is continuous at z_0. But $z_0 \in O$ was arbitrary, so f is continuous on O. (The reason for bringing out this fact at the outset is that we need the fact that f is integrable along paths in O.)

Next take $T = (z_1, z_2, z_3)$ so that the triangle T^* determined by T lies in O. The path $\partial T = [z_1, z_2] + [z_2, z_3] + [z_3, z_4]$, considered as a point set, is compact, so $\langle f_n \rangle$ converges uniformly on it. By Proposition 8.3.1, $\int_{\partial T} f(z)\,dz = \lim_{n\to\infty} \int_{\partial T} f_n(z)\,dz$. However, by the Cauchy-Goursat theorem for a triangle [Theorem 9.2.1], $\int_{\partial T} f_n(z)\,dz = 0$ for each n, $n = 0, 1, 2, \ldots$. Therefore, $\int_{\partial T} f(z)\,dz = 0$. It follows from Morera's theorem [Theorem 9.4.3] that since T is arbitrary, f is analytic on O.

It remains to prove the statement in the theorem about the uniform convergence of the sequences of derivatives.

Let E be a compact set in O, and let $2r = \text{dist}(E, O^c)$. By Theorem 5.6.1, $r > 0$. If $z_0 \in O^c$ and $z_E \in E$, then

$$|z_0 - z_E| \geq \text{dist}(E, O^c) = 2r. \tag{9.9.1}$$

Let $K = \bigcup_{z \in E} \Delta(z, r)$. We make the following assertions about $\bar{K}$.

(i) $\bar{K} \subset O$. For take $z_{\bar{K}} \in \bar{K}$. There exists $z_K \in K$ such that $|z_{\bar{K}} - z_K| < r$, by the definition of closure. But z_K lies in some one of the disks $\Delta(z, r)$, $z \in E$; say in $\Delta(z_E, r)$. We have $|z_{\bar{K}} - z_E| \leq |z_{\bar{K}} - z_K| + |z_K - z_E| < r + r = 2r$. Therefore, by (9.9.1), $z_{\bar{K}} \notin O^c$. Thus, $z_{\bar{K}} \in O$, and $\bar{K} \subset O$.

(ii) $\bar{K} \subset \bigcup_{z \partial E} \bar{\Delta}(z, r)$. For if $\Delta(z, r) \subset K$, by Proposition 5.1.1(vi), $\bar{\Delta}(z, r) \subset \bar{K}$, so the union of all such closed disks with centers in E must also be contained in $\bar{K}$.

(iii) $\bar{K}$ is compact. For since E is compact, there exists $\Delta(0, R)$, $R > 0$, containing E. Look at a typical $\Delta(z, r) \subset K$, and recall that $z \in E$. Take $z' \in \Delta(z, r)$. We have $|z'| = |z' - z + z| \leq |z' - z| + |z| < r + R$. This implies that $\Delta(z, r) \subset \Delta(0, R + r)$ and, therefore, $K \subset \Delta(0, R + r)$. So $\bar{K} \subset \bar{\Delta}(0, R + r) \subset \Delta(0, R')$ for any $R' > R + r$. Therefore, $\bar{K}$ is bounded, and since it is closed, it is compact.

It follows from the hypothesis of the theorem and from (i) and (iii) that $\langle f_n \rangle$ converges uniformly on $\bar{K}$. In the sequel, we fix the value of k in $f^{(k)}$. Take $\varepsilon > 0$. There exists $N = N_{K,\varepsilon,k}$ such that for $n \geq N$,

$$|f_n(z) - f(z)| < \frac{\varepsilon r^k}{k!}, \tag{9.9.2}$$

for every $z \in \bar{K}$. The function $f_n - f$ is analytic on O, so the Cauchy estimates

(9.5.3) can be applied to the kth derivative at the center of any one of the closed disks $\bar{\Delta}(z, r)$, $z \in E$. Such a closed disk is contained in $\bar{K}$, by (ii), so (9.9.2) is valid with restriction to the closed disk. Fix attention on such a disk with center $a \in E$. Using (9.5.3), and with $M_n(r) = \max\{f_n(z) - f(z)|$, $z \in C(a, r)$, we have for $n \geq N$,

$$|[f_n(a) - f(a)]^{(k)}| = |f_n^{(k)}(a) - f^{(k)}(a)|$$
$$\leq \frac{M_n(r)k!}{r^k} < \frac{\varepsilon r^k k!}{k!\,r^k} = \varepsilon.$$

But this inequality holds for every $a \in E$, so $\lim_{n\to\infty} f^{(k)}(a) = f^{(k)}(a)$ uniformly for $a \in E$. Q.E.D.

Remark. The conclusion of the theorem is in marked contrast with the situation on the real line. If a function f from an interval $[\alpha, \beta]$ into $\mathbf{R}$ is continuous on $[\alpha, \beta]$, then, according to a theorem of Weierstrass, there exists a sequence of real polynomials (each of which is infinitely differentiable) which converges uniformly on $[\alpha, \beta]$ to f (which may be nowhere differentiable).

Corollary Let O be an open set, let $f_j: O \to \mathbf{C}$, $j = 0, 1, 2, \ldots$, be analytic on O, and let $\sum_{j\in 0} f_j$ converge a.u. on O to a function $f: O \to \mathbf{C}$. Then f is analytic on O and $\sum_{j=0} f^{(k)}$ converges a.u. on O to $f^{(k)}$, for each k, $k = 1, 2, \ldots$.

The corollary follows at once from Theorem 9.9.1 because the sequence of partial sums $\langle \sum_{j=0}^{n} f_j \rangle$ satisfies the hypothesis of the theorem.

The remainder of this section is devoted to a few of the many applications of Theorem 9.9.1 and its corollary.

Proposition 9.9.1 (*Weierstrass Double-Series Theorem*). *Let the functions* $g_j: \Delta(a, R) \to \mathbf{C}$, $R > 0$, $j = 0, 1, 2, \ldots$, *be analytic, and let* $\sum_{j=0}^{\infty} g_j$ *converge a.u. on* Δ *to a function* f. *Then* f *is analytic on* Δ *and, therefore, has a power series representation* $f(z) = \sum_{k=0}^{\infty} b_k(z - a)^k$ *valid for every* $z \in \Delta$. *Let the power series for* $g_j(z)$ *be* $\sum_{k=0}^{\infty} a_{jk}(z - a)^k$, *again convergent on* Δ, $j = 0, 1, 2, \ldots$. *Then* $b_k = \sum_{j=0}^{\infty} a_{jk}$, $k = 0, 1, 2, \ldots$, *and, therefore, for every* $z \in \Delta(a, R)$,

$$f(z) = \sum_{j=0}^{\infty} g_j(z) = \sum_{j=0}^{\infty}\left[\sum_{j=0}^{\infty} a_{jk}(z - a)^k\right]$$
$$= \sum_{k=0}^{\infty}\left(\sum_{j=0}^{\infty} a_{jk}\right)(z - a)^k. \tag{9.9.3}$$

Proof. The analyticity of f is assured by the corollary of Theorem 9.9.1. The various power series representations assumed in the statement of the proposition are validated by Theorem 9.4.2. It only remains to show that $b_k = \sum_{j=0}^{\infty} a_{jk}$, $k = 0, 1, 2, \ldots$.

Now $f^{(k)}(a) = k!b_k$ and $g_j^{(k)}(a) = k!a_{jk}$; $k = 0, 1, 2, \ldots, j = 0, 1, 2, \ldots$.

(As usual, $f^{(0)}$ means f, $g_j^{(0)}$ means g_j.) By the corollary to Theorem 9.9.1, for each k, $k = 1, 2, \ldots$, the series $\sum_{j=0}^{\infty} g_j^{(k)}$ converges a.u. on Δ to $f^{(k)}$, and in particular, $\sum_{j=0}^{\infty} g_j^{(k)}(a) = f^{(k)}(a)$. Therefore, $\sum_{j=0}^{\infty} k!a_{jk} = k!b_k$, and division by $k!$ completes the proof. Q.E.D.

Remark. A formal sum of the type $\sum_{j=0}^{\infty} \sum_{h=0}^{\infty} c_{jk}$ is called a *double series*. For any fixed j, the series $\sum_{k=0}^{\infty} c_{jk}$ is called a row sum of the double series, and for any fixed k, the series $\sum_{j=0}^{\infty} c_{jk}$ is called a column sum. This nomenclature is motivated by the possibility of displaying the coefficients in a doubly infinite square matrix $[c_{jk}]$, where j is the row index and k is the column index. The conclusion of Proposition 9.9.1 can be loosely paraphrased by saying that the value given to the double series $\sum\sum a_{jk}(z-a)^k$ by summing its row sums is the same as that given by summing its column sums.

Proposition 9.9.2 *Let B be a bounded region, let the functions $f_n: \bar{B} \to \mathbf{C}$, $n = 0, 1, 2, \ldots$, be analytic on B and continuous on $\bar{B}$. Furthermore, let $\langle f_n \rangle$ converge uniformly on the boundary ∂B. Then $\langle f_n \rangle$ converges uniformly on $\bar{B}$ to a function f which is analytic on B, continuous on $\bar{B}$. Also $\langle f_n^{(k)} \rangle$ converges a.u. on B to $f^{(k)}$, $k = 1, 2, \ldots$.*

Remark. This is the stronger form of Exercise 9.7.2 promised in the remark following that exercise.

Proof. The uniform convergence of $\langle f_n \rangle$ on ∂B implies that $\langle f_n \rangle$ has the uniform Cauchy property on ∂B. (See Sec. 3.3; in particular, Theorem 3.3.1.) In arithmetic terms, this means that given $\varepsilon > 0$, there exists N such that if $n > m \geq N$, then $|f_n(z) - f_m(z)| < \varepsilon$ for every $z \in \partial B$. But by Corollary 2 of Theorem 9.7.1, $f_n - f_m$ (a function which is continuous on $\bar{B}$ and analytic on B) attains its maximum value on $\bar{B}$ at a point or at points on ∂B. Therefore, for $n > m \geq N$, $|f_n(z) - f_m(z)| < \varepsilon$ for every $z \in \bar{B}$. Thus, $\langle f_n \rangle$ has the uniform Cauchy property not only for $z \in \partial B$ but also for $z \in \bar{B}$. It follows that $\langle f_n \rangle$ converges uniformly to some function f on $\bar{B}$ [Theorem 3.3.1]. The function f is continuous on $\bar{B}$ [Theorem 3.3.3]. Furthermore, by the definition of uniform convergence, $\langle f_n \rangle$ converges uniformly to f on any subset of $\bar{B}$. Therefore, Theorem 9.9.1 is applicable, and it establishes the statements in the proposition concerning the analyticity of f and the convergence of $\langle f_n^{(k)} \rangle$. Q.E.D.

Remark (i). There is an obvious translation of Proposition 9.9.2 into the terminology of infinite series, as in the corollary of Theorem 9.9.1.

Remark (ii). With the stated hypotheses of Proposition 9.9.2, the conclusion concerning the behavior of the derivative sequences $\langle f_n^{(k)} \rangle$ cannot, in general, be strengthened. For example, the sequence $\langle \sum_{j=1}^{m} (z^j/j^2) \rangle_{n=1}^{\infty}$ converges uniformly on $\bar{B} = \bar{\Delta}(0, 1)$. The sequence of first derivatives is $\langle \sum_{j=1}^{n} (z^{j-1}/j) \rangle_{n=1}^{\infty}$, which diverges at $z = 1$, and does not converge uniformly on $\Delta(0, 1)$. But it does converge a.u. on $\Delta(0, 1)$ [Exercise 5.6.1].

Proposition 9.9.3 *Let O be an open set, and let $BA(O)$ denote the collection of functions $f: O \to \mathbf{C}$, each of which is bounded and analytic on O. Then $BA(O)$, taken with the usual operations of function addition, multiplication, and multiplication by complex constants, and normed with $\|f\| = \sup\{|f(z)|: z \in O\}$, is a Banach algebra over $\mathbf{C}$.*

Proof. The proof chiefly consists in bringing together certain facts established somewhat earlier in this book. According to Exercise 3.2.7, the collection of all bounded functions on O, taken with the usual algebraic function operations, is an algebra over $\mathbf{C}$. By Exercise 6.2.2, the subcollection of these bounded functions consisting of the bounded functions analytic on O also is an algebra. Exercise 3.2.11 establishes that $\|f\|$ as described previously is indeed a norm for these algebras. This norm defines in the usual way a metric. Recall that a Banach algebra is a Cauchy complete normed algebra, and this means in particular that each Cauchy sequence must have a limit which is a member of the algebra. A sequence $\langle f_n \rangle$ of functions in $BA(O)$ for which $\lim_{m,n\to\infty} \|f_n - f_m\| = 0$ is a sequence which has the uniform Cauchy property on O. Therefore, by Theorem 3.3.1, it converges uniformly on O to some function f. By Exercise 3.3.6, f is bounded on O. The only new information supplied by the present section is that according to Theorem 9.9.1, f is also analytic on O, and thus $f \in BA(O)$. Q.E.D.

Exercises 9.9

1. According to Exercise 7.6.7, the series $\sum_{n=1}^{\infty} n^{-z}$ (with principal logarithms used in evaluating the exponentials) converges uniformly on $\{z: \text{Re } z \geq p\}$ for any $p > 1$. Show that the sum of the series is an analytic function f on $H = \{z: \text{Re } z > 1\}$ and find a series representation for f' on H. *Answer.* $\sum_1^{\infty} n^{-z} \ln n$.
2. Show that $\sum_{n=1}^{\infty} e^{-n} \sin nz$ converges and represents an analytic function in the strip $\{z = x + iy: -1 < y < 1\}$. (*Hint*: $|\sin[n(x + iy)]| = [\sin^2 nx + \sinh^2 (ny)]^{1/2} \leq [1 + \sinh^2 ny + \cosh^2 ny]^{1/2} = [1 + (e^{2ny} + e^{-2ny})/2]^{1/2}$.)
3. Show that $\sum_{n=0}^{\infty} (z^n + z^{-n})/3^n$ converges and represents an analytic function on $\{z: (1/3) < |z| < 3\}$.

9.10 Introduction to the Laplace Transform

In this section, we present an application of Theorem 9.9.1 to the Laplace transform. This type of functional transform is useful in various problems in applied mathematics, and particularly in questions involving differential equations.

A Laplace transform is the integral of a complex-valued function along an interval in $\mathbf{R}$ which typically is the unbounded interval $[0, \infty)$ or $(-\infty, \infty)$, so we make some preliminary remarks about such integrals.

In the first place, we need a precise definition of the limit of a complex-valued function defined on the real line as the independent variable becomes infinite. Let f be a complex-valued function with domain $[\alpha, \infty) \subset \mathbf{R}$. The limit $\lim_{t\to\infty} f(t)$, if it exists, is defined to be a complex number L with the property that for any $\varepsilon > 0$ there exists $T \geq \alpha$ such that for every $t \geq T$, $|f(t) - L| < \varepsilon$. Similarly, if the domain of f is $(-\infty, \beta]$, $\lim_{t\to-\infty} f(t) = L$ is defined by the property that for any $\varepsilon > 0$ there exists $T \leq \beta$ such that for every $t \leq T$, $|f(t) - L| < \varepsilon$.

Now let the function $\phi\colon [\alpha, \infty) \to \mathbf{C}$ be bounded and Riemann integrable on every interval $[\alpha, \beta]$, $\beta > \alpha$. (The integral of ϕ over $[\alpha, \beta]$ is defined via the real and imaginary parts as in the case of a continuous complex-valued function; see equation (8.1.3).) When it exists, the ("improper") Riemann integral of ϕ along $[\alpha, \infty)$, written $\int_\alpha^\infty \phi(x)\,dx$, is defined by

$$\int_\alpha^\infty \phi(x)\,dx = \lim_{t\to\infty} \int_\alpha^t \phi(x)\,dx.$$

Similarly if $\phi\colon (-\infty, \beta] \to \mathbf{C}$ is bounded and Riemann integrable on every interval $[\alpha, \beta]$, $\alpha < \beta$, then by definition

$$\int_{-\infty}^\beta \phi(x)\,dx = \lim_{t\to-\infty} \int_t^\beta \phi(x)\,dx.$$

If $\phi\colon (-\infty, \infty) \to \mathbf{C}$ is bounded and Riemann integrable on every finite interval, then $\int_{-\infty}^{+\infty} \phi(x)\,dx$ should ordinarily be taken to mean $\int_0^\infty \phi(x)\,dx + \int_{-\infty}^0 \phi(x)\,dx$, with the implication that each of the latter two integrals must exist.

In studying an improper integral such as $\int_0^\infty \phi(x)\,dx$, it is sometimes convenient (as it is in this section) to relate it to the sequence $\langle \int_0^n \phi(x)\,dx \rangle_{n=1}^\infty$, or equivalently to the infinite series $\sum_{n=0}^\infty \int_j^{j+1} \phi(x)\,dx$. It is easy to see that if $I = \int_0^\infty \phi(x)\,dx$ exists, then $\sum_0^\infty \int_j^{j+1} \phi(x)\,dx$ must exist and be equal to I. The converse is not true; for example, $\sum_0^\infty \int_j^{j+1} (\cos 2\pi x)\,dx = \sum_0^\infty 0 = 0$, but $\int_0^{n+(1/4)} \cos 2\pi x\,dx = 1/2\pi$, $n = 1, 2, \ldots$, and $\int_0^\infty \cos 2\pi x\,dx$, therefore, does not exist. Perhaps the simplest criterion for the equivalence of the improper integral and the sequence or series is the following.

Proposition 9.10.1 *Let $\phi\colon [0, \infty) \to \mathbf{C}$ be bounded and Riemann integrable on every interval $[0, \beta]$, $\beta > 0$, and let $\lim_{x\to\infty} \phi(x) = 0$. Then*

$$\int_0^\infty \phi(x)\,dx = \sum_{j=0}^\infty \int_j^{j+1} \phi(x)\,dx, \tag{9.10.1}$$

with the implication that if one member of the equation exists, then so does the other.

Proof. We have already noted that if the left-hand member of (9.10.1) exists, then so does the right-hand member. Now suppose that the series in (9.10.1) converges to the sum I. For any $t > 0$, let $[t]$ denote the greatest integer in t. Let $d[t] = \int^{[t]} \phi(x)\,dx - I$. Then by hypothesis, $\lim_{t\to\infty} d[t] = 0$. We have

$$\begin{aligned}\left|\int_0^t \phi(x)\,dx - I\right| &= \left|\int_0^t \phi(x)\,dx - \int_0^{[t]} \phi(x)\,dx + d[t]\right| \\ &\le \int_{[t]}^t |\phi(x)|\,dx + |d[t]| \\ &\le \int_{[t]}^{[t]+1} |\phi(x)|\,dx + |d[t]| \\ &\le \sup_{x\ge[t]} |\phi(x)| + |d[t]|.\end{aligned}$$

The limit as $t \to \infty$ of each term in the last member of this inequality is 0, so the limit of the first member is also 0; and this is the same as stating that $\lim_{t\to\infty} \int_0^t \phi(x)\,dx = I$. Q.E.D.

We now introduce an integral known as the finite Laplace transform. Let $[\alpha, \beta]$ be a finite interval in **R**, and let $F: [\alpha, \beta] \to \mathbf{R}$ be bounded and Riemann integrable on $[\alpha, \beta]$. The finite Laplace transform of F is $\mathscr{L}(F; \alpha, \beta; z) = \int_\alpha^\beta e^{-zt}F(t)\,dt$. It is thereby defined for every $z \in \mathbf{C}$.

Proposition 9.10.2 *The transform $\mathscr{L}(z) = \mathscr{L}(F; \alpha, \beta; z)$ is an entire function of z. For each k, $k = 1, 2, 3, \ldots$, $L^{(k)}(z) = (-1)^k \int_\alpha^\beta t^k e^{-zt}F(t)\,dt$, for every $z \in \mathbf{C}$.*

Proof. Our method of proof does not involve Cauchy theory and is similar to that used for Theorem 8.4.1. We give only an outline. The series representation

$$\begin{aligned} e^{-zt}F(t) &= F(t) - ztF(t) + z^2t^2F(t)/2! \\ &\quad - z^3t^3F(t)/3! + \cdots, \end{aligned} \tag{9.10.2}$$

is valid for every $z \in \mathbf{C}$, $t \in \mathbf{R}$. Take $z = z_0$, and hold it fixed. If $\alpha \le t \le \beta$, then $|t| \le \max(|\alpha|, |\beta|)$. From this and the boundedness of F on $[\alpha, \beta]$, it is easy to set up an M-test for the uniform convergence of the series with the terms treated as functions of t alone. By Proposition 8.3.1, the series in (9.10.2) can be integrated term by term on $[\alpha, \beta]$, and the resulting series of integrals converges to the integral of the sum function. We obtain

$$\mathscr{L}(z_0) = \sum_{k=0}^{\infty} z_0^k \left(\frac{(-1)^k}{k!} \int_\alpha^\beta t^k F(t)\,dt \right),$$

but this is a power series in z_0, and z_0 is arbitrary, so $\mathscr{L}(z)$ is the sum function of a power series in powers of z, convergent for every $z \in \mathbf{C}$. Therefore, $\mathscr{L}$ is analytic on **C**.

To obtain the formulas for the derivatives of $\mathscr{L}$, first notice that $\mathscr{L}^{(k)}(0) = (-1)^k \int_\alpha^\beta t^k F(t)\, dt$. From this it is easy to establish the general formulas for $\mathscr{L}^{(k)}(z)$; the details are an exercise (Exercise 9.10.1). Q.E.D.

We pass to "the" Laplace transform, by which is usually meant the one-sided infinite Laplace transform now to be defined. Let $F: [0, \infty) \to \mathbf{R}$ be bounded and Riemann integrable on each interval $[0, \alpha]$, $\alpha > 0$. The Laplace transform of F is defined by

$$L(z) = L(F; z) = \int_0^\infty e^{-zt} F(t)\, dt \tag{9.10.3}$$

for any z for which the integral exists.

For the definition to be useful, the function F must, of course, be restricted in some way so that the transform exists for certain values of z. A natural restriction is given by the following definition. The function $F: [0, \infty) \to \mathbf{R}$ is said to be of *exponential type with index* λ if it is bounded and integrable on each interval $[0, \alpha]$, $\alpha > 0$, and if there exist $M_F > 0$, $\lambda \geq 0$ such that $|F(t)| \leq M_F e^{\lambda t}$ for every $t > 0$. Notice that a function bounded on $[0, \infty)$ is of exponential type with index 0. Another example: the function $e^{3x}/(1 + x^2)$ is of exponential type with index $\lambda = 3$.

Theorem 9.10.1 *If F is of exponential type with index λ, then the Laplace transform $L(z) = L(F; z)$ exists and is analytic on $H = \{z\colon \operatorname{Re} z > \lambda\}$. The derivatives $L^k(z)$, $k = 1, 2, \ldots$, are given by the formula*

$$L^{(k)}(z) = (-1)^k \int_0^\infty t^k e^{-zt} F(t)\, dt \tag{9.10.4}$$

for each $z \in H$.

Proof. According to Proposition 9.10.2, each of the terms of the series

$$\sum_{j=0}^\infty \int_j^{j+1} e^{-zt} F(t)\, dt \tag{9.10.5}$$

is an analytic function of z with domain $\mathbf{C}$. The series of kth derivatives is

$$(-1)^k \sum_{j=0}^\infty \int_j^{j+1} t^k e^{-zt} F(t)\, dt, \qquad k = 1, 2, \ldots. \tag{9.10.6}$$

We first prove that (9.10.5) converges a.u. on H to a function $f(z)$. By the corollary of Theorem 9.9.1, it follows that f is analytic on H and that the series (9.10.6) represents $f^{(k)}(z)$ for every $z \in H$.

Take a compact set $E \subset H$ and let $2r = \operatorname{dist}(E, H^c)$. Let $H_1 = \{z\colon \operatorname{Re} z \geq \lambda + r\}$; then $E \subset H_1$. We find an M-test for (9.10.5) valid for z on H_1. Let $z = x + iy$, $x, y \in \mathbf{R}$. We have, since $x \geq \lambda + r$,

$$\left|\int_j^{j+1} e^{-zt}F(t)\,dt\right| \le \int_j^{j+1} e^{-xt}|F(t)|\,dt$$
$$\le \int_j^{j+1} e^{-(\lambda+r)t}M_F e^{\lambda t}\,dt$$
$$= \int_j^{j+1} M_F e^{-rt}\,dt$$
$$\le \int_j^{j+1} M_F e^{-rj}\,dt = M_F(e^{-r})^j = \mu_j.$$

Since $r > 0$ and $0 < e^{-r} < 1$, the series $\sum \mu_j$ converges, and the M-test is established. Thus, (9.10.5) converges uniformly on H_1 and, in particular, on E.

It remains to show that for $z \in H$, $L(z) = L(F; z)$ exists and equals $f(z)$, and $L^{(k)}(z)$ as given by (9.10.4) exists and equals the sum of the series in (9.10.6). According to Proposition 9.10.1, this is accomplished if we can prove that $\lim_{t\to\infty} t^k e^{-z_0 t}F(t) = 0$ for any z_0 held fixed on H and for $k = 0, 1, 2, \ldots$.

Let $z_0 = x_0 + iy_0$, with $x_0 > \lambda$; write $x_0 = \lambda + \rho$, $\rho > 0$. Then with $t > 0$,

$$|t^k e^{-z_0 t}F(t)| \le t^k e^{-(\lambda+\rho)t}M_F e^{\lambda t} = \frac{M_F t^k}{e^{\rho t}}. \tag{9.10.7}$$

Now by looking at the power series for e^z, it is easily seen that with $t > 0$, $e^{\rho t} > (\rho t)^{k+1}/(k + 1)!$. After substitution with this into (9.10.7) it becomes apparent that the limit as $t \to \infty$ is zero. Q.E.D.

Extensive tables of Laplace transforms are available. See for example [1], where further references are given.

Exercises 9.10

1. Prove the formula for $\mathscr{L}^{(k)}(z)$ given in Proposition 9.10.2, starting with the fact that $\mathscr{L}^{(k)}(0) = (-1)^k \int_\alpha^\beta t^k F(t)\,dt$. (*Hint*: Let $w = z - a$, $z = w + a$, $\mathscr{L}(z) = \mathscr{L}(w + a) = \phi(w)$; $\phi^{(k)}(0) = \mathscr{L}^{(k)}(a)$. Identify $\phi(w)$ with $\mathscr{L}(G; \alpha, \beta; w)$, where $G(t) = e^{-at}F(t)$.)
2. Assuming that F and F' are both of exponential type with index λ and $F(0) = 0$, prove that for Re $z > \lambda$, $L(F'; z) = zL(F; z)$. (*Hint*: Integrate by parts in $\int_0^T e^{-zt}F'(t)\,dt$.)
3. By direct integration, show that if $F(t) \equiv 1$ on $[0, \infty)$, then $L(F; z) = 1/z$, valid for Re $z > 0$. Show that if $F(t) = t^n$ on $[0, \infty)$, then $L(F; z) = n!/z^{n+1}$, $n = 1, 2, \ldots$. (*Hint*: (9.10.4) applied to $F(t) = 1$.)
4. Let F: $[0, \infty) \to \mathbf{R}$ be continuous and such that there is a constant M with the

property that $|F(t)| < M/t$ for $t \geq 1$. Show that

$$f(z) = \int_0^\infty \frac{F(t)}{t - z}\, dt$$

exists for every $z \in \mathbf{C} \sim [0, \infty]$ and represents an analytic function on $\mathbf{C} \sim [0, \infty]$. (*Hint*: Follow the general method of the proof of Theorem 9.10.1. For the analyticity of $\int_j^{j+1} [F(t)/(t - z)]\, dt$, use Theorem 8.4.1.)

5. Let the function Γ be defined by

$$\Gamma(z) = \int_0^\infty t^{z-1} e^{-t}\, dt \tag{9.10.8}$$

for any z for which the improper integral exists. The principal logarithm is used in evaluating t^{z-1}. Carry out the following program (which is similar to that used in this section for the Laplace transform) to show that the integral in (9.10.8) exists and represents an analytic function on $H = \{z : \text{Re } z > 1\}$:

(a) Hold z fixed, and let j be any nonnegative integer. Expand the integrand of the integral $\int_j^{j+1} t^{z-1} e^{-t}\, dt$ in an exponential series with base t.

(b) Set up an M-test for the uniform convergence of this series for $j \leq t \leq j + 1$.

(c) Integrate the series term-by-term over $[j, j + 1]$. The resulting series is

$$\sum_{k=0}^{\infty} (-1)^k \frac{(j + 1)^{z+k} - j^{z+k}}{k!(z + k)} = \int_j^{j+1} t^{z-1} e^{-t}\, dt, \tag{9.10.9}$$

where $0! = 1$. (Cite pertinent authority for the validity of this equation.)

(d) Let E be an arbitrary compact subset of H. Set up an M-test for the uniform convergence of the series in (9.10.9) for $z \in E$. Each term of the series is clearly analytic on H. Cite an authority which establishes that the integral in (9.10.9) is analytic on H.

(e) Let $f(z) = \sum_{j=0}^{\infty} \int_j^{j+1} t^{z-1} e^{-t}\, dt$, $z \in H$. Let E again be an arbitrary compact subset of H. Set up an M-test for the uniform convergence of this series on E. Again cite an authority which shows that f is analytic on H.

(f) Finally, for $z \in H$ identify $f(z)$ with $\Gamma(z)$ as given by (9.10.8), by showing that for every $z \in H$, the hypothesis of Proposition 9.10.1 are satisfied by the integrand $t^{z-1}e^{-t}$.

(*Hint*: Assume these easily proved facts: A primitive of t^z, t real, $t \neq 0$, z complex $z \neq -1$, is $t^{z+1}/(z + 1)$; if $z = x + iy$, x, y real, $|t^z| = t^x$, $t > 0$; $\sum_{n=1}^{\infty} n^\alpha e^{-n}$, α real, converges [ratio test]; $\lim_{t \to \infty} t^w e^{-t} = 0$, t real.)

Remark. The integral in (9.10.8) is known as Euler's integral, and (9.10.8) is Euler's definition of the so-called gamma function. This function has a particularly rich formal theory, which was studied in depth by Euler and various nineteenth century mathematicians including Gauss. An introduction to the theory is given in [9, Section 8.8]. See [1, Chapter 6] for a comprehensive survey of the many formal relations involving the gamma function, and for tables of numerical values. Incidentally, with a little more attention to detail, it can be shown that the integral in (9.10.8) exists and is analytic on $\{z: \text{Re } z > 0\}$.

6. Let $\Gamma(1)$ be defined by the integral in (9.10.8) with $z = 1$. Show that $\Gamma(1)$ exists and equals 1, and that for every real z, say $z = x$, with $x \geq 1$, $\Gamma(x + 1) = x\Gamma(x)$. Thereby show that if n is any positive integer, $\Gamma(n + 1) = n!$. (*Hint*: Integrate by parts.)
7. (Continuation of Exercise 9.10.6.) Cite authorities to show that the "difference equation" $\Gamma(x + 1) = x\Gamma(x)$, $x \geq 1$, implies that $\Gamma(z + 1) \equiv z\Gamma(z)$, $z \in H = \{z: \text{Re } z > 1\}$. Thereafter show that this identity can be used as a definition to obtain an analytic continuation of $\Gamma(z)$ onto $\mathbf{C} \sim \{0, -1, -2, -3, \ldots\}$.

9.11 Evaluation of Improper Real Integrals by Using the Cauchy Integral Theorem

This is a classical and perhaps somewhat old-fashioned application of the Cauchy theory. Success in any given instance depends on the integrand being related to some analytic function, and also on the ingenuity of the problem solver in selecting a suitable analytic function and path of integration. The following example illustrates the general idea. Other examples are given in the exercises.

Example. Evaluate $\int_0^\infty (\sin \alpha t)/t \, dt$, $\alpha > 0$. If the integrand is assigned the value α at $t = 0$, then it is continuous in any interval $[0, R]$, $R > 0$, and $\int_0^R t^{-1} \sin \alpha t \, dt = \lim_{r \to 0} \int_r^R t^{-1} \sin \alpha t \, dt$, $0 < r < R$. The problem can be restated as follows: Evaluate $\lim_{R \to \infty} [\lim_{r \to 0} \int_r^R t^{-1} \sin \alpha t \, dt]$.

The function $e^{i\alpha z}/z$ is analytic on $\mathbf{C} \sim \{0\}$. Let γ be the closed path $\gamma_1 + \gamma_2 + \gamma_3 + \gamma_4$, where $\gamma_1 = [r, R]$, γ_2 is a positively oriented semicircle in the closed upper half-plane joining R to $-R$, $\gamma_3 = [-R, -r]$, and γ_4 is a negatively oriented semicircle joining $-r$ to r and lying in the closed upper half-plane. This path γ appeared in Exercise 9.3.2, and according to that exercise, $\int_\gamma (e^{i\alpha z}/z) \, dz = 0$.

We use the following parametric equations:

$$\begin{aligned} \gamma_1&: z = t, & r \leq t \leq R;\\ \gamma_2&: z = Re^{it}, & 0 \leq t \leq \pi;\\ -\gamma_3&: z = -t, & r \leq t \leq R;\\ -\gamma_4&: z = re^{it}, & 0 \leq t \leq \pi. \end{aligned}$$

We have

$$\begin{aligned}\int_{\gamma_1}\frac{e^{i\alpha z}}{z}\,dz+\int_{\gamma_3}\frac{e^{i\alpha z}}{z}\,dz&=\int_{\gamma_1}\frac{e^{i\alpha z}}{z}\,dz-\int_{-\gamma_3}\frac{e^{i\alpha z}}{z}\,dz\\&=\int_r^R\frac{e^{i\alpha t}}{t}\,dt-\int_r^R\frac{-e^{-i\alpha t}}{-t}\,dt\\&=\int_r^R\frac{e^{i\alpha t}-e^{-i\alpha t}}{t}\,dt=2i\int_r^R\frac{\sin\alpha t}{t}\,dt.\qquad(9.11.1)\end{aligned}$$

Also

$$\begin{aligned}\left|\int_{\gamma_2}\frac{e^{i\alpha z}}{z}\,dz\right|&=\left|\int_0^\pi\frac{e^{i\alpha R(\cos t+i\sin t)}iRe^{it}\,dt}{iRe^{it}}\right|\\&\le\int_0^\pi e^{-\alpha R\sin t}\,dt.\qquad(9.11.2)\end{aligned}$$

The inequality in (9.11.2) follows from the fact that for real x, y, $|e^{i(x+iy)}|=|e^{-y+ix}|=e^{-y}$. We show that the right-hand member of (9.11.2) tends to zero as $R\to\infty$. For later reference we prove a slightly more general result which includes this assertion.

Lemma 1 $\lim_{\beta\to\infty}\beta^p\int_0^{\pi/2}e^{-\beta\sin t}\,dt=\lim_{\beta\to\infty}\beta^p\int_0^{\pi/2}e^{-\beta\cos t}\,dt=0$, $0\le p<1$.

Proof. By Proposition 7.3.1, $\sin t\ge 2t/\pi$, $0\le t\le\pi/2$. Therefore, for $\beta>0$,

$$\begin{aligned}0<\beta^p\int_0^{\pi/2}e^{-\beta\sin t}&\le\beta^p\int_0^{\pi/2}e^{-2\beta t/\pi}\,dt\\&=\beta^p\left[-\frac{\pi}{2\beta}e^{-2\beta t/\pi}\right]_0^{\pi/2}\\&=\frac{\pi}{2\beta^{1-p}}(1-e^{-\beta}).\end{aligned}$$

The limit of the right-hand member as $\beta\to\infty$ is clearly zero, since $1-p>0$. The integral involving the cosine in the lemma has the same value as that of the integral involving the sine; this can easily be seen by making the change of variables $t=(\pi/2)=u$. Q.E.D.

The right-hand member of (9.11.2) can be written in the form

$$\begin{aligned}\int_0^{\pi/2}e^{-\alpha R\sin t}\,dt&+\int_{\pi/2}^{\pi}e^{-\alpha R\sin t}\,dt\\&=2\int_0^{\pi/2}e^{-\alpha R\sin t}\,dt;\qquad(9.11.3)\end{aligned}$$

this equation is obtained by making the change of variable $t=\pi-u$ in the second integral in the first member. We apply Lemma 1 to the last integral in (9.11.3), letting $\alpha R = \beta$ and $p = 0$, and thereby establish that

$$\lim_{R\to\infty} \int_{\gamma_2} \frac{e^{i\alpha z}}{z}\, dz = 0. \tag{9.11.4}$$

We now study $\int_{\gamma_4} z^{-1} e^{i\alpha z}\, dz$. We write this integral in the form

$$-\int_{-\gamma_4} \frac{e^{i\alpha z}}{z}\, dz = \int_{-\gamma_4} \frac{1 - e^{i\alpha z}}{z}\, dz - \int_{-\gamma_4} \frac{dz}{z}. \tag{9.11.5}$$

The second integral in the right-hand member can be evaluated explicitly by using the equation for $-\gamma_4$; the value is πi. To estimate the first integral, we use the following lemma.

Lemma 2 If a is any complex number and $0 < |z| \le 1$, then

$$\left|\frac{1 - e^{az}}{z}\right| \le e^{|a|}.$$

Proof.

$$\left|\frac{1 - e^{az}}{z}\right| = \left|\frac{1 - (1 + az + a^2z^2/2! + a^3z^3/3! + \cdots)}{z}\right|$$

$$\le |a| + \frac{|a|^2|z|}{2!} + \frac{|a|^3|z|^2}{3!} + \cdots$$

$$\le 1 + |a| + \frac{|a|^2}{2!} + \frac{|a|^3}{3!} + \cdots = e^{|a|}. \qquad \text{Q.E.D.}$$

Using Lemma 2 and (9.11.5), we obtain

$$\left|\int_{\gamma_4} \frac{e^{iz}}{z}\, dz + \pi i\right| = \left|\int_{-\gamma_4} \frac{1 - e^{iz}}{z}\right| \le e^{|i\alpha|}\pi r.$$

Hence

$$\lim_{r\to 0} \int_{\gamma_4} \frac{e^{i\alpha z}}{z}\, dz = -\pi i. \tag{9.11.6}$$

We now put the pieces of the problem together. Let $f(z) = e^{i\alpha z}/z$. By the Cauchy integral theorem as it appears in Exercise 9.3.2, we have

$$\left[\int_{\gamma_1} f(z)\, dz + \int_{\gamma_3} f(z)\, dz\right] + \int_{\gamma_2} f(z)\, dz + \int_{\gamma_4} f(z)\, dz = 0.$$

We apply (9.11.1) to the integrals in the square brackets and obtain

$$2i \int_r^R \frac{\sin \alpha t}{t}\, dt = -\int_{\gamma_2} f(z)\, dz - \int_{\gamma_4} f(z)\, dz,$$

or

$$\int_r^R \frac{\sin \alpha t}{t}\, dt = -\frac{1}{2i}\int_{\gamma_2} f(z)\, dz - \frac{1}{2i}\int_{\gamma_4} f(z)\, dz.$$

Using (9.11.6) and noting that the integral along γ_2 is independent of r, we obtain

$$\lim_{r\to 0}\int_r^R \frac{\sin \alpha t}{t}\, dt = -\frac{1}{2i}\int_{\gamma_2} f(z)\, dz - \frac{1}{2i}(-\pi i).$$

Finally, from (9.11.4),

$$\lim_{R\to 0}\left[\lim_{r\to 0}\int_r^R \frac{\sin \alpha t}{t}\, dt\right] = 0 + \frac{\pi}{2}.$$

Thus, the value of $\int_0^\infty (t^{-1}\sin \alpha t)\, dt$ is $\pi/2$ for any $\alpha > 0$. The integral is known as Dirichlet's integral.

Exercises 9.11

1. Evaluate $\int_0^\infty (\sin^2 \alpha x)/x^2\, dx$, $\alpha > 0$. *Answer*: $\pi\alpha/2$.
(*Hint*: Integrate $(1 - e^{2i\alpha z})/2z^2$ along the path γ used in the previous example. Show that the sum of the integrals along γ_1 and γ_3 is

$$2\int_t^R \frac{1-\cos 2\alpha t}{2t^2}\, dt = 2\int_r^R \frac{\sin^2 \alpha t}{t^2}\, dt.$$

Show that the absolute value of the integral along γ_2 is dominated by π/R. In the integral along γ_4, add and subtract $2i\alpha z$ to the numerator of the integrand. Proceed as in the example, but replace Lemma 2 by the estimate $|(1 - e^{az} + az)/z^2| \le e^{|a|}$, $0 < |z| \le 1$, which can be obtained by the method used in the proof of Lemma 2.)

2. Evaluate the so-called Fresnel integrals $\int_0^\infty \cos x^2\, dx$, $\int_0^\infty \sin x^2\, dx$, assuming that $\int_0^\infty e^{-x^2}\, dx = \sqrt{\pi}/2$. *Answer*: $\sqrt{\pi}/(2/\sqrt{2})$ for each integral.
(*Hint*: Integrate the function e^{-z^2} along the boundary γ of the sector $\{z: 0 \le |z| \le R, 0 \le \operatorname{Arg} z \le \pi/4\}$. With $\gamma = \gamma_1 + \gamma_2 + \gamma_3$, the suggested parametrizations are $\gamma_1 : z = te^{i\pi/4}, 0 \le t \le R; -\gamma_2 : z = Re^{it}, 0 \le t \le \pi/4; -\gamma_3 : z = t$, $0 \le t \le R$. Intuitively speaking, the integration is "counterclockwise." Use Lemma 1 in the example with $\beta = R^{1/2}$ to show that the integral along γ_2 tends to zero as $R \to \infty$. The goal of the analysis should be to establish that

$$\lim_{R\to\infty}\left[\left(\int_0^R \cos t^2\, dt + \int_0^R \sin t^2\, dt - \sqrt{\frac{\pi}{2}}\right) + i\left(\int_0^R \cos t^2\, dt - \int_0^R \sin t^2\, dt\right)\right] = 0.$$

This equation implies that $\int_0^R \cos t^2\, dt + \int_0^R \sin t^2\, dt = \sqrt{\pi/2} + \varepsilon(R)$, $\int_0^R \cos t^2\, dt - \int_0^R \sin t^2\, dt = \eta(R)$, where $\lim_{R\to\infty} \varepsilon(R) = \lim_{R\to\infty} \eta(R) = 0$. Treat these as simultaneous linear equations with the integrals as the unknowns, and solve the equations to establish both the existence and the values of Fresnel's integrals.)

3. With $\alpha > 0$, prove these equations:

$$\int_0^\infty e^{-x^2} \cos 2\alpha x\, dx = \frac{e^{-\alpha^2}\sqrt{\pi}}{2},$$

$$\int_0^\infty e^{-x^2} \sin 2\alpha x\, dx = e^{\alpha^2} \int_0^\alpha e^{t^2}\, dt.$$

Assume that $\int_0^\infty e^{-x^2}\, dx = \sqrt{\pi}/2$. (*Hint*: Integrate e^{-z^2} along the rectangular path $[i\alpha, R + i\alpha] + [R + i\alpha, R] + [R, 0] + [0, i\alpha]$. Suggested parametrizations:

$$\begin{aligned} [i\alpha, R + i\alpha]&: z = i\alpha + t, \quad 0 \le t \le R; \\ -[R + i\alpha, R]&: z = R + it, \quad 0 \le t \le \alpha; \\ -[R, 0]&: z = t, \quad 0 \le t \le R; \\ [0, i\alpha]&: z = it, \quad 0 \le t \le \alpha.) \end{aligned}$$

4. Show by an appropriate change of variables that

$$\left(\int_{-R}^{-r} + \int_r^R\right) \frac{e^{i\alpha t}}{t}\, dt = \int_r^R \frac{2i \sin \alpha t}{t}\, dt$$

and, consequently, that

$$\int_{-\infty}^\infty \frac{e^{i\alpha t}}{t}\, dt = \pi i,$$

provided that this improper integral is given a special interpretation.

10

Zeroes and Isolated Singularities of Analytic Functions

10.1 A Factorization of Analytic Functions

Let B be a region, and let $f: B \to \mathbf{C}$ be analytic on B. The function f has the LPS property on B [Theorem 9.4.2]. Then according to Corollary 3 of Theorem 6.5.1 the set of zeros of f on B (that is, the set $\{z : f(z) = 0, z \in B\}$) cannot have an accumulation point in B unless $f(z) \equiv 0$ on B. Henceforth in this chapter, functions which are identically equal to zero on their domains are ruled out, unless in a given instance the contrary is specifically stated.

It is, of course, possible for a function analytic on a region to have an infinite number of zeros in the region, whether or not the region is bounded. For example, the function $f(z) = \sin(2\pi/z)$ restricted to $\Delta(1, 1)$ has zeros at $z = 1/n$, $n = 2, 3, \ldots$. But this possibility disappears if the function is restricted to a compact domain. More formally, we have the following proposition.

Proposition 10.1.1 *Let $f: B \to \mathbf{C}$ be analytic on the region B, and let E be a compact subset of B. Then f can have at most a finite number of zeros in E.*

Proof. If the set of zeros of f in E were infinite, then by the Bolzano–Weierstrass theorem [Exercise 5.5.4], it would have an accumulation point in E and, thus, an accumulation point in B. But as was stated previously, this would be impossible unless $f(z) \equiv 0$ on B. Q.E.D.

Let $f: B \to \mathbf{C}$ be analytic on the region B, and let $f(a) = 0$, $a \in B$. There is a local power series representation $f(z) = \sum_{k=0}^{\infty} a_k(a)(z - a)^k$ converging in some neighborhood of a, and $a_0(a) = f(a) = 0$. If $a_0(a) = a_1(a) = \cdots =$

$a_{m-1}(a) = 0$ but $a_m(a) \neq 0$, $m \geq 1$, then f is said to have a zero at a of *order* m, or of *multiplicity* m. A zero of order 1 is also called a *simple* zero.

Proposition 10.1.2 *A necessary and sufficient condition for a function $f: B \to$ **C**, analytic on the region B, to have a zero of order m at $a \in B$, is that there exists a function $g: B \to$ **C**, analytic on B, with $g(z) \neq 0$ for every z in some neighborhood of a, and such that $f(z) = (z - a)^m g(z)$ for every $z \in B$.*

Proof. The function f has a local power series representation $\sum a_k(z - a)^k$ converging to $f(z)$ in a disk $\Delta(a, R) \subset B$, $R > 0$. If the function has a zero or order m at a, then by definition $a_0 = a_1 = \cdots = a_{m-1} = 0$ and $a_m \neq 0$. It follows from Theorem 4.3.1 that there exists a function $g: \Delta(a, R) \to$ **C** such that $f(z) = (z - a)^m g(z)$ for every $z \in \Delta(a, R)$. According to that theorem, $g(z) \neq 0$ in some neighborhood of a, and g is the sum function of a power series convergent in $\Delta(a, R)$. Therefore, g is analytic at least on $\Delta(a, R)$. We extend the definition of g to B simply by using the equation $g(z) = f(z)/(z - a)^m$, $z \in B \sim \{a\}$.

Conversely, suppose that a representation of the form $f(z) = (z - a)^m g(z)$ as described in Proposition 10.1.2 does exist. Let the local power series expansion of g at a be $g(z) = \sum_0^\infty b_k(z - a)^k$. Since by hypothesis $g(a) \neq 0$, it follows that $b_0 = g(a) \neq 0$. The local power series for $f(z) = (z - a)^m g(z)$ can be written as follows: $0 + 0(z - a) + 0(z - a)^2 + \cdots + 0(z - a)^{m-1} + b_0(z - a)^m + b_1(z - a)^{m+1} + \cdots$. Thus, by definition, f has a zero of order m at a. Q.E.D.

Proposition 10.1.3 *Let $f: B \to$ **C** be analytic on the region B, let f have a zero of order m at $a \in B$, and let the function $\phi: B \to$ **C** be analytic on B with $\phi(a) \neq 0$. Then $\phi \cdot f$ has a zero of order m at a.*

The proof is deferred to the exercises.

Theorem 10.1.1 *Let $f: B \to$ **C** be analytic on the region B, and let the set of zeros of f on B be the finite set $Z = \{z_1, z_2, \ldots, z_\mu\}$. Let the multiplicity of the zero z_k be m_k, $k = 1, 2, \ldots, \mu$. Then there exists a function $g: B \to$ **C** analytic on B, with $g(z) \neq 0$ for every $z \in B$, and such that*

(i) $f(z) = (z - z_1)^{m_1}(z - z_2)^{m_2} \cdots (z - z_\mu)^{m_\mu} g(z)$,

for every $z \in B$, and

(ii) $$\frac{f'(z)}{f(z)} = \frac{m_1}{z - z_1} = \frac{m_2}{z - z_2} + \cdots + \frac{m_\mu}{z - z_\mu} + \frac{g'(z)}{g(z)},$$

for every $z \in B \sim Z$.

Proof. (i) By Proposition 10.1.2 there exists a function $g_1: B \to$ **C** analytic on B with $g_1(z_1) \neq 0$, and such that for every $z \in B$, $f(z) = (z - z_1)^{m_1} g_1(z)$. Now $(z - z_1)^{-m}$ is analytic on $B \sim \{z_1\}$ and does not have a zero at z_2. By Proposition 10.1.3, the function $(z - z_1)^{-m} f(z)$, which is defined and equal to $g_1(z)$ on $B \sim \{z_1\}$, has a zero of order m_2 at z_2. That is to say, g_1, which is analytic everywhere on B, has a zero of order m_2 at z_2. Therefore, there exists a function $g_2 : B \to \mathbf{C}$, analytic on B, with $g_2(z_2) \neq 0$, such that $g_1(z) = (z - z_2)^{m_2} g_2(z)$ for every $z \in B$. Also $g_2(z_1) \neq 0$ because $g_1(z_1) \neq 0$. We have arrived at the factorization $f(z) = (z - z_1)^{m_1}(z - z_2)^{m_2} g_2(z)$.

The next step is to observe that $(z - z_1)^{-m_1}(z - z_2)^{-m_2}$ is analytic on $B \sim (\{z_1\} \cup \{z_2\})$ and nonzero at z_3. Therefore, $(z - z_1)^{-m_1}(z - z_2)^{-m_2} f(z) = g_2(z)$, has a zero of order m_3 at z_3; and so forth.

The process is continued until all the zeros of f are exhausted. The final factorization is $f(z) = (z - z_1)^{m_1}(z - z_2)^{m_2} \cdots (z - z_\mu)^{m_\mu} g(z)$, where $g(z_1) \neq 0$, $g(z_2) \neq 0, \ldots, g(z) \neq 0$ and g is analytic on B. If there were a point $z_0 \in B$, $z_0 \neq z_j$, $j = 1, \ldots, \mu$, such that $g(z_0) = 0$, then $f(z_0) = 0$, contrary to the hypotheses.

To prove part (ii), we let $p(z) = (z - z_1)^{m_1}(z - z_2)^{m_2} \cdots (z - z)^{m_\mu}$. According to Exercise 9.6.3,

$$\frac{p'(z)}{p(z)} = \frac{m_1}{z - z_1} + \frac{m_2}{z - z_2} + \cdots + \frac{m_\mu}{z - z_\mu}.$$

We have from (i): $f(z) \equiv p(z)g(z)$, $f'(z) \equiv p'(z)g(z) + p(z)g'(z)$. Thus,

$$\frac{f'(z)}{f(z)} = \frac{p'(z)}{p(z)} + \frac{g(z)}{g'(z)}.$$

The formula in (b) follows at once. Q.E.D.

Exercises 10.1

1. Let $f : B \to \mathbf{C}$ be analytic on the region B and have a zero at $a \in B$. Prove that a necessary and sufficient condition for the zero of f at a to be of order m is that f' has a zero of order $m - 1$ at a. (The case $m = 1$, $m - 1 = 0$ is to be construed as meaning that $f'(a) \neq 0$. *Hint*: First, use Proposition 10.1.2 to show that whenever f has a zero of order p at a, then f' must have a zero or order $p - 1$ at a, or $f'(a) \neq 0$ if $p = 1$. Then use the facts that here *by hypothesis f does have a zero at a*, and the order of a zero of an analytic function is uniquely defined.)
2. Let $f : B \to \mathbf{C}$ and $g : B \to \mathbf{C}$ be analytic on the set B, and let f and g have zeros at $a \in B$ of respective orders $m \geq 0$ and $n \geq 0$, where $m = 0$ or $n = 0$ is to be interpreted as indicating a nonzero. Show that $f \cdot g$ has a zero of order $m + n$ at a. (Proposition 10.1.3 is a special case.)

10.2 Integrals Related to the Set of Zeros of an Analytic Function

Theorem 10.2.1 *Let $f\colon B \to \mathbf{C}$ be analytic and not identically zero on the region B. Let the closed* disk $\bar{\Delta}(a, r)$ $r > 0$, *be contained in B, and let z_1, $z_2, \ldots, z_\nu$ be the zeros of f in $\Delta(a, r)$, assumed to exist and of respective orders $m_1, m_2, \ldots, m_\nu$. Suppose further that no zero of f lies on $C(a, r)$. Let γ denote the positively oriented circle with graph $C(a, r)$. Then*

(i) $\displaystyle \frac{1}{2\pi i}\int_\gamma \frac{f'(z)}{f(z)}\,dz = m_1 + m_2 + \cdots + m_\nu;$

(ii) if $h\colon B \to \mathbf{C}$ is analytic, then

$$\frac{1}{2\pi i}\int_\gamma h(z)\frac{f'(z)}{f(z)}\,dz = m_1 h(z_1) + m_2 h(z_2) + \cdots + m_\nu h(z_\nu). \quad (10.2.1)$$

Proof. We prove part (ii), since part (i) is the special case in which $h(z) \equiv 1$. Let R be any real number such that $r < R < \text{dist}\,(a, B^c)$; then $\bar{\Delta}(a, r) \subset \bar{\Delta}(a, R) \subset B$. By Proposition 10.1.1, f has only a finite number of zeros on $\bar{\Delta}(a, R)$ and, therefore, only a finite number of zeros on $\Delta(a, R)$, among which are the zeros $z_1, \ldots, z_\nu$ in $\Delta(a, r)$. We denote the zeros in $\Delta(a, R)$ by $z_1, \ldots, z_\nu, z_{\nu+1}, \ldots, z_\mu$, of respective multiplicities $m_1, \ldots, m_\nu, m_{\nu+1}, \ldots, m_\mu$. By Theorem 10.1.1, there exists $g\colon \Delta(a, R) \to \mathbf{C}$ analytic and nonzero on $\Delta(a, R)$ such that

$$h(z)\frac{f'(z)}{f(z)} = \sum_{j=1}^{\nu} \frac{m_j h(z)}{z - z_j} + \sum_{j=\nu+1}^{\mu} \frac{m_j h(z)}{z - z_j} + \frac{h(z)g'(z)}{g(z)}. \quad (10.2.2)$$

The disk $\Delta(a, R)$ is a starlike region, so the formulations of the Cauchy integral theorem and formula in Theorems 9.2.2 and 9.4.1 are applicable. The calculation can be displayed as follows:

$$\begin{aligned}\frac{1}{2\pi i}\int_\gamma \frac{h(z)f'(z)}{f(z)}\,dz &= \sum_{j=1}^{\nu} m_j \operatorname{Ind}_\gamma(z_j)h(z_j)\\ &\quad + \sum_{j=\nu+1}^{\mu} m_j \operatorname{Ind}_\gamma(z_j)h(z_j) + 0\\ &= \sum_{j=1}^{\nu} m_j h(z_j) + 0 + 0. \end{aligned} \quad (10.2.3)$$

To obtain the last equation, we used the computation of the index of a point with respect to a positively oriented circle which appears in Example (i) of Sec. 8.4. Q.E.D.

Remark. If f has no zeros in $\bar{\Delta}(a, r)$, then by the Cauchy integral theorem the integrals in parts (i) and (ii) of the conclusion of the theorem have the value zero.

Corollary Let $f\colon B \to \mathbf{C}$ be analytic, $f(z) \not\equiv 0$, on the region B. Let $\bar{\Delta}(a, r)$, $r > 0$, be contained in B. Suppose that f has no zeros on $C(a, r)$. Then $f(z) \neq 0$ for every $z \in \Delta(a, r)$ if and only if

$$\frac{1}{2\pi i}\int_\gamma \frac{f'(z)}{f(z)}\,dz = 0, \tag{10.2.4}$$

where γ is the positive or negatively oriented circle with graph $C(a, r)$.

Theorem 10.2.2 *Let $f\colon B \to \mathbf{C}$ be analytic on the region B, let $\bar{\Delta}(a, r)$, $r > 0$, be contained in B, and let w_0 be a complex number. Let γ be the positively oriented circle with graph $C(a, r)$, and let γ^* be the closed path $\gamma^* = f \circ \gamma$, with graph Γ^*. Suppose that $w_0 \notin \Gamma^*$. Then the number of zeros of $f(z) - w_0$ in $\Delta(a, r)$, counted according to multiplicities, is given by* $\operatorname{Ind}_{\gamma^*}(w_0)$.

Remark. The case in which there are no zeros of $f(z) - w_0$ in $\Delta(a, r)$ and $\operatorname{Ind}_{\gamma^*}(w_0) = 0$ is not excluded.

Proof of the Theorem. Let the equation of γ be $z = \gamma(t) = a + re^{it}$, $0 \le t \le 2\pi$. The equation of γ^* is $w = f[\gamma(t)]$, $0 \le t \le 2\pi$. The condition $w_0 \notin \Gamma^*$ means that there is no number $t \in [0, 2\pi]$ such that $f(\gamma(t)) = w_0$, so there is no zero of $f(z) - w_0$ on $C(a, r)$.

If $f(z)$ is a constant function on B, then since $f(z) = w_0 \neq 0$ on $C(a, r)$, there are no zeros of $f(z) - w_0$ in B. In this case, γ^* is a point arc, so it follows from the definition of complex line integral that $\operatorname{Ind}_{\gamma^*}(w_0) = 0$.

Now suppose that f is nonconstant. The derivative of $f(z) - w_0$ with respect to z is $f'(z)$. We use Theorem 8.3.2 twice in the following computation:

$$\begin{aligned}\frac{1}{2\pi i}\int_\gamma \frac{f'(z)\,dz}{f(z) - w_0} &= \frac{1}{2\pi i}\int_0^{2\pi} \frac{f'[\gamma(t)]\gamma'(t)\,dt}{f[\gamma(t)] - w_0}\\ &= \frac{1}{2\pi i}\int_{\gamma^*} \frac{dw}{w - w_0} = \operatorname{Ind}_{\gamma^*}(w_0).\end{aligned} \tag{10.2.5}$$

By the corollary of Theorem 10.2.1, if $f(z) - w_0$ has no zeros on $\Delta(a, r)$, then the left-hand member of (10.2.5) is 0, so in this case $\operatorname{Ind}_{\gamma^*}(w_0) = 0$. If $f(z) - w_0$ does have zeros on $\Delta(a, r)$, by Theorem 10.2.1(i) the left-hand member of (10.2.5) is equal to the sum of their multiplicities. Q.E.D.

Exercises 10.2

1. Use Theorem 10.2.1 to find the value of

$$\frac{1}{i}\int_\gamma \frac{z^m \cos \pi z}{\sin \pi z}\,dz,$$

where γ is the positively oriented circle with graph $C(0, n + 1/2)$ and n and m are each positive integers on zero.

Answer. $\sum_{k=-n}^{n} k^m$, with 0^0 interpreted as 1.

2. Let the function f be analytic, $f(z) \not\equiv 0$, on a region R containing the set $\bar{B} = \{z : z = a + sr(t)e^{it},\ 0 \le t \le 2\pi,\ 0 \le s \le 1\}$, where $r(t)$ is piecewise continuously differentiable, $r(t) > 0$, and $r(0) = r(2\pi)$. Let γ be the closed path with parametric equation $z = a + r(t)e^{it}, 0 \le t \le 2\pi$. Suppose that no zero of f lies on the graph of γ. Let $z_1, z_2, \ldots, z_\nu$ be the zeros of f which are interior points of $\bar{B}$, with respective multiplicities $m_1, \ldots, m_\nu$. Show that Theorem 10.2.1 and its corollary remain true with this γ replacing the circle γ in that theorem. (*Hint*: If a representation of hf'/f like that in (10.2.2) can be obtained, then Propositions 9.3.3 and 9.4.1 establish the validity of the computation in (10.2.3). The following device, which could have been used in the proof of Theorem 10.2.1, achieves this end. By Exercise 7.5.11, $\bar{B}$ is compact, so $\text{dist}(\bar{B}, R^c) > 0$ [Theorem 5.6.1]. Choose $\rho, 0 < \rho < \text{dist}(\bar{B}, R^c)$. Let Γ be the graph of γ. Cover each point $z \in \Gamma$ with a disk $\Delta(z, \rho)$. Let B^* be the union of all these disks and the set $B = \{z : z = a + sr(t)e^{it}, 0 \le t \le 2\pi, 0 \le s < 1\}$. According to Exercise 7.5.10, B is open, so B^* is open. Show that there are only a finite number of zeros of f on B^* and complete the proof.)

10.3 Rouché's Theorem with Applications

The number of zeros of an analytic function in a region can sometimes be inferred from the known number of zeros of a "nearby" function. This idea is developed in this section.

Theorem 10.3.1 (*Rouché's Theorem*). *Let B be a region; let $f: B \to \mathbf{C}$ and $g: B \to \mathbf{C}$ be functions analytic on B; let $\bar{\Delta}(b, r)$, $r > 0$, be contained in B; and finally suppose that $|g(z) - f(z)| < |f(z)|$ for every $z \in C(b, r)$. Then f and g have the same number of zeros in $\Delta(b, r)$, with zeros counted according to multiplicities.*

Proof. Let the parametric equation of $C(b, r)$ be $z = \gamma(t) = b + re^{it}$, $0 \le t \le 2\pi$, and let $F(z, \sigma) = f(z) + \sigma[g(z) - f(z)]$, $F_1(z, \sigma) = f'(z) + \sigma[g'(z) - f'(z)] = \partial F/\partial z$, $0 \le \sigma \le 1$. Clearly, F and F_1 as functions of z are analytic on B for each fixed $\sigma \in [0, 1]$. It is easily verified that $F[\gamma(t), \sigma]$ and, therefore, $|F[\gamma(t), \sigma]|$ are continuous functions of the real variables t and σ on the rectangle $R = \{(t, \sigma) : 0 \le t \le 2\pi, 0 \le \sigma \le 1\}$. Now R is a compact set in $\mathbf{R}^2$ [Theorem 5.5.1], so by Corollary 1 of Theorem 5.5.3, F has a minimum value m on R. The condition $|f(z)| > |g(z) - f(z)|$ implies that $m > 0$. Similarly, $F_1[\gamma(t), \sigma]$ is continuous on R. These facts together imply that the quotient $F_1[\gamma(t), \sigma]/F[\gamma(t), \sigma]$ is continuous on R.

We now set up the integral

$$I(\sigma) = \frac{1}{2\pi i}\int_\alpha^\beta \frac{F_1[\gamma(t), \sigma]}{F[\gamma(t), \alpha]}\gamma'(t)\,dt$$
$$= \frac{1}{2\pi i}\int_\gamma \frac{f'(z) + \sigma[g'(z) - f'(z)]}{f(z) + \sigma[g(z) - f(z)]}\,dz.$$

According to Exercise 8.3.7, $I(\sigma)$ is continuous on $[0, 1]$. But according to Theorem 10.2.1(i), for each $\sigma \in [0, 1]$, $I(\sigma)$ is the number of zeros of $f(z) + \sigma[g(z) - f(z)]$ in $\Delta(b, r)$, counted according to multiplicities; therefore, $I(\sigma)$ is an integer for each $\sigma \in [0, 1]$. Moreover $I(0)$ is the number of zeros of $f(z)$ in $\Delta(b, r)$ and $I(1)$ is the number of zeros of $f(z) + [g(z) - f(z)] = g(z)$ in $\Delta(b, r)$.

Suppose now that $I(0) \neq I(1)$. Then by the intermediate value theorem [Proposition 5.3.2], if c is any noninteger lying between these two integers, then for some value of $\sigma \in [0, 1]$, $I(\sigma) = c$, which contradicts the fact that $I(\sigma)$ is always an integer. Q.E.D.

Remark. Although the hypotheses rules out the posibility that either f or g is identically equal to zero on B, the argument does cover cases in which either f or g or both are constant functions on B. In any of these cases, under the conditions of the theorem, neither function has any zeros in $\Delta(b, r)$.

Example. We show that the equation $e^{z-\alpha} = z^n$, where n is a positive integer and $\alpha > 1$, has exactly n roots in $\Delta(0, 1)$, counted according to multiplicities. (By "root" of an equation $\phi(z) = \psi(z)$, we mean a zero of $\phi(z) - \psi(z)$.) In Rouché's theorem, let $g(z) = e^{z-\alpha} - z^n$ and $f(z) = -z^n$. Then f clearly has a zero of order n in $\Delta(0, 1)$, and no other zeros. We have, for $z = x + iy$, $|z| = 1$,

$$|g(z) - f(z)| = |e^{z-\alpha} - z^n - (-z)^n|$$
$$= |e^{z-\alpha}| = e^{x-\alpha} < 1 = |-z^n|,$$

since $x \le 1 < \alpha$ if $(x^2 + y^2)^{1/2} = |z| = 1$. Thus, by Rouché's theorem, $e^{z-\alpha} - z^n$ has exactly n roots in $\Delta(0, 1)$.

Applications of Rouché's theorem to various questions in complex analysis occur in the sequel. Several of them are found in the exercises at the end of this section. An application of sufficient importance to qualify for a formal statement is contained in the following proposition.

Theorem 10.3.2 (*Hurwitz*). *Let B be a region, let $\langle f_n\rangle_0^\infty$ be a sequence of functions, each of which is analytic on B, and let the sequence converge almost uniformly on B to the limit function f. Let $\bar{\Delta}(a, r)$, $r > 0$, be contained in B. If f has no zeros on $C(a, r)$ and has $p \ge 0$ zeros in $\Delta(a, r)$ counted according to multiplicities, then there exists $N > 0$ such that for every $n \ge N$, f_n has exactly p zeros in $\Delta(a, r)$ counted according to multiplicities.*

Remark. For "almost uniform" convergence, refer to Sec. 9.9.

Proof. According to Theorem 9.9.1, f is analytic on B. Then $|f|$ is continuous on the compact set $C(a, r)$, so by Corollary 1 of Theorem 5.5.2, $|f|$

attains a minimum value m on $C(a, r)$. By hypothesis, $m > 0$. The sequence converges uniformly to f on the compact set $C(a, r)$, so there exists an integer N such that if $n \geq N$, $|f_n(z) - f(z)| < m \leq |f(z)|$ for every $z \in C(a, r)$. Reference to Rouché's theorem completes the proof. Q.E.D.

Corollary 1 Let B be a region, let $f_n: B \to \mathbf{C}$ be analytic, $n = 0, 1, 2, \ldots$, and let $\langle f_n \rangle$ converge a.u. on B to a function f. If E is a compact subset of B on which f has no zeros, then there exists $N_E > 0$ such that for $n \geq N_E$, $f_n(z) \neq 0$ on E.

The proof is an exercise.

Corollary 2 Let $B \subset \mathbf{C}$ be a region, let $f_n: B \to \mathbf{C}$ be analytic, $n = 0, 1, 2, \ldots$, and let $\langle f_n \rangle$ converge a.u. on B to a function f. If $f_n(z) \neq 0$ for every z on B, $n = 0, 1, 2, \ldots$, then either $f(z) = 0$ for every $z \in B$, or $f(z)$ is never zero on B.

Proof. By Theorem 9.9.1, f is analytic on B. Suppose that f is not identically zero on B, but $f(a) = 0$, $a \in B$. Since a must be an isolated point of the set of zeros of f, there must exist $\Delta(a, 2r) \subset O$, $r > 0$, such that $f(z) \neq 0$ for every $z \in \Delta(a, 2r) \sim \{a\}$, and in particular on $C(a, r)$. But then by Theorem 10.3.2, for all sufficiently large N, f_n would have a zero or zeros in $\Delta(a, r)$, contrary to hypothesis. Q.E.D.

Example. The members of the sequence $\langle e^z/n \rangle_{n=1}^{\infty}$ have no zeros on $\mathbf{C}$, but the sequence converges a.u. on $\mathbf{C}$ to the limit zero.

Exercises 10.3

1. Show that the equation $e^z = \alpha z^n$, n a positive integer, $\alpha > e$, has exactly n roots in $\Delta(0, 1)$, counted according to multiplicities. (*Hint*: In Rouché's theorem, take $f(z) = \alpha z^n$.)
2. Prove by using Rouché's theorem that the polynomial $a_0 + a_1 z + a_2 z + \cdots + a_n z^n$, $a_n \neq 0$, has exactly n zeros in $\mathbf{C}$, counted according to multiplicities. (*Hint*: In Rouché's theorem, take $f(z) = a_n z^n$.)
3. Let the function h be analytic on a region containing $\Delta(0, 1)$, and suppose that $|h(z)| < 1$ for every $z \in C(0, 1)$. Prove that the function $h(z) - z$ has a simple zero in $\Delta(0, 1)$ and that it is the only zero of $h(z) - z$ in $\Delta(0, 1)$. (*Hint*: Let $g(z) = h(z) - z$ and $f(z) = -z$ in Rouché's theorem.)
4. Let $f: B \to \mathbf{C}$ and $g: B \to \mathbf{C}$ be analytic on the region B. Let $\bar{\Delta}(b, r)$, $r > 0$, be contained in B. Prove that if $|g(z) - f(z)| < |f(z) - w_0|$ for every $z \in C(b, r)$, then the equations $f(z) = w_0$ and $g(z) = w_0$ have the same number of roots in $\Delta(b, r)$ counted according to multiplicities.
5. Let $P_n(z) = 1 + z + (z^2/2!) + \cdots + (z^n/n!)$. Given any bounded set $E \subset \mathbf{C}$, show that there exists N_E such that if $n \geq N_E$, $P_n(z)$ has no zeros on E. (*Hint*: Hurwitz's theorem with $f(z) = e^z$.)
6. Let $B \subset \mathbf{C}$ be a region, let $f_n: B \to \mathbf{C}$ be analytic on B, and let $\langle f_n \rangle$ converge a.u. on B to a function f which is not identically zero. Given $a \in B$, prove that $f(z) = 0$ if

and only if a is an accumulation point of the set of all zeros of the family $\{f_n\}$. (*Hint*: If a is such an accumulation point, first show that given any $\Delta(a, r) \subset B$, $r > 0$, there exists an infinite subsequence of the positive integers, say $\langle n_j \rangle_1^\infty$ such that for each j, f_{n_j} has a zero, say z_{n_j}, in $\Delta(a, r)$. Then look at

$$|f(a)| \leq |f(a) - f(z_{n_j})| + |f(z_{n_j}) - f_{n_j}(z_{n_j})|.$$

The converse follows easily from Hurwitz's theorem.)

7. Prove Corollary 1 of Theorem 10.3.2. (*Hint*: Cover E with disks in each of which Hurwitz's theorem is applicable.)
8. Let $B \subset \mathbf{C}$ be a region; let $f_n: B \to \mathbf{C}$ be analytic *and univalent*, $n = 0,1,2, \ldots$; and let $\langle f_n \rangle$ converge a.u. on B to a function f. Prove that f either is a constant function on B or is univalent. (*Hint*: Suppose that f is nonconstant and that there exist $z_1 \in B$, $z_2 \in B$, such that $f(z_1) = f(z_2) = b$, but $z_1 \neq z_2$. The assumption implies that $f(z) - b$ (to which $\langle f_n(z) - b \rangle$ converges a.u. on B) has zeros at z_1 and at z_2. Take $r > 0$ so that $\Delta(z_1, r) \subset B$, $\Delta(z_2, r) \subset B$, and $\Delta(z_1, r) \cap \Delta(z_2, r) = \phi$. Use Hurwitz's theorem to show that there exist $N > 0$, and $z_3 \in \Delta(z_1, r)$, and $z_4 \in \Delta(z_2, r)$, such that $f_N(z_3) - b = 0$, $f_N(z_3) - b = 0$.)
9. Show that the formulation in the text of Rouché's theorem can be generalized by replacing the disk $\bar{\Delta}(b, r)$ by the set $\bar{B}$ of Exercise 10.2.2 and the circle $C(b, r)$ by the graph of the closed path γ of that exercise.

10.4 The Open Mapping Theorem

A complex-valued function f which is analytic on an open set is also continuous on that set. Thus, according to Theorem 5.2.1, if O is an open set in $\mathbf{C}$ then $f^{-1}(O)$ is open. A remarkable property of a nonconstant analytic function, to be established in this section, is that it also maps open sets in its *domain* onto open sets in its range. This phenomenon goes under the name of the *open mapping property* of analytic functions. The basic theorem can be formulated as follows.

Theorem 10.4.1 *Let B be a region, let the function f: $B \to \mathbf{C}$ be analytic and nonconstant on B, and let $w_0 = f(z_0)$ for a given $z_0 \in B$. Suppose further that the zero of $f(z) - w_0$ at $z = z_0$ is of order m. Then there exist open disks $\Delta(z_0, r) \subset B$ and $\Delta(w_0, \rho) \subset f(B)$, such that for each point $w^* \in \Delta(w_0, \rho) \sim \{w_0\}$ there are exactly m distinct points $z_1(w^*), z_2(w^*), \ldots, z_m(w^*)$ in $\Delta(z_0, r) \sim \{z_0\}$ with $f(z_j(w^*)) = w^*, j = 1, 2, \ldots, m$.*

Proof. Since the analytic function f is nonconstant on B, its derivative f' cannot be identically zero on B [Theorem 6.2.1]. Also $f(z) - w_0$ cannot be identically zero in B. Since both f' and $f - w_0$ are analytic on B, the combined set of zeros of these functions must be a set of isolated points. Therefore, there exists $\Delta(z_0, 2r) \subset B$, $r > 0$, such that $f(z) - w_0 \neq 0$ and $f'(z) \neq 0$ for every $z \in \Delta(z_0, 2r) \sim \{z_0\}$. Then the closed disk $\bar{\Delta}(z_0, r) = \Delta(z_0, r) \cup C(z_0, r)$ lies in B.

Now $|f(z) - w_0|$ is continuous on the compact set $C(z_0, r)$ so it assumes a minimum value μ on this circle [Corollary 1 of Theorem 5.5.3]. The open disks referred to in the conclusion of the theorem are the disks $\Delta(z_0, r)$ of the first paragraph of this proof, and $\Delta(w_0, \rho)$ with $0 < \rho < \mu$. We now show that they have the asserted properties.

Take any point $w^* \in \Delta(w_0, \rho) \sim \{w_0\}$. With $z \in C(z_0, r)$ we have

$$|f(z) - w_0| > \rho > |w_0 - w^*| = |[f(z) - w^*] - [f(z) - w_0]|.$$

Therefore, by Rouché's theorem [Theorem 10.3.1], the functions $f(z) - w^*$ and $f(z) - w_0$ have the same number of zeros in $\Delta(z_0, r)$, namely m. Clearly the zeros of $f(z) - w^*$ all lie in $\Delta(z_0, r) \sim \{z_0\}$, because if one of them were to coincide with z_0, we would have the contradiction $f(z_0) = w^*$ and $f(z_0) = w_0$. Now $f'(z) = d[f(z) - w^*]/dz \neq 0$ in $\Delta(z_0, r) \sim \{z_0\}$ so by Exercise 10.1.1 the m zeros of $f(z) - w^*$ are each of multiplicity one. Thus, they are distinct points of $\Delta(z_0, r) \sim \{z_0\}$. Finally, since we have shown that *each* point $w \in \Delta(w_0, \rho)$ has one or more preimage points in $\Delta(z_0, r)$, it must follow that $\Delta(w_0, \rho) \subset f(B)$. Q.E.D.

Remark (i). Part (i) of the conclusion of Theorem 10.4.1 can be restated in a somewhat less precise manner by saying that every point of $f(B)$ is an interior point of $f(B)$. In other words, $f(B)$ is an open set in **C**.

Remark (ii). (Continuation.) By definition of region, B is connected. Since f is continuous on B, it follows from Proposition 5.3.1 that $f(B)$ is connected and, therefore, is a region.

Remark (iii). (Continuation.) Another way to state the content of Theorem 10.4.1(i) is to assert that a complex-valued function analytic on a region maps that region either onto another region or onto a point. (The "point" case, of course, is that in which the function is a constant function.)

Remark (iv). We can restate Theorem 10.4.1 as follows. Let B be a region, let $f: B \to \mathbf{C}$ be analytic and nonconstant on B, and for a given $z_0 \in B$, let $f(z) - f(z_0)$ have a zero of order m at $z = z_0$. Then there exist neighborhoods O_z of z_0 and O_w of $f(z_0)$ such that f gives an m-to-one mapping of $O_z \sim \{z_0\}$ onto $O_w \sim \{f(z_0)\}$, and $O_w = f(O_z)$. (In the notation of Theorem 10.4.1, $O_w = \Delta(w_0, \rho)$ and $O_z = \Delta(z_0, r) \cap f^{-1}[\Delta(w_0, \rho)]$.)

Corollary 1 (The Open Mapping Theorem). Let B be a region, let $f: B \to \mathbf{C}$ be analytic and nonconstant on B, and let E be an open subset of B. Then $f(E)$ is open in **C**.

Proof. Let C be a component of E. By Corollary 2 of Theorem 5.4.1, C is open in **C**. Then f is nonconstant on C, because if it were constant on C, then by Corollary 2 of Theorem 6.5.1 it would be constant on B, contrary to hypothesis. If follows from Remark (i) that $f(C)$ is open in **C**. Now $f(E) = \bigcup f(C)$, where the union is taken over all components of E. Thus, $f(E)$ is open in **C**. Q.E.D.

Corollary 2 Let B be a region, let the function $f: B \to \mathbf{C}$ be analytic and nonconstant on B, and let $w_0 = f(z_0)$ for a given $z_0 \in B$. If $f'(z_0) \neq 0$, then there exist sets O_z and O_w open in $\mathbf{C}$, with $z_0 \in O_z \subset B$ and $w_0 \in O_w = f(O_z) \subset f(B)$, and such that f gives a homeomorphism of O_z onto O_w. The local inverse function is analytic on O_w.

Proof. The sets O_z and O_w are constructed as indicated in Remark (iv). Since $f'(z_0) \neq 0$, by Exercise 10.1.1, the zero of $f(z) - w_0$ at $z = z_0$ is of order 1. Therefore, by Remark (iv), the mapping of O_z onto O_w given by $w = f(z)$ is one-to-one. Let $g: O_w \to O_z$ be the inverse function. If $E \subset O_z$ is open, then $f(E) = g^{-1}(E)$ is open, by Corollary 1, and therefore g is continuous on O_w [Theorem 5.2.1]. We have now shown that f gives a homeomorphism of O_z onto O_w.

As for the analyticity of g, we first observe that the construction of O_z is such that $f'(z) \neq 0$ on O_z. It follows at once from Theorem 6.1.3 (the inverse function law of differentiation) that for every $w \in O_w$, $g'(w)$ exists and equals $1/f'(z)$, where $w = f(z)$. Q.E.D.

A converse of sorts to *Corollary 2* is contained in the following proposition.

Proposition 10.4.1 *Let B be a region, and let $f: B \to \mathbf{C}$ be analytic on B.*

(i) *If for a given $z_0 \in B$ there exist neighborhoods N_z of z_0 and N_w of $w_0 = f(z_0)$ such that f gives a one-to-one mapping of N_z onto N_w, then $f'(z_0) \neq 0$.*
(ii) *If f is univalent on B, then $f'(z) \neq 0$ for every $z \in B$, and the inverse function f^{-1} is analytic on $f(B)$.*

Proof. (i) The hypothesis implies that f is nonconstant on B. As in the proof of Theorem 10.4.1, let $\bar{\Delta}(z_0, r) \subset B$ be such that $f(z) - w_0 \neq 0$ and $f'(z) \neq 0$ for every z in $\bar{\Delta}(z_0, r) \sim \{z_0\}$. But in the present case, adjust r downward (if necessary) so that $\Delta(z_0, r)$ is contained in N_z. Define O_w and O_z as indicated in Remark (iv). Since O_z so defined is a subset of $\Delta(z_0, r)$, it follows that $O_z \subset N_z$. The hypothesis of Proposition 10.4.1(i) implies that for any $w \in N_w \cap O_w$ there exists one and only one $z \in N_z$ such that $f(z) - w = 0$.

Now suppose that $f'(z_0) = 0$. Then $f(z) - w_0$ must have a zero of order $m \geq 2$ at z_0 [Exercise 10.1.1]. Therefore, according to Theorem 10.4.1(ii), for any $w \in N_w \cap O_w$, $w \neq w_0$, there exist two or more distinct zeros of $f(z) - w$ in $O_z \sim \{z_0\}$; but this is a contradiction of the statement in the last sentence of the preceding paragraph. Therefore, $f'(z_0) \neq 0$.

(ii) The hypothesis that f is univalent means that for any $z_0 \in B$ the mapping given by f of a disk $\Delta(z_0, r) \subset B$, $r > 0$, onto $f[\Delta(z_0, r)]$ is one-to-one. By the open mapping theorem, $f[\Delta(z_0, r)]$ is open, and so it constitutes a neighborhood of $w_0 = f(z_0)$. Thus by part (i) of the proposition, $f'(z_0) \neq 0$.

In other words, f' can have no zeros on B. By Corollary 2 of Theorem 10.4.1, the inverse function f^{-1} is analytic in a neighborhood of each point in $f(B)$, which, of course, means that it is analytic on $f(B)$. Q.E.D.

Remark (v). Notice that Proposition 10.4.1(ii) is a global statement about univalence and the nonvanishing of the derivative, whereas Corollary 2 of Theorem 10.4.1 (a corresponding statement in the opposite direction) is only a local statement. It is possible for a function $f: B \to \mathbf{C}$, analytic on the region B, to satisfy the condition $f'(z) \neq 0$ for every $z \in B$ and still *not* be univalent. For example, consider $f(z) = z^2$, $B = \mathbf{C} \sim \{0\}$. Here $f'(z) = 2z \neq 0$ on B, but f is not univalent on B [Exercise 1.11.11].

Remark (vi). We can now recognize an unnecessary condition in the hypotheses of two propositions in Section 9.8. In both Propositions 9.8.1 and 9.8.2, the hypotheses state that a certain function analytic and univalent on a certain region also satisfies the further conditions that it possesses an analytic inverse. Proposition 10.4.1(ii) shows that the assumption as to an analytic inverse is fulfilled automatically.

Exercises 10.4

1. Let $f(z) = z^3 + 2$; let $z_0 = 0$, $w_0 = f(z_0) = 2$. By following the proof of Theorem 10.3.1, show that the equation $f(z) - w_1 = 0$ has three distinct zeros $z_1(w_1)$, $z_2(w_1)$, $z_3(w_1)$ in $\Delta(0, 1)$ for each $w_1 \in \Delta(2, 1) \sim \{2\}$.
2. Use the open mapping theorem to establish the maximum modulus principle in the following formulation. Let B be a region, let $f: B \to \mathbf{C}$ be analytic and nonconstant on B. Then f can have no local maximum on B; that is, given $z_0 \in B$, in any disk $\Delta(z_0, r) \subset B$, $r > 0$, there exists $z_1 \in \Delta(z_0, r)$ such that $|f(z_1)| > |f(z_0)|$. (*Hint*: Assume $w_0 = f(z_0) \neq 0$; otherwise the result is trivial. By the open mapping theorem, $E = f[\Delta(z_0, r)]$ is open in $\mathbf{C}$, so there exists a disk $\Delta = \Delta[w_0, r'] \subset E$, $r' > 0$. Find a point $w_1 \in \Delta$ for which $|w_1| > |w_0|$.)
3. By using Theorem 10.2.2 a proof can be given for Theorem 10.4.1, restated with a neighborhood N_w of $w_0 = f(z_0)$ replacing $\Delta(w_0, \rho)$. First choose $\bar{\Delta}(z_0, r)$ as in the proof of Theorem 10.4.1 in the text. Let $z = \gamma(t)$, $0 \leq t \leq 2\pi$, be a parametric equation of the positively oriented circle with graph $C(z_0, r)$. Let Γ be the graph of the closed path γ^* with parametric equation $w = f[\gamma(t)]$, $0 \leq t \leq 2\pi$. Provide arguments and authorities to support the following statements:

   (a) w_0 lies in Γ^c.

   (b) The components of Γ^c are regions. (Let A be the component which contains w_0.)

   (c) $m = \mathrm{Ind}_{\gamma^*}(w_0) = \mathrm{Ind}_{\gamma^*}(w^*)$ for every $w^* \in A$, so $f(z) - w^*$ has m zeros in $\Delta(z_0, r) \sim \{z_0\}$.

   (d) The zeros of $f(z) - w^*$ in $\Delta(z_0, r) \sim \{z_0\}$ are distinct points.
4. With reference to Theorem 10.4.1 for notation, show that the sums of the powers of the preimages $z_1(w^*)$, $z_2(w^*)$, $\ldots$, $z_m(w^*)$ of $w^* \in \Delta(w_0, \rho)$ have the following integral representation:

$$S_k(w^*) = [z_1(w^*)]^k + [z_2(w^*)]^k + \cdots + [z_m(w^*)]^k$$
$$= \frac{1}{2\pi i}\int_\gamma \frac{\zeta^k f'(\zeta)\,d\zeta}{f(\zeta) - w^*}, \qquad k = 0,1,2,\ldots,$$

and that $S_k(w^*)$ is analytic on $\Delta(w_0, \rho)$. (*Hint*: Theorems 10.2.1(ii) and 8.4.1.)

Remark. In the case $m = 1$, which is the subject of Corollary 2 of Theorem 10.4.1, the exercise shows that the local inverse $g(w)$ is given by

$$g(w) = \frac{1}{2\pi i}\int_\gamma \frac{\zeta f'(\zeta)\,d\zeta}{f(\zeta) - w}.$$

Theorem 8.4.1 provides an alternative proof that $g(w)$ is analytic on $\Delta(w_0, \rho)$.

5. In Exercise 10.4.1, find the integral representations and the power series representation (in powers of $w - 2$) for $S_k(w) = [z_1(w)]^k + [z_2(w)]^k + [z_3(w)]^k$, $k = 1,2,\ldots$, for $w \in \Delta(2, 1)$. (*Hint*: Exercise 10.4.4, Theorem 8.4.1(ii).) *Answers*: $S_{3j} = 3(w-2)^j$, $j = 0,1,2,\ldots$; $S_{3j+1} = 0$, $S_{3j+2} = 0$, $j = 0,1,2,\ldots$.

10.5 Isolated Singularities

Let Ω be an open set in **C**, and let E be a subset of Ω with no accumulation points in Ω. Then E consists only of isolated points. Let $f: \Omega \sim E \to \mathbf{C}$ be analytic on $\Omega \sim E$. Such a function is said to have an *isolated singularity* at each point in E. This classical terminology is slightly peculiar because it seems to imply that E both does and does not lie in the domain of f. The intended meaning is that E does *not* lie in the original domain of definition of f, as is indicated by the notation $\Omega \sim E$ in the definition.

Let K be a compact subset of Ω. If K were to contain an infinite subset of the set E discussed in the preceeding paragraph, then by the Bolzano–Weierstrass theorem [Exercise 5.5.4], there would be an accumulation point of this infinite subset in K and, therefore, in Ω. This is contrary to the description of E. Thus, any compact subset of Ω can contain at most a finite number of points of E.

As might be expected, the classification and characterization of an isolated singularity of a function f at a point $a \in E$ depend only on the local behavior of f in a neighborhood of a. Therefore, without loss of generality, for such a study we can take as our starting point the simplified hypothesis that f is a function which is analytic on a set $O \sim \{a\}$, where O is open and $a \in O$. The singularity of f at a must fall into one of the following three classes:

(i) *Removable singularities.* The function f has a *removable singularity* at a (also called an *artificial singularity*), if f can be assigned a value at a in such a way that the thereby extended function is analytic on O.

(ii) *Poles.* The function f is said to have a *pole of order p* at a, if there exists a finite series $Q(z) = \sum_{k=1}^{p} c_k(z-a)^{-k}$, $c_p \neq 0$, such that $f(z) - Q(z)$

has a removable singularity at a. The function Q is called the *principal part* of f at a, and the number c_1 is called the *residue* of f at a. If $p = 1$, the pole is said to be *simple*.

(iii) *Essential singularities.* The function f is said to have an *essential singularity* at a, if it has an isolated singularity at a which is neither a pole nor a removable singularity. We supplement this rather vapid negative definition by affirmative characterizations in Theorem 10.5.5 and Proposition 10.6.3.

Theorem 10.5.1 *Let O be an open set containing the point a; let $f\colon O \sim \{a\} \to \mathbf{C}$ be analytic. The following statements are equivalent:*

(i) *f has a removable singularity at a;*
(ii) *f can be defined at a so that f is continuous on O;*
(iii) *There exists a punctured* disk $\Delta'(a, r) = \Delta(a, r) \sim \{a\} \subset O$, $r > 0$ *such that f is bounded in $\Delta'(a, r)$;*
(iv) $\lim_{z \to a} f(z)(z - a) = 0$.

Proof. The equivalences can be proved by showing that (i) ⇒ (ii) ⇒ (iii) ⇒ (iv) ⇒ (i). The first three implications here are relatively obvious in view of the definition of removable singularity, since they seem to pass from stronger to weaker statements. The proofs are deferred to the exercises. The interesting implication is (iv) ⇒ (i), which we now prove.

Define

$$h(z) = \begin{cases} (z - a)^2 f(z), & z \neq a, \quad z \in O, \\ 0, & z = a. \end{cases}$$

We claim that $h'(a)$ exists and is zero. For with $z \neq a$,

$$\delta_a(z) = \frac{h(z) - h(a)}{z - a} = \frac{(z - a)^2 f(z) - 0}{z - a} = (z - a)f(z).$$

Thus, by hypothesis and the definition of derivative, $\lim_{z \to a} \delta_a(z)$ exists and is zero.

This implies that h is analytic on O. It therefore has a local power series at a, which looks like this: $h(z) = 0 + 0(z - a) + a_2(z - a)^2 + a_3(z - a)^3 + \cdots$. The series converges in some disk $\Delta(a, \rho) \subset O$, $\rho > 0$. Then at least for every $z \in \Delta(a, \rho) \sim \{a\}$, $f(z) = (z - a)^{-2}h(z) = a_2 + a_3(z - a) + \cdots$. But the power series in the right-hand member of this equation converges everywhere on $\Delta(a, \rho)$ [Corollary 2 of Theorem 4.1.2] and represents an analytic function there. We assign the value a_2 to f at $z = a$. Then f so extended is analytic on $\Delta(a, \rho)$ and, therefore, on O. Q.E.D.

Corollary Let O be open, $a \in O$, and let $f\colon O \sim a \to \mathbf{C}$ and $g\colon O \sim a \to \mathbf{C}$

each have a removable singularity at a. Then $f + g$ and $f \cdot g$ each have a removable singularity at a.

The proof is an exercise.

In concrete examples, the criterion in Theorem 10.5.1(ii) is sometimes the easiest one to apply, and it provides automatically the proper definition of f at a.

Examples

(i) Let O in Theorem 10.5.1 be $\Delta(0, 1)$, $a = 0$, and let $f(z) = z/\sin z$. The function is analytic on $\Delta(0,1) \sim \{0\}$. By Exercise 6.1.6 (L'Hospital's rule), $\lim_{z \to 0} (z/\sin z) = \lim_{z \to 0} (1/\cos z) = 1$. The function has a removable singularity at $z = 0$, which can be removed by letting $f(0) = 1$.

(ii) Let O be $\mathbf{C} \sim (\{1\} \cup \{2\})$, $a = 2$, and let

$$f(z) = \frac{z^3 + 1}{z^2 - 3z + 2} - \frac{9}{z - 2}. \tag{10.5.1}$$

By straightforward algebraic manipulation it can be shown that for $z \neq 2$, $z \neq 1$,

$$f(z) = \frac{z^3 - 9z + 10}{(z - 2)(z - 1)} = \frac{z^2 + 2z - 5}{z - 1}.$$

Thus $\lim_{z \to 2} f(z)$ exists and equals 3, which is the value f should be given at $z = 2$ to remove the singularity. With even less algebra, it is easy to show that f as given by (10.5.1) satisfies $\lim_{z \to 2} (z - 2)f(z) = 0$, but although that computation shows that f satisfies the condition in Theorem 10.5.1(iv) at $z = 2$, it fails to indicate the appropriate value for f at $z = 2$.

(iii) Let $f: O \to \mathbf{C}$ be analytic on the open set O, and take $a \in O$. The difference quotient

$$\delta(z) = \frac{f(z) - f(a)}{z - a}, \quad z \in O \sim \{a\},$$

introduced in Section 6.1, is analytic on $O \sim \{a\}$, and $\lim_{z \to a} \delta(z) = f'(a)$. Thus, δ has a removable singularity at $z = a$, and the appropriate value for $\delta(a)$ is $f'(a)$.

Theorem 10.5.2 *Let O be an open set containing the point a; and let f: $O \sim a \to \mathbf{C}$ be analytic. The following statements are equivalent*:

(i) *f has a pole of order p at a;*

(ii) *$(z - a)^p f(z) \equiv g(z)$ has a removable singularity at a, and $\lim_{z \to a} g(z) \neq 0$;*

(iii) *There exists a function g_1 analytic on O, with $g_1(a) \neq 0$, such that $f(z) \equiv (z - a)^{-p} g_1(z)$ for every $z \in O \sim \{a\}$;*

(iv) *There exists* $\Delta(a, r) \subset O$, $r > 0$, *such that* $f(z) \neq 0$ *for every* $z \in \Delta(a, r) \sim \{a\}$, *and such that the function*

$$h(z) = \begin{cases} \dfrac{1}{f(z)}, & z \in \Delta(a, r) \sim \{a\} \\ 0, & z = a, \end{cases}$$

is analytic on $\Delta(a, r)$ *and has a zero of order* p *at* a.

Proof. The plan is to show that (i) $\Rightarrow$ (ii) $\Rightarrow$ (iii) $\Rightarrow$ (iv) $\Rightarrow$ (iii) $\Rightarrow$ (i).

(i) $\Rightarrow$ (ii). If f has a pole of order p at a, let the principal part be $Q(z) = \sum_{j=1}^{p} c_j(z-a)^{-j}$, where $c_p \neq 0$. Then $H(z) = f(z) - Q(z)$ has a removable singularity at a. For every $z \in O \sim \{a\}$, $(z-a)^p f(z) = (z-a)^p H(z) + \sum_{j=1}^{p} c_j(z-a)^{p-j}$. By Theorem 10.5.1(iv), $\lim_{z\to a} (z-a)^p H(z) = 0$, so $\lim_{z\to a} (z-a)^p f(z) = 0 + c_p \neq 0$. Therefore, by Theorem 10.5.1(ii), $g(z) = (z-a)^p f(z)$ has a removable singularity at $z = a$, and the appropriate value for $g(z)$ at $z = a$ is c_p. We have shown that (i) $\Rightarrow$ (ii).

(ii) $\Rightarrow$ (iii). This is an immediate consequence of (ii) in which $g_1(z) = (z-a)^p f(z)$ for $z \in O \sim \{a\}$ and $g(a) = c_p$.

The proofs of the implications (iii) $\Rightarrow$ (iv) and (iv) $\Rightarrow$ (iii) are deferred to the exercises.

(iii) $\Rightarrow$ (i). The analytic function g_1 has a local power series representation $g_1(z) = \sum_0^\infty a_j(z-a)^j$, $a_0 \neq 0$, valid in a disk $\Delta(a, R) \subset O$, $R > 0$. For every $z \in \Delta(a, R) \sim \{a\}$, $f(z) = a_0(z-a)^{-p} + a_1(z-a)^{-p+1} + \cdots + a_p + a_{p+1}(z-a) + a_{p+2}(z-a)^2 + \cdots$. Let $Q(z) = a_0(z-a)^{-p} + a_1(z-a)^{-p+1} + \cdots + a_{p-1}(z-a)^{-1}$; this function is analytic on $O \sim \{a\}$. For every $z \in \Delta(a, R) \sim a, f(z) - Q(z) = \sum_{j=0}^\infty a_{j+p}(z-a)^j$. The right-hand member is a function defined and analytic on $\Delta(a, R)$, so if $f(a) - Q(a)$ is defined to be a_p, then $f - Q$ is analytic on O. Therefore, by definition, $f - Q$ has a removable singularity at $z = a$, and thus by definition of pole, f has a pole of order p at a.

Corollary 1 Let O be an open set containing the point a. Let $f\colon O \sim \{a\} \to \mathbf{C}$ and $g\colon O \sim \{a\} \to \mathbf{C}$ be analytic, let f have a pole of order p_1 at a, and let g have a pole of order p_2 at a. Then $f \cdot g$ has a pole of order $p_1 + p_2$ at a, and if $p_1 < p_2$, $f + g$ has a pole of order p_2 at a.

The proof is an exercise.

Corollary 2 Let O be an open set containing a; let $f\colon O \sim \{a\} \to \mathbf{C}$ have a removable singularity at a and let $g\colon O \sim \{a\} \to \mathbf{C}$ have a pole of order p at a. Then $f + g$ has a pole of order p at a, and the principal part of $f + g$ at a is the same as that of g. Also if the appropriate value $f(a) \neq 0$, then $f \cdot g$ has a pole of order p at a.

The proof again is an exercise.

Corollary 3 Let O be an open set, let $P = \{\zeta_1, \zeta_2, \ldots, \zeta_\nu\}$ be a set of distinct points contained in O, and let $f\colon O \sim P \to \mathbf{C}$ be analytic and have poles at $\zeta_1, \zeta_2, \ldots, \zeta_\nu$ of respective orders $p_1, p_2, \ldots, p_\nu$. There exists a function $h\colon O \to \mathbf{C}$ analytic on O such that for every $z \in O \sim P$,

$$f(z) = (z - \zeta_1)^{-p_1}(z - \zeta_2)^{-p_2} \cdots (z - \zeta_\nu)^{-p_\nu}h(z), \qquad (10.5.2)$$

and $h(\zeta_k) \neq 0$, $k = 1, \ldots, \nu$.

Proof. By Theorem 10.5.2(ii), $(z - \zeta_1)^{p_1}f(z)$ has a removable singularity at ζ_1, and since $(z - \zeta_1)^{p_1}$ is analytic at $z = \zeta_2$, the function $(z - \zeta_1)^{p_1}f(z)$ has a pole of order p_2 at ζ_2 [Corollary 2 of Theorem 10.5.2]. Then $(z - \zeta_2)^{p_2}$ $(z - \zeta_1)^{p_1}f(z)$ has a removable singularity at ζ_2 and a pole of order p_3 at ζ_3. And so on. We finally arrive at a function

$$h^*(z) = (z - \zeta_\nu)^{p_\nu}(z - \zeta_{\nu-1})^{p_{\nu-1}} \cdots (z - \zeta_1)^{p_1}f(z), \qquad (10.5.3)$$

defined and analytic for every $z \in O \sim P$, and with a removable singularity at each of the points of P. We complete the definition of h^* appropriately at each point of P and obtain a function $h\colon O \to \mathbf{C}$ analytic on O. Solving for $f(z)$ in (10.5.3), we obtain (10.5.2). Theorem 10.5.2(iii) applied to the representation

$$f(z) = (z - \zeta_k)^{-p_k}[(z - \zeta_1)^{-p_1} \cdots (z - \zeta_{k-1})^{-p_{k-1}}(z - \zeta_{k+1})^{-p_{k+1}} \cdots (z - \zeta_\nu)^{p_\nu}h(z)]$$

indicates that $h(\zeta_k) \neq 0$. Q.E.D.

Examples

(iv) Let $f(z) = z/\sin z$. By Exercise 7.4.2, the only zeros of $\sin z$ are at $z = \pi k$, $k = 0$, $\pm 1, \pm 2, \ldots$, and clearly the function f is analytic on $\mathbf{C}$ except at these points. In Example (i), it was shown that f has a removable singularity at $z = 0$. Since $\cos z = (\sin z)' = \pm 1 \neq 0$ at $z = \pi k$, the zero of $\sin z$ at $z = \pi k$, $k \neq 0$, is of order 1 [Exercise 10.1.1]. By Theorem 10.5.2(iv), $1/\sin z$ has a pole at $z = \pi k$ of order 1, and by Corollary 2, $z/\sin z$ also has a pole of order 1 at $z = \pi k$, $k = \pm 1, \pm 2, \ldots$.

(v) This example illustrates the use of Theorem 10.5.2(iv) and Corollary 1 of Theorem 10.5.2. Let $O = \Delta(0, 2\pi)$, and let $f\colon O \sim \{0\} \to \mathbf{C}$ be given by $f(z) = z^{-m}$ $(e^z - 1)^{-1}$, where m is a positive integer. By our definition of a pole, $f_1(z) = z^{-m}$ has a pole of order m at $z = 0$. (It is its own principal part at 0.) By Theorem 5.7.1, $h(z) = e^z - 1$ has a zero at $z = 0$ and no other zero in $\Delta(0, 2\pi)$. In fact, $e^z - 1 = z + (z^2/2!) + (z^3/3!) + \cdots$, valid for every $z \in \mathbf{C}$, so by the definition of the order of a zero in Section 10.1, h has a simple zero at $z = 0$. Let $f_2(z) = (e^z - 1)^{-1}$. Then $h(z) = 1/f_2(z)$, $z \in \Delta(0, 2\pi) \sim \{0\}$, so by Theorem 10.5.2(iv), f_2 has a simple pole at $z = 0$. Therefore, by Corollary 1, $f(z) = f_1(z)f_2(z)$ has a pole of order $m + 1$ at $z = 0$.

(vi) Recall that a rational function is the quotient of two polynomial functions, one or both of which can be constant functions, with the denominator not the zero polynomial. *We henceforth always assume when dealing with a rational function that the two polynomials have no zeros in common.* Let $r(z) = p(z)/q(z)$ be a rational function, and let the distinct zeros of q be $\zeta_1, \zeta_2, \ldots, \zeta_\nu$ of respective multiplicities $p_1, p_2, \ldots, p_\nu$. The function r is analytic on $\mathbf{C} \sim \{\zeta_1, \zeta_2, \ldots, \zeta_\nu\}$ [Exercise 6.1.4]. We write q in the factored form $q(z) = c(z-\zeta_1)^{p_1}(z-\zeta_2)^{p_2}\cdots(z-\zeta_\nu)^{p_\nu}$ [corollary of Theorem 9.6.2]. To classify the isolated singularity at ζ_k, $1 \le k \le \nu$, we arrange the function r as follows:

$$r(z) = (z-\zeta_k)^{-p_k}R(z),$$

$$R(z) = \frac{p(z)}{c(z-\zeta_1)^{p_1}\cdots(z-\zeta_{k-1})^{p_{k-1}}(z-\zeta_{k+1})^{p_{k+1}}\cdots(z-\zeta_\nu)^{p_\nu}}.$$

The rational function R is analytic on the open set $\mathbf{C} \sim \{\zeta_1, \ldots, \zeta_{k-1}, \zeta_{k+1}, \ldots, \zeta_\nu\}$, and $R(\zeta_k) \neq 0$. Therefore by Theorem 10.5.2(iii), r has a pole of order p_k at ζ_k. *The only singularities in* **C** *of a rational function are poles.* If there are no singularities, the function must be a polynomial. A converse to this statement appears in Theorem 10.7.2.

We now turn to a characterization of essential singularities.

Theorem 10.5.3 *Let O be an open set containing the point a, and let $f: O \sim \{a\} \to \mathbf{C}$ be analytic. The isolated singularity of f at a is an essential singularity if and only if for every* disk $\Delta(a, r) \subset O$, $r > 0$, *$f[\Delta'(a, r)]$ is dense in* **C**, *where $\Delta'(a, r) = \Delta(a, r) \sim \{a\}$.*

Remark (i). This theorem is often called the Casorati–Weierstrass theorem, but the first proof was published in 1859 by C. A. A. Briot and J. C. Bouquet.

Remark (ii). The condition that $f[\Delta'(a,r)]$ is dense in **C** means that for any $\varepsilon > 0$ and any complex number c, there exists $z_c \in \Delta'(a, r)$ such that $|f(z_c) - c| < \varepsilon$. In other words, $f(z)$ comes arbitrarily close to any complex number c for z in any punctured disk with center at a.

Proof of the Theorem. First, suppose that for every $\Delta(a, r) \subset O$, $r > 0$, $f(\Delta'(a, r))$ is dense in **C**. We show that this rules out the possibility that a is a removable singularity or a pole. If a were removable, then by Theorem 10.5.1(iii), there would exist a punctured disk $\Delta'(a, r) \subset O$, $r > 0$, such that f would be bounded in $\Delta'(a, r)$; say $|f(z)| \le M$, $M > 0$, for every $z \in \Delta'(a, r)$. But the hypothesis implies that there exists $z_0 \in \Delta'(a, r)$ such that $|f(z_0) - 2M| < M$. It is easily seen that this implies that $|f(z_0)| > M$, a contradiction.

If a were a pole, then according to Theorem 10.5.2(iv), there would exist $\Delta'(a, r) \subset O$, $r > 0$, such that

$$h(z) = \begin{cases} \dfrac{1}{f(z)}, & z \in \Delta'(a, r) \\ 0, & z = a \end{cases}$$

is analytic on $\Delta(a, r)$. Therefore, $h(z)$ would be bounded on $\bar{\Delta}(a, r/2)$; say $|h(z)| \leq M$, $M > 0$, $z \in \bar{\Delta}(a, r/2)$. Then $|f(z)| \geq 1/M$ for every $z \in \bar{\Delta}(a, r/2)$. But by hypothesis there exists $z_0 \in \Delta(a, r/2)$ such that $|f(z_0) - 0| < 1/M$, again a contradiction.

Now conversely suppose that the isolated singularity of f at a is neither a removable singularity nor a pole and, therefore, is an essential singularity. Also suppose that there were to exist a disk $\Delta(a, r) \subset O$, $r > 0$, such that $f[\Delta'(a, r)]$ is not dense in **C**. This would mean that for some $\varepsilon > 0$ and some $c \in \mathbf{C}$, $|f(z) - c| \geq \varepsilon$ for every $z \in \Delta'(a, r)$. Let $g(z) = 1/[f(z) - c]$ for $z \in \Delta'(a, r)$. This function would be analytic in $\Delta'(a, r)$ and $|g(z)| \leq 1/\varepsilon$ for $z \in \Delta'(a, r)$, so by Theorem 10.5.1(iii), g would have a removable singularity at a. We could remove the singularity by continuity and call the extended function g^*. Two cases need to be examined.

(1) $g^*(a) \neq 0$. Since g^* would be continuous at a, if m is such that $0 < m < |g^*(a)|$, there would exist a disk $\Delta(a, r_1) \subset \Delta(a, r)$, $r_1 > 0$, such that $|g^*(z)| \geq m$ for every $z \in \Delta(a, r_1)$ [Exercise 3.2.3(b)]. Then $f(z) = c + [1/g^*(z)]$ and $|f(z)| \leq |c| + (1/m)$ for every $z \in \Delta'(a, r_1)$. But then by Theorem 10.5.1(iii), f would have a removable singularity at a, which would be a contradiction.

(2) $g^*(a) = 0$. The zero of g^* at a would be isolated because $g(z)$ as already defined cannot be identically zero on $\Delta'(a, r)$, so there would again exist a disk $\Delta(a, r_1) \subset \Delta(a, r)$ such that $g^*(z) \neq 0$ in $\Delta'(a, r_1)$. By Theorem 10.5.2(iv), $1/g^*(z)$ would have a pole at a, and by Corollary 1 of that theorem, $f(z) = c + [1/g^*(z)]$ would also have a pole at a. Again this would be a contradiction. Q.E.D.

Corollary Let O be an open set containing the point a; let $f\colon O \sim \{a\} \to \mathbf{C}$ be analytic. The function f has a pole at a if and only if for every $M > 0$ there exists $\Delta(a, r_M) \subset O$, $r_M > 0$ such that $|f(z)| > M$ for every $z \in (a, r_M) \sim \{a\}$.

The proof is an exercise.

Exercises 10.5

1. Let $f\colon 0 \sim \{a\} \to \mathbf{C}$ be analytic, where O is open and $a \in O$. Furthermore, let f have a pole at a. As in the definition of pole at the beginning of this section, let c_1 be the residue of f at a. Prove the following statements:
   (a) Whatever the order of the pole, $c_1 = \int_\gamma f(z)\, dz/2\pi i$, where γ is a positively oriented circle with graph $C(a, r)$, and $\bar{\Delta}(a, r) \subset O$.
   (b) If the pole is simple, then $c_1 = \lim_{z \to a} (z - a) f(z)$.
2. Let $f\colon O \sim a \to \mathbf{C}$ be analytic, where O is open, $a \in O$, and let f have a pole of order

p at a. Let the residue of f at a be c_1. According to Theorem 10.5.2(iii), there exists a function g analytic on O such that $f(z) = (z-a)^{-p}g(z)$, $z \in O \sim \{a\}$. Show that if γ is a positively oriented circle with graph $C(a, r)$ such that $\bar{\Delta}(a, r) \subset O$, then the residue c_1 of f at a is given by

$$c_1 = \frac{1}{2\pi i}\int \frac{g(z)\,dz}{(z-a)^p} = \frac{g^{(p-1)}(a)}{(p-1)!}.$$

(*Hint*: Exercise 10.5.1(a) and Theorem 9.4.2(ii).)

3. Show that each of the singularities in **C** of each of the following functions is a pole. Find the location and order of each pole and the residue of the function at the pole. Use Exercise 10.5.1 and 10.5.2 and L'Hospital's rule [Exercise 6.1.6] as needed.

(a) $\sin z/z^2$. *Answer*. 0,1,1.

(b) $\cot z$. *Answer*. $k\pi$, 1, 1; $k = 0, \pm 1, \pm 2, \ldots$.

(c) $1/z \sin z$. *Answer*. 0, 2, 0; $k\pi$, 1, $(-1)^k/k\pi$, $k = \pm 1, \pm 2, \ldots$.

(d) $(e^{2z} - 1)^{-1}$. *Answer*. $k\pi i$, 1, 1/2, $k = 0, \pm 1, \pm 2, \ldots$.

(e) $z^{-3}e^{z^2} \cdot /2$. *Answer*. 0,3,1/2.

(f) $z^2/[(z+1)^2(z-2)]$. *Answer*. -1, 2, 5/9; 2, 1, 4/9.

4. For Theorem 10.5.1, prove that (i) ⇒ (ii) ⇒ (iii) ⇒ (iv).

5. Prove the corollary of Theorem 10.5.1.

6. For Theorem 10.5.2, prove that (iii) ⇔ (iv). (*Hint*: In the notation of that theorem, let $h(z) = (z-a)^m/g_1(z)$, and show that this function is defined and analytic in some neighborhood of a. Use Proposition 10.1.2 to show that (iv) ⇒ (iii).)

7. Prove Corollaries 1 and 2 of Theorem 10.5.2.

8. Prove the corollary of Theorem 10.5.3. (*Hint*: If a is a pole of f, the conclusion follows at once from Theorem 10.5.2(iii). Show that the converse hypothesis rules out both an essential singularity [Theorem 10.5.3] and a removable singularity.)

9. Let $f: O \sim \{a\} \to \mathbf{C}$ be analytic, O be an open set, $a \in O$, and let f have a pole at a. Show that the principal part is the *unique polynomial* p in $w = 1/(z-a)$ such that $f - p$ has a removable singularity at a.

10. Let $f: O \sim \{a\} \to \mathbf{C}$ be analytic, where O is an open set and $a \in O$. Prove that if there exists $\Delta(a, r) \subset O$, $r > 0$, and a real M such that either Re $f(z) < M$ or Re $f(z) > M$ for every $z \in \Delta(a, r) \sim \{a\}$, then f has a removable singularity at a. (*Hint*: First, show that if Re $f(z) < M$, then $\phi(z) = e^{f(z)}$ has a removable singularity at a. If the definition of ϕ is properly extended to $z = a$, then ϕ is analytic and nonzero on $\Delta(a, r)$. Therefore ϕ has an analytic logarithm g on $\Delta(a, r)$ [corollary of Proposition 9.4.2]. Show that g can be used to complete the definition of f at a so that f is analytic on $\Delta(a, r)$.)

11. Let $r = p/q$ be a rational function; let the zeros of p be $z_1, z_2, \ldots, z_\mu$ with respective multiplicities $m_1, m_2, \ldots, m_\mu$; and let the zeros of q be $\zeta_1, \zeta_2, \ldots, \zeta_\nu$

with respective multiplicities $p_1, p_2, \ldots, p_\nu$. Show that for every z not a zero of p or q,

$$\frac{r'(z)}{r(z)} = \sum_{j=1}^{\mu} \frac{m_j}{z - z_j} - \sum_{k=1}^{\nu} \frac{p_j}{z - \zeta_j}.$$

12. Show that the function $w = e^{1/z}$ maps any punctured disk $\{z: 0 < |z| < r\}$, $0 < r < 1$ onto $\mathbf{C} \sim \{0\}$; therefore, by Theorem 10.5.3, $z = 0$ is an essential singularity of the function. (*Hint*: Let $\zeta = 1/z$; consult Theorem 7.5.1.)
13. Let $f: B \sim \{a\} \to \mathbf{C}$ and $g: B \sim \{a\} \to \mathbf{C}$ be analytic, where B is a region and $a \in B$. Suppose that f has an essential singularity at a and g does *not* have an essential singularity at a. Prove that $f + g$ and $f \cdot g$ each have an essential singularity at a. (*Hint*: The case $g \equiv 0$ is ruled out in this chapter. If, say, $f + g$ were to have a removable singularity or a pole at a, which are the alternatives to an essential singularity, then there would be functions $h: B \to \mathbf{C}$ analytic on B, and a nonnegative integer m, such that $f(z) + g(z) = (z - a)^{-m}h(z)$. There is a similar representation for g itself.)
14. Let $f: B \sim \{a\} \to \mathbf{C}$ be analytic on the region $B \sim \{a\}$ and have an essential singularity at a. Furthermore, suppose that $f(z) \neq 0$ for every z in some deleted neighborhood of a. Show that $1/f(z)$ has an essential singularity at a. (*Hint*: See the hint for Exercise 10.5.13, and refer to Theorem 10.5.2(iii).)

10.6 Laurent Series

Our starting point in this section is a generalization of the Cauchy integral formula for a circle [corollary of Theorem 9.4.1].

Theorem 10.6.1 *Let $f: O \sim \{a\} \to \mathbf{C}$ be analytic, where O is an open set and $a \in O$. Let $R = \text{dist}(a, O^c) \leq \infty$. Let γ_1 and γ_2 be positively oriented circles with respective graphs $C(a, r_1)$ and $C(a, r_2)$, where $0 < r_1 < r_2 < R$. Then*

$$f(z) = \frac{1}{2\pi i} \int_{\gamma_2} \frac{f(\zeta)\, d\zeta}{\zeta - z} - \frac{1}{2\pi i} \int_{\gamma_1} \frac{f(\zeta)\, d\zeta}{\zeta - z} \tag{10.6.1}$$

for every $z \in A = \{z: r_1 < |z - a| < r_2\}$.

Proof. The proof is similar to that of the original version of the Cauchy integral formula, Theorem 9.4.1. Hold z fixed in O. Let

$$\delta(\zeta) = \begin{cases} \dfrac{f(\zeta) - f(z)}{\zeta - z}, & \zeta \neq z, \zeta \in O. \\ f'(z), & \zeta = z. \end{cases}$$

The function $\delta(\zeta)$ is an analytic function of ζ on O except at $\zeta = z$, where δ

is continuous. Therefore, according to Exercise 9.3.3,

$$\frac{1}{2\pi i}\int_{\gamma_1}\frac{f(\zeta)-f(z)}{\zeta - z}\,d\zeta = \frac{1}{2\pi i}\int_{\gamma_2}\frac{f(\zeta)-f(z)}{\zeta - z}\,d\zeta. \tag{10.6.2}$$

Now choose any z in the annulus A. Then $\mathrm{Ind}_{\gamma_1}(z) = 0$ and $\mathrm{Ind}_{\gamma_2}(z) = 1$, and (10.6.2) becomes

$$\frac{1}{2\pi i}\int_{\gamma_1}\frac{f(\zeta)\,d\zeta}{\zeta - z} - f(z)\cdot 0 = \frac{1}{2\pi i}\int_{\gamma_2}\frac{f(\zeta)\,d\zeta}{\zeta - z} - f(z)\cdot 1. \tag{10.6.3}$$

This equation is equivalent to (10.6.1). Q.E.D.

Remark. For any fixed z in O, and in particular in A, the remark following Exercise 9.3.3 indicates that the integrals in (10.6.2) are independent of r_1 and r_2, $0 < r_1 < r_2 < R$. Taken with (10.6.3), this implies that each of the integrals in (10.6.1) is constant with respect to r_1 and r_2, $0 < r_1 < r_2 < R$, provided r_1 and r_2 are varied only so that z always remains either in the annulus $\{z : r_1 < |z - a| < r_2\}$ or in the set $\{z : |z - a| < r_1 \text{ and } |z - a| > r_2\}$.

The Laurent expansion of an analytic function in the neighborhood of an isolated singularity a involves both positive and negative powers of $(z - a)$. By way of preparation, we first present a new series development of an integral of the type first introduced in Theorem 8.4.1.

Proposition 10.6.1 *Let γ_r be a positively oriented circle with graph $C(a, r)$, $r > 0$, and let $g\colon C(a, r) \to \mathbf{C}$ be continuous on $C(a, r)$. The function*

$$\psi(z) = -\int_{\gamma_r}\frac{g(\zeta)\,d\zeta}{\zeta - z}, \qquad z \notin C(a, r),$$

has a series representation $\psi(z) = \sum_{j=1}^{\infty} C_j(z - a)^{-j}$ converging absolutely for every $z \in \{z\colon |z - a| > r\}$ and uniformly on any set $\{z\colon |z - a| \geq r_1\}$, $r_1 > r$. The function is analytic on $[C(a, r)]^c$.

Proof. The analyticity of ψ on $[C(a, r)]^c$ is assured by Theorem 8.4.1.

To establish the series representation, we proceed as in the proof of Theorem 8.4.1. Choose any z in $\bar{\Delta}(a, r)^c = \{z\colon |z - a| > r\}$ and then hold z fixed. The "independent variable" in the steps immediately following is ζ, restricted to $C(a, r)$. With this restriction, $|\zeta - a|/|z - a| = r/|z - a| < 1$, so the following computation is valid:

$$\begin{aligned}\frac{-g(\zeta)}{\zeta - z} &= \frac{g(\zeta)}{(z - a)[1 - (\zeta - a)/(z - a)]}\\ &= \frac{g(\zeta)}{z - a}\sum_{j=0}^{\infty}\left(\frac{\zeta - a}{z - a}\right)^j\\ &= \sum_{j=0}^{\infty}\frac{g(\zeta)(\zeta - a)^j}{(z - a)^{j+1}}.\end{aligned} \tag{10.6.4}$$

Let M be the maximum value of the continuous function $|g|$ on $C(a, r)$. Then

$$\left|\frac{g(\zeta)(\zeta - a)^j}{(z - a)^{j+1}}\right| \leq \frac{Mr^j}{|z - a|\,|z - a|^j} = M_j.$$

The series $\sum_0^\infty M_j$ converges because $r/|z - a| < 1$, so the series in (10.6.4) converges uniformly for $\zeta \in C(a, r)$, by the Weierstrass M-test [Theorem 3.4.1]. It can, therefore, be integrated term-by-term with respect to ζ along γ_r [corollary of Theorem 8.3.1]. We obtain

$$\begin{aligned}\psi(z) &= -\int_{\gamma_r} \frac{g(\zeta)\,d\zeta}{\zeta - z} = \sum_{j=0}^{\infty} (z - a)^{-j-1} \int_{\gamma_r} g(\zeta)(\zeta - a)^j\,d\zeta \\ &= \sum_{j=1}^{\infty} c_j(z - a)^{-j}, \end{aligned} \qquad (10.6.5)$$

where

$$c_j = \int_{\gamma_r} g(\zeta)(\zeta - a)^{j-1}\,d\zeta, \quad j = 1, 2, \ldots.$$

Now we view z as being the "variable" in the situation. With $r_1 > r$, let $\Delta = \{z\colon |z - a| \geq r_1\}$. We set up a new M-test, this time for the series in (10.6.5). For $z \in \Delta$, we have

$$\left|\frac{\int_{\gamma_r} g(\zeta)(\zeta - a)^j\,d\zeta}{(z - a)^{j+1}}\right| \leq \frac{2\pi r M\cdot r^j}{r_1 r_1^j} = M'_j,$$

and since $r/r_1 < 1$, $\sum M_j$ converges. Therefore, the series in (10.6.5) converges uniformly and absolutely on Δ. For any point $z_0 \in \bar{\Delta}(a, r)^c$, if r_1 is chosen so that $r < r_1 < |z_0 - a|$, then $z_0 \in \Delta$, so the series converges absolutely at every point in $\bar{\Delta}(a, r)^c$. Q.E.D.

We are now ready for the Laurent expansion.

Theorem 10.6.2 *Let $f\colon O \sim \{a\} \to \mathbf{C}$ be analytic, where O is an open set and $a \in O$. Let $R =$ dist $(a, O^c) \leq \infty$.*

(i) *There exists a series representation of f as follows:*

$$f(z) = \sum_{k=0}^{\infty} b_k(z - a)^k + \sum_{j=1}^{\infty} c_j(z - a)^{-j}, \qquad (10.6.6)$$

valid for every $z \in \Delta(a, R) \sim \{a\}$. The first series in (10.6.6) (the power series) converges absolutely and almost uniformly on $\Delta(a, R)$ and represents a function analytic on $\Delta(a, R)$. The second series in (10.6.6) converges absolutely on $\mathbf{C} \sim \{a\}$ and uniformly on any closed set in $\mathbf{C} \sim \{a\}$ and represents a function analytic on $\mathbf{C} \sim \{a\}$.

(ii) *The coefficients in (10.6.6) are given by the formulas*

$$b_k = \frac{1}{2\pi i}\int_{\gamma_\rho} \frac{f(\zeta)\,d\zeta}{(\zeta - a)^{k+1}}, \qquad k = 0, 1, 2, \ldots, \tag{10.6.7}$$

$$c_j = \frac{1}{2\pi i}\int_{\gamma_\rho} f(\zeta)(\zeta - a)^{j-1}\,d\zeta, \qquad j = 1, 2, \ldots, \tag{10.6.8}$$

where γ_ρ is any positively oriented circle with graph $C(a, \rho)$, $0 < \rho < R$.

(iii) *The series representation in (10.6.6) is unique in the sense that if for some ρ', $0 < \rho' < R$, series $\sum_0^\infty b_k^*(z - a)^k$ and $\sum_1^\infty c_j^*(z - a)^{-j}$ are found which converge uniformly on $C(a, \rho')$ to respective sum functions f_2^* and f_1^* such that $f(z) = f_2^*(z) + f_1^*(z)$ for every $z \in C(a, \rho')$ then $b_k^* = b_k$, $k = 0, 1, 2, \ldots$, and $c_j^* = c_j$, $j = 1, 2, \ldots$.*

Remark (i). Part (iii) of the Theorem implies that the coefficients b_k and c_j given by (10.6.7) and (10.6.8) are independent of the choice of ρ, $0 < \rho < R$.

Remark (ii). The pair of series in (10.6.6) are usually written in the condensed form $\sum_{k=-\infty}^{\infty} a_k(z - a)^k$, and together are called "the Laurent series" or "the Laurent expansion" of f in the punctured disk $\Delta(a, R) \sim \{a\}$; or more loosely, the "Laurent expansion of f about a."

Remark (iii). The function represented by the first series (the power series) in (10.6.6) is sometimes called the *analytic part of f at a*, and the function represented by the second series is called the *principal part* or *singular part of f at a*. (Proposition 10.6.2 indicates that the use of the term "principal part" here is consistent with the use of this term in the general definition of pole in Sec. 10.5.)

Proof of the Theorem. First choose ρ, $0 < \rho < R$. This number is held fixed throughout the proof. Let γ_1 and γ_2 be positively oriented circles with respective graphs $C(a, r_1)$ and $C(a, r_2)$, where $0 < r_1 < \rho < r_2 < R$. Let

$$f_1(z) = -\frac{1}{2\pi i}\int_{\gamma_1} \frac{f(\zeta)\,d\zeta}{\zeta - z},$$

$$f_2(z) = \frac{1}{2\pi i}\int_{\gamma_2} \frac{f(\zeta)\,d\zeta}{\zeta - z}.$$

Then by Theorem 10.6.1, $f(z) = f_2(z) + f_1(z)$ for every $z \in A(r_1, r_2) = \{z : r_1 < |z - a| < r_2\}$.

According to Theorem 8.4.1, f_2 can be represented in $\Delta(a, r_2)$ by a power series $\sum_{k=1}^{\infty} b_k(z - a)^k$. This series converges absolutely at each point of $\Delta(a, r_2)$ and uniformly on any closed subset of $\Delta(a, r_2)$ [Theorem 4.1.2, Exercise 5.6.1]. It represents a function analytic on $\Delta(a, r_2)$.

By Proposition 10.6.1 (with $g(z) = f(z)/(2\pi i)$ and $\psi(z) = f_1(z)$), f_1 has a series representation $f_1(z) = \sum_{j=1}^{\infty} c_j(z - a)^{-j}$ converging absolutely for every $z \in \{z : |z - a| > r_1\}$ and uniformly on any set $\{z : |z - a| \geq r_1' > r_1\}$. The function f_1 is analytic on $\{z : |z - a| > r_1\}$.

To derive the formulas (10.6.7) and (10.6.8) for the coefficients of these series, we observe that both series converge uniformly on $C(a, \rho)$. They remain uniformly convergent after term-by-term multiplication by $(z - a)^{J-1}$, $J = 0, \pm 1, \pm 2, \ldots$. The individual terms of the series and the sum functions are continuous on $C(a, \rho)$. Therefore, by the corollary of Proposition 8.3.1, the following computation is valid:

$$\begin{aligned}\int_{\gamma_\rho} f(z)(z-a)^{J-1}\,dz &= \int_{\gamma_\rho} f_2(z)(z-a)^{J-1}\,dz + \int_{\gamma_\rho} f_1(z)(z-a)^{J-1}\,dz \\ &= \sum_{k=0}^{\infty} b_k \int_{\gamma_\rho} (z-a)^{k+J-1}\,dz \\ &\quad + \sum_{j=1}^{\infty} c_j \int_{\gamma_\rho} (z-a)^{-j+J-1}\,dz. \qquad (10.6.9)\end{aligned}$$

If $J = 1, 2, 3, \ldots$, the integrals in the first series in (10.6.9) are each zero, and each of the terms in the second series is zero except possibly when $j = J$. The value of the corresponding term is $2\pi i\, c_J$. Similarly if $J = -K$, $K = 0, 1, 2, \ldots$, the only possible nonzero term in the two series in (10.6.9) is $2\pi i\, b_K$. Formulas (10.6.7) and (10.6.8) follow at once.

Part (iii) of the theorem can be proved by essentially the same technique used here to establish part (ii). Given ρ', $0 < \rho' < R$, r_1 and r_2 can be chosen so that $0 < r_1 < \min(\rho, \rho')$, $\max(\rho, \rho') < r_2 < R$. The equation (10.6.9) is valid with γ_ρ replaced by $\gamma_{\rho'}$. It is also valid with γ_ρ replaced by $\gamma_{\rho'}$, f_2 and f_1 replaced, respectively, by f_2^* and f_1^*, and b_k and c_j replaced, respectively, by b_k^* and c_j^*. After carrying out the integrations, it is found that b_k and b_k^* are both given by (10.6.7) with $\gamma_{\rho'}$ replacing γ_ρ, and c_j and c_j^* are both given by (10.6.8) with $\gamma_{\rho'}$ replacing γ_ρ.

It is now clear that the coefficients b_k and c_j in (10.6.6) do not depend on r_2 and r_1, provided that $0 < r_1 < \rho < r_2 < R$. Given any point $z_0 \in \Delta(a, R) \sim a$, we can adjust r_1 and r_2 (without changing the coefficients) so that $z_0 \in A(r_1, r_2)$. Therefore, both of the series in (10.6.6) converge absolutely at z_0. The radius of convergence of the power series in (10.6.6) must then be greater than or equal to R, so this power series converges almost uniformly on $\Delta(a, R)$ [Exercise 5.6.1]. Let E be any closed set contained in $\mathbf{C} \sim \{a\}$, and let $d = \operatorname{dist}(a, E)$. By Theorem 5.6.1, we have $d > 0$. We take $r_1 = \min(d/2, \rho/2)$, and it then follows that $E \subset \{z\colon |z - a| \geq d > r_1\} = D$ and $r_1 < \rho$. According to the third paragraph in the proof, the series $\sum_1^\infty c_j (z - a)^{-j}$ converges uniformly on D and, therefore, on E.

The sum function of the power series in (10.6.6) is, of course, analytic on $\Delta(a, R)$ [Theorem 6.4.1]. The terms of the second series in (10.6.6) are all analytic on $\mathbf{C} \sim \{a\}$ and the series converges a.u. on this domain. Therefore, the sum function is analytic on $\mathbf{C} \sim \{a\}$ [corollary of Theorem 9.9.1]. Q.E.D.

Examples

(i) To find the Laurent expansion of $f(z) = z^{-2}e^z$ in $\mathbf{C} \sim \{0\}$, we multiply the Maclauren series for e^z (known to converge a.u. on $\mathbf{C}$) by z^{-2}, and again obtain a series converging a.u. on $\mathbf{C}$. In particular, the latter series converges uniformly on $C(0, r)$ for any $r > 0$, so it must be the authentic Laurent expansion, by uniqueness. The result is

$$z^{-2}e^z = \frac{1}{z^2} + \frac{1}{z} + \frac{1}{2!} + \frac{z}{3!} + \frac{z^2}{4!} + \cdots.$$

(ii) To find the Laurent expansion of $f(z) = (z-1)^{-2}(z-2)^{-1}$ in $\Delta(1, 1) \sim \{1\}$, we first use "partial fractions" to rewrite f in the form

$$f(z) = -\frac{1}{(z-1)^2} - \frac{1}{z-1} + \frac{1}{z-2}. \tag{10.6.10}$$

(The technique of representing a complex rational function by partial fractions in the same as that used in the theory of formal integration in elementary calculus. We discuss it further in Sec. 11.2.) Since this Laurent expansion is a series containing positive and negative powers of $z - 1$, the first two terms in (10.6.10) are a part of the series. We write $1/(z-2) = 1/[(z-1) - 1] = -1/[1 - (z-1)]$ $= 1 + (z-1) + (z-1)^2 + (z-1)^3 + \cdots$. This power series converges a.u. on $\Delta(1, 1)$. The desired Laurent expansion is

$$\frac{1}{(z-2)^2(z-1)} = -\frac{1}{(z-1)^2} - \frac{1}{(z-1)} + \sum_0^\infty (z-1)^k.$$

If the Laurent expansion is readily available in a given case, as in the foregoing examples, it provides a convenient tool for identifying types of isolated singularities. The basic result is Proposition 10.6.2.

Proposition 10.6.2 *Let $f: O \sim a \to \mathbf{C}$ be analytic, where O is an open set and $a \in O$, and let $R =$ dist (a, O^c). Let $\sum_{-\infty}^{\infty} a_k(z-a)^k$ be the Laurent expansion of f in $\Delta(a, R) \sim \{a\}$. Then*

(i) *f has a removable singularity at a if and only if $a_k = 0$, $k = 1, 2, \ldots$.*
(ii) *f has a pole of order p at a if and only if $a_{-p} \neq 0$ and $a_{-p-1} = a_{-p-2} = \cdots = 0$. The principal part of f at a as defined in Sec. 10.5 is identical with the principal part of the Laurent series.*
(iii) *f has an essential singularity at a if and only if there is an infinite subsequence of nonzero coefficients in the coefficient sequence $\langle a_{-k}\rangle_1^\infty$.*

Proof. Part (iii) of the conclusion follows by default, as it were, from parts (i) and (ii).

(i) If f has a removable singularity at a, then by definition, a value can be assigned to f at a in such a way that the so extended function is analytic on O.

By Theorem 9.4.2, there is a local power series representation $\sum_1^\infty a_k(z-a)^k$ convergent in $\Delta(a, R)$ for the extended functions. This series converges a.u. on $\Delta(a, R)$ and represents f in $\Delta(a, R) \sim \{a\}$. By the uniqueness property of the Laurent expansion [Theorem 10.6.2(iii)], this power series must be the Laurent expansion of f in $\Delta(a, R) \sim \{a\}$, and in that Laurent expansion, $a_{-k} = 0, k = 1\ 2, \ldots$. Conversely, if the negative powers of $(z - a)$ in the Laurent expansion of f are absent, the expansion is a power series which represents a function which is analytic in $\Delta(a, R)$, and in particular analytic at a.

(ii) If f has a pole of order p at a, then by definition there exists a finite series $Q(z) = \sum_{k=-p}^{-1} a_k(z-a)^k$, $a_{-p} \neq 0$, such that $H(z) = f(z) - Q(z)$ has a removable singularity at a. Let $\sum_{k=0}^\infty a_k(z-a)^k$ denote the local power series representation of H, convergent a.u. in $\Delta(a, R)$. (Its existence is assured by part (i)). Then $f = Q + H$ is represented by the series $\sum_{k=-p}^\infty a_k(z-a)^k$, which converges a.u. on $\Delta(a, R) \sim \{a\}$ and, therefore, must be the Laurent expansion. Conversely, if $\sum_{k=-p}^\infty a_k(z-a)^k$ is the Laurent expansion of f in $\Delta(a, R) \sim \{a\}$, then the function $H(z) = f(z) - \sum_{k=-p}^{-1} a_k(z-a)^k = \sum_{k=0}^\infty a_k(z-a)^k$, $z \neq a$, has a removable singularity at a, so by definition, a is a pole if f of order p. Q.E.D.

Examples

(iii) By examining the Laurent series in Example (i), it is evident that $z^{-2}e^z$ has a pole of order 2 at $z = 0$.

(iv) By examining the Laurent series in Example (ii), it is at once evident that $(z-1)^{-2}(z-2)^{-1}$ has a pole of order 2 at $z = 1$.

In these two examples, the nature of the singularity is made obvious by Theorem 10.5.2 without appeal to the Laurent series. The next example involves a situation in which the identification of the singularity is really facilitated by the Laurent series.

Example

(v) The Laurent expansion for $f(z) = z^3 \sin(1/z)$ in $C(0, \infty) \sim \{0\}$ is obtainable as follows: The power series representation $\sin w = w - w^3/3! + w^5/5! - \cdots$ converges a.u. on $\mathbf{C}$, so the series $z^{-1} - z^{-3}/3! + z^{-5}/5! - + \cdots$ converges a.u. to $\sin(1/z)$ on $\mathbf{C} \sim \{0\}$, and therefore (by uniqueness) the Laurent expansion of $z^3 \sin(1/z)$ in $\mathbf{C} \sim \{0\}$ is $z^2 - 1/3! + z^{-2}/5! - z^{-5}/7! + \cdots$. Clearly, there is an infinite number of nonzero coefficients of negative powers of z, so the function $z^3 \sin(1/z)$ has an essential singularity at $z = 0$.

Remark. Suppose that Ω is an open set in $\mathbf{C}$, and E is a subset of Ω consisting only of isolated points, and $f: \Omega \sim E \to \mathbf{C}$ is analytic on the indicated domain. Furthermore, suppose that the annulus $A(r_1, r_2) = \{z: r_1 < |z-a| < r_2\}, 0 < r_1 < r_2$, is contained in $\Omega \sim E$, and that a series representation

$$f(z) = \sum_{k=0}^\infty b_k(z-a)^k + \sum_{j=1}^\infty c_j(z-a)^{-j} \tag{10.6.11}$$

has been found in which the first series converges a.u. on $\Delta(a, r_2)$ and the second series converges a.u. on $\{z : |z - a| > r_1\}$. If this series is to be used with Proposition 10.6.3 to classify singularities of f in $E \cap \bar{\Delta}(a, r_1)$, then to insure a correct indication, care must be taken to verify that there is *only one point of E in* $\bar{\Delta}(a, r_1)$ *and that point is a*. Under these circumstances, (10.6.11) is identical with the Laurent expansion of f in some punctured circular disk with center at a. The following example illustrates this remark.

Example

(vi) Let $f(z) = 1/(z - 1)$, which is analytic on $\mathbf{C} \sim \{1\}$ with a simple pole at $z = 1$. By writing this function in the form $(1/z)/[1 - (1/z)]$ and using the geometric series, we obtain a series expansion $f(z) = z^{-1} + z^{-2} + z^{-3} + \cdots$, converging a.u. in $A = \{z : 1 < |z| < \infty\}$. This is an expansion of the form $\sum_{-\infty}^{+\infty} a_k(z - 0)^k$, and if Proposition 10.6.3 were to be applied blindly to it, there would be a false indication of an essential singularity at $z = 0$ and no indication of the pole at $z = 1$. The same technique applied to $f(z)/z = 1/[z(z - 1)]$ yields the series expansion $f(z)/z = z^{-2} + z^{-3} + z^3 + \cdots$, again converging a.u. in the annulus A. The function actually has simple poles at $z = 0$ and $z = 1$, but the expansion gives only a false indication of an essential singularity at $z = 0$.

Exercises 10.6

1. Find the Laurent expansion of $1/[z(z - 1)]$ in $\Delta(0, 1) \sim \{0\}$, and also Laurent expansion of this function in $\Delta(1, 1) \sim \{1\}$. (*Hint*: In finding Laurent expansions for explicitly given functions, it is usually easier to think of some way of using the standard known power series. Use the integral formulas (10.6.7) and (10.6.8) only as a last resort. For the second expansion here, write $1/[z(z - 1)] = [1/(z - 1)]\{1/[1 - (z - 1)]\}$ and expand the function in braces in a geometric series.)
2. Let $f : \Delta(0, R) \sim \{0\} \to \mathbf{C}$, $R > 0$, be analytic and have an essential singularity at $z = 0$. Show that $f(z^m)$ also has an essential singularity at $z = 0$, where m is any positive integer. (*Hint*: Proposition 10.6.2.)
3. Let $f : O \sim \{a\} \to \mathbf{C}$ be analytic, where O is open and $a \in O$, and let $\sum_{-\infty}^{\infty} a_k(z - a)^k$ be the Laurent expansion of f in the punctured disk $\Delta(a, R) \sim \{a\}$, where $R = \operatorname{dist}(a, O^c)$. Show without using Theorem 10.5.1(iv) that if $\lim_{z \to a} (z - a)f(z) = 0$, then $a_k = 0$, $k = -1, -2, \ldots$. (Taken with Proposition 10.6.2, this exercise provides another proof of Theorem 10.5.1(iv). *Hint*: Use the integral formulas for a_k, $k = -1, -2, \ldots$, which are displayed in (10.6.8). You can take an arbitrarily small ρ in those formulas.)
4. Let $f : O \sim \{a\} \to \mathbf{C}$ be analytic, where O is open and $a \in O$. Use the Laurent expansion of f about a to prove that f has a pole of order p at a if and only if f' has a pole of order $(p + 1)$ at a and the residue of f' at a is zero. (*Hint*: The Laurent expansion of f' can be derived from that of f by using the corollary of Theorem 9.9.1.)
5. Let $f : \mathbf{C} \sim \{0\} \to \mathbf{C}$ be analytic, and suppose that $\overline{f(\bar{z})} = f(1/z)$ for every

$z \in \mathbf{C} \sim \{0\}$. Let $\sum_{-\infty}^{\infty} a_k z^k$ be the Laurent expansion of f in $\mathbf{C} \sim \{0\}$. Show that $\bar{a}_k = a_{-k}, k = 1,2, \ldots$, and that a_0 is real. What conclusion can you draw about the values of f on $C(0, 1)$? (*Hint*: The question hinges on the uniqueness of a Laurent expansion. Explicit formulas for the coefficients a_k are not needed.)

10.7 Zeros and Singularities at ∞

In Sec. 1.15 we introduce the extended complex plane $\hat{\mathbf{C}} = \mathbf{C} \cup \{\infty\}$ and discuss the geometric model obtained by stereographic projection onto the Riemann sphere. It is now convenient to use the setting $\hat{\mathbf{C}}$ for our further development of the theory of zeros and singularities of analytic functions. Recall these algebraic conventions: for $z \in \mathbf{C}$, $z \pm \infty = \infty \pm z = \infty$ and $z/\infty = 0$; for $z \in \hat{\mathbf{C}}$ but $z \neq 0$, $z \cdot \infty = \infty \cdot z = \infty$ and $z/0 = \infty$.

We extend the family of open sets in $\mathbf{C}$ to a family of open sets in $\hat{\mathbf{C}}$ as follows. For any real r, $0 \leq r$, let $\Delta'(\infty, r) = \{z : z \in \mathbf{C}, |z| > r\}$, and let $\Delta(\infty, r) = \Delta'(\infty, r) \cup \{\infty\}$. We call $\Delta(\infty, r)$ a disk with center at ∞ and $\Delta'(\infty, r)$ a punctured disk with center at ∞. A neighborhood of the point at infinity is defined to be the union of $\{\infty\}$ with an open set of $\mathbf{C}$ containing a punctured disk $\Delta'(\infty, r)$. The family of open sets in $\hat{\mathbf{C}}$ consists of the open sets in $\mathbf{C}$ and the neighborhoods of ∞. (The space $\hat{\mathbf{C}}$ together with the open sets in $\hat{\mathbf{C}}$ is known in topology as the *one-point conpactification of* $\mathbf{C}$.)

The open sets in $\mathbf{C}$ are defined in Chapter 5 by regarding $\mathbf{C}$ as a metric space $(\mathbf{C}, d)$ in which d is the usual metric for $\mathbf{C}$ given by $d(z_1, z_2) = |z_1 - z_2|$. This is not a useful metric for $\hat{\mathbf{C}}$ because of problems in measuring the distance from $z \in \mathbf{C}$ to ∞. As indicated in Sec. 2.2, the chordal distance $\rho(z_1, z_2)$ between two points in $\hat{\mathbf{C}}$ given by (1.15.5) and (1.15.6) satisfies the postulates for a metric in $\mathbf{C}$; it is the "usual" metric for $\hat{\mathbf{C}}$. It is obvious from inspection of (1.15.6) that a disk $\Delta(\infty, r)$ as defined is an open ball in $(\hat{\mathbf{C}}, \rho)$ with center at ∞. In fact, by using the homeomorphic character of the transformation (1.15.1), it is easy to verify that an open set in $(\hat{\mathbf{C}}, \rho)$ not containing ∞ is also an open set in $(\mathbf{C}, d)$, and conversely. For the limited purposes of this book it is not necessary to endow $\hat{\mathbf{C}}$ with a metric; it suffices to work with the usual metric in $\mathbf{C}$ and the open sets in $\hat{\mathbf{C}}$ as defined in the preceding paragraph.

Let E be a subset of $\hat{\mathbf{C}}$, let a be a point in E, and let f be a function from E into $\hat{\mathbf{C}}$. The function f is defined to be continuous at a if and only if for any ε, $0 < \varepsilon < \infty$, there exists δ_ε, $0 < \varepsilon < \infty$, such that if $z \in \Delta(a, \delta_\varepsilon) \cap E$, then $f(z) \in \Delta(f(a), \varepsilon)$. (Notice that the notation used here includes the cases in which $a = \infty$, or $f(a) = \infty$, or $a = f(a) = \infty$.) In particular, if $a = \infty$ and if these conditions are fulfilled, then f is said to be continuous at ∞ even if $f(a) = \infty$.

Let $E \subset \hat{\mathbf{C}}$ have an accumulation point a, which may or may not be an element of E, and which can be ∞. Let $g: E \to \hat{\mathbf{C}}$ be such that $g(z) \neq \infty$, $z \in E \sim \{a\}$. An element $L \in \hat{\mathbf{C}}$ is defined to be the limit of g as z in E approaches a, written $L = \lim_{z \to a} g(a)$, $z \in E$, if there exists a function $f: E \cup \{a\} \to \hat{\mathbf{C}}$ continuous at a, with $f(a) = L$ and $f(z) = g(z)$ for every $z \in E \sim \{a\}$.

These definitions are basically the same as those in Sec. 5.2 with $\hat{\mathbf{C}}$ replacing $\mathbf{C}$.

Examples

(i) The function from $\hat{\mathbf{C}}$ onto $\hat{\mathbf{C}}$ given by $w = 1/z$ is a homeomorphism from $\mathbf{C} \sim \{0\}$ onto $\mathbf{C} \sim \{0\}$. By the present definitions, it is continuous at $z = 0$ ($w = \infty$) and at $z = \infty$ ($w = 0$). Thus, the function is a homeomorphism of $\hat{\mathbf{C}}$ onto $\hat{\mathbf{C}}$.

(ii) According to these definitions and the corollary of Theorem 10.5.3, if O is an open set, $a \in O$, and $f: O \sim \{a\} \to \mathbf{C}$ has a pole of any order at $z = a$, then if f is assigned the value ∞ at a, f becomes continuous at a, and $\lim_{z \to a} f(z) = \infty$.

(iii) Theorem 10.5.3 shows that it is not useful to try to assign a complex value or infinity to a function at a point at which it has an essential singularity. For instance, it follows from Proposition 10.7.4 that $f(z) = e^z$ has an essential singularity at ∞. Let $z = x + iy$. Then $\lim_{x \to \infty} e^x = \infty$ and $\lim_{x \to -\infty} e^x = 0$. Thus, $\lim_{z \to \infty} e^z$, $z \in \mathbf{C}$, does not exist.

In the definition of continuity, if $a = \infty$ and $f(a) \neq \infty$, the definition is equivalent to the usual definition of the continuity of $f(1/w)$ at $w = 0$, where the domain of $f(1/w)$ is taken to be the image of E under the reciprocation $w = 1/z$. The use of this reciprocation provides a successful approach to the definitions of zeros and isolated singularities at ∞.

Specifically, let O be a neighborhood of infinity containing $\Delta'(\infty, r)$, $0 \leq r < \infty$, and let O' be the image of O under the reciprocation $w = 1/z$. Then O' is a neighborhood of $w = 0$, which contains $\Delta(0, 1/r)$. We have the following definitions.

(1) *Isolated singularities at* ∞. Let $f: O \sim \{\infty\} \to \mathbf{C}$ be analytic on the indicated domain. Then f is said to have an *isolated singularity* at ∞. Let $\phi: O' \sim \{0\} \to \mathbf{C}$ be given by $\phi(w) = f(1/w)$. Then ϕ has an isolated singularity at $w = 0$. The singularity of f at ∞ is said to be a *removable singularity*, a *pole of order p*, or an *essential singularity*, if and only if the singularity of ϕ at $w = 0$ is, respectively, a removable singularity, a pole of order p, or an essential singularity.

(2) *A zero at infinity*. Let $f: O \sim \{\infty\}$ have a removable singularity at ∞. Then $\phi: O' \sim \{0\} \to \mathbf{C}$ given by $\phi(w) = f(1/w)$ has a removable singularity at $w = 0$, and when the definition of ϕ is extended to $w = 0$ by continuity, ϕ is analytic on O'. The function f is said to have a zero of order m at ∞ if and only if ϕ, with definition so extended, has a zero of order m at $w = 0$.

Examples

(iv) Let $O = \hat{\mathbf{C}}$, and let $p_n(z) = a_0 + a_1 z + a_2 z^2 + \cdots + a_n z^n$, $a_n \neq 0$, $z \in \hat{\mathbf{C}} \sim \{\infty\} = \mathbf{C}$. Then $p_n(1/w) = a_0 + a_1 w^{-1} + a_2 w^{-2} + \cdots + a_n w^{-n}$, and according to the definition of pole, this function has a pole of order n at $w = 0$. Therefore, $p_n(z)$ has a pole of order n at ∞. Notice that the number of zeros of $p_n(z)$ in $\hat{\mathbf{C}}$, counted according to multiplicities, equals the number of poles of $p_n(z)$ in $\hat{\mathbf{C}}$. In Exercise 10.7.6, it is shown that this statement is true for any rational function.

(v) In Example (v) of Section 10.6, it is shown that $\phi(w) = w^3 \sin(1/w)$ has an essential singularity at $w = 0$. Therefore, $f(z) = z^{-3} \sin z$ has a essential singularity at $z = \infty$.

(vi) Let $O = \hat{\mathbf{C}}$, and let $f: O \to \hat{\mathbf{C}}$ be given by $f(z) = 1/z^m$, m a positive integer. Then $f(\infty) = 0$, f is continuous at ∞, and ϕ given by $\phi(w) = f(1/w) = w^m$ is obviously continuous (and, in fact, analytic) at $w = 0$. The zero of ϕ at $w = 0$ is of order m, so f has a zero of order m at ∞.

We define differentiation at ∞ as follows. Let O be a neighborhood of infinity, and let $f: O \sim \{\infty\} \to \mathbf{C}$ be analytic on the indicated domain. Then the derivative function f' exists at every point $z \in O \sim \{\infty\}$, and f' is analytic in this domain. If f' has a removable singularity at ∞, and if $f'(\infty)$ denotes the value for f' at ∞ indicated by continuity, then f is said to have a derivative at ∞ which has the value $f'(\infty)$. Notice that according to this definition, $f'(\infty) \neq \infty$.

Example

(vii) The function $w = f(z) = (1/2)[Rz + (1/Rz)]$, $R > 1$, introduced in Exercise 1.14.16, gives a one-to-one mapping of $\Delta(\infty, 1)$ onto the "outside" of a certain ellipse in the w-plane. Since $f'(z) = (1/2)[R - (1/Rz^2)]$ and f' is analytic in a neighborhood of infinity and has a removable singularity at ∞ (because $f'(1/\zeta)$ has a removable singularity at $\zeta = 0$), the value of f' indicated by continuity at ∞ is $f'(\infty) = R/2$.

We now introduce the Laurent expansion of an analytic function in a neighborhood of ∞.

Theorem 10.7.1 *Let O be a neighborhood of ∞ containing the punctured* disk $\Delta'(\infty, R)$, $R \geq 0$, *and let $f: O \sim \{\infty\} \to \mathbf{C}$ be analytic on $O \sim \{\infty\}$.*

(i) *There exists a series representation of f as follows:*

$$f(z) = \sum_{k=0}^{\infty} b_k z^{-k} + \sum_{j=1}^{\infty} c_j z^j, \tag{10.7.1}$$

valid for every $z \in \Delta'(\infty, R)$. The first series in (10.7.1) converges absolutely and almost uniformly on $\Delta'(\infty, R)$ and represents a function analytic on $\Delta'(\infty, R)$. The second series converges absolutely and almost uniformly on $\mathbf{C}$ and represents a function analytic on $\mathbf{C}$.

(ii) *The coefficients in (10.7.1) are given by the formulas*

$$b_k = \frac{1}{2\pi i}\int_{\gamma_\rho} f(\zeta)\zeta^{k-1}\,d\zeta, \qquad k = 0, 1, 2, \ldots, \tag{10.7.2}$$

$$c_j = \frac{1}{2\pi i}\int_{\gamma_\rho} \frac{f(\zeta)}{\zeta^{j+1}}\,d\zeta, \qquad j = 1, 2, \ldots, \tag{10.7.3}$$

where γ_ρ is any positively oriented circle with graph $C(0, \rho)$, $\rho > R$. The coefficients are independent of ρ, $\rho > R$.

(iii) *The series representation in (10.7.1) is unique in the sense that if for some r, $r > R$, two series $\sum_{k=0}^{\infty} b_k^* z^{-k}$ and $\sum_{j=1}^{\infty} c_j^* z^j$ are found which converge uniformly on $C(0, r)$, and the sum of their sum functions is f, then $b_k^* = b_k$, $k = 0, 1, 2, \ldots$, and $c_j^* = c_j$, $j = 1, 2, \ldots$.*

Remark (i). The pair of series in (10.7.1) are usually written in the condensed form $\sum_{-\infty}^{\infty} a_k z^k$ and together are called the Laurent expansion of f in a neighborhood of ∞, or "at ∞."

Remark (ii). The function $g_1 : \Delta'(\infty, R) \to \mathbf{C}$ represented by the first series in (10.7.1) has a removable singularity at ∞, and the appropriate definition of g_1 at ∞ is $g_1(\infty) = b_0$. The function g_1 is sometimes called the *analytic part* of f at ∞. The function g_2 represented by the second series in (10.7.1) is called the *principal part* or *singular part of f at* ∞.

Proof of the Theorem. (i) As usual, let $\phi(w) = f(1/w)$; ϕ is analytic on $O' \sim \{0\}$, where O' is the image of O under the reciprocation $w = 1/z$. Notice that $\Delta(0, 1/R) \subset O'$. According to Theorem 10.6.2, there exist sequences $\langle b_k\rangle_0^\infty$ and $\langle c_j\rangle_1^\infty$ such that for every $z \in \Delta(0, 1/R) \sim \{0\}$,

$$\phi(w) = \sum_{k=0}^{\infty} b_k w^k + \sum_{j=1}^{\infty} c_j w^{-j}. \tag{10.7.4}$$

The first series in (10.7.4) converges absolutely and almost uniformly on $\Delta(0, 1/R)$. Therefore, $\sum_{k=0}^{\infty} b_k z^{-k}$ converges absolutely on $\Delta'(\infty, R)$. Let E be a compact subset of $\Delta'(\infty, R)$, and let E' be its image under $w = 1/z$. By Exercise 5.2.1, E' is closed; and if $R > 0$, then since $E' \subset \Delta(0, 1/R)$, E' is bounded and so is compact. If $R = 0$, let $d = \text{dist}(0, E)$, which is positive by Theorem 5.6.1 and the fact that $\Delta'(\infty, 0)$ does not contain 0. Then if $z \in E$, so $w = 1/z \in E'$, we have $|z| \geq d$ and $|w| \leq 1/d$. Thus, again E' is bounded and, therefore, compact. The first series in (10.7.4) converges uniformly on E', so the series $\sum_{k=0}^{\infty} b_k z^{-k}$ converges uniformly on E.

Again according to Theorem 10.6.2, the second series in (10.7.4) converges absolutely on $\mathbf{C} \sim \{0\}$, so $\sum_{k=1}^{\infty} c_j z^j$ converges absolutely on $\mathbf{C} \sim \{0\}$. The latter series is a power series, and its radius of convergence must be infinite, so it converges a.u. on $\mathbf{C}$. The statements in (i) concerning the analyticity of the sum functions of the two series in (10.7.1) when the coefficients in (10.7.1) are identified with those in (10.7.4), are now implied by the corollary of Theorem 9.9.1.

(ii) The series in (10.7.1) remain uniformly convergent on $C(0, \rho)$ after multiplication by $z^{J-1}/2\pi i$, $J = 0, \pm 1, \pm 2, \ldots$. Therefore, after being so multiplied, each of the series can be integrated term-by-term along the path γ_ρ, and the sum functions of the series so obtained are the integrals of the respective sum functions along γ_ρ [corollary of Proposition 8.3.1]. In this way, we obtain

$$\frac{1}{2\pi i}\int_{\gamma_\rho} z^{J-1}f(z)\,dz = \sum_{k=0}^{\infty} b_k \int_{\gamma_\rho} \frac{z^{-k+J-1}}{2\pi i}\,dz + \sum_{j=1}^{\infty} c_j \int_{\gamma_\rho} \frac{z^{j+J-1}}{2\pi i}\,dz$$
$$= \begin{cases} b_J, & J = 0, 1, 2, \ldots. \\ c_{-J}, & J = -1, 2, \ldots. \end{cases}$$

These equations are equivalent to the formulas (10.7.2) and (10.7.3).

(iii) The proof of the uniqueness is essentially the same as the above proof of part (ii), with b_k^* and c_j^* replacing b_k and c_j, respectively, and γ_r replacing γ_ρ. Q.E.D.

In many cases, the definitions of singularities and zeros at ∞ already given in terms of the reciprocation $w = 1/z$, taken with the theorems in Sec. 10.1 and 10.5, provide sufficient information for classifying zeros and singularities at ∞ of explicitly given functions. However, for convenience in reference, we list here the main criteria translated into terms involving the behavior of a function in a neighborhood of ∞.

In each of the following four propositions, O is a neighborhood of ∞, and $f: O \sim \{\infty\} \to \mathbf{C}$ is analytic on the domain $O \sim \{\infty\}$.

Proposition 10.7.1 *The following statements are equivalent:*

(i) *f has a zero of order m at ∞.*
(ii) *There exists a function $g: O \to \mathbf{C}$ analytic on $O \sim \{\infty\}$ and continuous and finite valued at ∞, with $g(\infty) \neq 0$, and such that $f(z) = z^{-m}g(z)$ for every $z \in O \sim \{\infty\}$.*
(iii) *In the Laurent expansion $\sum_{-\infty}^{\infty} a_k z^k$ in a neighborhood of ∞, $a_k = 0$, $k = 0, 1, 2, \ldots,$ and $k = -1, -2, \ldots, -(m-1)$, and $a_{-m} \neq 0$.*

Proposition 10.7.2 *The following statements are equivalent:*

(i) *f has a removable singularity at ∞.*
(ii) *f can be assigned a finite value at ∞ so that f is continuous at ∞.*
(iii) *In the Laurent expansion $\sum_{-\infty}^{\infty} a_k z^k$ of f in a neighborhood of ∞, $a_k = 0$, $k = +1, +2, \ldots.$*
(iv) *There exists a punctured* disk $\Delta'(\infty, r) \subset O$ *such that f is bounded in* $\Delta'(\infty, r)$.
(v) $\lim_{z\to a} f(z)/z = 0$.

Proof. The proposition follows at once from Theorem 10.5.1, Proposition 10.6.2(i), and the definition of removable singularity at ∞.

Proposition 10.7.3 *The following statements are equivalent:*

(i) *f has a pole of order p at ∞, where p is a positive integer.*
(ii) *There exists a polynomial $P(z)$ of degree p such that $f(z) - P(z)$ has a removable singularity at ∞.*
(iii) *In the Laurent expansion $\sum_{-\infty}^{\infty} a_k z^k$ of f in a neighborhood of ∞, $a_p \neq 0$ and $a_{p+1} = a_{p+2} = \cdots = 0$.*
(iv) *$h(z) = z^{-p}f(z)$ is analytic on $O \sim \{\infty\}$ and has a removable singularity at ∞, and $h(\infty)$ defined by continuity is nonzero.*
(v) *There exists $h(z): O \to \mathbf{C}$ analytic on $O \sim \{\infty)$ and continuous at ∞, with $h(\infty) \neq 0$, finite, and such that $f(z) = z^p h(z)$ for every $z \in O \sim \{\infty\}$.*
(vi) *There exists a* disk $\Delta(\infty, r) \subset O$ *such that $f(z) \neq 0$ for every $z \in \Delta'(\infty, r)$, and*

$$h(z) = \begin{cases} \dfrac{1}{f(z)}, & z \in \Delta'(\infty, r), \\ 0, & z = \infty, \end{cases}$$

is analytic on $\Delta'(\infty, r)$, is continuous at ∞, and has a zero of order p at ∞.

Proof. The relevant authorities are Theorem 10.5.2, Proposition 10.6.2, and the definition of pole at ∞.

Similarly, the following proposition is a translation of Theorem 10.5.3 and Proposition 10.6.2 into the language of singularities of ∞.

Proposition 10.7.4 *The following statements are equivalent:*

(i) *f has an essential singularity at ∞.*
(ii) *In the Laurent expansion $\sum_{-\infty}^{\infty} a_k z^k$ of f in a neighborhood of ∞, there is an infinite subsequence of nonzero coefficients in the sequence $\langle a_k \rangle_1^{\infty}$.*
(iii) *For each punctured* disk $\Delta'(\infty, r) \subset O$, *$f(\Delta'(\infty, r))$ is dense in $\mathbf{C}$.*

We conclude this section with a sufficient condition for an analytic function to be a rational function.

Theorem 10.7.2 *Let $f: \hat{\mathbf{C}} \to \hat{\mathbf{C}}$ be continuous on $\hat{\mathbf{C}}$ and analytic on $\mathbf{C}$ except possibly at a finite number of points $\zeta_1, \zeta_2, \ldots, \zeta_n$ in $\mathbf{C}$, at which the function has poles of respective orders $p_1, p_2, \ldots, p_n$. Furthermore, let the isolated singularity at ∞ be removable or a pole. Then f is a rational function of the form $P(z)/[(z - \zeta_1)^{p_1}(z - \zeta_2)^{p_2} \cdots (z - \zeta_n)^{p_n}]$, where $P(z)$ is a polynomial. If the function f is analytic on $\mathbf{C}$, then f is a polynomial.*

Proof. Since ∞ is an isolated singularity of f, there exists $\Delta(\infty, R)$ such that f is analytic on $\Delta(\infty, R) \sim \{\infty\}$. If the singularity of f at ∞ is a pole, let the order be p. According to Proposition 10.7.3(v), in this case there exists $h\colon \Delta(\infty, R) \to \mathbf{C}$ analytic on this domain and continuous at ∞, such that $f(z) = z^p h(z)$, $z \in \Delta(\infty, R) \sim \{\infty\}$. By Proposition 10.7.2(ii), if ∞ is a removable singularity of f, this representation of f is still valid with $p = 0$. By Proposition 10.7.2(iv), there exists $\Delta(\infty, r)$, $r > R$, and $M > 0$, such that $|h(z)| \leq M$ for every $z \in \Delta(\infty, r)$.

If f has no singularities in $\mathbf{C}$, then since $|f(z)| \leq M|z|^p$ for $z \in \Delta(\infty, r) \sim \{\infty\}$, it follows from Exercise 9.6.1 that f is a polynomial of degree at most p. (The case $p = 0$ is simply Liouville's theorem.)

If f has poles in $\mathbf{C}$ as described in the theorem, then by Corollary 3 of Theorem 10.5.2 there exists $g\colon \mathbf{C} \to \mathbf{C}$ analytic on $\mathbf{C}$ and such that for every $z \in \mathbf{C}$, $z \neq \zeta_j$, $j = 1, \ldots, n$,

$$f(z) = (z - \zeta_1)^{-p_1}(z - \zeta_2)^{-p_2} \cdots (z - \zeta_n)^{-p_n} g(z). \qquad (10.7.5)$$

These poles must all lie in $[\Delta(\infty, R)]^c$ and in particular in $[\Delta(\infty, r)]^c$, where these are the disks described in the first paragraph of the proof. Then for every $z \in \Delta(\infty, r) \sim \{\infty\}$ we have

$$\begin{aligned} g(z) &= (z - \zeta_1)^{p_1}(z - \zeta_2)^{p_2} \cdots (z - \zeta_n)^{p_n} f(z) \\ &= (z - \zeta_1)^{p_1}(z - \zeta_2)^{p_2} \cdots (z - \zeta_n)^{p_n} z^p h(z), \end{aligned}$$

$$\begin{aligned} |g(z)| &\leq |z|^m \left(1 + \left|\frac{\zeta_1}{z}\right|\right)^{p_1} \left(1 + \left|\frac{\zeta_2}{z}\right|\right)^{p_2} \cdots \left(1 + \left|\frac{\zeta_n}{z}\right|\right)^{p_n} \cdot M \\ &\leq |z|^m \left(1 + \frac{R}{r}\right)^{p_1 + p_2 + \cdots + p_n} \cdot M, \end{aligned}$$

where $m = p + p_1 + p_2 + \cdots + p_n$. Therefore, again by Exercise 9.6.1, g is a polynomial of degree at most m. Reference to (10.7.5) completes the proof. Q.E.D.

Corollary Let $f\colon \hat{\mathbf{C}} \to \hat{\mathbf{C}}$ be analytic on $\hat{\mathbf{C}}$ except for poles. Then f is a rational function.

Proof. The statement follows at once from the theorem provided that we can show that there is at most a finite number of poles of f in $\hat{\mathbf{C}}$. This is partly a matter of semantics: A pole is by definition an *isolated* singularity; that is, an isolated point at which f is not analytic. If f has a pole at $\{\infty\}$, then there exists $\Delta'(\infty, R)$ in which f is analytic. Suppose that f were to have an infinite number of poles in $\bar{\Delta}(0, R)$. Then by the Bolzano–Weierstrass theorem [Exercise 5.5.4] there would be an accumulation point of these poles, say a. If f were defined and analytic at a, it would be continuous at a [corollary of Proposition 6.1.1] and would be bounded in some $\Delta(a, r)$, $r > 0$ [Exercise 3.2.3(a)]. But if $\zeta \in \Delta(a, r)$, $\zeta \neq a$, were one of the poles of f, then according to the corollary of Theorem 10.5.3, f would be unbounded in a neighborhood of ζ. Thus, f cannot be defined and analytic at a. But a cannot be one of the poles of f because a is not an isolated singularity. The conclusion is that f can have only a finite number of poles in $\hat{\mathbf{C}}$. Q.E.D.

Exercises 10.7

1. Discuss the singularities of $\sin z/[z(z-1)]$ in $\hat{\mathbf{C}}$.
2. Show that the linear fractional transformation $w = (az+b)/(cz+d)$, $ad - bc \neq 0$, is a homeomorphism of $\hat{\mathbf{C}}$ onto $\hat{\mathbf{C}}$.
3. Is ∞ an isolated singularity of $\tan z$? of $1/(e^{2z}-1)$? of $1/e^{2z-1}$? In case it is an isolated singularity, find the Laurent expansion.
4. Recall from Sec. 9.6 that a function $f:\mathbf{C} \to \mathbf{C}$ analytic on $\mathbf{C}$ is called an *entire function*. Show that a nonconstant entire function either is a polynomial or else has an essential singularity at ∞. (*Hint*: What is the Laurent expansion of an entire function in a neighborhood of ∞?)
5. Let $q(z)$ be a polynomial of degree $n \geq 1$. In Example (v) of Sec. 10.5, it is shown that if the distinct zeros of $q(z)$ are $\zeta_1, \zeta_2, \ldots, \zeta_\nu$ of respective multiplicities $p_1, p_2, \ldots, p_\nu$, $\sum_1^\nu p_j = n$, then $1/q(z)$ has poles at $\zeta_1, \zeta_2, \ldots, \zeta_\nu$ of respective orders $p_1, p_2, \ldots, p_\nu$. Show that the number of zeros of $1/q(z)$ in $\hat{\mathbf{C}}$, counted according to multiplicities, is equal to the sum of the orders of the poles.
6. (Continuation of 5) Let $r(z) = p(z)/q(z)$, where p and q are polynomials, $q(z) \not\equiv 0$. Show that the number of zeros of $r(z)$ in $\hat{\mathbf{C}}$, counted according to multiplicities, is equal to the sum of the orders of the poles. (Assume as usual that p and q have no zeros in common. *Hint*: Let $p(z) = a_0 + a_1 z + \cdots + a_m z^m$, $a_m \neq 0$, and let $q(z) = b_0 + b_1 z + \cdots + b_n z^n$, $b_n \neq 0$. The number of zeros and number of poles in $\mathbf{C}$, of course, follows from the fundamental theorem of algebra, with the poles counted as in Exercise 10.7.5. To study the order of a possible pole or zero at infinity, write $r(z)$ in the form:

$$r(z) = z^{m-n}\left[\frac{a_0/z^m + a_1/z^{m-1} + \cdots + a_m}{b_0/z^n + b_1/z^{n-1} + \cdots + b_n}\right].$$

Note. The *order* of a rational function is the common number of its zeros and poles, counted according to multiplicities and orders. Thus, in the notation of the "Hint," the order of $r(z)$ is $\max\{m, n\}$.
7. Let O be a neighborhood of ∞, and let $f: O \sim \{\infty\} \to \mathbf{C}$ be analytic. Show that f has a removable singularity at ∞ if and only if $f'(\infty)$ exists and *is* zero, and that f has a simple pole at ∞ if and only if $f'(\infty)$ exists and is *not* zero. (*Hint*: Use the corollary of Theorem 9.9.1 to derive the Laurent expansion for f' in a neighborhood of infinity from that of f. Then use Propositions 10.7.2 and 10.7.3.)
8. Let γ be a path with graph Γ, and let $h:\Gamma \to \mathbf{C}$ and $g:\Gamma \to \mathbf{C}$ be continuous on Γ. There exists $\Delta(0, R)$ such that $h(\Gamma) \subset \Delta(0, R)$. According to Theorem 8.4.1, the

function

$$f(z) = \int_\gamma \frac{g(\zeta)d\zeta}{h(\zeta) - z}$$

is analytic on $\Delta'(\infty, R)$ (indeed, on $[h(\Gamma)]^c$). Show that the Laurent expansion of $f(z)$ in $\Delta'(\infty, R)$ is of the form $\sum_1^\infty b_k z^{-k}$ and that, therefore, f has a simple zero at ∞. Also show that

$$b_k = -\int_\gamma g(\zeta)[h(\zeta)]^{k-1} d\zeta, \qquad k = 1,2,\ldots.$$

(*Hint*: Look at the proof of Proposition 10.6.2.)

9. Let O be a neighborhood of ∞ containing the disk $\Delta(\infty, R)$, let $f: O \sim \{\infty\} \to \mathbf{C}$ be analytic, let $g_1(z)$ be the analytic part of f at ∞, and let $g_1(\infty)$ be the appropriate value of g_1 at ∞. Prove the following version of the Cauchy integral formula (D. R. Curtiss). Let γ be a positively oriented circle with graph $C(0, r)$, $r > R$. Then for every $z \in \Delta(\infty, r)$,

$$g_1(z) - g_1(\infty) = \frac{1}{2\pi i}\int_{-\gamma} \frac{f(\zeta)d\zeta}{\zeta - z}. \tag{10.7.6}$$

(Thus, if f has a zero at ∞, then the integral represents $f(z)$ itself in $\Delta(\infty, r)$. *Hint*: Let f be represented by its Laurent series as in Theorem 10.7.1, with the coefficients of the analytic part given in (10.7.2). Expand the integral in (10.7.6) in a Laurent series as indicated in Exercise 10.7.8, and look at the coefficients.)

10. Let $O, \Delta(\infty, R), f,$ and g_1 be given as in Exercise 10.7.9. Let γ be any closed path in $\Delta(\infty, R)$. Show that if g_1 has a zero at ∞ of order at least 2, then $\int_\gamma f(z)\, dz = 0$.

11. The purpose of this exercise and the next two is to introduce the reader to the Faber polynomials, for which many applications have been found in complex analysis. (See [5].) Let $\psi: \Delta(\infty, \rho) \to \hat{\mathbf{C}}$ be analytic and nonconstant on $\Delta'(\infty, \rho)$, $\rho \geq 0$ and have a simple pole at ∞. Assume that $\psi[\Delta'(\infty, \rho)] \neq \mathbf{C}$. For any $z \in \mathbf{C} \sim \psi[\Delta'(\infty, r)] = E_r$, where $\rho \leq r < \infty$, show that the function given by

$$\psi_z(w) = \frac{w\psi'(w)}{\psi(w) - z}$$

is an analytic function of w on $\Delta'(\infty, r)$ and has a removable singularity at ∞. Let the Laurent expansion of ψ_z at ∞ be $\psi_z(w) = \sum_0^\infty p_n w^{-n}$, $|w| > r$ [Proposition 10.7.2(iii)]. Show that for each n, $n = 0,1,2,\ldots$, the coefficient p_n is a polynomial in z of degree n, with $p_0(z) \equiv 1$ and that these polynomials do not depend on r. (The polynomials p_n are called Faber polynomials belonging to ψ. *Hint*: Let the Laurent expansion of ψ at ∞ be $\psi(w) = bw + w_0 + b_1 w^{-1} + b_2 w^{-2} + \cdots$, $b \neq 0$ [Proposition 10.7.3(iii)]. It converges a.u. on $\Delta'(\infty, \rho)$, and so it can be differentiated term-by-term to find the Laurent expansion of ψ' at ∞ [corollary of Theorem 9.9.1]. Cite authorities to establish that ψ' has a removable singularity at ∞; that $\psi(w) - z$ has a simple pole at ∞; that $[\psi(w) - z]^{-1}$ has a simple zero at

∞; and finally that $\psi_z(w)$ has a removable singularity at ∞. To show that the coefficients p_n are polynomials in z, refer to the absolute convergence of the Laurent series for $\psi_z(w)$ and $\psi(w) - z$ [Theorem 10.7.1] to justify deriving the Laurent expansion of $\psi_z(w)[\psi(w) - z]$ by forming the Cauchy product series [Propositions 2.7.2 and 2.7.3], and cite an authority for assuming that this product series is identical with the Laurent expansion of $w\psi'(w)$ at ∞. Thus, the coefficients p_n can be found by the method of undetermined coefficients. Thereby find p_0 and p_1 explicitly, and then show that p_n is a polynomial in z of degree n by induction on n.)

12. (Continuation of Exercise 10.7.11.) Prove the following statements concerning the polynomials $p_n(z)$ of Exercise 10.7.11.

(a) Let r and R be such that $\rho \le r < R$. Then for every $z \in \mathbf{C} \sim \psi[\Delta'(\infty, r)]$,

$$p_n(z) = \frac{1}{2\pi i} \int_{\gamma R} \frac{\zeta^n \psi'(\zeta)\, d\zeta}{\psi(\zeta) - z}, \qquad n = 1,2, \ldots . \tag{10.7.7}$$

where γ_R is a positively oriented circle with graph $C(0, R)$.

(b) If the ψ in Exercise 10.7.11 is univalent on $\Delta(\infty, \rho)$ as well as analytic on $\Delta'(\infty, \rho)$, and R_1 is such that $\rho < R_1$, and $z = \psi(w)$, then for every $w \in \Delta(\infty, R_1)$, $z = \psi(w) \in \psi[\Delta(\infty, R_1)]$, we have

$$p_n(z) = p_n[\psi(w)] = w^n + \frac{1}{2\pi i} \int_{\gamma R_1} \frac{\zeta^n \psi'(\zeta)\, d\zeta}{\psi(\zeta) - \psi(w)}, \qquad n = 1,2, \ldots, \tag{10.7.8}$$

where γ_{R_1} is the positively oriented circle with graph $C(0, R_1)$.

(c) Let r and R_1 be such that $\rho < R_1 < r$. There exists $M = M(r, R_1) > 0$ such that for every w with $|w| = r$, and with $z = \psi(w)$, we have

$$p_n(z) = p_n[\psi(w)] = w^n + I_n(w), \; |I_n(w)| \le M R_1^n, \qquad n = 1,2, \ldots . \tag{10.7.9}$$

Also there exists $M_1 = M_1(r, R_1) > 0$ such that for every w, $|w| = r$,

$$|p_n(z)| = |p_n \psi(w)| \le M_1 r^n, \qquad n = 1,2, \ldots . \tag{10.7.10}$$

(d) With $z = \psi(w)$, we have

$$\lim_{n \to \infty} \frac{p_{n+1}(z)}{p_n(z)} = w$$

and

$$\lim_{n \to \infty} |p_n(z)|^{1/n} = |w|, \tag{10.7.11}$$

uniformly for $z \in \psi[C(0, r)]$, $r > \rho$.

(*Hint*: For (a), refer to (10.7.2). For (b) first choose any w with $|w| > R_1$ and then choose the R in part (a) so that $R > |w|$. Define a function D by $D(\zeta) = [\psi(\zeta) - \psi(w)]/(\zeta - w)$ for $\zeta \in \Delta'(\infty, \rho)$, $\zeta \ne w$, and $D(w) = \psi'(w)$. This function D is analytic and nonzero on $\Delta'(\infty, \rho)$ (why?), so the function χ defined by $\chi(\zeta$

$= \zeta^n\psi'(\zeta)/D(\zeta)$ is also analytic on $\Delta'(\infty, \rho)$. Notice that $\chi(w) = w^n$. By Theorem 10.6.1, we have

$$\chi(w) = w^n = \frac{1}{2\pi i}\int_{\gamma R}\frac{\chi(\zeta)\,d\zeta}{\zeta - w} - \frac{1}{2\pi i}\int_{\gamma R_1}\frac{\chi(\zeta)\,d\zeta}{\zeta - w}. \tag{10.7.12}$$

Now use (10.7.7). The problem in (c) is only that of estimating the absolute value of the integral in (10.7.8). Since $\psi[C(0, R_1)]$ and $\psi[C(0, r)]$ are compact sets [Theorem 5.5.2], the distance d between them is positive [Theorem 5.6.1] and $|\psi(\zeta) - \psi(w)|^{-1} \geq d$ for every $\zeta \in C(0, R_1)$ and $w \in C(0, r)$. The remainder of the proof of (c) is straightforward. Both of the asymptotic formulas in part (d) stem from (10.7.9). First, show that the second part of (10.7.9) implies that the sequence $\langle w^{-m}I_m(w)\rangle_1^\infty$ converges uniformly to zero for $w \in C(0, r)$. For (10.7.11), use the fact (easily derived by logarithms) that for any bounded sequence of complex numbers $\langle z_n\rangle_1^\infty$, $\lim_{n\to\infty} |1 + z_n|^{1/n} = 1$.)

13. (Continuation of Exercises 10.7.11 and 10.7.12.) Let the function ψ introduced in Exercise 10.7.11 be univalent on $\Delta(\infty, \rho)$ as well as analytic on $\Delta'(\infty, \rho)$. Prove that:

(a) $p_n[\psi(w)]$ has a Laurent expansion at ∞ of the following type:

$$p_n[\psi(w)] = w^n + \sum_{k=1}^{\infty}\frac{\alpha_{nk}}{w^k}, \quad |w| > \rho; \tag{10.7.13}$$

(b) if $P_1(z), P_2(z), \ldots, P_n(z), \ldots$, is any sequence of polynomials such that the degree of P_n is exactly n, $n = 1,2, \ldots$, and such that $P_n[\psi(w)]$ has a Laurent expansion at ∞ of the type $w^n + \sum_{k=1}^{\infty}\beta_{nk}w^{-k}$, $|w| > \rho$, $n = 1,2, \ldots$, then $P_n(z) \equiv p_n(z)$, where p_n is the nth Faber polynomial belonging to ψ. (The numbers α_{nk} in (10.7.13) are called Faber coefficients. *Hint*: (a) The function $p_n[\psi(w)] - w^n$ is certainly analytic on $\Delta'(\infty, \rho)$, and the problem here is merely to determine the form of its Laurent expansion at ∞. As in the proof of Exercise 10.7.11(b), choose any $|w| > \rho$; then choose R_1 so that $\rho < R_1 < |w|$, and set up the function χ which appears in that proof. Equation (10.7.8) may be rewritten as follows:

$$p_n[\psi(w)] - w^n = \frac{1}{2\pi i}\int_{\gamma R_1}\frac{\chi(\zeta)\,d\zeta}{\zeta - w}, \quad n = 1,2, \ldots.$$

To deduce the type of the Laurent expansion at ∞ of the right-hand member, refer to Proposition 10.6.1. Then cite an authority which assures the permanence of equation (10.7.12) throughout $\Delta(\infty, \rho)$. The hypotheses of (b), taken with (10.7.13), imply that $Q_n(z) = p_n(z) - P_n(z)$ is a polynomial of degree $\leq n$ such that $Q_n[\psi(w)]$ has a Laurent expansion at ∞ of the type $\sum_{k=1}^{\infty} a_k w^{-k}$. The problem is to show that such a function Q_n is the zero polynomial. Let $Q_n(z) = A_n z^n + A_{n-1}z^{n-1} + \cdots + A_0$. By Proposition 10.7.4(v), there exists $h(w)$ analytic on $\Delta'(\infty, \rho)$, continuous at ∞, with $c = h(\infty) \neq 0, \neq \infty$, and such that $\psi(w) = wh(w)$, $z \in \Delta(\infty, \rho)$. Therefore, by

Propositions 10.7.1 and 10.7.2, in the Laurent expansion at ∞ of $[\psi(w)]^n = w^n[h(w)]^n$, the term containing the greatest positive power of w is $c^n w^n$, and thus the term in the Laurent expansion at ∞ of $Q_n[\psi(w)]$ with the greatest positive power of w is $A_n c^n w^n$. But by hypothesis and the uniqueness of Laurent expansions, the coefficient of w^n in this term is zero, so we have $A_n = 0$. Proceed inductively to show that $A_{n-1} = A_{n-2} = \cdots = A_0 = 0$.)

14. Let O be a neighborhood of ∞; let f be analytic on $O \sim \{\infty\}$ and have a simple pole at ∞. Show that $f'(\infty)$ exists, is nonzero, and $f'(\infty) = \lim_{z \to \infty} [f(z)/z] = a_1$, where a_1 is the coefficient of z in the Laurent expansion of f about ∞. (*Hint*: Use the corollary of Theorem 9.9.1 to justify term-by-term differentiation of the Laurent expansion of f about ∞. Proposition 10.7.2(iii) identifies the type of singularity of f' at ∞. Use the reciprocation $w = 1/z$ to transform the Laurent expansions of $f'(z)$ and $f(z)/z$ into power series, and refer to Theorem 4.2.1(i) to justify taking limits as $z \to \infty$.)

15. Prove that f is continuous on a set $E \subset \hat{\mathbf{C}}$ if and only if for each set O open in $\hat{\mathbf{C}}$, $f^{-1}(O) = O' \cap E$, where the set O' is open in $\hat{\mathbf{C}}$. (*Hint*: Imitate the proof of Theorem 5.2.1.)

11

Residues and Rational Functions

11.1 The Residue Theorem

Let O be an open set in $\mathbf{C}$, and let $f\colon O \sim a \to \mathbf{C}$ be analytic, where $a \in O$. Recall from Sec. 10.5 that if f has a pole at a, then by the definition of a pole there exists a polynomial Q in $1/(z-a)$ without a constant term, called the principal part of f at a, and such that $f - Q$ has a removable singularity at a. The coefficient of $1/(z-a)$ in Q is called the residue of f at a. Henceforth, we use the notation Res $(f; a)$ for the residue of f at a. In Exercise 10.5.1, it is shown that if $(1/2\pi i)f(z)$ is integrated over a suitably chosen circle with center at a, the result is Res $(f; a)$. This fact is the motivation for the word "residue": The "residue" is "what is left" after integration.

The following is a special form of what is known as the residue theorem. A more general formulation appears below in Theorem 12.6–(11.1.1).

Theorem 11.1.1 (*Residue Theorem*). *Let B be a starlike region, let $P = \{\zeta_1, \zeta_2, \ldots, \zeta_\nu\}$ be a set of distinct points in B, and let $f\colon B \sim P \to \mathbf{C}$ be analytic and have a pole at ζ_j with residue* Res $(f; \zeta_j)$, $j = 1, 2, \ldots, \nu$. *Let γ be a closed path in B such that no point of P lies on its graph. Then*

$$\frac{1}{2\pi i}\int_\gamma f(z)\,dz = \sum_{j=1}^{\nu} \operatorname{Res}(f; \zeta_j)\,\operatorname{Ind}_\gamma(\zeta_j). \qquad (11.1.1)$$

Proof. Let the principal parts of f at $\zeta_1, \zeta_2, \ldots, \zeta_\nu$ be denoted, respectively, by $Q_1, Q_2, \ldots, Q_\nu$. The function $f - Q_1$ has a removable singularity at ζ_1. Since Q_1 is analytic at $\zeta_2, \zeta_3, \ldots, \zeta_\nu$, by Corollary 2 of Theorem 10.5.2, $f - Q_1$ has poles at $\zeta_2, \zeta_3, \ldots, \zeta_\nu$ *with the same principal parts* as those of f. The function $(f - Q_1) - Q_2$ has a removable singularity at ζ_1 [corollary of Theorem 10.5.1]; it also has another removable singularity at ζ_2, and poles at $\zeta_3, \ldots, \zeta_\nu$ with the same principal parts as those of f. Proceeding in this way until all the poles in P are exhausted, we finally arrive at a function

$$G(z) = f(z) - \sum_{j=1}^{\nu} Q_j(z), \qquad z \in B \sim P, \tag{11.1.2}$$

which is analytic on B except for removable singularities at $\zeta_1, \ldots, \zeta_\nu$. We now assume that the definition of G has been suitably extended at these points so that G is analytic on B.

The typical principal part Q_j has the form

$$Q_j(z) = \frac{\text{Res}\,(f; \zeta_j)}{z - \zeta_j} + q_j(z),$$

where q_j is a polynomial in $1/(z - \zeta_j)$ with no constant term and no first degree term. Therefore, by the definition of $\text{Ind}_\gamma\,(\zeta_j)$ and by Example (ii) of Sec. 8.5 (which assures the existence of a primitive for q_j on $B \sim P$),

$$\frac{1}{2\pi i}\int_\gamma Q_j(z)\,dz = \text{Res}\,(f; \zeta_j)\,\text{Ind}_\gamma\,(\zeta_j) + 0.$$

Using (11.1.2) with the Cauchy integral theorem for a starlike region, we obtain

$$0 = \frac{1}{2\pi i}\int_\gamma G(z)\,dz = \frac{1}{2\pi i}\int_\gamma f(z)\,dz - \sum_{j=1}^{\nu} \text{Res}\,(f; \zeta_j)\,\text{Ind}_\gamma\,(\zeta_j).$$

Q.E.D.

Exercises 11.1

1. Evaluate the integral

$$\int_\gamma \frac{z^2 + 3z + 1}{(z - 1)(z + 4)(z - 5i)}\,dz,$$

where γ is a positively oriented circle with graph $C(0, 6)$. (*Hint*: In this problem and in other exercises in this set involving residues, use Exercises 10.5.1 and 10.5.2 as needed.)

2. For each of the functions in Exercise 10.5.3, find the value of its integral along the positively oriented circle with graph $C(0, 4)$. Use the residues in the answers to that exercise.

3. Let γ be a closed path with graph Γ not containing the origin, and let the point a

lie in the component of Γ^c containing the origin. Show by means of the residue theorem that

$$\int_\gamma \frac{dz}{z^j(z-a)} = 0,$$

where j is a positive integer. (See the hint in Exercise 11.1.1.)

4. Let O be a neighborhood of infinity containing $\Delta(\infty, R)$, let $f: O \sim \{\infty\} \to \mathbf{C}$ be analytic and have a pole of order n at ∞, and let γ be a circle with graph $C(0, r)$, $R < r$. Use the Laurent series of f at infinity and the preceding exercise to show that the integral

$$\int_\gamma \frac{f(z)\, dz}{z-a}$$

represents a polynomial in a of degree n for $a \in \Delta(0, r)$.

5. Let O be a neighborhood of infinity containing $\Delta(\infty, R)$, and let $f: O \sim \{\infty\} \to \mathbf{C}$ be analytic. The negative of the coefficient of z^{-1} in the Laurent expansion of f at ∞ is called the residue of f at ∞, and is given (see (10.7.2)) by

$$\operatorname{Res}(f, \infty) = -\frac{1}{2\pi i}\int_{\gamma_\rho} f(\zeta)\, d\zeta,$$

where γ_ρ is any positively oriented circle with graph $C(0, \rho)$, $\rho > R$. Show that if the function $F: \mathbf{C} \to \mathbf{C}$ is analytic on $\mathbf{C}$ except for a finite number of poles, then the sum of the residues of F in $\hat{\mathbf{C}}$ is zero. (Thus, for every rational function, the sum of the residues is zero.)

6. Let O be a neighborhood of infinity containing $\Delta(\infty, R)$, and let $f: O \sim \{\infty\} \to \mathbf{C}$ be analytic except for poles at $\{\zeta_1, \zeta_2, \ldots, \zeta_n\} \subset \Delta'(\infty, R)$. Let f have a pole or a removable singularity at ∞. Let γ be the positively oriented circle with graph $C(0, \rho)$, where $R < \rho < \min_j |\zeta_j|$. Prove that

$$-\frac{1}{2\pi i}\int_\gamma f(z)\, dz = \operatorname{Res}(f; \infty) + \sum_{j=1}^{n} \operatorname{Res}(f; \zeta_j).$$

(*Hint*: The key to the problem is the equation (11.1.2). Notice that by the Cauchy integral theorem, $\int_\gamma Q_j(z)\, dz = 0$; the notation is that of the equation. After removal of artificial singularities, G is analytic on $\Delta'(\infty, R)$, and the only possible nonzero contribution to the integral of G along γ comes from the coefficient of $1/z$ in the Laurent expansion of G at ∞.)

7. Let γ be a closed path given by the parametric equation $z = a + r(t)e^{it}$, $0 \le t \le 2\pi$, where $r(t) > 0$ for every $t \in [0, 2\pi]$, $r(0) = r(2\pi)$, and $r(t)$ is piecewise continuously differentiable. Let $B = \{z : z = a + sr(t)e^{it},\ 0 \le t \le 2\pi,\ 0 \le s < 1\}$. Let $f: \bar{B} \to \mathbf{C}$ be analytic on B except for a pole at each point of the set $\{\zeta_1,$

$\zeta_2, \ldots, \zeta_\nu\} \subset B$, and let f be continuous on $\bar{B}$. Show that

$$\frac{1}{2\pi i}\int_\gamma f(z)\,dz = \sum_{j=1}^{\gamma} \mathrm{Res}(f;\zeta_j).$$

(*Hint*: Imitate the proof of Theorem 11.1.1. See Sec. 8.4, Example (iii), and Proposition 9.3.3.)

11.2 A General Cauchy Integral Theorem for Rational Functions

The residue theorem leads to a Cauchy integral theorem for rational functions which dispenses with the starlike restriction in Theorem 9.2.2, and which is used in the next chapter to obtain a much more general form of the Cauchy theory than that which our methods have heretofore made available. The theorem in question is Theorem 11.2.1.

First, a remark is in order concerning the technique used to prove Theorem 11.1.1 when applied to a rational function. Let $R(z)$ be a rational function with its poles at $\zeta_1, \zeta_2, \ldots, \zeta_\nu$ and possibly at ∞, and let the principal parts at the finite poles be, respectively, $Q_1(z), Q_2(z), \ldots, Q_\nu(z)$. We apply the reduction used to derive equation (11.1.2), and obtain $G(z) = R(z) - \sum_{j=1}^{\nu} Q_j(z)$. But now after removal of artificial singularities, G is analytic on $\mathbf{C}$, and the singularity at ∞ is either removable or a pole. Therefore, by Theorem 10.7.2, G with definition so completed is a polynomial. The resulting representation for R,

$$R(z) = G(z) + \sum_{j=1}^{\nu} Q_j(z) \tag{11.2.1}$$

valid on $\mathbf{C} \sim \{\zeta_1, \zeta_2, \ldots, \zeta_\nu\}$, is called the *partial fraction decomposition* of R. There are various devices for obtaining the coefficients of this decomposition in given cases; some of these are discussed in the undergraduate calculus textbooks. The coefficients of G can always be found by long division, and when the pole ζ_j is simple, the only coefficient of Q_j is Res (R, ζ_j) $= \lim_{z\to\zeta_j} (z - \zeta_j)R(z)$ [Exercise 10.5.1].

It is useful here to introduce a type of region which is a generalization of the starlike type and to which most of the Cauchy theory valid with the starlike restriction still applies. Recall that a region $B \subset \mathbf{C}$ is an open connected set. A region $B \subset \mathbf{C}$, $B \neq \mathbf{C}$, is called *simply connected* (in $\mathbf{C}$) if and only if every component of $B^c = \mathbf{C} \sim B$ is unbounded. By definition, $\mathbf{C}$ is simply connected. For example, if B is bounded and B^c has only one component, then B is simply connected [Proposition 5.4.2]. The interior region of a closed rectangular path is simply connected [Sec. 8.4, Example (ii)].

A starlike region B with star center a is simply connected, because (briefly) if $z \in B^c$, then when the line segment $[a, z]$ is indefinitely continued beyond z with initial point a, the continuation can never intersect B, by the definition of "starlike." Thus, z lies in an unbounded component of B^c. (It is possible but unnecessary here to argue that there can be only one component of B^c in the starlike case.)

A final definition: $\text{Ind}_\gamma(\infty) = 0$ for any path γ.

Theorem 11.2.1 *Let O be an open set in $\mathbf{C}$, and let $R(z)$ be a rational function with all its poles in $\hat{\mathbf{C}} \sim O$.*

(i) *If γ is a closed path with graph in O, such that* $\text{Ind}_\gamma(z) = 0$ *for every $z \in \hat{\mathbf{C}} \sim O$, then $\int_\gamma R(\zeta)\, d\zeta = 0$.*
(ii) *If γ_1 and γ_2 are closed paths with graphs in O, such that* $\text{Ind}_{\gamma_1}(z) = \text{Ind}_{\gamma_2}(z)$ *for every $z \in \hat{\mathbf{C}} \sim O$, then $\int_{\gamma_1} R(z)\, dz = \int_{\gamma_2} R(z)\, dz$.*
(iii) *If O is a simply connected region, then $\int_\gamma R(z)\, dz = 0$ for every closed path γ with graph in O.*

Proof. Clearly, in all three parts of the theorem, there are no poles of $R(z)$ on the paths of integration. If $O = \mathbf{C}$, then all three parts of the theorem are merely special cases of the Cauchy integral theorem for a starlike region [Theorem 9.2.2.]. Henceforth, we assume $O \neq \mathbf{C}$. For (i) in Theorem 11.1.1, we take the starlike region B to be $\mathbf{C}$. The topological indices in (11.1.1) are all zero, by hypothesis. Similarly for (ii), the right-hand members of (11.1.1) are the same for γ_1 and γ_2. For (iii), let Γ be the graph of γ. Since $\Gamma \subset O$, $O^c = \mathbf{C} \sim O \subset \Gamma^c$. Every component of O^c is unbounded, by the definition of simple connectivity. Let A be such a component. By Theorem 5.4.1, it lies in a component of Γ^c, and this component must be unbounded, so it must be *the* unbounded component of Γ^c [Proposition 5.4.2]. By Theorem 8.4.2, $\text{Ind}_\gamma(z) = 0$ for every z in the unbounded component of Γ^c. Thus, $\text{Ind}_\gamma(z) = 0$ for every $z \in \hat{\mathbf{C}} \sim O$, and we are back to part (i) of the theorem. Q.E.D.

Exercises 11.2

1. In the partial fraction decomposition (11.2.1), show that if R has a pole of order m at ∞, then G is a polynomial of degree m, and if R has a removable singularity at ∞ with the appropriate value $R(\infty) = g$, then $G(z) = g$ for every $z \in \mathbf{C}$.
2. Show that the partial fraction decomposition (11.2.1) is unique, in the sense that if a polynomial $G^*(z)$ (which may be a constant) is found, and polynomials $Q_j^*(w)$ without constant terms are found, $j = 1, \ldots, \nu$, such that

$$R(z) = G^*(z) + \sum_{j=1}^{\nu} Q_j^* \left(\frac{1}{z - \zeta_j}\right)$$

for every $z \in \mathbf{C}$, $z \neq \zeta_1, \zeta_2, \ldots, \zeta_\nu$, then the coefficients of G^* and Q_j^* are equal to the corresponding coefficients of G and Q_j in (11.2.1). (*Hint*: See Exercise 10.5.9.)

11.3 The Principle of the Argument and Jensen's Formula

A classical application of residue theory which generalizes Theorem 10.2.1 follows.

Theorem 11.3.1 *Let B be a starlike region, and let $Z = \{z_1, z_2, \ldots, z_\mu\}$ and $P = \{\zeta_1, \zeta_2, \ldots, \zeta_\nu\}$ be contained in B, with $Z \cap P = \varnothing$. Let $f\colon B \sim P \to \mathbf{C}$ be analytic and have a pole at ζ_k of order p_k, $k = 1, \ldots, \nu$, and a zero at z_j of multiplicity m_j, $j = 1, \ldots, \mu$. Let γ be a closed path with graph $\Gamma \subset B$ and such that $\Gamma \cap (P \cup Z) = \varnothing$. Then*

(i) $\displaystyle \frac{1}{2\pi i}\int_\gamma \frac{f'(z)}{f(z)}\,dz = \sum_{j=1}^{\mu} m_j \operatorname{Ind}_\gamma(z_j) - \sum_{k=1}^{\nu} p_k \operatorname{Ind}_\gamma(\zeta_k);$

(ii) if $h\colon B \to \mathbf{C}$ is analytic, then

$$\frac{1}{2\pi i}\int_\gamma h(z)\frac{f'(z)}{f(z)}\,dz = \sum_{j=1}^{\mu} m_j h(z_j)\operatorname{Ind}_\gamma(z_j) - \sum_{k=1}^{\nu} p_k h(\zeta_k)\operatorname{Ind}_\gamma(\zeta_k).$$

Remark. If γ is such that $\operatorname{Ind}_\gamma(z_j) = 1, j = 1, \ldots, \nu$, and $\operatorname{Ind}_\gamma(\zeta_k) = 1, k = 1, \ldots, \nu$, then the integral in (i) obviously represents the surplus (perhaps negative) of the number of zeros over the number of poles of f in B, with multiplicities and orders taken into account. This is the case which is known classically as the principle of the argument. We do not attempt to motivate this terminology rigorously, but the general idea is this. A local primitive of f'/f along Γ is $\log f(z) = \ln|f(z)| + i\arg[f(z)]$. (Recall that according to Proposition 9.4.2 and its corollary, if f were analytic and nonzero everywhere in B, then f'/f would have a global primitive in B called the analytic logarithm of f, and the integrals in (i) would be zero. But this would follow anyway for the integrals in (i) and (ii) by the Cauchy integral theorem.) When z describes Γ once from some chosen initial point to the coincident terminal point, $\ln|f(z)|$ returns to its original value. Intuitively, therefore, the integral in (i) measures in units of 2π the change in $\arg[f(z)]$ as z describes Γ.

Proof of the Theorem. If R is a rational function with Z as its zero set and P as its set of poles, then according to Exercise 10.5.11,

$$\frac{R'(z)}{R(z)} = \sum_{j=1}^{\mu}\frac{m_j}{z - z_j} - \sum_{k=1}^{\nu}\frac{p_k}{z - \zeta_k} \tag{11.3.1}$$

for every $z \in \mathbf{C} \sim (Z \cup P)$. By Corollary 3 of Theorem 10.5.2, there exists a function $H\colon B \to \mathbf{C}$ analytic in B, with $H(\zeta_k) \neq 0$, $k = 1, \ldots, \nu$, and such that for every $z \in B \sim P$,

$$f(z) = (z - \zeta_j)^{-p_1}(z - \zeta_2)^{-p_2}\cdots(z - \zeta_\nu)^{-p_\nu}H(z). \tag{11.3.2}$$

By Proposition 10.1.3, H has a zero of multiplicity m_j at z_j, $j = 1, \ldots, \mu$, and since f has no other zeros, H has no other zeros. Then by Theorem 10.1.1, there exists a function $G: B \to \mathbf{C}$ analytic in B, with $G(z) \neq 0$ for every $z \in B$, and such that

$$H(z) = (z - z_1)^{m_1}(z - z_2)^{m_2} \cdots (z - z_\mu)^{m_\mu}G(z).$$

Substitution in (11.3.2) yields the representation $f(z) = R(z)G(z)$, where R is a rational function with zero set Z and set of poles P. Notice that $f' = RG' + R'G$, and $f'/f = (G'/G) + (R'/R)$. Thus, using (11.3.1), we obtain

$$h(z)\frac{f'(z)}{f(z)} = \sum_{j=1}^{\mu} \frac{m_j h(z)}{z - z_j} - \sum_{k=1}^{\nu} \frac{p_k h(z)}{z - \zeta_k} \frac{h(z)G'(z)}{G(z)} \tag{11.3.3}$$

for every $z \in B \sim (Z \cup P)$. This equation shows that the only singularities of hf'/f are simple poles which occur at some or all of the points of $Z \cup P$. (If h has a zero at one of these points, then hf'/f has a removable singularity at this point. We assume that the definition of hf'/f has been suitably completed at such points.) If z_j is a pole of hf'/f, then by Exercise 10.5.1(b), $\operatorname{Res}(hf'/f; z_j) = \lim_{z \to z_j} (z - z_j)h(z)f'(z)/f(z) = m_j h(z_j)$. A similar calculation applies at a pole ζ_k. Reference to the residue theorem—or more simply, to the Cauchy integral theorem and formula—now establishes part (ii) of this theorem. Part (i) is the special case in which $h(z) = 1$ for every $z \in B$. Q.E.D.

Ordinarity the choice $h(z) = \log z$ is not permissible in Theorem 11.3.1(ii) because of the multivalued character of $\log z$, but when γ has the circular graph $C(0, r)$, this specialization can be handled in such a way as to derive a result known as Jensen's formula, which is important in the modern theory of analytic functions of which the only singularities in $\mathbf{C}$ are poles. We do not attempt to prove Jensen's formula via this route (the reader if interested may consult [9, pp. 256–257]), but instead we give a proof based on the factorization $f(z) = R(z)G(z)$ whereby we proved Theorem 11.3.1.

Theorem 11.3.2 (*Jensen's Formula*). *Let $f: \Delta(0, R) \to \mathbf{C}$ be analytic except for poles at the points $P = \{\zeta_1, \zeta_2, \ldots, \zeta_\nu\} \subset \Delta(0, R)$, and let the order of the pole at ζ_k be p_k, $k = 1, \ldots, \nu$. Let the zero set of f in $\Delta(0, R)$ be $Z = \{z_1, z_2, \ldots, z_\mu\}$, with z_j of multiplicity m_j, $j = 1, \ldots, \mu$. Furthermore, suppose that $Z \cup P$ does not contain the point 0. Finally, let r, $0 < r < R$, be such that $(Z \cup P) \subset \bar{\Delta}(0, r)$. Then*

$$\frac{1}{2\pi}\int_0^{2\pi} \ln |f(re^{i\theta})|\, d\theta = \ln |f(0)| + \sum_{j=1}^{\mu} \ln \frac{r}{|z_j|} - \sum_{k=1}^{\nu} \ln \frac{r}{|\zeta_k|}. \tag{11.3.4}$$

Remark (i). The formula, of course, remains valid (with obvious missing terms) when Z or P or both are empty.

Remark (ii). If there are points of $Z \cup P$ on $C(0, r)$, the integral in (11.3.4) is an improper integral. It seems worthwhile to state explicitly here the definition of the improper integral which is being used. Let α and β be real numbers, $\alpha < \beta$, and let $\{\alpha = t_0 < t_1 < \cdots < t_n = \beta\}$ be a partition of $[\alpha, \beta]$. Furthermore, let $\langle u_k \rangle$ be a finite sequence of real numbers satisfying the inequalities $t_0 < u_1 < t_1 < u_2 < t_3 < \cdots < t_{n-1} < u_n < t_n$. Let $\phi : \bigcup_1^n (t_{k-1}, t_k) \to \mathbf{C}$ be continuous on the indicated domain. With $u < \tau$, we use the symbol $\int_u^{\tau-} \phi(t)\, dt$ as an abbreviation for

$$\lim_{\substack{s \to \tau \\ u < s < \tau}} \int_u^s \phi(t)\, dt,$$

with the implied assumption that this limit exists. Similarly, the symbol $\int_{\tau+}^u \phi(t)\, dt$ denotes

$$\lim_{\substack{s \to \tau \\ \tau < s < u}} \int_s^u \phi(t)\, dt.$$

Then if each of the individual implied limits in the right-hand member of the following equation exists, we define the improper Riemann integral of ϕ over the interval $[\alpha, \beta]$ by

$$\int_\alpha^\beta \phi(t)\, dt = \int_{t_0+}^{u_1} \phi(t)\, dt + \sum_{k=1}^{n-1} \left[\int_{u_k}^{t_k-} \phi(t)\, dt + \int_{t_k+}^{u_{k+1}} \phi(t)\, dt \right] + \int_{u_n}^{t_n-} \phi(t)\, dt. \tag{11.3.5}$$

It follows from this definition, the additive property of the definite integral, and elementary limit theory, that if $\phi_1, \phi_2, \ldots, \phi_N$ are each continuous on $\bigcup_1^n (t_{k-1}, t_k)$ and $\int_\alpha^\beta \phi_h(t)\, dt$, $h = 1, \ldots, N$, exist, then $\int_\alpha^\beta \sum_{h=1}^N \phi(t)\, dt$ exists and equals $\sum_{h=1}^N \int_\alpha^\beta \phi_h(t)\, dt$. It is also noted that if ϕ in (11.3.5) happens to be continuous at a point t_k, the corresponding limit or limits of its integral in (11.3.5) automatically exist [Proposition 8.1.1(d)]. The implication of (11.3.4) is that if there are points in the interval $[0, 2\pi]$ at which $\ln|f(re^{i\theta})|$ is discontinuous (and these correspond to poles and zeros of f on $C(0, r)$), then nevertheless the integral in (11.3.4) does exist in the sense just now specified.

Remark (iii). The proof given here of the theorem depends on certain results of independent interest concerning the integrals of logarithms along circular paths. They appear in the following three lemmas.

Lemma 1 Let $g : \Delta(0, R) \to \mathbf{C}$ be analytic and never zero in $\Delta(0, R)$; let r be such that $0 < r < R$. Then

$$\ln|g(0)| = \frac{1}{2\pi} \int_0^{2\pi} \ln|g(re^{i\theta})|\, d\theta.$$

Proof of the Lemma. By the corollary of Proposition 9.4.2, g possesses an analytic logarithm $\lambda(z)$ in $\Delta(0, R)$, and by the definition of analytic logarithm, $e^\lambda = g$, $|e^\lambda| = e^{\mathrm{Re}\,\lambda} = |g|$, $\mathrm{Re}\,\lambda = \ln|g|$. By the Gauss mean value theorem (9.5.2),

$$\lambda(0) = \frac{1}{2\pi}\int_0^{2\pi} \lambda(re^{i\theta})\,d\theta,$$

and the conclusion follows when the real parts of the members of this equation are equated.

Lemma 2 For $r > 0$,

$$\frac{1}{2\pi}\int_0^{2\pi} \ln(r|1 - e^{i\theta}|)\,d\theta = \ln r. \tag{11.3.6}$$

Proof of the Lemma. The implication of (11.3.6) is that for any $\beta \in (0, 2\pi)$, limits $\int_{0+}^{\beta} \ln(r|1 - e^{i\theta}|)\,d\theta$ and $\int_{\beta}^{2\pi-} \ln(r|1 - e^{i\theta}|)\,d\theta$ both exist and their sum is $\ln r$. A standard proof involves integration of $\ln|1 - z|$ along a path consisting of $C(0, r)$ with an indentation to avoid $z = 1$, but this leads conveniently only to the existence and value of the integral in (11.3.6) as a principal value defined by the above limits when $\theta_2 = 2\pi - \theta_1$. (See Sec. 11.4 for a discussion of the principal value concept.) We proceed by power series methods.

The polynomial $1 - z$ has an analytic logarithm $L(z)$ in $\{z: \mathrm{Re}\,z < 1\}$, and $\mathrm{Re}\,L(z) = \ln|1 - z|$. (See the proof of Lemma 1.) Also $L'(z) = -1/(1 - z)$, from which, as in Exercise 7.7.2, we easily deduce that L has a power series representation $L(z) = -z - (z^2/2) - (z^3/3) - \cdots$ convergent in $\Delta(0, 1)$. Also $L(z)/(iz) = -(1/i) - (z/2i) - (z^2/3i) - \cdots = \sum_1^\infty a_k z^k$, $z \in \Delta(0, 1)$. The coefficient sequence $\langle a_k \rangle$ satisfies the conditions $\lim_{k\to\infty} a_k = 0$ and $\sum_1^\infty |a_k - a_{k+1}| = \sum_1^\infty |(1/k) - [1/(k + 1)]| \le \sum_1^\infty (1/k^2) < \infty$, so by Proposition 4.2.1 the series for $L(z)/(iz)$ converges uniformly on $\overline{\Delta(0, 1)} \sim \Delta(1, \rho)$, $0 < \rho < 2$. It may, therefore, be integrated term-by-term along any subarc of $C(0, 1) \sim \{1\}$ [Proposition 8.3.1]. Choose any $\beta \in (0, 2\pi)$ and $\theta_1 \in (0, \beta)$, $\theta_2 \in (\beta, 2\pi)$. Let γ_1 and γ_2 be the arcs with equations $z = e^{i\theta}$, $\theta_1 \le \theta \le \beta$ and $z = e^{i\theta}$, $\beta \le \theta \le \theta_2$. Then integrating by primitives we obtain

$$\int_{\gamma_1} \frac{L(z)}{iz}\,dz = \sum_{k=1}^{\infty} \frac{e^{i\beta k} - e^{ik\theta_1}}{ik^2} = \int_{\theta_1}^{\beta} L(e^{i\theta})\,d\theta,$$

$$\int_{\gamma_2} \frac{L(z)}{iz}\,dz = \sum_{k=1}^{\infty} \frac{e^{ik\theta_2} - e^{ik\beta}}{ik^2} = \int_{\beta}^{\theta_2} L(e^{i\theta})\,d\theta.$$

A Weierstrass M-test with $M_k = 2/k^2$ shows that the first of these series converges uniformly for $\theta_1 \in \mathbf{R}$, and the second converges uniformly for $\theta_2 \in \mathbf{R}$.

They therefore represent continuous functions of θ_1 and θ_2, respectively. Thus the following limit calculations are valid:

$$\int_{0+}^{\beta} L(e^{i\theta})\,d\theta = \sum_{k=1}^{\infty} \frac{e^{i\beta k} - 1}{ik^2},$$

$$\int_{\beta}^{2\pi-} L(e^{i\theta})\,d\theta = \sum_{k=1}^{\infty} \frac{1 - e^{i\beta k}}{ik^2},$$

$$\int_{0+}^{\beta} L(e^{i\theta})\,d\theta + \int_{\beta}^{2\pi-} L(e^{i\theta})\,d\theta = 0;$$

$$\begin{aligned} 0 &= \operatorname{Re}\int_{0+}^{\beta} L(e^{i\theta})\,d\theta + \operatorname{Re}\int_{\beta}^{2\pi-} L(e^{i\theta})\,d\theta \\ &= \int_{0+}^{\beta} \ln|1 - e^{i\theta}|\,d\theta + \int_{\beta}^{2\pi-} \ln|1 - e^{i\theta}|\,d\theta; \end{aligned}$$

$$\begin{aligned} &\int_{0+}^{\beta} (\ln r + \ln|1 - e^{i\theta}|)\,d\theta + \int_{\beta}^{2\pi-} (\ln r + \ln|1 - e^{i\theta}|)\,d\theta \\ &\qquad = 2\pi \ln r + \int_{0}^{2\pi} \ln|1 - e^{i\theta}|\,d\theta = 2\pi\ln r + 0. \end{aligned}$$

This establishes (11.3.6).

Lemma 3 For $a \in \mathbf{C}$, $r > 0$,

$$\frac{1}{2\pi}\int_0^{2\pi} \ln|re^{i\theta} - a|\,d\theta = \begin{cases} \ln|a|, & |a| > r. \\ \ln r, & |a| \le r. \end{cases} \tag{11.3.7}$$

Proof of the Lemma. If $|a| > r$, $z - a$ is analytic and nonzero in $\Delta(0, |a|)$, so by Lemma 1 the integral has the value $\ln|0 - a| = \ln|a|$. If $|a| \le r$, we first notice that

$$\begin{aligned} \ln|re^{i\theta} - a| &= \ln|\overline{re^{i\theta} - a}| = \ln(|e^{i\theta}|\,|re^{-i\theta} - \bar{a}|) \\ &= \ln|r - \bar{a}e^{i\theta}| = \ln\left(r\left|1 - \frac{\bar{a}e^{i\theta}}{r}\right|\right). \end{aligned} \tag{11.3.8}$$

If $|a| < r$, then $r(1 - (\bar{a}/r)w)$ is analytic in w and nonzero for $|w| < r/|a|$, so by Lemma 1 and (11.3.8) the integral in (11.3.7) has the value $\ln|r(1 - 0)| = \ln r$. If $|a| = r$, let $a/r = e^{-i\alpha}$, $0 \le \alpha < 2\pi$. If $\alpha = 0$, the integral from 0 to 2π of the right-hand member of (11.3.8) is the same as that which appears in Lemma 2, so again the integral in (11.3.7) has the value $\ln r$.

Finally, suppose that $0 < \alpha < 2\pi$. According to our definition of improper integrals, we have

$$\begin{aligned} \int_0^{2\pi} \ln|re^{i\theta} - a|\,d\theta &= \int_0^{2\pi} \ln(r|1 - e^{-i\alpha}e^{i\theta}|)\,d\theta \\ &= \lim_{\substack{\eta\to 0\\ \eta>0}} \int_0^{\alpha-\eta} \ln(r|1 - e^{i-\alpha}e^{i\theta}|)\,d\theta + \lim_{\substack{\eta'\to 0\\ \eta'>0}} \int_{\alpha+\eta'}^{2\pi} \ln(r|1 - e^{-i\alpha}e^{i\theta}|)\,d\theta, \end{aligned}$$

provided the limits in the third member of this equation both exist. We make the changes of variables $t = 2\pi - \alpha + \theta$ in the first integral in the third member of the equation, and $t = \theta - \alpha$ in the second integral. The transformed sum of limits of integrals is

$$\left(\int_{2\pi-\alpha}^{2\pi-} + \int_{0+}^{2\pi-\alpha}\right) \ln (r|1 - e^{it}|)\, dt$$
$$= \int_0^{2\pi} \ln (r|1 - e^{it}|)\, dt = 2\pi \ln r,$$

by Lemma 2. As we remarked at the beginning of the proof of Lemma 2, this equation implies the existence of all the individual limits involved in this part of the proof of Lemma 3.

This concludes the proofs of the three lemmas.

We now turn to the proof of the theorem. Proceeding as in the proof of Theorem 11.3.1, we factor the function f as follows:

$$f(z) = \frac{(z - z_1)^{m_1}(z - z_2)^{m_2} \cdots (z - z_\mu)^{m_\mu}}{(z - \zeta_1)^{p_1}(z - \zeta_2)^{p_2} \cdots (z - \zeta_\nu)^{p_\nu}} G(z), \qquad z \in \Delta(0, R) \sim P, \tag{11.3.9}$$

where G is analytic and nonzero on $\Delta(0, R)$. Then for $z \in C(0, r) \sim (Z \cup P)$,

$$\ln |f(z)| = \sum_{j=1}^{\mu} m_j \ln |z - z_j| + \sum_{k=1}^{\nu} (-p_k) \ln (z - \zeta_k| + \ln |G(z)|. \tag{11.3.10}$$

Now according to Lemma 3, whether or not $z_j \in C(0, r)$, the integral $\int_0^{2\pi} m_j \ln |re^{i\theta} - z_j|\, d\theta$ exists, at least in the sense of Remark (ii), and has the value $m_j \ln r$. The statement holds true for $j = 1, \ldots, \mu$ and with ζ_k and p_k replacing z_j and m_j, $k = 1, \ldots, \nu$. Also, $\ln |G(re^{i\theta})|$ is continuous. Therefore, by using (11.3.10) and the additivity property described in Remark (ii), we can establish that the integral on the left-hand side of the following equation exists and is given by the right-hand side:

$$\frac{1}{2\pi}\int_0^{2\pi} \ln |f(re^{i\theta})|\, d\theta$$
$$= \sum_{j=1}^{\mu} m_j \ln r - \sum_{k=1}^{\nu} p_k \ln r + \frac{1}{2\pi}\int_0^{2\pi} \ln |G(re^{i\theta})|\, d\theta \tag{11.3.11}$$

But by Lemma 1 and (11.39),

$$\frac{1}{2\pi}\int_0^{2\pi} \ln |G(re^{i\theta})|\, d\theta = \ln |G(0)|$$
$$= \ln \left| \frac{(-\zeta_1)^{p_1}(-\zeta_2)^{p_2} \cdots (-\zeta_\nu)^{p_\nu}}{(-z_1)^{m_1}(-z_2)^{m_2} \cdots (-z_\mu)^{m_\mu}} \cdot f(0) \right|$$
$$= -\sum_{j=1}^{\mu} m_j \ln |z_j| + \sum_{k=1}^{\nu} p_k \ln |\zeta_k| + \ln |f(0)|. \tag{11.3.12}$$

Substituting (11.3.12) into (11.3.11) and collecting terms, we obtain (11.3.4). Q.E.D.

Corollary Let f: $\Delta(0, R) \to \mathbf{C}$ be analytic, $f(0) \neq 0$. Then for any r, $0 < r < R$,

$$\frac{1}{2\pi}\int_0^{2\pi} \ln|f(re^{i\theta})|\, d\theta \geq \ln|f(0)|.$$

The inequality is a strong one if there is a zero of f in $\Delta(0, r)$.

Proof. The conclusion is obvious from inspection of (11.3.4). Q.E.D.

Exercises 11.3

1. Use Lemma 2 of Theorem 11.3.2, with $r = 1$, to show that $\int_0^{\pi} \ln(\sin\theta)\, d\theta = -\pi \ln 2$. (*Hint*: See Exercise 7.2.3.)
2. Let $f: \Delta(0, R) \to \mathbf{C}$ be analytic except at points $\zeta_1, \zeta_2, \zeta_3$, where f has poles of orders respectively 1, 2, and 3. Let γ be a positively oriented circle or radius r, $0 < r < R$ and with graph $C(0, r)$, such that $\{\zeta_1, \zeta_2, \zeta_3\} \subset \Delta(0, r)$. Given that $\int_\gamma [f'(z)/f(z)]\, dz = -8\pi i$, how many zeros, counted by multiplicities, does f have in $\Delta(0, r)$? *Answer* 2.
3. Suppose that the region B of Theorem 11.3.1 contains a closed disk $\bar{\Delta}(0, R)$, $R > 0$, and that $(Z \cup P) \subset \Delta(0, R)$. Let γ be the positively oriented circle with parametric equation $z = Re^{it}$, $0 \leq t \leq 2\pi$. Show that with $z = Re^{it}$,

$$\frac{1}{2\pi}\int_0^{2\pi} \mathrm{Re}\left(z\frac{f'(z)}{f(z)}\right) dt = \sum_{j=1}^{\mu} m_j - \sum_{k=1}^{\mu} 0_k.$$

From this show that

$$\min_{|z|=R} \mathrm{Re}\left(z\frac{f'(z)}{f(z)}\right) \leq \sum_{j=1}^{\mu} m_j - \sum_{k=1}^{\nu} p_k \leq \max_{|z|=R} \mathrm{Re}\left(z\frac{f'(z)}{f(z)}\right).$$

4. Let $p(z) = z^n + a_1 z^{n-1} + a_2 z^{n-2} + \cdots + a_n$, where $a_1, a_2, \ldots, a_n$ are complex numbers. Show that for all polynomials of this type,

$$\frac{1}{2\pi}\int_0^{2\pi} \ln|p(e^{i\theta})|\, d\theta \geq 0,$$

and that for every degree n there is at least one polynomial of this type for which the integral attains its lower bound. (*Hint*:

$$\begin{aligned}|p(z)| &= |z|^n|1 + a_1 z^{-1} + a_2 z^{-2} + \cdots + a_n z^{-n}|,\\ |p(e^{i\theta})| &= |1 + a_1 e^{-i\theta} + a_2 e^{-2i\theta} + \cdots + a_n e^{-ni\theta}|\\ &= |\overline{p(e^{i\theta})}| = |1 + \bar{a}_1 e^{i\theta} + \bar{a}_2 e^{2i\theta} + \cdots + \bar{a}_n e^{ni\theta}|.\end{aligned}$$

Use the corollary of Theorem 11.3.2.)

5. Show that if the hypotheses of Theorem 11.3.2 are modified to permit f to have a zero of order m at $z = 0$, then the term $\ln|f(0)|$ in (11.3.4) must be replaced by $\ln|f^{(m)}(0)/m!| + m \ln r$, but the formula is otherwise unchanged. (*Hint*: $f(z)/z^m$ is nonzero at $z = 0$.)

11.4 Evaluation of Definite Integrals by Residues

Applications of complex function theory to the evaluation of real integrals were introduced in Sec. 9.11. The residue theorem provides a far-reaching extension of these methods, but the reservations expressed at the beginning of Sec. 9.11 as to the assurance of success in any given case still apply. These methods are covered at great length in some of the textbooks which emphasize applications of analytic function theory rather than the theory itself (see for example [15, pp. 253–279]). The treatment here is confined to a few general theorems, each of which covers a substantial class of specific integration problems.

Proposition 11.4.1 *Let $R(s, t)$ be a function of the form*

$$R(s, t) = \frac{\sum_{j=0}^{m} a_j s^j t^{m-j}}{\sum_{k=0}^{n} b_k s^k t^{n-k}},$$

where the coefficients a_j and b_k are complex numbers. Let $f(z) = (iz)^{-1} R(2^{-1}(z + z^{-1}), (2i)^{-1}(z - z^{-1}))$. Let the poles of the rational function $f(z)$ in $\Delta(0, 1)$ be $\zeta_1, \zeta_2, \ldots, \zeta_\nu$, and let f have no pole on $C(0, 1)$. Then

$$\int_0^{2\pi} R(\cos\theta, \sin\theta)\, d\theta = 2\pi i \sum_{k=1}^{\nu} \operatorname{Res}(f; \zeta_k).$$

Proof. By Theorem 11.1.1, if γ is the positively oriented circle with parametric equation $z = e^{i\theta}$, $0 \le \theta \le 2\pi$, we have

$$2\pi i \sum_{h=1}^{\nu} \operatorname{Res}(f; \zeta_h) = \int_\gamma f(z)\, dz = \int_0^{2\pi} f(e^{i\theta}) i e^{i\theta}\, d\theta$$

$$= \int_0^{2\pi} R\left[\frac{1}{2}(e^{i\theta} + e^{-i\theta}), \frac{1}{2i}(e^{i\theta} - e^{-i\theta})\right] d\theta.$$

Q.E.D.

Example

(i) We evaluate

$$\int_0^{2\pi} \frac{d\theta}{\alpha + \beta \sin\theta}, \qquad \alpha > \beta > 0. \qquad (11.4.1)$$

Here

$$f(z) = \frac{1}{iz} \cdot \frac{1}{\alpha + \beta[(z - 1/z)/2i]} = \frac{1}{\beta(z^2 + 2irz - 1)}, \quad r = \frac{\alpha}{\beta} > 1.$$

The zeros of the denominator are $i[-r \pm \sqrt{r^2 - 1})$. Clearly, $-r - \sqrt{r^2 - 1} < -1$, and it is easily verified that the zero at $z = i(-r + \sqrt{r^2 - 1})$ lies in $\Delta(0, 1)$. The residue of f at this simple pole can be calculated by Exercise 10.5.1(b), and it turns out to be $(i\beta\sqrt{r^2 - 1})^{-1} = (i\sqrt{\alpha^2 - \beta^2})^{-1}$. Therefore, by Theorem 11.4.1, the value of the integral (11.4.1) is $2\pi/\sqrt{\alpha^2 - \beta^2}$.

Other applications of Theorem 11.4.1 appear in Exercise 11.4.1.

Application of residue calculus to the evaluation of an improper real integral often leads most conveniently to a principal value determination of the integral, which may exist when the improper integral as defined in Sec. 9.10 and in Remark (ii) of Sec. 11.3 does not exist. Specifically, let α and β be real numbers, $\alpha < \beta$, let $\{\alpha = t_0 < t_1 < \cdots < t_n = \beta\}$ be a partition of $[\alpha, \beta]$, and let $\langle u_k \rangle_0^{n+1}$ be a finite sequence of real numbers satisfying the inequalities $u_0 < t_0 < u_1 < t_1 < \cdots < t_{n-1} < u_n < t_n < u_{n+1}$. Let ϕ be a complex-valued function continuous on the domain $(-\infty, t_0) \cup (\bigcup_{j=1}^{n} t_{j-1}, t_j) \cup (t_n, \infty)$. Then if each of the limits in the following sum exists:

$$\lim_{R\to\infty}\left(\int_{-R}^{u_0} + \int_{u_{n+1}}^{R}\right)\phi(t)\,dt + \sum_{k=0}^{n} \lim_{\substack{\delta\to 0\\ \delta>0}}\left(\int_{u_k}^{t_k-\delta} + \int_{t_k+\delta}^{u_{k+1}}\right)\phi(t)\,dt, \quad (11.4.2)$$

the sum is taken to be the definition of the principal value of the integral of ϕ along $(-\infty, \infty)$, and is written $PV\int_{\infty}^{\infty} \phi(t)\,dt$. It is sometimes possible to show from symmetry properties of the integrand that if the principal value exists, then the integral exists within the more restrictive definitions of Sec. 9.10 and 11.3. If it is known that $\int_{-\infty}^{\infty} \phi(t)\,dt$ exists within the definitions of Sec. 9.10 and 11.3, then, of course, the value of the integral as so defined coincides with the principal value. There are various more or less obvious modifications of (11.4.2) when some or all of the implied discontinuities of ϕ are absent; for example, if ϕ is continuous on $(-\infty, \infty)$, then $PV\int_{-\infty}^{\infty} \phi(t)\,dt = \lim_{R\to\infty} \int_{-R}^{R} \phi(t)\,dt$.

Proposition 11.4.2 *Let $R(z)$ be a rational function with a zero at infinity, and let α be a positive real number.*

(i) *If R has no poles in the half-plane $\{z\colon \operatorname{Im} z \geq 0\}$, then*

$$PV\int_{-\infty}^{\infty} R(z)e^{i\alpha x}\,dx = 0.$$

(ii) *If $\{\zeta_1, \zeta_2, \ldots, \zeta_\nu\}$ is the set of poles of R in $\{z\colon \operatorname{Im} z > 0\}$ and if there is no pole of R on the real axis, then*

$$PV\int_{-\infty}^{\infty} R(x)e^{i\alpha x}\,dx = 2\pi i \sum_{k=1}^{\nu} \operatorname{Res}\,[R(z)e^{i\alpha z}; \zeta_k].$$

(iii) *If* $Z = \{\zeta_1, \zeta_2, \ldots, \zeta_\nu\}$ *is the set of poles of* R *in* $\{z\colon \operatorname{Im} z > 0\}$ *and if* R *has a simple pole at* $z = \beta$ *on the real axis but no other poles on the real axis, then*

$$PV \int_{-\infty}^{\infty} R(x)e^{i\alpha x}\,dx = \pi i e^{i\alpha\beta} \operatorname{Res}(R; \beta) + 2\pi i \sum_{k=1}^{\nu} \operatorname{Res}[R(z)e^{i\alpha z}; \zeta_k]. \quad (11.4.3)$$

If Z *is empty, the second term in the right-hand member of (11.4.3) is absent.*

Proof. We discuss only part (iii) in detail, since the method of proof can be made essentially to cover parts (ii) and (i) by assigning zero residues to missing poles. Actually, part (i) can be established by using only the Cauchy integral theorem with an argument which is similar to that which was used for the example in Sec. 9.11.

Choose $\rho_0 > 0$ and $\delta_0 > 0$ so that $-\rho_0 < \beta - \delta_0 < \beta < \beta + \delta_0 < \rho_0$, and so that all the poles of R in **C** are contained in $\Delta(0, \rho_0)$ and the only pole of R in $\bar{\Delta}(\beta, \delta_0)$ is $z = \beta$. These numbers ρ_0 and δ_0 henceforth are held fixed. We further introduce the following variable points on the real axis: $\rho > \rho_0$, $-\rho < -\rho_0$; and $\beta - \delta$, $\beta + \delta$, $0 < \delta < \delta_0$, so that $\beta - \delta_0 < \beta - \delta < \beta < \beta + \delta < \beta + \delta_0$. The plan is to integrate $R(z)e^{i\alpha z}$ along a path γ consisting of the following arcs, here specified by means of their graphs: γ_1, a semicircle in $\{z\colon \operatorname{Im} z \geq 0\}$, center at 0, joining ρ to $-\rho$; $\gamma_2 = [-\rho, \beta - \delta]$; γ_3, a semicircle in $\{z\colon \operatorname{Im} z \geq 0\}$ with center at β, joining $\beta - \delta$ to $\beta + \delta$, and $\gamma_4 = [\beta + \delta, \rho]$. See Fig. 1, in which for simplicity $C(0, \rho_0)$ is not shown.

Since $e^{i\alpha z}$ is analytic and nonzero in **C**, by Corollary 2 of Theorem 10.5.2 the set of poles of $R(z)e^{i\alpha z}$ is the same as the set of poles of $R(z)$. To use the residue theorem, it is necessary to compute $\operatorname{Ind}_\gamma(\zeta)$ for each pole ζ of R. To do this, we complete the circular paths γ_1 and γ_3 as shown in Fig. 1. Let Γ be the graph of γ, and Γ' the graph of $\gamma_6 - \gamma_4 + \gamma_5 - \gamma_2$. Now Γ lies in $H = \{z\colon \operatorname{Im} z \geq 0\} \sim \{\beta\}$, and H^c is clearly connected and unbounded. Therefore, H^c is contained in the unbounded component of Γ. All the poles of R not in $Z \cup \{\beta\}$ lie in H^c, by hypotheses, so $\operatorname{Ind}_\gamma(\zeta) = 0$ for every pole $\zeta \notin Z \cup \{\beta\}$. Also $\operatorname{Ind}_\gamma(\beta) = 0$, since $\beta \in H^c$. By the same reasoning, Z is contained in the unbounded component of Γ', so for each pole $\zeta_k \in Z$, $0 = \operatorname{Ind}_{\gamma_6 - \gamma_4 + \gamma_5 - \gamma_2}(\zeta_k) = \operatorname{Ind}_{-\gamma_6 + \gamma_2 - \gamma_5 + \gamma_4}(\zeta_k)$. We use the familiar calculation of topological indices for circular paths to obtain $\operatorname{Ind}_{\gamma_1 + \gamma_6}(\zeta_k) = 1$, $\operatorname{Ind}_{\gamma_5 + \gamma_3}(\zeta_k) = 0$, $k = 1, \ldots, \nu$. Now $(\gamma_1 + \gamma_6) + (\gamma_5 + \gamma_3) + (-\gamma_6 + \gamma_2 - \gamma_5 + \gamma_4) = \gamma_1 + \gamma_2 + \gamma_3 + \gamma_4 = \gamma$. Adding the topological indices, corresponding to the terms in parentheses, we obtain $\operatorname{Ind}_\gamma(\zeta_k) = 1 + 0 + 0 = 1$, $k = 1, \ldots, \nu$.

The residue theorem, Theorem 11.1.1, now gives us the equation $\int_\gamma R(z)e^{i\alpha z}\,dz = \sum_{k=1}^{\gamma} \operatorname{Res}(R(z)e^{i\alpha z}; \zeta_k)$, since the terms in the summation

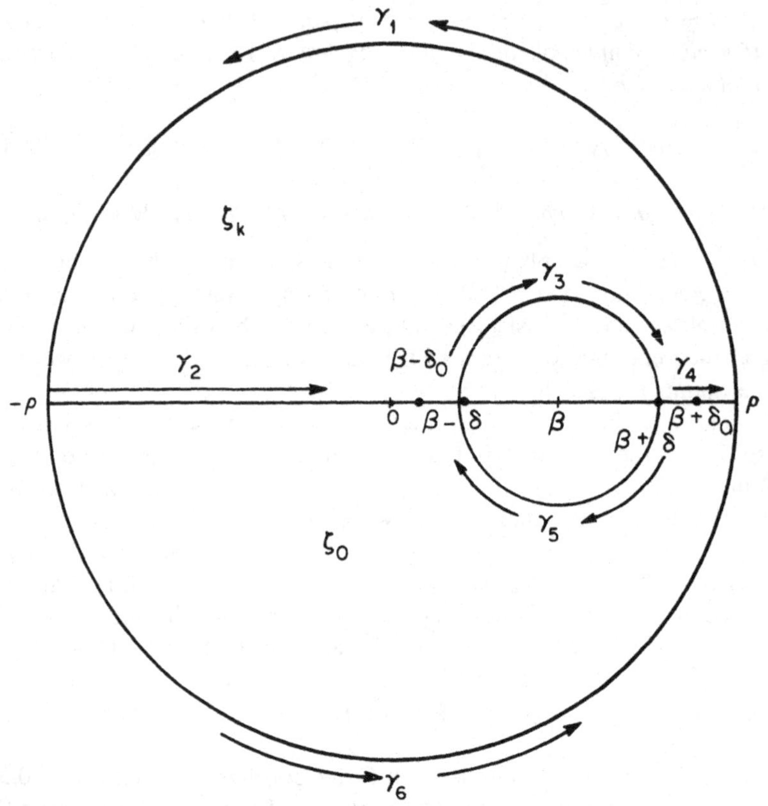

Fig. 1

corresponding to poles not in Z are zero. To analyze the integral with a view toward using the definition (11.4.2), we introduce the additional notation $I_1(\rho) = \int_{\gamma_1} R(z)e^{i\alpha z}\,dz$, $I_2(\rho) = [\int_{-\rho}^{\beta-\delta_0} + \int_{\beta+\delta_0}^{\rho}]R(x)e^{i\alpha x}\,dx$, $I_3(\delta) = \int_{\gamma_3} R(z)e^{i\alpha z}\,dz$, and $I_4(\delta) = [\int_{\beta-\delta_0}^{\beta-\delta} + \int_{\beta+\delta}^{\beta+\delta_0}]R(x)e^{i\alpha x}\,dx$. Then

$$\int_\gamma R(z)e^{i\alpha z}\,dz = I_1(\rho) + I_2(\rho) + I_3(\delta) + I_4(\delta)$$

$$= \sum_{k=1}^{\nu} \operatorname{Res}\,[R(z)e^{i\alpha z}; \zeta_k] = S. \qquad (11.4.4)$$

We now evaluate $\lim_{\rho\to\infty} I_1(\rho)$ and $\lim_{\delta\to 0} I_3(\delta)$.

First, we look at $I_1(\rho)$. Since R has a zero at ∞ and is analytic in $\Delta'(\infty, \rho_0) = \Delta(\infty, \rho_0) \sim \{\infty\}$, by Proposition 11.7.1(ii) there exists $g\colon \Delta(\infty, \rho_0) \to \mathbf{C}$ analytic on $\Delta'(\infty, \rho_0)$, continuous on $\Delta(\infty, \rho_0)$, and such that $f(z) = z^{-1}g(z)$, $z \in \Delta(\infty, \rho_0)$. Let $M = \max |g(z)|$, $z \in \bar{\Delta}(\infty, 2\rho_0)$.

Let the parametric equation of γ_1 be $z = \rho e^{i\theta}, 0 \le \theta \le \pi$. Then for $\rho \ge 2\rho_0$,

$$|I_1(\rho)| = \left| \int_0^\pi \frac{g(\rho e^{i\theta})}{\rho e^{i\theta}} \cdot e^{i\rho(\cos\theta + i\sin\theta)} i\rho e^{i\theta}\, d\theta \right|$$

$$\le \int_0^\pi |g(\rho e^{i\theta})| e^{-\rho\sin\theta}\, d\theta$$

$$\le 2M \int_0^{\pi/2} e^{-\rho\sin\theta}\, d\theta.$$

By Lemma 1 in Sec. 9.11, the limit of the last integral as $\rho \to \infty$ is zero. Hence, $\lim I_1(\rho) = 0$.

For $I_3(\delta)$, we introduce a lemma of some interest for itself alone.

Lemma Let $g: \Delta(a, r) \sim \{a\} \to \mathbf{C}$ be analytic and have a simple pole at $a \in \mathbf{C}$. Let $\gamma(\delta)$ be a circular arc with parametric equation $z = a + \delta e^{i\theta}$, $-\pi < \theta_1 \le \theta \le \theta_2 \le \pi$, $0 < \delta < r$. Then $\lim_{\delta \to 0} \int_{\gamma(\delta)} g(z)\, dz = i(\theta_2 - \theta_1)$ Res $(g; a)$.

Proof of the Lemma. By the definition of a pole, there exists $G: \Delta(a, r) \to \mathbf{C}$ analytic on $\Delta(a, r)$, and such that $g(z) = G(z) + b(z - a)^{-1}$ for every $z \in \Delta(a, r) \sim \{a\}$, where $b = \text{Res}\,(g; a)$. Then

$$\int_{\gamma(\delta)} g(z)\, dz = \int_{\theta_1}^{\theta_2} G(a + \delta e^{i\theta}) i\delta e^{i\theta}\, d\theta + b \int_{\theta_1}^{\theta_2} \frac{i\delta e^{i\theta}\, d\theta}{\delta e^{i\theta}}$$

$$= \int_{\theta_1}^{\theta_2} G(a + \delta e^{i\theta}) i\delta e^{i\theta}\, d\theta + i(\theta_2 - \theta_1)b.$$

Choose δ_1, $0 < \delta_1 < r$, and let $M_0 = \max |G(z)|$, $|z - a| \le \delta_1$. Then for $\delta < \delta_1$, we have

$$\left| \int_{\gamma(\delta)} g(z)\, dz - i(\theta_2 - \theta_1)b \right| \le \delta M_0(\theta_2 - \theta_1) \to 0$$

as $\delta \to 0$. This concludes the proof of the lemma.

We apply the lemma to $-I_3(\delta)$ with $\gamma_\delta = -\gamma_3$, $g(z) = R(z)e^{i\alpha z}$, and $\theta_2 - \theta_1 = \pi - 0$. The residue of $R(z)e^{i\alpha z}$ at the simple pole β can be calculated as follows [Exercise 10.5.1(b)]: Res $(R(z)e^{i\alpha z}; \beta) = \lim_{z \to \infty} (z - \beta)$ $R(z)e^{i\alpha z} = \text{Res}\,(R; \beta)e^{i\alpha\beta}$. The lemma then yields $\lim_{\delta \to 0} (-I_3(\delta)) = i\pi e^{i\alpha\beta}$ Res $(R; \beta)$.

Armed with these limit calculations, we return to (11.4.4). The sum $I(\rho) = I_1(\rho) + I_2(\rho)$ must be constant for $\rho > \rho_0$, because if $I(\rho_1) \ne I(\rho_2)$ for two numbers $\rho_1 > \rho_0$, $\rho_2 > \rho_0$, then the constraint (11.4.4) forces the function $I_3(\delta) + I_4(\delta)$ to have two different values for each δ, $0 < \delta < \delta_0$.

It follows then from the fact that $\lim_{\rho\to\infty} I_1(\rho) = 0$ that $\lim_{\rho\to\infty} I_2(\rho) = I_2(\infty)$ exists, and $0 + I_2(\infty) + I_3(\delta) + I_4(\delta) = S$. But from this it is clear that since $\lim_{\delta\to 0} I_3(\delta)$ exists and equals $-i\pi e^{i\alpha\beta}\,\mathrm{Res}\,(R;\beta)$, it must follow that $\lim_{\delta\to 0} I_4(\delta) = I_4(0)$ exists and satisfies $0 + I_2(\infty) - i\pi e^{i\alpha\beta}\,\mathrm{Res}\,(R;\beta) + I_4(0) = S$. However, by the definition (11.4.2), $PV\int_{-\infty}^{\infty} R(x)e^{i\alpha x}\,dx = I_2(\infty) + I_4(0)$, so (11.4.3) has now been established. Q.E.D.

An obvious extension of this proof yields the following corollary.

Corollary Let $R(z)$ be a rational function with a zero at infinity. Let $Z = \{\zeta_1, \ldots, \zeta_\nu\}$ be the set of poles of R in $\{z: \mathrm{Im}\, z > 0\}$, and let R have simple poles at $\beta_1, \beta_2, \ldots, \beta_\mu$ on the real axis and no other poles on the real axis. Then for $\alpha > 0$,

$$PV\int_{-\infty}^{\infty} R(x)e^{i\alpha x}\,dx = \pi i \sum_{j=1}^{\mu} e^{i\alpha\beta_j}\,\mathrm{Res}\,(R;\beta_j) + 2\pi i \sum_{k=1}^{\nu} \mathrm{Res}\,[R(z)e^{i\alpha z};\zeta_k].$$

Example

(ii) Evaluate

$$I = \int_0^{\infty} \frac{x \sin \alpha x}{x^2 + \beta^2}\,dx, \qquad \alpha > 0, \qquad \beta > 0.$$

Solution. If $R(z) = z/(z^2 + \beta^2)$, then

$$\int_{-\rho}^{\rho} \frac{xe^{i\alpha x}}{x^2 + \beta^2}\,dx = \int_{-\rho}^{\rho} R(x)[\cos \alpha x + i \sin \alpha x]\,dx$$

$$= 2i\int_0^{\rho} R(x) \sin \alpha x\,dx \to 2iI$$

as $\rho \to \infty$. Here we use the fact that $R(x)\cos \alpha x$ is an odd function and $R(x)\sin \alpha x$ is an even function. Now $R(x)$ has simple poles at $\pm i\beta$. Therefore, by Proposition 11.4.2,

$$2iI = 2\pi i\,\mathrm{Res}(R(z)e^{i\alpha z}; i\beta),$$

where

$$\mathrm{Res}(R(z)e^{i\alpha z}; i\beta) = \lim_{z\to i\beta} [(z - i\beta)R(z)e^{i\alpha z}]$$

$$= \lim_{z\to i\beta} \frac{ze^{i\alpha z}}{z + i\beta} = \frac{i\beta e^{i\alpha i\beta}}{2i\beta} = \frac{e^{-\alpha\beta}}{2}.$$

Therefore, $I = (\pi/2)e^{-\alpha\beta}$.

Example
(iii) Evaluate

$$PV\int_0^{\infty} \frac{\cos(\pi x/2)}{x^2-1}\,dx = I.$$

Solution. The integrand is an even function, so

$$2I = PV\int_{-\infty}^{\infty} \frac{\cos(\pi x/2)}{x^2-1}\,dx = \mathrm{Re}\left[PV\int_{-\infty}^{\infty} \frac{e^{(i\pi x)/2}}{x^2-1}\,dx\right].$$

The residues of $e^{i\pi z/2}/(z^2-1)$ at the poles $z = +1$ and $z = -1$ are, respectively, $e^{i\pi/2}/2 = i/2$ and $e^{-i\pi/2}/(-2) = i/2$. By the corollary of Proposition 11.4.2,

$$2I = \mathrm{Re}\left[\pi i\left(\frac{i}{2}+\frac{i}{2}\right)\right] = -\pi.$$

Thus, $I = -\pi/2$.

Proposition 11.4.3 *Let $R(z)$ be a rational function with a zero of multiplicity at least 2 at infinity. Let R have simple poles at $\beta_1, \beta_2, \ldots, \beta_\mu$ on the real axis, and no other poles on the real axis. Let $\zeta_1, \zeta_2, \ldots, \zeta_k$ be the set of poles of R in $\{z\colon \mathrm{Im}\, z > 0\}$. Then*

$$PV\int_{-\infty}^{\infty} R(x)\,dx = \pi i\sum_{j=1}^{\mu} \mathrm{Res}\,(R;\beta_j) + 2\pi i\sum_{k=1}^{\nu} \mathrm{Res}\,(R;\zeta_k).$$

The proof is an exercise.

Example
(iv) Evaluate

$$I = \frac{1}{\pi}PV\int_{-\infty}^{\infty} \frac{1}{x(x-a)}\,dx, \qquad a = \alpha + i\beta.$$

Solution. Let $R(z) = 1/[\pi z(z-a)]$. If $\beta = 0$: $\mathrm{Res}\,(R;0) = -1/\pi\alpha$, $\mathrm{Res}(R;\alpha) = 1/\pi\alpha$; $I = \pi i[(-1/\pi\alpha) + (1/\pi\alpha)] = 0$.
If $\beta > 0$; $\mathrm{Res}(R;\,0) = -1/\pi a$, $\mathrm{Res}(R;\,a) = 1/\pi a$; $I = \pi i(-1/\pi a) + 2\pi i(1/\pi a) = 1/a$.
If $\beta < 0$: the only pole in $\{z\colon \mathrm{Im}\, z \geq 0\}$ is at $z = 0$, $\mathrm{Res}(R;0) = -1/\pi a$. $I = \pi i(-1/\pi a) = -i/a$.
To summarize, if $a = \alpha + i\beta$, α, β real:

$$\frac{1}{\pi}PV\int_{-\infty}^{\infty} \frac{1}{x(x-a)}\,d = \begin{cases} -\dfrac{i}{a}, & \beta < 0; \\ 0, & \beta = 0; \\ \dfrac{i}{a}, & \beta > 0. \end{cases}$$

The integral is important in the theory of the Hilbert transform of a continuous function $f:\mathbf{R}\to\mathbf{R}$:

$$H(f;\, z) = \frac{1}{\pi} PV \int_{-\infty}^{\infty} \frac{f(x)}{z-x}\, dx.$$

Exercises 11.4

Remark. Exercises 10.5.1 and 10.5.2 are useful in calculating the residues in these problems.

1. Evaluate:

(a) $\displaystyle\int_0^{\pi/2} \frac{d\theta}{\alpha + \sin^2\theta}$, $\alpha > 0$. *Answer.* $\dfrac{\pi}{2\sqrt{\alpha^2+\alpha}}$.

(b) $\displaystyle\int_0^{\pi/2} \frac{d\theta}{\alpha^2\cos^2\theta + \beta^2\sin^2\theta}$, $\beta \geq \alpha > 0$. *Answer.* $\dfrac{\pi}{2\alpha\beta}$.

(*Hint*: Use (a).)

(c) $\displaystyle\int_0^{\pi} \frac{\sin^2\theta\, d\theta}{\alpha + \beta\sin\theta}$, $\alpha > \beta > 0$. *Answer.* $\dfrac{\pi}{\beta^2}(\alpha - \sqrt{\alpha^2-\beta^2})$.

(d) $\displaystyle\int_0^{2\pi} \frac{e^{in\theta}}{1 + 2r\cos\theta + r^2}$, $-1 < r < 1$, $n = 0,1,2,\ldots$.

Answer. $\dfrac{(-r)^n 2\pi}{1-r^2}$.

(e) $\displaystyle\int_0^{2\pi} (\sin\theta)^{2n}\, d\theta$. *Answer.* $\dfrac{2\pi(2n)!}{(2^n n!)^2}$.

2. Evaluate:

(a) $\displaystyle\int_0^{\infty} \frac{\cos x}{x^2+\alpha^2}\, dx$, α real. *Answer.* $\dfrac{\pi}{2|\alpha|} e^{-|\alpha|}$.

(b) $\displaystyle\int_0^{\infty} \frac{x\sin x}{(x^2+1)(x^2+4)}$. *Answer.* $\dfrac{(e-1)\pi}{6e^2}$.

(c) $\displaystyle\int_0^{\infty} \frac{\cos^2 x}{(x^2+1)^2}\, dx$. *Answer.* $\dfrac{\pi}{8}(1+3e^{-2})$.

3. Evaluate by residues:

(a) $\displaystyle\int_0^{\infty} \frac{x^2}{(x^2+1)(x^2+4)}\, dx$. *Answer.* $\dfrac{\pi}{6}$.

(b) $\displaystyle\int_0^\infty \frac{x^{2m}}{1+x^{2n}}\,dx$, $\qquad m, n$ integers, $0 \le m < n$.

Answer. $\dfrac{\pi}{2n\sin\left[\dfrac{(2m+1)\pi}{2n}\right]}$.

(c) $\displaystyle\int_{-\infty}^\infty \frac{dx}{(1+x^2)^n}$. $\qquad$ *Answer.* $\dfrac{1\cdot 3\cdot 5\cdots(2n-3)}{2\cdot 4\cdot 6\cdots(2n-2)}\pi$.

4. Prove Proposition 11.4.3.
5. Show that in Proposition 11.4.2(i) and (ii) the use of the principal value definition of the improper integrals is unnecessary. (*Hint*: Use a rectangular path of integration.)
6. Show that in Proposition 11.4.3 the use of the principal value definition of the improper integral is unnecessary when there are no poles on the real axis. (*Hint*: Use a rectangular path of integration.)
7. Let $f(z) = z^{-2m}\pi\cot\pi z$, with m a positive integer. Let

$$J(n) = \frac{1}{2\pi i}\int_{\gamma(n)} f(z)\,dz,$$

where $\gamma(n)$ is a positively oriented square path joining $n + (1/2) + i(n + (1/2))$ to $-n - (1/2) + i(n + (1/2))$ to $-n - (1/2) - i(n + (1/2))$ to $n + 1/2 - i(n + (1/2))$, and where n is a positive integer. Show that

$$J(n) = 2\sum_{k=1}^{n}\frac{1}{k^{2m}} + \text{Res}(f; 0);$$

show also that $\lim_{n\to\infty} J(n) = 0$, and that

$$\sum_{k=1}^{\infty}\frac{1}{k^{2m}} = -\frac{1}{2}\,\text{Res}(f; 0).$$

(*Hint*: In applying the residue theorem to a square path, see Sec. 8.4, Example(ii).)
8. Use the preceding exercise to show that $\sum_{k=1}^{\infty} k^{-2} = \pi^2/6$ and $\sum_{k=1}^{\infty} k^{-4} = \pi^4/900$. (*Hint*: Write the function $f(z)$ of Exercise 11.4.7 in the form $z^{-2m-1}\pi z\cot\pi z = z^{-2m-1}g(z)$; then g has a removable singularity at $z = 0$ and $g(0) \ne 0$. By Exercise 10.5.2, the desired residue is $g^{(2m)}(0)/(2m)!$ The coefficients of the Maclauren series for g are perhaps most easily found by applying the method of undertermined coefficients, using the known series for $\cos\pi z$ and $\sin\pi z/(\pi z)$.)

12

Approximation of Analytic Functions by Rational Functions and Generalizations of the Cauchy Theory

12.1 General Remarks

It is shown in Sec. 9.4 that a function f analytic in an open set O in $\mathbf{C}$ can be locally represented at each point $a \in O$ by a power series convergent in a neighborhood of a. This means that for each $a \in O$, there exists a sequence of polynomials (the partial sums of the local power series for f) which converges almost uniformly to $f(z)$ on some disk with center at a. But in general, the sequence of polynomials changes with a. The question on which this chapter is based is that of whether or not a single sequence of polynomials, or at least of rational functions, can be found which converges pointwise and almost uniformly to $f(z)$ in O.

If O is a region, the behavior of f throughout O is determined by its value and those of its derivatives at a single point in O, or by the values of f on a sequence of points in O having an accumulation point in O [Theorem 6.5.1 and corollaries]. But if O has two or more components, the behavior of f in any one of the components clearly is independent of the behavior of f in any other component. The question in the preceding paragraph is going to be answered in the affirmative in this chapter for approximation by rational functions. It is remarkable, therefore, that a function consisting of diverse "pieces" can be approximated pointwise in its domain by a single sequence of rational functions. The basic formulation of this approximation process is due to Carl Runge, who published it in 1885. Aside from the intrinsic interest in such a process, a further motivation for studying it here is to achieve thereby a wide generalization of parts of the Cauchy theory as presented in Chapters 9 and 11.

Even if O is only one region, a global almost uniform approximation by polynomials to a function analytic in O is generally impossible. For example, let $O = \Delta(0, 2) \sim \{0\}$ and $f(z) = 1/z$. If $\langle p_n \rangle$ is a sequence of polynomials converging a.u. to f on O, we would have $\lim \int_\gamma p_n(z)\,dz = 0$ for every closed path γ in O. However, if γ is a positively oriented circle with graph $C(0, 1)$, then $\int_\gamma f(z)\,dz = 2\pi i$. But by Proposition 8.3.1, the sequence of integrals of $\langle p_n \rangle$ along any closed path in O must converge to the integral of f. So a.u. approximation of $1/z$ by polynomials in this region O is impossible. Of course, $1/z$ is perfectly approximated on O by the rational function $1/z$.

The basic idea in Runge's original proof consists in representing the function f by the Cauchy integral formula and then approximating the integral by means of the sums which appear in the definition of the complex line integral given in Sec. 8.2. These sums in the case of the Cauchy integral formula assume the form of rational functions in the parameter z (see (9.4.1)). The idea is a simple one, but in direct application it involves a number of more or less hidden topological difficulties, which we surmount or bypass here by a somewhat roundabout approach.

Our first step, taken in the next section, consists of deriving a general Cauchy-type integral representation for a function analytic at each point of an arbitrary compact set. Recall from Sec. 6.2 that such a function is by definition analytic in a neighborhood of each point of its domain. The union of these neighborhoods is open and covers the domain, so we might as well start with the hypothesis that the function is analytic in an open set which contains the compact set in question. This hypothesis appears frequently in the sequel.

12.2 A Cauchy-type Integral Representation of a Function Analytic on a Compact Set

With $a \in \mathbf{C}$ and $h > 0$, let γ be a positively oriented closed square path with vertices a, $a + h$, $a + h + ih$, $a + ih$, to be taken in that order. In Example (ii) of Sec. 8.4 it is shown that the complement of the graph of γ is the union of a bounded region S (the *interior region* of γ) and an unbounded region U, and that $\operatorname{Ind}_\gamma(z) = 1$ for every $z \in S$. Of course, $\operatorname{Ind}_\gamma(z) = 0$ for every $z \in U$. Therefore, if f is a function analytic on a disk Δ containing S, by Theorem 9.4.1 we have the representation

$$\frac{1}{2\pi i}\int_\gamma \frac{f(\zeta)\,d\zeta}{\zeta - z} = \begin{cases} f(z), & z \in S. \\ 0, & z \in \bar{S}^c = U. \end{cases} \tag{12.2.1}$$

We call the closure $\bar{S}$ of the bounded region S determined by a closed square path (such as this γ) a *square*, and the line segments in γ are called the *edges* of $\bar{S}$. (Since $\bar{S}$ is the convex hull of the vertices, this nomenclature is consistent with that in Chapter 9, where the convex hull of three points was called a triangle.)

According to Exercise 5.6.3, diam (S) = diam $(\bar{S}) = h\sqrt{2}$. Notice that if $b \in \bar{S}$ and $z \in \bar{S}$, then $|z - b| \le h\sqrt{2}$. This implies that if $r > h\sqrt{2}$, then the disk $\Delta(b, r)$ contains $\bar{S}$.

With these preliminaries out of the way, we take up the result described in the title of this section.

Theorem 12.2.1 *Let $O \subset \mathbf{C}$ be a nonempty open set, and let F be a nonempty compact subset of O. There exists a finite set of line segments $\{\gamma_1, \gamma_2, \ldots, \gamma_n\}$, each parallel to either the real or the imaginary axis, and with graphs in $O \sim F$, such that for every function $f: O \to \mathbf{C}$ analytic on O,*

$$f(z) = \sum_{j=1}^{n} \frac{1}{2\pi i} \int_{\gamma_j} \frac{f(\zeta)\, d\zeta}{\zeta - z} \tag{12.2.2}$$

for every $z \in F$.

Proof. The paths of integration in this proof are all unions of line segments. We have been using the same notation for the arc and its graph in the case of line segments, and it causes no confusion to adopt this convention for the paths in this proof.

Let 2η = dist (F, O^c); by Theorem 5.6.1, $2\eta > 0$. Construct a grid of lines parallel to the real and imaginary axes so that the distance between adjacent lines is η. Let $\bar{S}_1, \bar{S}_2, \ldots, \bar{S}_m$ be those squares formed by the grid (each with edges of length η) which intersect F, and let $S_1, S_2, \ldots, S_m$ be the corresponding interior regions. There is certainly at least one such square. Notice that $F \subset \bigcup_k \bar{S}_k$.

Now diam $(\bar{S}_k) = \eta\sqrt{2} < 2\eta$, $k = 1, \ldots, m$. Take $b_k \in F \cap \bar{S}_k$; then $\bar{S}_k \subset \Delta(b_k, 2\eta)$, $k = 1, \ldots, m$. (See the remarks preceding the statement of the theorem.) Also, by definition of dist (F, O^c), $2\eta \le |z - w|$ for every $z \in F$ and $w \in O^c$ (see (5.6.1)), so each disk $\Delta(b_k, 2\eta)$, and, therefore, each $\bar{S}_k$ is contained in O.

Let z_k, $z_k + \eta$, $z_k + \eta + i\eta$, $z_k + i\eta$ denote the vertices of $\bar{S}_k$, and let $\gamma(S_k)$ be the positively oriented closed square path determined by these points. Since f is analytic in the disk $\Delta(b_k, 2\eta)$, by (12.2.1) we have

$$\frac{1}{2\pi i} \int_{\gamma(S_k)} \frac{f(\zeta)\, d\zeta}{\zeta - z} = \begin{Bmatrix} f(z), & z \in S_k \\ 0, & z \in \bar{S}_k^c \end{Bmatrix}, \qquad k = 1, 2, \ldots, m. \tag{12.2.3}$$

Thus, in particular, for a given k the value of this integral is zero when $z \in S_h$, $h \ne k$.

When any two of the $\bar{S}_k$'s have an edge in common, one of the line segments in $\bigcup_k \gamma(S_h)$ appears twice, but with opposite orientation. Discard the line segments which appear twice, and let $\gamma_1, \gamma_2, \ldots, \gamma_n$ denote the remaining ones. If an edge of $\bar{S}_k$ intersects F, then it is also the edge of an adjacent member of the set $\{\bar{S}_1, \bar{S}_2, \ldots, \bar{S}_m\}$. It therefore has been discarded in selecting $\gamma_1, \gamma_2, \ldots, \gamma_n$. Therefore, $\gamma_1 \cup \gamma_2 \cup \cdots \cup \gamma_n \subset O \sim F$. We now exploit the cancellations of integrals along discarded edges of squares and use (12.2.3) to obtain, for $z \in S_K$, $1 \leq K \leq m$,

$$\sum_{j=1}^{m} \frac{1}{2\pi i} \int_{\gamma_j} \frac{f(\zeta)\, d\zeta}{\zeta - z} = \sum_{k=1}^{m} \frac{1}{2\pi i} \int_{\gamma(S_k)} \frac{f(\zeta)\, d\zeta}{\zeta - z}$$

$$= \frac{1}{2\pi i} \int_{\gamma(S_K)} \frac{f(\zeta)\, d\zeta}{\zeta - z} = f(z). \tag{12.2.4}$$

Thus, (12.2.2) is valid for every $z \in \bigcup_k S_k$ and, in particular, for every $z \in F \cap [\bigcup_k S_k]$. Now take $z' \in F$, $z' \notin \bigcup_k S_k$. Then $z' \in \bigcup_k \bar{S}_k = \overline{\bigcup_k S_k}$. By Exercise 5.1.2, there exists a sequence $\langle z_N \rangle_1^\infty$ with $z_N \in \bigcup_k S_k$ and $\lim_{N\to\infty} z_N = z'$. Since $\bigcup_j \gamma_j \subset O \sim F$, $z' \notin \bigcup_j \gamma_j$. By Theorem 8.4.1, the integrals in (12.2.2) are each continuous at z', and so is their sum. Also f is continuous at z', and the equation (12.2.2) holds for each z_N, $N = 1, 2, \ldots$. Therefore, by Exercise 5.2.3, the equation persists at z'. Q.E.D.

It is important for certain applications of Theorem 12.2.1 to show that *the set of line segments* $\{\gamma_j, j = 1, 2, \ldots, n\}$ *appearing in the theorem can be chosen so that the line segments form a set of disjoint closed polygonal paths.* We develop this idea in two propositions which follow.* The proofs require some preliminary explanations.

Recall that each γ_j is one of the four line segments of a positively oriented square path, so this γ_j itself has a well-defined orientation and has an initial and a terminal point. Also each γ_j is an edge of a square in the special system of squares $S = \{\bar{S}_1, \bar{S}_2, \ldots, \bar{S}_m\}$ described in the proof of the theorem. Since each endpoint of any γ_j is a vertex of one or more of the squares in S, each endpoint must be the common endpoint of at least two edges belonging to a square, or to squares, in S. However, there is only one square in S having this γ_j among its edges, because if there had been two squares with the edge γ_j in common, this edge would have been discarded in the selection of the set $\{\gamma_j\}$. We henceforth call the members of $\{\gamma_j\}$ *boundary edges* of S, and their endpoints, *boundary vertices* of S.

The reader can assure himself by looking at the possible cases (see Fig. 1) that if q_j is, say, the terminal point of γ_j, then at least one of the other edges of squares in S on which q_j lies is itself a boundary edge with the *initial* point

†The discussion in the remainder of this section is not directly relevant to Runge's theorem and is referred to only in the last part of the proof of Theorem 12.8.1. Reading may, therefore, be postponed.

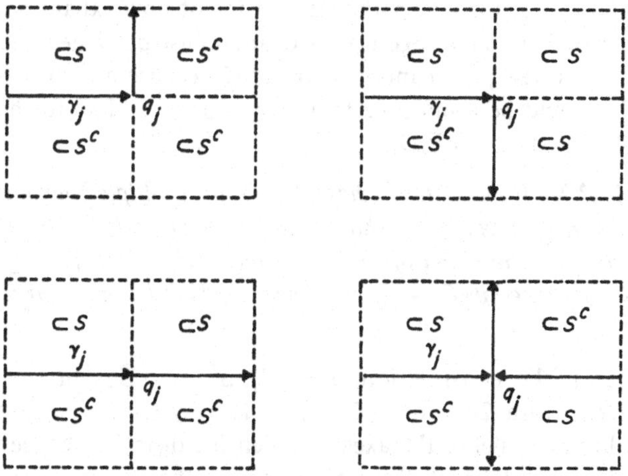

Fig. 1

q_j. (A more formal proof is given in [19, pp. 33–34].) Furthermore, the two possible cases are that (i) q_j is the common endpoint of only two boundary edges and (ii) q_j is the common endpoint of four boundary edges. In the latter case, q_j is simultaneously the initial point of two of the edges and the terminal point of two edges, one of which, of course, is γ_j. Similarly, if p_j is the initial point of γ_j, then either (i) it is the common endpoint of only two boundary edges and is the terminal point of the edge other than γ_j or else (ii) p_j is the common endpoint of four boundary edges. If a boundary vertex is the common endpoint of four boundary edges, we call it a *multiple boundary vertex* of the system S.

Proposition 12.2.1 *If the system S has no multiple boundary vertices, then the set of line segments $\{\gamma_j\}$ constructed in the proof of Theorem 12.2.1 are such that their union consists of a finite set of disjoint closed polygonal paths.*

Proof. Choose an arbitrary member of $\{\gamma_j\}$, say γ_J, and form a chain of boundary edges $G = \gamma_J + \gamma_{J_1} + \gamma_{J_2} + \cdots$ in which the terminal point of each is the initial point of its successor. There is only one possible choice of the next boundary edge at each boundary vertex because there are no multiple boundary vertices. If the terminal point q of a boundary edge in G were to coincide with an endpoint of a *preceding* boundary edge other than the initial point of γ_J, then q would also be the terminal point of one of the boundary edges in G and would, therefore, be a multiple boundary point. Since this is impossible by hypothesis, and since the set $\{\gamma_j\}$ is finite, eventually a terminal point of an edge in G must coincide with the initial point of γ_J.

Thus, G is a closed polygonal path. If $\{\gamma_j\}$ is not now exhausted, we start again with a new boundary edge not in G and construct a new closed polygonal path. This new path cannot intersect G because a common point of the two paths would be a multiple boundary vertex, and so forth, until the set $\{\gamma_j\}$ is exhausted. Q.E.D.

Proposition 12.2.2 *If the system S has one or more multiple boundary vertices, there is a system of squares S' which contains S and which has no multiple boundary vertices. The union of the set of boundary edges $\{\gamma'_1, \gamma'_2, \ldots, \gamma'_{n'}\}$ of S' is again contained in $O \sim F$, and Theorem 12.2.1 remains true with $\{\gamma'_j\}$ replacing $\{\gamma_j\}$.*

Proof. Recall the construction $S = \{\bar{S}_1, \bar{S}_2, \ldots, \bar{S}_m\}$ in the proof of Theorem 12.2.1: with $2\eta = \operatorname{dist}(F, O^c)$, S is a system of squares formed by a grid of lines parallel to the axes in which the distance between adjacent lines is η. The members of S were chosen by the requirement that each intersect F. Now if $b_k \in F \cap \bar{S}_k$, then $|z - b_k| \le \eta\sqrt{2}$ for every $z \in \bar{S}_k$, but $|b_k - w| \ge 2\eta$ for every $w \in O^c$. From these inequalities, it follows that $\operatorname{dist}(\bigcup_k \bar{S}_k, O^c) \ge (2 - \sqrt{2})\eta > \eta/2$. Thus, if σ is a square with edges of length $\eta/4$ and with a vertex contained in any square in S, then $\sigma \subset O$. (The argument is the same as that which established that $\bigcup_k \bar{S}_k \subset O$.) Now construct a refinement of the original grid of lines parallel to the axes which contains the old grid and in which the distance between adjacent lines is $\eta/4$. Each of the squares of S is subdivided into sixteen smaller squares by this grid. To this new system of smaller squares contained in the squares of S, adjoin each square formed by the new grid which contains a multiple boundary vertex of S. The resulting system of squares $S' = \{\bar{S}'_1, \bar{S}'_2, \ldots, \bar{S}'_{m'}\}$ contains no multiple boundary vertices, and $\bigcup_k \bar{S}'_k \subset O$. We select the new set of boundary edges $\{\gamma'_1, \gamma'_2, \ldots, \gamma'_{n'}\}$ of S' by discarding multiple edges in S', as in the proof of Theorem 12.2.1. Once again $\bigcup \gamma'_j \subset O \sim F$, because each γ'_j is either contained in $\{\gamma_j\}$ or belongs to a square in S' which is not contained in S. The calculation in (12.2.4) and the ensuing last part of the proof of Theorem 12.2.1 are valid with $\{\gamma'_j\}$ replacing $\{\gamma_j\}$ and S' replacing S. Q.E.D.

Exercises 12.2

In these exercises, O is an open set in $\mathbf{C}$, F is a nonempty compact subset of O, and $\gamma_1, \gamma_2, \ldots, \gamma_n$ are the line segments which appear in the conclusion of Theorem 12.2.1.

1. Show that if $g: O \to \mathbf{C}$ is analytic on O, then $\sum_{j=1}^{n} \int_{\gamma_j} g(\zeta)\, d\zeta = 0$. (*Hint*: Take any $z_0 \in F$; then $g(z)(z - z_0)$ is analytic on O and $g(z_0) = 0$. Substitute into (12.2.2).)
2. Show that

$$\sum_{j=1}^{n} \frac{1}{2\pi i} \int_{\gamma_j} \frac{d\zeta}{\zeta - z} = 1$$

for every $z \in F$.

3. Show that if m is a positive integer, $m \geq 2$, then

$$\sum_{j=1}^{n} \frac{1}{2\pi i} \int_{\gamma_j} \frac{d\zeta}{(\zeta - z)^m} = 0$$

for every $z \in F$. (*Hint*: According to Theorem 8.4.1, each of the integrals in the summation in Exercise 12.2.2 is an analytic function of z. Look at the formulas for the derivatives in (8.4.2).)

4. Let $P = \{\zeta_1, \zeta_2, \ldots, \zeta_\nu\}$ be a set of distinct points contained in F, and let $f: O \sim P \to \mathbf{C}$ be analytic and have a pole at ζ_k with residue $\mathrm{Res}(f; \zeta_k)$, $k = 1, \ldots, \nu$. Show that

$$\sum_{j=1}^{n} \frac{1}{2\pi i} \int_{\gamma_j} f(\zeta)\, d\zeta = \sum_{k=1}^{\nu} \mathrm{Res}(f; \zeta_k).$$

(*Hint*: Follow the proof of Theorem 11.1.1.)

5. Let $Z = \{z_1, z_2, \ldots, z_\mu\}$ and $P = \{\zeta_1, \zeta_2, \ldots, \zeta_\nu\}$ be contained in F, with $Z \cap P = \phi$. Let $f: B \sim P \to \mathbf{C}$ be analytic and have a pole at ζ_k of order p_k, $k = 1, \ldots, \nu$, and a zero at z_h of multiplicity m_k, $h = 1, \ldots, \mu$, and no other zeros in O. Show that

$$\sum_{j=1}^{n} \frac{1}{2\pi i} \int_{\gamma_j} \frac{f'(\zeta)}{f(\zeta)}\, d\zeta = \sum_{h=1}^{\mu} m_h - \sum_{k=1}^{\nu} p_k.$$

(*Hint*: Proof of Theorem 11.3.1.)

12.3 Approximation of a Rational Function by Another Rational Function

Our plan of attack on Runge's theorem involves the approximation of the integrals in (12.2.2) by rational functions with poles lying on a prescribed set. To carry out this program, two propositions covering a technique sometimes known as the "translation of poles" are needed. These are set forth in this section.

Proposition 12.3.1 *Let F be a closed set in $\mathbf{C}$, and let $\zeta_1 \in \mathbf{C} \sim F$ and $\zeta_2 \in \hat{\mathbf{C}} \sim F$ be given. Let R be a rational function with a pole at ζ_1 and no other poles in $\hat{\mathbf{C}}$.*

(i) *If* $\zeta_2 \neq \infty$ *and* $|\zeta_2 - \zeta_1| \leq (1/2)\,\mathrm{dist}\,(\zeta_2, F)$, *then for any* $\varepsilon > 0$ *there exists a rational function* Q_ε *with a pole at* ζ_2 *and no other pole in* $\hat{\mathbf{C}}$, *such that* $|R(z) - Q_\varepsilon(z)| < \varepsilon$ *for every* $z \in F$.
(ii) *If* $\zeta_2 = \infty$ *and if* $|z| \leq (1/2)|\zeta_1|$ *for every* $z \in F$, *then there exists a polynomial* Q_ε *such that* $|R(z) - Q_\varepsilon(z)| < \varepsilon$ *for every* $z \in F$.

Proof. We first prove the proposition for the special case in which $R(z) = c/(z - \zeta_1)^k$, where $c \in \mathbf{C}$ and k is a positive integer.

To establish part (i) in this case, we proceed as follows. We have the identity

$$\frac{c}{(z-\zeta_1)^k} = \frac{c}{(z-\zeta_2)^k[1-(\zeta_1-\zeta_2)/(z-\zeta_2)]^k} = \frac{c}{(z-\zeta_2)^k}\left[1 + \frac{\zeta_2-\zeta_1}{z-\zeta_2}\right]^{-k}. \tag{12.3.1}$$

Recall that according to Exercise 7.7.3, if $w \in \mathbf{C}$, $a \in \mathbf{C}$, and the principal logarithm is used in the determination of $(1 + w)^a$, then the Maclaurin series for $(1 + w)^a$ converges almost uniformly to this function on $\Delta(0, 1)$, and takes the form of the so-called binomial series:

$$(1+w)^a = 1 + \sum_{j=1}^{\infty} \binom{a}{j} w^j; \qquad \binom{a}{j} = \frac{a(a-1)\cdots(a-j+1)}{j!}.$$

In particular, it converges uniformly on $\bar{\Delta}(0, 1/2)$. Now by hypothesis, $|\zeta_2 - \zeta_1| \leq (1/2)|z - \zeta_2|$ for every $z \in F$. Therefore, for $z \in F$, we can represent the left-hand member of (12.3.1) with a binomial series as follows:

$$\begin{aligned}\frac{c}{(z-\zeta_1)^k} &= \frac{c}{(z-\zeta_2)^k}\left[1 + \sum_{j=1}^{\infty}\binom{-k}{j}\left(\frac{\zeta_2-\zeta_1}{z-\zeta_2}\right)^j\right]\\ &= \frac{c}{(z-\zeta_2)^k} + \sum_{j=1}^{\infty}\binom{-k}{j}\frac{(\zeta_2-\zeta_1)^j}{(z-\zeta_2)^{j+k}},\end{aligned}$$

and the series converges uniformly for $z \in F$.

Now take $\varepsilon > 0$. By the definition of uniform convergence, there exists N_ε such that for every $z \in F$,

$$\left|\frac{c}{(z-\zeta_1)^k} - \left[\frac{c}{(z-\zeta_2)^k} + \sum_{j=1}^{N_\varepsilon}\binom{-k}{j}\left(\frac{(\zeta_2-\zeta_1)^j}{(z-\zeta_2)^{k+j}}\right)\right]\right| < \varepsilon. \tag{12.3.2}$$

The rational function $Q_\varepsilon(z)$ promised in part (i) of the proposition for this special R is the function in the square brackets in (12.3.2).

To prove part (ii) for the special case in which $R(z) = c/(z - \zeta_1)^k$, we use the hypothesis $|z| \leq (1/2)|\zeta_1|$ for every $z \in F$, together with an obvious identity, to obtain another binomial series representation as follows:

$$\begin{aligned}\frac{c}{(z-\zeta_1)^k} &= \frac{(-1)^k c}{\zeta_1^k(1-z/\zeta_1)^k}\\ &= \frac{(-1)^k c}{\zeta_1^k}\left[1+\sum_{j=1}^{\infty}\binom{-k}{j}\left(-\frac{z}{\zeta_1}\right)^j\right]\\ &= \frac{(-1)^k c}{\zeta_1^k}+\sum_{j=1}^{\infty}\binom{-k}{j}(-1)^{k+j}c\frac{z^j}{\zeta_1^{k+j}}.\end{aligned} \tag{12.3.3}$$

This series converges uniformly for $z \in F$. As in the proof of part (i), given any $\varepsilon > 0$, we let the polynomial $Q_\varepsilon(z)$ promised in the proposition be a suitably chosen partial sum of the series which appears in the right-hand member of (12.3.3).

The extension to a more general rational function R with a single pole in $\mathbf{C}$ at ζ_1 is immediate. Such a function R has a partial fraction decomposition (11.2.1) of the form

$$R(z) = c + \sum_{k=1}^{p} \frac{c_k}{(z-\zeta_1)^k},$$

where c is a complex constant which may be zero. Let Q_k be a rational function (part (i)) or a polynomial (part (ii)) such that for every $z \in F$,

$$\left|\frac{c}{(z-\zeta_1)^k} - Q_k(z)\right| < \frac{\varepsilon}{p}, \qquad k = 1, 2, \ldots, n.$$

Then let $Q_\varepsilon(z) = c + \sum_{k=1}^{p} Q_k(z)$. Clearly, by the triangle inequality, $|R(z) - Q_\varepsilon(z)| < p(\varepsilon/p) = \varepsilon$. Q.E.D.

In our further discussion of the translation of poles, questions arise relating to the connectivity of subsets of the extended plane $\hat{\mathbf{C}}$. Since we have not metricized $\hat{\mathbf{C}}$, and since in this book the concept of connectedness has heretofore been placed in a setting of metric spaces, a digression on elementary topology seems to be in order here. By reviewing Sec. 5.3, the reader can assure himself that the definition of a connected set in a space X, and the properties of such a set (insofar as they do not depend on an algebraic structure imposed of X), depend only on the concept of an *open set* in X. If X is equipped with a metric, the metric furnishes a convenient way to define a family of open sets with all the properties that open sets intuitively are supposed to have. But these properties can easily be divorced from the metric space concept, and the result is the more general concept of a *topological space*. Specifically, a topological space is an ordered pair (X, Φ) in which X is a nonempty space and Φ is a family of subsets of X which contains $\varnothing$ and X, and which also contains the intersection of any finite collection of members of Φ and the union of any finite or infinite collection of members of Φ. The family Φ is said to be a *topology* in (or on) X and the members of Φ are arbitrarily declared to be the *open sets* in X. A *closed set* in a topological space is then defined to be the complement with respect to X of an open set. If E is any nonempty subset of X, the family of sets $\{O \cap E: O \in \Phi\}$ constitutes a topology which is called the *relative topology* in (or on) E. The family of open intervals in $\mathbf{R}$ is the "usual" or "standard" topology in $\mathbf{R}$, or just "the" topology in $\mathbf{R}$. The family of open sets in $\mathbf{C}$ as previously defined via open disks is the "usual" topology or "the" topology in $\mathbf{C}$.

A *separation* of a subset E of a topological space is a pair of disjoint nonempty sets O_1, O_2 belonging to the relative topology on E and such that $E = O_1 \cup O_2$. The set E is *connected* if and only if there is no such separation. A *component* of a subset A of a topological space is a maximal connected subset of A.

In Sec. 10.7, a certain family of open sets in the extended complex plan $\hat{\mathbf{C}}$ was introduced. It consists of the usual topology in the finite complex plane $\mathbf{C}$ augmented by a collection of sets, each of which consists of the union of a set in the usual topology in $\mathbf{C}$ and some disk $\Delta(\infty, r)$ with center at ∞. As indicated in that section, it is easily verified that this is a topology in $\hat{\mathbf{C}}$; it is the "usual" topology, or "the" topology. The only topologies in $\mathbf{R}$, $\mathbf{C}$, and $\hat{\mathbf{C}}$ which are used in this text are the usual topologies. It is customary to employ abbreviations when the usual topology is implied, such as "O is open in $\mathbf{C}$", "O is open in $\hat{\mathbf{C}}$." Notice that if O is open in $\hat{\mathbf{C}}$ and contains ∞, then O necessarily contains a disk $\Delta(\infty, r)$, and $O \sim \{\infty\}$ is an open set in $\mathbf{C}$.

Now let A be a subset of $\hat{\mathbf{C}}$ with $\infty \in A$. Then we claim that *the components of A in $\hat{\mathbf{C}}$ consist of the bounded components of A in $\mathbf{C}$ and a set A_∞ which is the union of $\{\infty\}$ and all the unbounded components of A in $\mathbf{C}$.* To see that the unbounded component A_∞ of A in $\hat{\mathbf{C}}$ is correctly defined here, we have only to show that it is connected in $\hat{\mathbf{C}}$. Suppose O_1 and O_2 are open sets in $\hat{\mathbf{C}}$ such that $O_1 \cap A_\infty$ and $O_2 \cap A_\infty$ is a separation of A_∞. One, but only one, of the sets O_1 and O_2 must contain ∞; say it is O_1. Then O_2 is open in $\mathbf{C}$, and O_1 contains some disk $\Delta(\infty, r)$, $r > 0$. Now $O_2 \cap A_\infty \subset \hat{\mathbf{C}} \sim \Delta(\infty, r)$, so $O_2 \cap A_\infty$ is bounded. An unbounded component of A in $\mathbf{C}$ must be contained entirely either in $O_1 \sim \{\infty\}$ or in O_2, because otherwise we would have a separation in $\mathbf{C}$ of this unbounded component. But such an unbounded component cannot be a subset of the *bounded* set $O_2 \cap A_\infty$, so it must be contained in O_1. Therefore, $O_2 \cap A_\infty$ is empty, so no such separation of A_∞ in $\hat{\mathbf{C}}$ exists. Thus, by definition, A_∞ is connected in $\mathbf{C}$.

In Sec. 11.2, we defined a *simply connected region* $B \subset \mathbf{C}$ to be a region B such that either every component of $\mathbf{C} \sim B$ is unbounded or $B = \mathbf{C}$. It is clear that an equivalent definition in the present terminology is that a region

$B \subset \mathbf{C}$ is simply connected in $\mathbf{C}$ if and only if $\hat{\mathbf{C}} \sim B$ has *only one component* in $\hat{\mathbf{C}}$, or in other words, $\hat{\mathbf{C}} \sim B$ is connected in $\hat{\mathbf{C}}$.

Proposition 12.3.2 *Let F be a closed subset of $\mathbf{C}$, and let $\zeta \in \mathbf{C} \sim F$, $\zeta' \in \hat{\mathbf{C}} \sim F$ be given with ζ and ζ' in the same component of $\hat{\mathbf{C}} \sim F$. Let R be a rational function with its only pole in $\hat{\mathbf{C}}$ at ζ.*

(i) *If $\zeta' \neq \infty$ and ζ, ζ' are contained in the same component of $\mathbf{C} \sim F$, then for any $\varepsilon > 0$ there exists a rational function Q_ε with its only pole in $\hat{\mathbf{C}}$ at ζ', such that $|R(z) - Q_\varepsilon(z)| < \varepsilon$ for every $z \in F$.*
(ii) *If $\zeta' = \infty$ and F is compact, then for any $\varepsilon > 0$ there exists a polynomial Q_ε such that $|R(z) - Q_\varepsilon(z)| < \varepsilon$ for every $z \in F$.*

Proof. (i) The component B of $\mathbf{C} \sim F$ in which ζ and ζ' lie is an open connected set in $\mathbf{C}$ [Corollary 2 of Theorem 5.4.1], and so it is arcwise connected by polygonal arcs [Theorem 5.3.2]. Let $L \subset B$ be the graph of a polygonal arc joining ζ to ζ', and let $d = \text{dist}(L, F)$. By Theorem 5.6.1, $d > 0$. Let $\{\zeta = z_0, z_1, \ldots, z_n = \zeta'\}$ be a finite subset of the points of L chosen so that $|z_k - z_{k-1}| \leq d/2$, $k = 1, 2, \ldots, n$. For each pair z_{k-1}, z_k, we have the inequality $|z_k - z_{k-1}| \leq (1/2)\,\text{dist}(z_k, F)$, so z_{k-1} and z_k satisfy the hypotheses of Proposition 12.3.1(i) with $\zeta_1 = z_{k-1}$ and $\zeta_2 = z_k$. Thus, for any $\varepsilon > 0$, there exists a chain of rational functions $Q_1(z)$ (only pole in $\hat{\mathbf{C}}$ at z_1), $Q_2(z)$ (only pole in $\hat{\mathbf{C}}$ at z_2), $\ldots$, $Q_n(z)$ (only pole in $\hat{\mathbf{C}}$ at $z_n = \zeta'$) such that for every $z \in F$, $|R(z) - Q_1(z)| < \varepsilon/n$, $|Q_1(z) - Q_2(z)| < \varepsilon/n, \ldots,$ $|Q_{n-1}(z) - Q_n(z)| < \varepsilon/n$. We add these inequalities and obtain $|R(z) - Q_n(z)| < \varepsilon$ for every $z \in F$. This proves part (i) with $Q_n = Q_\varepsilon$.

(ii) If $\zeta' = \infty$, then by hypothesis ζ lies in an unbounded component of of $\mathbf{C} \sim F$. By Proposition 5.4.2, there is only one unbounded component of $\mathbf{C} \sim F$, and if r is such that $F \subset \bar{\Delta}(0, r)$, this unbounded component contains the punctured disk $\Delta'(\infty, r) = \Delta(\infty, r) \sim \{\infty\}$. Choose $\zeta'' \in \Delta'(\infty, 2r) \subset \Delta'(\infty, r)$. Then $|z| \leq (1/2)|\zeta''|$ for every $z \in F$. By part (i) of this proposition, for any $\varepsilon > 0$, there exists a rational function Q with its only pole in $\hat{\mathbf{C}}$ at ζ'', such that $|R(z) - Q(z)| < \varepsilon/2$ for every $z \in F$. Then by Proposition 12.3.1(ii), there exists a polynomial Q_ε such that $|Q(z) - Q_\varepsilon(z)| < \varepsilon/2$ for every $z \in F$. Adding these inequalities, we find that $|R(z) - Q_\varepsilon(z)| < \varepsilon$ for every $z \in F$. Q.E.D.

Exercise 12.3

Let the set $E \subset \hat{\mathbf{C}}$ be connected, and let $f: E \to \hat{\mathbf{C}}$ be continuous on E in the extended sense described in Sec. 10.7. Prove that $f(E)$ is connected in $\hat{\mathbf{C}}$. (*Hint*: Imitate the proof of Theorem 5.3.1.)

12.4 Approximation of a Function Analytic on a Compact Set by a Rational Function with Poles on a Prescribed Set

We now use the integral representation (12.2.2) to obtain a rational function approximation to an arbitrary analytic function with a compact domain. We first establish a result on approximation of Cauchy-type integrals, which was the basis of Runge's proof of the rational function approximation and which has been used in various explicit constructions of rational function approximations in the literature.

Proposition 12.4.1 *Let γ be a path with graph Γ and parametric equation $z = \gamma(t)$, $\alpha \le t \le \beta$. Let $f\colon \Gamma \to \mathbf{C}$ be continuous on Γ, and let F be a closed set such that $F \cap \Gamma = \varnothing$. Let $\langle \mathscr{P}_n \rangle_1^\infty$ be a sequence of partitions of $[\alpha, \beta]$ in which $\mathscr{P}_n$ has the form $\mathscr{P}_n = \{\alpha = t_{n0} < t_{n1} < t_{n2} < \cdots < t_{nn} = \beta\}$. Let $\zeta_{nk} = \gamma(t_{nk})$, $k = 1, \ldots, n$, and let $\|\mathscr{P}_n\| = \max\limits_k (t_{nk} - t_{n,k-1})$. Finally, let R_n be the rational function with poles only on Γ given by*

$$R_n(z) = \sum_{k=1}^{n} \frac{f(\zeta_{nk})(\zeta_{nk} - \zeta_{n,k-1})}{\zeta_{nk} - z}. \tag{12.4.1}$$

Then if $\lim\limits_{n\to\infty} \|\mathscr{P}_n\| = 0$, *it follows that*

$$\lim_{n\to\infty} R_n(z) = \int_\gamma \frac{f(\zeta)\, d\zeta}{\zeta - z}$$

uniformly for $z \in F$.

Proof. Without the uniformity condition, the theorem would follow at once from Theorem 8.3.2 and the definition of the complex line integral given in Sec. 8.2. However, the uniformity is important in the applications, and it seems to be worthwhile to give the technical details.

Let $\gamma_{nk} = \gamma|[t_{n,k-1}, t_{nk}]$. In the computations which follow, for typographical simplicity we drop the first subscript in t_{nk}, ζ_{nk}, γ_{nk}. With $z \in F$ fixed, we have, using (8.3.3),

$$\begin{aligned} \left| \int_\gamma \frac{f(\zeta)\, d\zeta}{\zeta - z} - R_n(z) \right| &= \left| \sum_{k=1}^{n} \int_{\gamma_k} \left[\frac{f(\zeta)}{\zeta - z} - \frac{f(\zeta_k)}{\zeta_k - z} \right] d\zeta \right| \\ &\le \sum_{k=1}^{n} M_k(z) L(\gamma_k), \end{aligned} \tag{12.4.2}$$

where $L(\gamma_k) = \int_{t_{n-1}}^{t_n} |\gamma'(t)|\, dt$ and

$$M_k = M_{nk}(z) = \max_{t_{k-1} \le t \le t_k} \left| \frac{f(\gamma(t))}{\gamma(t) - z} - \frac{f(\gamma(t_k))}{\gamma(t_k) - z} \right|, \quad k = 1, 2, \ldots, n. \tag{12.4.3}$$

The maximum $M_k(z)$ exists because the function in the absolute value signs in (12.4.3) is clearly a continuous function of t on the closed interval $[t_{k-1}, t_k]$.

If $f(z)$ is identically zero on Γ, the proposition is trivially true. We then assume that $\max_{z\in\Gamma} |f(z)| = \mu > 0$. Let $d = \text{dist}(\Gamma, F)$. By Theorem 5.6.1, $d > 0$. The functions γ and $f \circ \gamma$ are *uniformly* continuous on $[\alpha, \beta]$ [Theorem 5.5.3]. Now take $\varepsilon > 0$. Let $L(\gamma) = \int_\alpha^\beta |\gamma'(t)|\, dt$, and let $\delta_\varepsilon > 0$ be chosen so that for every t', $t \in [\alpha, \beta]$ with $|t' - t| < \delta_\varepsilon$, we have the inequalities

$$|\gamma(t') - \gamma(t)| < \frac{\delta^2 \varepsilon}{2\mu L(\gamma)},$$

$$|f[\gamma(t')] - f[\gamma(t)]| < \frac{d\varepsilon}{2L(\gamma)}.$$

By hypothesis, there exists $N > 0$ such that for every $n \geq N$, $\|\mathscr{P}_n\| < \delta_\varepsilon$. Then for every $t \in [t_{k-1}, t_k]$, every $z \in F$, and $n \geq N$, we have

$$\begin{aligned}
&\left|\frac{f[\gamma(t)]}{\gamma(t) - z} - \frac{f[\gamma(t_k)]}{\gamma(t_k) - z}\right| \\
&\quad = \left| f[\gamma(t)]\left[\frac{1}{\gamma(t) - z} - \frac{1}{\gamma(t_k) - z}\right] + \frac{1}{\gamma(t_k) - z}[f[\gamma(t)] - f[\gamma(t_k)]]\right| \\
&\quad \leq \left| f[\gamma(t)]\frac{\gamma(t_k) - \gamma(t)}{(\gamma(t) - z)(\gamma(t_k) - z)}\right| + \left|\frac{1}{\gamma(t_k) - z}[f[\gamma(t)] - f[\gamma(t_k)]]\right| \\
&\quad < \mu \cdot \frac{d^2\varepsilon}{2\mu L(\gamma)} \cdot \frac{1}{d^2} + \frac{1}{d} \cdot \frac{d\varepsilon}{2L(\gamma)} = \frac{\varepsilon}{L(\gamma)}.
\end{aligned}$$

Therefore, if $n \geq N$, we have $M_k(z) \leq \varepsilon / L(\gamma)$, $k = 1, 2, \ldots, n$, for every $z \in F$. We return to (12.4.2) with this information and find that the right-hand member of (12.4.2) satisfies the inequality

$$\sum_{k=1}^{n} M_k(z) L(\gamma_k) \leq \frac{\varepsilon}{L(\gamma)} \sum_{k=1}^{n} L(\gamma_k) = \frac{\varepsilon}{L(\gamma)} \cdot L(\gamma) = \varepsilon$$

for every $z \in F$. Q.E.D.

Theorem 12.4.1 *Let $O \subset \mathbf{C}$ be an open set and F a nonempty compact subset of O. Let the function $f\colon O \to \mathbf{C}$ be analytic on O. Let $P \subset \hat{\mathbf{C}}$ be a set which contains ∞ and at least one point of each component of $\hat{\mathbf{C}} \sim F$. For any $\varepsilon > 0$, there exists a rational function R_ε with all its poles in P, such that $|f(z) - R_\varepsilon(z)| < \varepsilon$ for every $z \in F$.*

Proof. Take $\varepsilon > 0$. By Theorem 12.2.1, there exist line segments γ_1, $\gamma_2, \ldots, \gamma_n$ contained in $O \sim F$ such that

$$f(z) = \sum_{j=1}^{n} \frac{1}{2\pi i} \int_{\gamma_j} \frac{f(\zeta)\, d\zeta}{\zeta - z}$$

for every $z \in F$. According to Proposition 12.4.1, for each j there exists a rational function R_{n_j} with the following properties (see (12.4.1)): $R_{n_j} = r_{j1} + r_{j2} + \cdots + r_{jn_j}$, where r_{jk} has only one pole ζ_{jk} in $\hat{\mathbf{C}}$ and that pole lies on γ_j, $k = 1, 2, \ldots, n_j$; and

$$D_j = \left| \frac{1}{2\pi i} \int_{\gamma_j} \frac{f(\zeta)\, d\zeta}{\zeta - z} - \sum_{k=1}^{n} r_{jh}(z) \right| < \frac{\varepsilon}{2n}$$

for every $z \in F$. (The function f appearing in Proposition 12.4.1 in this application is $f/2\pi i$.) Now ζ_{jk} must lie in some component of $\hat{\mathbf{C}} \sim F$. This component contains at least one point of P, which we here call z_{jk}. If ζ_{jk} lies in the unbounded component of $\hat{\mathbf{C}} \sim F$ and there is no point of P in the component of $\mathbf{C} \sim F$ which contains ζ_{jk}, then it must follow that $z_{jk} = \infty$. (It is to be understood that in so designating points of P by a double-subscript notation, different pairs of subscripts may refer to the same point of P.)

In any case, by Proposition 12.3.2, there exists a rational function $Q_{jh}(z)$ with its only pole in $\hat{\mathbf{C}}$ at z_{jk}, such that $|r_{jk}(z) - Q_{jk}(z)| < \varepsilon/(2nn_j)$ for every $z \in F$. Let $Q_j = \sum_{k=1}^{n_j} Q_{jk}$. Then for each j, $j = 1, \ldots, n$,

$$\begin{aligned} &\left| \frac{1}{2\pi i} \int_{\gamma_j} \frac{f(\zeta)\, d\zeta}{\zeta - z} - Q_j(z) \right| \\ &\qquad \leq \left| \frac{1}{2\pi i} \int_{\gamma_j} \frac{f(\zeta)\, d\zeta}{\zeta - z} - \sum_{k=1}^{n_j} r_{jk}(z) \right| + \left| \sum_{k=1}^{n_j} [r_{jk}(z) - Q_{jk}(z)] \right| \\ &\qquad < D_j + \frac{n_j \varepsilon}{2nn_j} < \frac{\varepsilon}{2n} + \frac{\varepsilon}{2n} = \frac{\varepsilon}{n}. \end{aligned}$$

Now, finally, let $R_\varepsilon(z) = \sum_{j=1}^{n} Q_j(z)$. Then

$$|f(z) - R_\varepsilon(z)| = \left| \sum_{j=1}^{n} \left[\frac{1}{2\pi i} \int_{\gamma_j} \frac{f(\zeta)\, d\zeta}{\zeta - z} - Q_j(z) \right] \right| < \frac{n\varepsilon}{n} = \varepsilon.$$

Q.E.D.

Corollary 1 If in the hypotheses of Theorem 12.4.1 it is further specified that $\hat{\mathbf{C}} \sim F$ is connected in $\hat{\mathbf{C}}$, then $R_\varepsilon(z)$ in the conclusion can be taken to be a polynomial.

Proof. The connectivity of $\hat{\mathbf{C}} \sim F$ implies that there is only one component of $\hat{\mathbf{C}} \sim F$, and the boundedness of F implies that this component of $\hat{\mathbf{C}} \sim F$ contains ∞. Thus each of the poles z_{jk} in the proof of the theorem can be taken to be ∞, and then, by Theorem 10.7.2, R_ε is a polynomial. Q.E.D.

Corollary 2 The hypothesis that P contains ∞ can be replaced in Theorem 12.4.1 by the hypothesis that P does not contain ∞ but does contain one point of each component of $\mathbf{C} \sim F$. The approximating R_ε is then not a polynomial and has a simple zero at ∞.

Proof. This is again obvious from the proof of the theorem, taken with Proposition 12.3.2(i).

Exercises 12.4

1. Let $f: C(0, 1) \to \mathbf{C}$ be continuous, let $\zeta_{nk} = e^{2\pi ik/n}$, $k = 1, \ldots, n$, $n = 1, 2, \ldots$, and let

$$R_n(z) = \frac{1}{n} \sum_{k=1}^{n} \frac{f(\zeta_{kn})}{\zeta_{kn} - z}.$$

Show that

$$\lim_{n \to \infty} R_n(z) = \frac{1}{2\pi i} \int_\gamma \frac{f(\zeta) d\zeta}{\zeta(\zeta - z)},$$

almost uniformly for $z \in [C(0, 1)]^c$, where γ is the positively oriented circle with graph $C(0, 1)$. Thereafter, use residues to show that if f is analytic on a disk $\Delta(0, R)$, $R > 1$, then

$$\lim_{n \to \infty} R_n(z) = \begin{cases} \dfrac{f(z) - f(0)}{z}, & z \in \Delta(0, 1) \sim \{0\}; \\ f'(0), & z = 0; \\ -\dfrac{f(0)}{z}, & 1 < |z| < R. \end{cases}$$

(*Hint*: R_n can be thrown into the form (12.4.1) by multiplying and dividing the summand by $\zeta_{nk} - \zeta_{n,k-1} = \zeta_{nk}(1 - e^{-2\pi i/n})$. It is easy to show that $\lim_{n \to \infty} n(1 - e^{-2\pi i/n}) = 2\pi i$.)

2. Let $f: \bar{\Delta}(0, 1) \to \mathbf{C}$ be continuous, and also analytic on $\Delta(0, 1)$. With ζ_{nk} as in Exercise 12.4.1, let

$$Q_n(z) = \frac{1}{n} \sum_{k=1}^{n} \frac{\zeta_{kn} f(\zeta_{kn})}{\zeta_{kn} - z}.$$

Show that $\lim_{n \to \infty} Q_n(z) = f(z)$ a.u. on $\Delta(0, 1)$. (*Hint*: Exercise 12.4.1 and Proposition 9.4.1.)

3. Let $\zeta_1, \zeta_2, \ldots, \zeta_n$ be distinct points in $\mathbf{C}$, and let $\omega(z) = (z - \zeta_1)(z - \zeta_2)$

$\cdots (z - \zeta_n)$. Define functions $\lambda_k\colon \mathbf{C} \to \mathbf{C}$, $k = 1,2, \ldots, n$, as follows:

$$\lambda_k(z) = \begin{cases} \dfrac{\omega(z)}{(z - \zeta_k)\omega'(\zeta_k)}, & z \neq \zeta_k; \\ 1, & z = \zeta_k. \end{cases}$$

Show that $\lambda_k(z)$ is a polynomial of degree $n - 1$. Then for given complex numbers $w_1, w_2, \ldots, w_k$, show that $L(z) = w_1\lambda_1(z) + w_2\lambda_2(z) + \cdots + w_n\lambda_n(z)$ is the unique polynomial of degree less than n which assumes the value w_k at $z = \zeta_k$, $k = 1, 2, \ldots, n$. Also show that if $P(z)$ is any polynomial of degree less than n, then $P(z) = P(\zeta_1)\lambda_1(z) + P(\zeta_2)\lambda_2(z) + \cdots + P(\zeta_n)\lambda_n(z)$ for every $z \in \mathbf{C}$. (The polynomial $L(z)$ is known as the Lagrange interpolation polynomial for the points ζ_k and values w_k. *Hint*: For the uniqueness assertion, the fundamental theorem of algebra can be used.)

4. (Fejér, Walsh.) Let $f\colon C(0, 1) \to \mathbf{C}$ be continuous, let $\zeta_{nk} = e^{2\pi ik/n}$, $k = 1, 2, \ldots, n$, $n = 1, 2, \ldots$. Let $L_n(z)$ be the Lagrange interpolation polynomial (see Exercise 12.4.3) determined by the conditions $L_n(\zeta_{nk}) = f(\zeta_{nk})$, $k = 1, 2, \ldots, n$. Prove that

$$\lim_{n\to\infty} L_n(z) = \frac{1}{2\pi i}\int_\gamma \frac{f(\zeta)\,d\zeta}{\zeta - z}$$

almost uniformly on $\Delta(0, 1)$, where γ is the positively oriented circle with graph $C(0, 1)$. Thereafter, show that if f is continuous on $\bar{\Delta}(0, 1)$ and analytic on $\Delta(0, 1)$, then $\lim_{n\to\infty} L_n(z) = f(z)$ a.u. on $\Delta(0, 1)$. (*Hint*: From Exercise 12.4.3, we see that the general formula for the Lagrange interpolation polynomial which assumes respective values $f(\zeta_k)$ at n distinct points $\zeta_k \in \mathbf{C}$, $k = 1, 2, \ldots, n$, is

$$L(z) = L_n(z) = \sum_{k=1}^{n} f(\zeta_k)\frac{\omega_n(z)}{(z - \zeta_k)\omega_n'(\zeta_k)}, \tag{12.4.4}$$

where $z = \zeta_k$, $k = 1, \ldots n$, and $\omega_n(z) = (z - \zeta_1)(z - \zeta_2)\cdots(z - \zeta_n)$. In this case, the numbers $\zeta_k = \zeta_{nk}$ are the n roots of the equation $z^n - 1 = 0$, so $\omega_n(z) = z^n - 1$ and $\omega_n'(\zeta_{nk}) = n\zeta_{nk}^{n-1} = n\zeta_{nk}^{-1}$. Substitute into (12.4.4) and use Exercise 12.4.1 and the fact that $\lim_{n\to\infty} z^n = 1$ a.u. on $\Delta(0, 1)$. For the last sentence, refer to Proposition 9.4.1.)

5. Show that there exists a sequence of polynomials $\langle P_n(z)\rangle_1^\infty$ (where n is not necessarily the degree of P_n), such that $\lim_{n\to\infty} P_n(z) = 0$ for every $z \neq 0$, and $\lim_{n\to\infty} P_n(0) = \infty$. (*Hint*: In Corollary 1 of Theorem 12.4.1, let $f(z) = n/z$ and $F = F_n = \bar{\Delta}(0, n) \sim [\Delta(0, 1/n) \cup \{z\colon \operatorname{Re} z > 0, 0 < \operatorname{Im} z < 1/n\}]$. Show that according to the corollary, for each n there exists a polynomial Q_n such that $|(n/z) - Q_n(z)| < 1/n$ for every $z \in F_n$. Then look at the sequence $\langle n - zQ_n(z)\rangle_1^\infty$. This construction was suggested by P.J. O'Hara, Jr.)

12.5 Almost Uniform Approximation of a Function Analytic on an Open Set by Rational Functions

The theorem in this section has been the goal of previous work in this chapter. We need the following elementary topological proposition.

Proposition 12.5.1 *Let O be a nonempty open subset of* $\mathbf{C}$. *There exists a sequence $\langle K_n \rangle_0^\infty$ of compact sets such that:*

(i) *If F is a compact subset of O, then for some integer $N = N(F)$, $F \subset K_n$, $n = N, N+1, \ldots$;*
(ii) $O = \bigcup_n K_n$;
(iii) *K_n is contained in the interior of K_{n+1}, $n = 1, 2, \ldots$;*
(iv) *Each component of $\hat{\mathbf{C}} \sim K_n$ contains a component of $\hat{\mathbf{C}} \sim O$, and the unbounded component of $\hat{\mathbf{C}} \sim K_n$ contains the unbounded component of $\hat{\mathbf{C}} \sim O$, $n = 1, 2, \ldots$.*

Proof. If $O = \mathbf{C}$, then obviously we can let $K_n = \bar{\Delta}(0, n)$ and the proposition becomes trivial. Henceforth, we assume that $O \neq \mathbf{C}$.

Let $O^c = \mathbf{C} \sim O$, and for each positive integer n, let $V_n = \Delta(\infty, n) \cup [\bigcup_{z \in O^c} \Delta(z, 1/n)]$, and $K_n = \mathbf{C} \sim (V_n \sim \{\infty\}) = \hat{\mathbf{C}} \sim V_n$. (This construction is given in Rudin [18, p. 286].) We prove that the sequence $\langle K_n \rangle_1^\infty$ has the properties described in the proposition.

It is readily verified that if K_n is not empty, then

$$K_n = \left\{z\colon |z| \le n \text{ and dist}\,(z; O^c) \ge \frac{1}{n}\right\}. \tag{12.5.1}$$

Clearly, $V_n \sim \{\infty\}$ is open in $\mathbf{C}$, so K_n is closed in $\mathbf{C}$. Equation (12.5.1) implies that K_n is bounded, so K_n is compact. It also implies that $K_n \subset O$. Now let F be any compact subset of O, and suppose that r_F is such that $F \subset \Delta(0, r_F)$. By Theorem 5.6.1, dist $(F, O^c) = d_F > 0$. Then for every $n \ge \max(r_F, 1/d_F)$, we have $F \in K_n$. This proves part (i); and since $\bigcup_n K_n \subset O$, it also follows that $\bigcup_n K_n = O$, as stated in part (ii).

To establish (iii), recall that the interior of a set E, int E, is the maximal open set contained in E [Sec. 5.1]. To show that $K_n \subset \text{int}\, K_{n+1}$, we must show that for every $z_0 \in K_n$, there exists a disk $\Delta(z_0, r) \subset K_{n+1}$, $r = r_z > 0$. We do this by proving that for every $z_0 \in K_n$, $\Delta = \Delta(z_0, 1/[n(n+1)]) \subset K_{n+1}$. Take any $z' \in \Delta$. According to (12.5.1), it suffices to show that $|z'| \le n + 1$ and dist $(z', O^c) \ge 1/(n+1)$. Applying the triangle inequality and (12.5.1) with $z = z_0$, we have

$$|z'| \le |z' - z_0| + |z_0| < \frac{1}{n(n+1)} + n < n + 1;$$

and for any $w \in O_c$,

$$\frac{1}{n} \le |z_0 - w| \le |z_0 - z'| + |z' - w| \le \frac{1}{n(n+1)} + |z' - w|,$$

$$|z' - w| \ge \frac{1}{n} - \frac{1}{n(n+1)} = \frac{1}{n+1}.$$

The proof of (iii) is now complete.

To prove (iv) we observe that $\hat{\mathbf{C}} \sim K_n = V_n$ and that by construction, $O^c = \mathbf{C} \sim O \subset V_n \sim \{\infty\}$. Let A be a component of V_n, and take $z \in A$. Then z satisfies one or both of these conditions: (a) $z \in O^c$, (b) $z \in \Delta(\infty, n)$. If (a) is satisfied, then z must lie in some component of O^c which is contained entirely in A [Theorem 5.4.1]. If (b) is satisfied, then z lies in the unbounded component U of V_n. We now show that U must contain the unbounded component U_0 of $\hat{\mathbf{C}} \sim O$. If $U_0 = \{\infty\}$, there is nothing to prove. If U_0 contains an unbounded component U_1 of O^c, then since $O^c \subset V_n \sim \{\infty\}$, U_1 must be contained entirely in U [Theorem 5.4.1], and, therefore, by definition of unbounded component, $U_0 \subset U$. Q.E.D.

Theorem 12.5.1 (*Runge's Theorem*) *Let $O \subset \mathbf{C}$ be a nonempty open set, let the function f: $O \to \mathbf{C}$ be analytic in O, and let $P \subset \hat{\mathbf{C}} \sim O$ be a set which contains ∞ and at least one point of each component of $\hat{\mathbf{C}} \sim O$. Then*:

(i) *There exists a sequence $\langle R_n \rangle_1^\infty$ of rational functions with poles only on P and which converges almost uniformly to f on O.*

(ii) *If $\hat{\mathbf{C}} \sim O$ is connected (and in particular if O is a simply connected region, then there exists a sequence of polynomials $\langle P_n \rangle_1^\infty$ converging almost uniformly to f on O.*

Proof. Let $\langle K_n \rangle_1^\infty$ be the sequence of compact sets described in Proposition 12.5.1. According to part (iv) of that proposition, each component of $\hat{\mathbf{C}} \sim K_n$ contains a component of $\hat{\mathbf{C}} \sim O$, so by hypothesis there is at least one point of P in each component of $\hat{\mathbf{C}} \sim K_n$. Thus, by Theorem 12.4.1, for each n, $n = 1, 2, \ldots$, there exists a rational function R_n with all its poles in P, such that $|f(z) - R_n(z)| < 1/n$ for every $z \in K_n$. Now let F be any compact subset of O. According to Proposition 12.5.1(iv), there exists $N = N(F)$ such that $F \subset K_n$, $n = N, N+1, N+2, \ldots$. Given $\varepsilon > 0$, take $N_\varepsilon = \max(N(F), 1/\varepsilon)$. Then for every integer $n \ge N_\varepsilon$, we have $|f(z) - R_n(z)| < 1/n \le 1/N_\varepsilon \le \varepsilon$ for every $z \in F$. This proves part (i) of the theorem.

As for (ii), if $\hat{\mathbf{C}} \sim O$ is connected, then by the remarks preceding Proposition 12.3.2, the one and only component of $\hat{\mathbf{C}} \sim O$ is the unbounded component. We can therefore let $P = \{\infty\}$ in part (i), and then by Theorem 10.7.2 each R_n is a polynomial. (The subscript n here does not necessarily coincide with the degree of P_n.) Q.E.D.

Remark (i). Of course, the approximating sequences $\langle R_n \rangle$ and $\langle P_n \rangle$ in the theorem are far from unique. The reader can verify, by tracing the thread of the argument from the proof of Theorem 12.2.1 through that of Theorem 12.5.1, that our proof here of the existence of these sequences is essentially a constructive one, but also that the exact details of the implied construction in concrete examples are usually complicated and difficult to formulate explicitly. There is a large literature devoted to straightforward constructions of approximations by rational functions and polynomials. A comprehensive exposition of results obtained to 1956 was given by J. L. Walsh in [21]. Two simple examples appear in Exercises 12.4.2 and 12.4.4.

Remark (ii). In contrast to Corollary 2 of Theorem 12.4.1, the hypotheses of Theorem 12.5.1 concerning the pole set P cannot be replaced by the hypothesis that P does not contain ∞ but does contain at least one point of each component of $\mathbf{C} \sim O$. For example, let $O = \{z : |z| > 0\}$, and let $f(z) = ze^{1/z}$. The sole component of $\mathbf{C} \sim O$ is $\{0\}$, and if a pole at ∞ is disallowed, R_n would have to take the form of a polynomial in $1/z$ (not necessarily of degree n). Suppose that such a sequence $\langle R_n \rangle$ were to converge a.u. to $ze^{1/z}$ on this O. Let γ be a positively oriented circle with graph $C(0, 1)$. Then we would have $\lim_{n\to\infty} [R_n(z)/z^2] = e^{1/z}/z$ uniformly on $C(0, 1)$. Now for each n, $\int^{\gamma}[R_n(z)/z^2]\,dz = 0$, but $\int_{\gamma}(e^{1/z}/z)\,dz = \int_{\gamma}[(1/z + 1 + (z/2!) + (z^2/3!) + \cdots]dz = 2\pi i$. Thus, Proposition 8.3.1. would be violated.

12.6 Generalized Formulations in the Cauchy Theory

Theorem 12.6.1 (*A General Gauchy Integral Theorem*). *Let O be an open set in $\mathbf{C}$, and let $f\colon O \to \mathbf{C}$ be analytic on O.*

(i) *If γ is a closed path with graph in O, and such that* $\operatorname{Ind}_\gamma(a) = 0$ *for every $a \in \hat{\mathbf{C}} \sim O$, then $\int_\gamma f(z)\,dz = 0$.*
(ii) *If γ_1 and γ_2 are closed paths with graphs in O and are such that* $\operatorname{Ind}_{\gamma_1}(a) = \operatorname{Ind}_{\gamma_2}(a)$ *for every $a \in \hat{\mathbf{C}} \sim O$, then $\int_{\gamma_1} f(z)\,dz = \int_{\gamma_2} f(z)\,dz$.*
(iii) *If O is a simply connected region, then $\int_\gamma f(z)\,dz = 0$ for every closed path γ with graph in O.*

Proof. According to Theorem 11.2.1, all parts of this theorem are true for a rational function with all its poles in $\hat{\mathbf{C}} \sim O$. By Theorem 12.5.1, there exists a sequence $\langle R_n \rangle_1^\infty$ of such functions with the property that $\lim_{n\to\infty} P_n(z) = f(z)$ uniformly on any compact subset of O, and in particular on the graph Γ of any arc such that $\Gamma \subset O$. The theorem follows at once from these facts and Proposition 8.3.1. For example, to establish part (ii), we have

$$\int_{\gamma_1} f(z)\,dz = \lim_{n\to\infty} \int_{\gamma_1} R_n(z)\,dz = \lim_{n\to\infty} \int_{\gamma_2} R_n(z)\,dz = \int_{\gamma_2} f(z)\,dz.$$

Q.E.D.

This theorem generally permits the substitution of the hypothesis of a simply connected region for that of a starlike region wherever the latter appears in Chapters 9, 10, and 11. Some of the resulting statements furnish necessary and sufficient conditions that a region be simply connected, which is basically a topological concept. We defer these to Sec. 12.8. (See Theorem 12.8.1.) Others seem to lie properly in the province of classical complex analysis, and the remainder of this section is devoted to results of this type. The numbering refers to the earlier versions for starlike regions.

Theorem 12.6–(9.4.1) (*Cauchy Integral Formula*). *Let B be a simply connected region, let $f\colon B \to \mathbf{C}$ be analytic on B, and let γ be a closed path with graph Γ contained in B. Then*

$$\operatorname{Ind}_\gamma(z) f(z) = \frac{1}{2\pi i}\int_\gamma \frac{f(\zeta)\,d\zeta}{\zeta - z} \tag{12.6.1}$$

for every $z \in \Gamma^c \cap B$.

Proof. The function

$$\delta(\zeta) = \begin{cases} \dfrac{f(\zeta) - f(z)}{\zeta - z}, & \zeta \neq z, \quad \zeta \in B, \\ f'(z), & \zeta = z \end{cases}$$

as a function of ζ is analytic in B except for a removable singularity at $\zeta = z$, and the appropriate value at $\zeta = z$ which makes $\delta(\zeta)$ analytic on B is $\delta(z) = f'(z)$ [Theorem 10.5.1(ii)]. The remainder of the proof is exactly like that of Theorem 9.4.1. Q.E.D.

Corollary Let B be a simply connected region, let $f\colon B \to \mathbf{C}$ be analytic on B, and let γ be a closed path with graph $\Gamma \subset B$ and with the further properties that Γ^c has only two components G_0 and G_1, and $\operatorname{Ind}_\gamma(z) = 1$ for every z on the bounded component, say G_0. Then

$$f(z) = \frac{1}{2\pi i}\int_\gamma \frac{f(\zeta)\,d\zeta}{\zeta - z}$$

for every $z \in G_0$.

Proof. The implication in the corollary that one of the two components of Γ^c is bounded and the other is unbounded is valid because Γ is compact [Proposition 5.4.2]. To apply Theorem 12.6–(9.4.1), we have only to show that $G_0 \subset B$. Suppose there existed a point $z_0 \in G_0 \cap (\mathbf{C} \sim B)$. Then z_0 is contained in a component V of $\mathbf{C} \sim B$, and by Proposition 5.4.1, $G_0 \cup V$ is connected in $\mathbf{C}$. Recall that all components of $\mathbf{C} \sim B$ are unbounded. Now $G_0 \cup V \subset \Gamma^c$, so $G_0 \cup V$ is contained in a component of Γ^c [Theorem 5.4.1]. Since V is unbounded, this component of Γ^c is to be G_1. Of course, this is a contradiction. Q.E.D.

Remark (i). This corollary of Theorem 12.6–(9.4.1) may be regarded as the formulation, in the language of this text, of the classical presentation of the Cauchy integral formula. In that presentation, the following assumptions are made explicitly or implicitly: (a) the path γ is homeomorphic to a positively oriented circle (such a curve is called a *simple closed curve* or *Jordan curve*); (b) the path, therefore, separates the plane into a bounded and an unbounded region (this is the so-called Jordan curve theorem); and (c) $\text{Ind}_\gamma(z) = 1$ for every z in the bounded region. Rigorous proofs of (b) and (c) require rather deep topological considerations.

Remark (ii). According to Ahlfors [2, p. 150] a path γ is said to "bound" a region G if and only if $\text{Ind}_\gamma(z) = 1$ for every $z \in G$ and $\text{Ind}_\gamma(z) = 0$ or is undefined on G^c. Thus, the path γ in the corollary bounds G_0 in this sense.

Theorem 12.6–(10.6.1) *Let f be analytic on the annulus $A = \Delta(a, R) \sim \bar{\Delta}(a, r)$, $0 \le r < R$. Let γ_1 and γ_2 be positively oriented circles with respective graphs $C(a, r_1)$ and $C(a, r_2)$, where $r < r_1 < r_2 < R$. Then*

$$\int_{\gamma_1} f(\zeta)\, d\zeta = \int_{\gamma_2} f(\zeta)\, d\zeta, \tag{12.6.2}$$

and

$$f(z) = \frac{1}{2\pi i}\int_{\gamma_2} \frac{f(\zeta)\, d\zeta}{\zeta - z} - \frac{1}{2\pi i}\int_{\gamma_1} \frac{f(\zeta)\, d\zeta}{\zeta - z} \tag{12.6.3}$$

for every $z \in A_1 = \{z: r_1 < |r - a| < r_2\}$.

Proof. Equation (12.6.2) is established in Exercise 9.3.3 in a somewhat artificial manner (necessitated by the starlike restriction on the Cauchy integral theorem), and only for the special case in which $A = \Delta(a, R) \sim \{a\}$. The more general result is now seen to be an obvious consequence of Theorem 12.6.1(ii) with $O = A$, because $\text{Ind}_{\gamma_1}(z) = \text{Ind}_{\gamma_2}(z) = 1$ for $z \in \bar{\Delta}(a, r)$ and $\text{Ind}_{\gamma_1}(z) = \text{Ind}_{\gamma_2}(z) = 0$ for $z \in \{z: |z - a| \ge R\}$. Thereafter the proof of (12.6.3) is exactly the same as that of Theorem 10.6.1. Q.E.D.

We now formulate a generalization of the residue theorem, Theorem 11.1.1. First, we define the term *meromorphic function.* Let O be an open set in $\hat{\mathbf{C}}$. A function $f: O \to \hat{\mathbf{C}}$ is said to be meromorphic on O if it is continuous on O (in the extended sense introduced in Sec. 10.7) and is analytic on $O \sim \{\infty\}$ except possibly for isolated singularities at the points of a set $P \subset O$, each of which is a pole of f. Notice that the definition implies that any subset of $O \sim \{\infty\}$ which is compact in $\mathbf{C}$ can contain only a finite number of points of P. (See the second paragraph of Sec. 10.5.) Therefore, by any of various devices, such as considering the number of points of P in $\bar{\Delta}(0, n) \sim \Delta(0, n - 1)$, $n = 2, 3, \ldots$, the points of P can be put into one-to-one correspondence with the integers $1, 2, \ldots$, and we can write P in the form $\{\zeta_1, \zeta_2, \zeta_3, \ldots\}$. (In more technical language, P is "countable.") The case $P = \varnothing$ is not excluded, so the class of functions analytic on O is contained in the class of functions meromorphic on O. Notice that the corollary of Theorem 10.7.2 states that a function meromorphic on $\hat{\mathbf{C}}$ is a rational function.

Theorem 12.6–(11.1.1) *(A General Form of the Residue Theorem). Let $O \subset \mathbf{C}$ be open, and let $f: O \to \hat{\mathbf{C}}$ be meromorphic in O with pole set $P = \{\zeta_1, \zeta_2, \ldots\}$. Let γ be any closed path with graph $\Gamma \subset O \sim P$ such that* $\operatorname{Ind}_\gamma(z) = 0$ *for every $z \in \hat{\mathbf{C}} \sim O$. Then*

$$\frac{1}{2\pi i}\int_\gamma f(z)\,dz = \sum_{n=1}^{\infty} \operatorname{Res}(f; \zeta_n) \operatorname{Ind}_\gamma(\zeta_n). \tag{12.6.4}$$

Proof. Let $P_0 = \{\zeta_n: \operatorname{Ind}_\gamma(\zeta_n) \neq 0, n = 1, 2, \ldots\}$. By Theorem 8.4.2, P_0 is contained in the union U of the *bounded* components of Γ^c, and by Proposition 5.4.2 there exists $\bar{\Delta}(0, R)$, $R > 0$, such that $U \subset \bar{\Delta}(0, R)$. Thus, $P \cap U$ and $P_0 \cap U$ are finite point sets; so let $P_0 = \{\zeta_{n1}, \zeta_{n2}, \ldots, \zeta_{nk}\}$. Also let $P_1 = P \sim P_0$; and let $Q_1, Q_2, \ldots, Q_k$, respectively, denote the principal parts of f at the poles $\zeta_{n1}, \zeta_{n2}, \ldots, \zeta_{nk}$. We proceed as in the proof of Theorem 11.1.1 to obtain a function

$$G(z) = f(z) - \sum_{j=1}^{k} Q_j(z)$$

analytic on $O \sim P_1$. Since $\operatorname{Ind}_\gamma(z) = 0$ for every $z \in P_1$, and for every $z \in \hat{\mathbf{C}} \sim O$, it follows that $\operatorname{Ind}_\gamma(z) = 0$ for every $z \in \hat{\mathbf{C}} \sim (O \sim P_1)$. Therefore, Theorem 12.6.1(i) is applicable to G and γ, and we have

$$0 = \int_\gamma G(z)\,dz = \int_\gamma f(z)\,dz - \sum_{j=1}^{k}\int_\gamma Q_j(z)\,dz. \tag{12.6.5}$$

For each j, the computation of the line integral of Q_j along γ proceeds exactly as in the proof of Theorem 11.1.1, and we obtain, as before,

$$\frac{1}{2\pi i}\int_\gamma Q_j(z)\,dz = \operatorname{Res}(f; \zeta_{nj}) \operatorname{Ind}_\gamma(\zeta_{nj}).$$

Substitution into (12.6.5) concludes the proof. Q.E.D.

Remark. Notice that the infinite series in (12.6.4) is illusory in the sense that with zero terms deleted, it is only a finite sum.

Exercises 12.6

1. (General Laurent Expansion.) Let f be analytic on the annulus $A = \Delta(a, R) \sim \bar{\Delta}(a, r)$, $0 \leq r < R$. Use Theorem 12.6–(10.6.1), Theorem 8.4.1, and Proposition 10.6.1 to show that

   (i) there exists a unique series representation of f as follows:

$$f(z) = \sum_{k=0}^{\infty} b_k(z-a)^k + \sum_{j=1}^{\infty} c_j(z-a)^{-j},$$

valid for every $z \in A$, and such that the first series converges absolutely and almost uniformly on $\Delta(a, R)$ to a function analytic on $\Delta(a, R)$, and the second series converges absolutely and uniformly on any closed set in $\mathbf{C} \sim \bar{\Delta}(a, r)$;

(ii) $$b_k = \frac{1}{2\pi i}\int_{\gamma_\rho} \frac{f(\zeta)\,d\zeta}{(\zeta - a)^{k+1}},\ k = 0, 1, 2, \ldots,$$

$$c_j = \frac{1}{2\pi i}\int_{\gamma_\rho} f(\zeta)(\zeta - a)^{j-1}\,d\zeta,\ j = 1, 2, \ldots,$$

where γ_ρ is any positively oriented circle with graph $C(a, \rho)$, $r < \rho < R$.

Remark. The reader's attention is invited to the Remark following Example (v) in Sec. 10.6 concerning the limitations of this generalized Laurent expansion with regard to the problem of characterizing singularities of f.

2. (Generalized Principle of the Argument.) Let $O \subset \mathbf{C}$ be open, and let $f: O \to \mathbf{C}$ be meromorphic in O with pole set $P = \{\zeta_1, \zeta_2, \ldots\}$ and zero set $Z = \{z_1, z_2, \ldots\}$. Assume that Z has no accumulation point in O. Let the order of the pole at ζ_k be p_k, $k = 1, 2, \ldots$, and the multiplicity of the zero at z_j be m_j, $j = 1, 2, \ldots$. Let γ be any closed path with graph $\Gamma \subset O \sim (P \cup Z)$ such that $\text{Ind}_\gamma z = 0$ for every $z \in \hat{\mathbf{C}} \sim O$. Then

$$\frac{1}{2\pi i}\int_\gamma \frac{f'(z)}{f(z)}\,dz = \sum_{j=1}^{\infty} m_j \,\text{Ind}_\gamma(z_j) - \sum_{k=1}^{\infty} p_k \,\text{Ind}_\gamma(\zeta_k). \tag{12.6.6}$$

(*Hint*: Use the argument in the proof of Theorem 12.6–(11.1.1) to show that the sums in (12.6.6) are really finite sums after deletion of zeros, and then look at (11.3.3).)

3. Let B be a simply connected region, let $f: B \to \mathbf{C}$ be analytic, and let $\zeta_1, \zeta_2, \ldots, \zeta_n$ be complex numbers, not necessarily distinct, such that the corresponding points are contained in B. Let γ be a closed path with graph $\Gamma \subset B$ containing none of the points ζ_k. Let $\omega(z) = (z - \zeta_1)(z - \zeta_2)\cdots(z - \zeta_n)$. Show that the Hermite–Lagrange interpolation formula:

$$H(z) = \frac{1}{2\pi i}\int_\gamma \frac{f(\zeta)}{\zeta - z}\left(1 - \frac{\omega(z)}{\omega(\zeta)}\right) d\zeta$$

$$= \frac{1}{2\pi i}\int_\gamma \frac{f(\zeta)}{\omega(\zeta)}\left[\frac{\omega(\zeta) - \omega(z)}{\zeta - z}\right] d\zeta$$

is a polynomial in z of degree at most $n - 1$ such that $H(\zeta_k) = f(\zeta_k)\,\text{Ind}_\gamma(\zeta_k)$. Show also that if $\omega(z)$ has a zero of order m at ζ_K, then $H^{(p)}(\zeta_K) = f^{(p)}(\zeta_K)\text{Ind}_\gamma(\zeta_K)$, p

$= 1, 2, \ldots, m - 1$. (*Hint*: For every z lying in the component of Γ^c in which ζ_K lies, we can write

$$H(z) = f(z)\text{Ind}_\gamma(\zeta_K) - \frac{\omega(z)}{2\pi i}\int_\gamma \frac{[f(\zeta)/\omega(\zeta)]}{\zeta - z}\,d\zeta.$$

By Theorem 8.4.1, the integral in the right-hand member is an analytic function of z. By Exercise 10.1.1 and the definition of a zero of order m, $\omega(\zeta_K) = \omega'(\zeta_K) = \cdots = \omega^{m-1}(\zeta_K) = 0$. Use Leibniz's formula for the repeated differentiation of the product of two functions.)

4. Show why $1/\sin z$ is meromorphic in **C** but not meromorphic in $\hat{\mathbf{C}}$.
5. Show that if B is a simply connected region, and $f: B \to \mathbf{C}$ is analytic, and γ is a closed path with graph $\Gamma \subset B$, then there exists a sequence of rational functions $\langle R_n \rangle_1^\infty$ with all poles contained in Γ, such that $\langle R_n(z) \rangle$ converges a.u. to $\text{Ind}_\gamma(z) f(z)$ on $\Gamma^c \cap B$. (*Hint*: Proposition 12.4.1 and the Cauchy integral formula.)

12.7 Homology, Homotopy, and a Generalization of the Complex Line Integral

Let O be an open set in **C**. A closed path γ with graph in O is said to be *O-homologous to zero* if and only if $\text{Ind}_\gamma(z) = 0$ for every $z \in \hat{\mathbf{C}} \sim O$. (Thus, the path γ in Theorem 12.6.1(i) is O-homologous to zero, and every closed path in a simply connected region B is B-homologous to zero [Theorem 12.6.1(iii)].) Two closed paths γ_0 and γ_1 with graphs in O are said to be O-homologous if and only if $\text{Ind}_{\gamma_1}(z) = \text{Ind}_{\gamma_2}(z)$ for every $z \in \hat{\mathbf{C}} \sim O$. (Thus, the paths in Theorem 12.6.1(ii) are O-homologous.)

We now define the concept of homotopy within the limited context of the topology in **C**. This concept plays the role of giving precision to the idea of continuously deforming one curve into another.

Let $O \subset \mathbf{C}$ be an open set, and let γ_0 and γ_1 be closed curves with graphs contained in O and with parametric equations respectively $z = \gamma_0(t)$ and $z = \gamma_1(t)$, $\alpha \le t \le \beta$. These two curves are said to be *O-homotopic* if and only if there exists a continuous mapping $\phi: ([\alpha, \beta] \times [0, 1]) \to O$ such that $\phi(t, 0) = \gamma_0(t)$, $\alpha \le t \le \beta$, $\phi(t, 1) = \gamma_1(t)$, $\alpha \le t \le \beta$, and $\phi(\alpha, s) = \phi(\beta, s)$, $0 \le s \le 1$. (The last equation is the *closed curve* condition.) Then $\gamma_s(t) = \phi(t, s)$, $\alpha \le t \le \beta$ represents a one-parameter family of closed curves which can be regarded intuitively as a continuous succession of deformations of γ_0 into γ_1. If $\gamma_0(t)$ is constant on $[\alpha, \beta]$, so its graph is a single point, then γ_1 is said to be *null-homotopic* in O. Intuitively, in this case, the graph of γ_1 can be shrunk continuously in O to a single point.

Examples

(i) Let B be a starlike region with star center a. Then every closed curve with graph in B is null-homotopic in B. For let γ be such a closed curve with equation $z = \gamma(t)$, $\alpha \le t \le \beta$. Let $\phi(t, s) = (1 - s)a + s\gamma(t)$, $\alpha \le t \le \beta$, $0 \le s \le 1$. This function is obviously continuous on $[\alpha, \beta] \times [0, 1]$, and $\phi(t, 0) = a$, $\phi(t, 1) = \gamma(t)$, $\alpha \le t \le \beta$, and $\phi(\alpha, s) = (1 - s)a + s\gamma(\alpha) = (1 - s)a + s\gamma(\beta) = \phi(\beta, s)$. For any fixed t, $z = (1 - s)a + s\gamma(t), 0 \le s \le 1$, is the equation of a straight line joining a to $\gamma(t)$, and by the definition of starlike region, every point on this straight line lies in B.

(ii) Let B be a convex region. Then any two closed curves γ_0 and γ_1 with common domain $[\alpha, \beta]$ and with graphs in B are B-homotopic. The mapping which verifies this statement is given by $\phi(t, s) = (1 - s)\gamma_0(t) + s\gamma_1(t)$, $\alpha \le t \le \beta$, $0 \le s \le 1$.

A basic connection between the concepts of homology and homotopy is now presented. We need a preliminary result which shows that a closed path can be perturbed to a certain extent without changing the topological index of the path with respect to a point.

Proposition 12.7.1 *Let $\gamma_1 \colon [\alpha, \beta] \to \mathbf{C}$ be a closed path, and let $\gamma_2 \colon [\alpha, \beta] \to \mathbf{C}$ be another closed path such that for a certain complex a,*

$$|\gamma_2(t) - \gamma_1(t)| < |\gamma_1(t) - a| \tag{12.7.1}$$

for every $t \in [\alpha, \beta]$. Then $\mathrm{Ind}_{\gamma_1}(a) = \mathrm{Ind}_{\gamma_2}(a)$.

Proof. Notice first that the strong inequality in (12.7.1) implies that a does not lie on the graph of either γ_1 or γ_2. Now define a new closed path γ by the parametric equation

$$\gamma(t) = \frac{\gamma_2(t) - a}{\gamma_1(t) - a} + a, \qquad \alpha \le t \le \beta.$$

Let the graph of γ be Γ. We claim that a lies in the unbounded component of Γ^c. To see this, rewrite (12.7.1) as follows:

$$|\gamma_2(t) - a - (\gamma_1(t) - a)| < |\gamma_1(t) - a|, \qquad \alpha \le t \le \beta;$$

$$\left|\frac{\gamma_2(t) - a}{\gamma_1(t) - a} - 1\right| < 1, \qquad \alpha \le t \le \beta;$$

$$|\gamma(t) - (a + 1)| < 1, \qquad \alpha \le t \le \beta.$$

The last inequality implies that $\Gamma \in \Delta(a + 1, 1)$. However, $|a - (a + 1)| \nless 1$, so $a \in [\Delta(a + 1, 1)]^c$. Now $[\Delta(a + 1, 1)]^c$ is contained in the unbounded component of Γ^c [Proposition 5.4.2], so in particular, a lies in that unbounded component.

A simple calculation shows that

$$\frac{\gamma'(t)}{\gamma(t) - a} = \frac{\gamma_2'(t)}{\gamma_2(t) - a} - \frac{\gamma_1'(t)}{\gamma_1(t) - a}.$$

Then by Theorem 8.4.2, the definition of a topological index, and Theorem 8.3.2, we have

$$\begin{aligned} 0 = \operatorname{Ind}_\gamma(a) &= \frac{1}{2\pi i}\int_\alpha^\beta \frac{\gamma'(t)\,dt}{\gamma(t) - a} \\ &= \frac{1}{2\pi i}\int_\alpha^\beta \frac{\gamma_2'(t)\,dt}{\gamma_2(t) - a} - \frac{1}{2\pi i}\int_\alpha^\beta \frac{\gamma_1'(t)\,dt}{\gamma_1(t) - a} \\ &= \operatorname{Ind}_{\gamma_2}(a) - \operatorname{Ind}_{\gamma_1}(a). \end{aligned}$$

Q.E.D.

Theorem 12.7.1 *Let O be an open set in* **C**, *and let γ_0 and γ_1 be closed paths with graphs in O. If γ_0 and γ_1 are O-homotopic, then they are O-homologous. If γ_1 is null-homotopic in O, then γ_1 is O-homologous to zero.*

Proof. If $O = \mathbf{C}$, then every closed path in O is homologous to zero, so the theorem becomes trivial. We henceforth assume that $O \neq \mathbf{C}$. Choose $a \in O^c$. Let γ_0 and γ_1 be O-homotopic with parametric equations $z = \gamma_0(t)$, $z = \gamma_1(t)$, $\alpha \le t \le \beta$. There exists a continuous mapping $\phi\colon [\alpha, \beta] \times [0, 1] \to O$ such that $\phi(t, 0) = \gamma_0(t)$, $\phi(t, 1) = \gamma_1(t)$, $\alpha \le t \le \beta$ and $\phi(\alpha, s) = \phi(\beta, s)$, $0 \le s \le 1$. For each $s \in [0, 1]$, let γ_s be the closed curve with parametric equation $z = \gamma_s(t)$, $\alpha \le t \le \beta$. The general idea of the proof is to take a monotonically increasing sequence of value of s in $[0, 1]$, starting with $s = 0$ and ending with $s = 1$, such that if $s < s'$ are any two adjacent members of the sequence then $|\gamma_{s'}(t) - \gamma_s(t)| < |\gamma_s(t) - a|$, so by Proposition 12.7.1, $\operatorname{Ind}_{\gamma_{s'}}(a) = \operatorname{Ind}_{\gamma_s}(a)$.

But unfortunately for this program, the curves γ_s, aside from γ_0 and γ_1, are not necessarily *paths* or even rectifiable curves, and there is no assurance that the complex line integral used to define $\operatorname{Ind}_{\gamma_s}(a)$ exists for such a curve. We therefore approach the problem via an approximation method. This method is of importance for itself alone in opening the way to a generalization of the concept of complex line integrals of analytic functions, as will be shown later in this section.

Lemma Let O be an open set in **C**, let γ be an arc with graph $\Gamma \subset O$ and equation $z = \gamma(t)$, $\alpha \le t \le \beta$, and let $\rho = \operatorname{dist}(\Gamma, O^c)$. For any ε, $0 < \varepsilon \le 1$, there exists an infinite family P_ε of polygonal arcs with the following properties: If $\lambda_\varepsilon \in P_\varepsilon$, then the graph of λ_ε is contained in O, the vertices of λ_ε are contained in Γ, the initial and terminal points of λ_ε are the same as those of γ, and $|\gamma(t) - \lambda_\varepsilon(t)| < \varepsilon\rho$ for every $t \in [\alpha, \beta]$, where $z = \lambda_\varepsilon(t)$, $\alpha \le t \le \beta$, is the equation of λ_ε.

Proof. Take ε, $0 < \varepsilon \le 1$, and notice that $0 < \rho \le \infty$ [Theorem 5.6.1]. For any $z \in \Gamma$, $\Delta(z, \varepsilon\rho) \subset O$. Since γ is uniformly continuous on $[\alpha, \beta]$ [Theorem 5.5.3], there exists $\delta > 0$ such that for every t, $t' \in [\alpha, \beta]$ with

$|t' - t| < \delta$, $|\gamma(t') - \gamma(t)| < \varepsilon\rho$. Let $\{\alpha = t_0 < t_1 < \cdots < t_n = \beta\}$ be any partition of $[\alpha, \beta]$ with the property that $\max_k (t_k - t_{k-1}) < \delta$. We construct $\lambda_\varepsilon(t)$ for this partition by requiring that its restriction to $[t_{k-1}, t_k]$ be given by

$$\lambda_\varepsilon(t) = \frac{t_k - t}{t_k - t_{k-1}}\gamma(t_{k-1}) + \frac{t - t_{k-1}}{t_k - t_{k-1}}\gamma(t_k), \qquad t_{k-1} \le t \le t_k.$$

The graph of $\lambda_\varepsilon(t)$ as so restricted is a line segment joining $\gamma(t_{k-1})$ to $\gamma(t_k)$, and since $|\gamma(t_{k-1}) - \gamma(t_k)| < \varepsilon\rho$, this line segment must be contained in $\Delta[\gamma(t_{k-1}), \varepsilon\rho] \subset O$, $k = 1, 2, \ldots, n$. Then for $t \in [t_{k-1}, t_k]$, $k = 1, 2, \ldots, n$, we have

$$\begin{aligned} |\lambda_\varepsilon(t) - \gamma(t)| &= \left|\frac{t_k - t}{t_k - t_{k-1}}[\gamma(t_{k-1}) - \gamma(t)] + \frac{t - t_{k-1}}{t_k - t_{k-1}}[\gamma(t_k) - \gamma(t)]\right| \\ &< \frac{t_k - t}{t_k - t_{k-1}}\varepsilon\rho + \frac{t - t_{k-1}}{t_k - t_{k-1}}\varepsilon\rho = \varepsilon\rho. \end{aligned}$$

The fact that the family $P_\varepsilon = \{\lambda_\varepsilon\}$ is infinite (in fact, uncountably so), of course, stems from the extremely loose restriction on the partitions of $[\alpha, \beta]$.

We return to the proof of the theorem. Let $K \subset O$ be the image of $S = [\alpha, \beta] \times [0, 1]$ under the mapping ϕ. Since K is compact [Theorem 5.5.2], $d = \operatorname{dist}(K, O^c) > 0$ [Theorem 5.6.1]. Since ϕ is uniformly continuous on S [Theorem 5.5.3], there exists δ such that $|\phi(t, s') - \phi(t, s)| < d/4$ for every (t, s), $(t, s') \in S$ with $|s' - s| < \delta$. Let $\{0 = s_0 < s_1 < \cdots < s_m = 1\}$ be a partition of $[0, 1]$ such that $\max_j (s_j - s_{j-1}) < \delta$. To simplify the notation slightly, let $\gamma^{(j)} = \gamma_{s_j}$, $j = 0, 1, 2, \ldots, m$. (Here $\gamma^{(0)} = \gamma_0$, $\gamma^{(m)} = \gamma_1$.) We have the inequalities

$$\begin{aligned} &|\gamma^{(j)}(t) - \gamma^{(j-1)}(t)| = |\phi(t, s_j) - \phi(t, s_{j-1})| < \frac{d}{4}, \\ &|\gamma^{j-1}(t) - a| > d, \qquad j = 1, 2, \ldots, m; \qquad t \in [\alpha, \beta]. \end{aligned}$$

Let $\Gamma^{(j)}$ be the graph of $\gamma^{(j)}$, and let $\rho_j = \operatorname{dist}(\Gamma^{(j)}, O^c)$, $j = 0, 1, \ldots, m$. Notice that since $\Gamma^{(j)} \subset K$, we have $\rho_j \ge d$, $j = 0, 1, \ldots, m$. Now for each $\gamma^{(j)}$, $j = 1, \ldots, m - 1$, let λ_j be the approximating polygonal arc promised by the lemma with $\varepsilon = \varepsilon_j = d/4\rho_j$, so that $|\gamma^{(j)}(t) - \lambda_j(t)| < d/4$. Also let $\lambda_0 = \gamma_0$, $\lambda_m = \gamma_1$. Each λ_j is a closed path, so Proposition 12.7.1 can be applied to these polygonal closed curves. We have

$$\begin{aligned} |\lambda_j(t) - \lambda_{j-1}(t)| &= |[\lambda_j(t) - \gamma^{(j)}(t)] \\ &\quad + [\gamma^{(j)}(t) - \gamma^{(j-1)}(t)] + [\gamma^{(j-1)}(t) - \lambda_{j-1}(t)]| \\ &< \frac{d}{4} + \frac{d}{4} + \frac{d}{4} = \frac{3d}{4}, \end{aligned} \tag{12.7.2}$$

$$|\lambda_{j-1}(t) - a| = |[\gamma^{(j-1)}(t) - a] - [\gamma^{(j-1)}(t) - \lambda_{j-1}(t)]|$$
$$> d - \frac{d}{4} - \frac{3d}{4}, \qquad j = 1, 2, \ldots, m; \quad t \in [\alpha, \beta]. \quad (12.7.3)$$

Thus $|\lambda_j(t) - \lambda_{j-1}(t)| < |\lambda_{j-1}(t) - a|$, $j = 1, 2, \ldots, m$, $t \in [\alpha, \beta]$. From Proposition 12.7.1 we obtain the chain $\text{Ind}_{\gamma_0}(a) = \text{Ind}_{\gamma_1}(a) = \text{Ind}_{\gamma_2}(a) = \cdots = \text{Ind}_{\gamma_{m-1}}(a) = \text{Ind}_{\gamma_1}(a)$, so γ_0 and γ_1 are O-homologous. Q.E.D.

Corollary Let O be an open set in **C**, and let $f\colon O \to \mathbf{C}$ be analytic on O. Let γ_0 and γ_1 be O-homotopic closed paths. Then $\int_{\gamma_0} f(z)\,dz = \int_{\gamma_1} f(z)\,dz$. If γ_1 is null-homotopic in O, then $\int_{\gamma_1} f(z)\,dz = 0$.

Proof. In the light of Theorem 12.7.1, this is merely a translation of Theorem 12.6.1(i) and (ii) into the language of this section. Q.E.D.

Remark (i). It can be shown by the methods of algebraic topology that the converse of Theorem 12.7.1 is false.

Remark (ii). The lemma of Theorem 12.7.1 can be used to extend, in a consistent and meaningful way, the definition of the complex line integral of an analytic function along a closed curve to the case in which the curve is merely a continuous mapping of a real interval, and is not even rectifiable. The following proposition contains one approach to this idea.

Proposition 12.7.2 *Let O be an open set in* **C**, *and let γ be a closed curve with graph $\Gamma \subset O$ and equation $z = \gamma(t)$, $\alpha \le t \le \beta$. Let $\rho = \text{dist}(\Gamma, O^c)$. For any r, $0 < r \le 1/3$, there exists an infinite family P_r of polygonal closed paths with the following properties: If $\lambda_r \in P_r$, then the graph of λ_r is contained in O; the vertices of λ_r are contained in Γ; and $|\gamma(t) - \lambda_r(t)| < r\rho$ for every $t \in [\alpha, \beta]$, where $z = \lambda_r(t)$, $\alpha \le t \le \beta$, is the equation of λ_r. Moreover if $\lambda_r \in P_r$, $\lambda_{r'} \in P_{r'}$, $0 < r, r' \le 1/3$, then λ_r and $\lambda_{r'}$ are O-homologous, and so if $f\colon O \to$* **C** *is any function analytic on O, then $\int_{\lambda_r} f(z)\,dz = \int_{\lambda_{r'}} f(z)\,dz$.*

Proof. In the light of Theorem 12.6.1(ii) and the lemma of Theorem 12.7.1, all that remains to be proved is the statement that the restriction $0 < r \le 1/3$ and the "proximity" of λ_r to γ forces any $\lambda_r, \lambda_{r'} \in \bigcup P_r$ to be O-homologous. The verification of this fact is similar to (but easier than) the proof of Theorem 12.7.1 and is left as an exercise.

Suppose now that Γ is the graph of a closed curve γ and f is a function which is analytic on Γ. Then f by implication is analytic on an open set O containing Γ, since for any $z \in \Gamma$, f is (by definition) analytic in an open disk with center at z. We simply *define* $\int_\gamma f(z)\,dz$ to have the value $\int_{\gamma_r} f(z)\,dz$, where λ_r is any member of the O-homologous family of polygonal closed paths related to O and Γ described in Proposition 12.7.2. This definition leads to a slightly more satisfying formulation of the corollary of Theorem 12.7.1, which follows.

Proposition 12.7.3 *Let O be an open set in* $\mathbf{C}$, *and let* $f\colon O \to \mathbf{C}$ *be analytic on* O. *Let* γ_0 *and* γ_1 *be* O*-homotopic closed curves (not necessarily paths). Then* $\int_{\gamma_0} f(z)\,dz = \int_{\gamma_1} f(z)\,dz$. *If* γ_1 *is null-homotopic in* O, *then* $\int_{\gamma} f(z)\,dz = 0$. *The integrals here are taken in the extended sense.*

The proof is an exercise.

Notice that our method of eliminating smoothness requirements in the definition of the line integral of an analytic function seems to apply only to integration along closed curves. Another approach which eliminates this restriction is presented in [17].

Exercises 12.7

1. Prove Proposition 12.7.2. (*Hint*: Imitate in simplified form the computations in (12.7.2) and (12.7.3).)
2. For any closed curve γ with graph Γ, let $\operatorname{Ind}_{\gamma}(a)$, $a \notin \Gamma$, be defined in terms of the families of approximating polygonal curves described in Proposition 12.7.2. Prove this generalization of Proposition 12.7.1.: If γ_1 and γ_2 are any two closed curves with parametric equations $z = \gamma_1(t)$, $\alpha \le t \le \beta$, and $z = \gamma_2(t)$, $\alpha \le t \le \beta$, and if for $a \in \mathbf{C}$, $|\gamma_1(t) - \gamma_2(t)| < |\gamma_1(t) - a|$. then $\operatorname{Ind}_{\gamma_1}(a) = \operatorname{Ind}_{\gamma_2}(a)$. (*Hint*: In Proposition 12.7.2, let $O = \mathbf{C} \sim \{a\}$. The problem is to show that if λ_1 and λ_2 are suitably chosen polygonal arcs which approximate γ_1 and γ_2, respectively, in the sense of Proposition 10.7.1, then
$$|\lambda_1(t) - \lambda_2(t)| < |\lambda_1(t) - a|, \qquad \alpha \le t \le \beta; \tag{12.7.4}$$
for this implies, according to Proposition 12.7.2 and the definitions, that $\operatorname{Ind}_{\gamma_1}(a) = \operatorname{Ind}_{\lambda_1}(a) = \operatorname{Ind}_{\lambda_2}(a) = \operatorname{Ind}_{\gamma_2}(a)$. In the usual notation, let $\rho_1 = \operatorname{dist}(a, \Gamma_1)$, $\rho_2 = \operatorname{dist}(a, \Gamma_2)$, and let $m = \min[|\gamma_1(t) - a| - |\gamma_2(t) - \gamma_1(t)|]$. Show that $m > 0$. Derive an explicit upper bound β for r in terms of m, ρ_1, ρ_2, such that if $r < \beta$ and $|\lambda_1(t) - \gamma_1(t)| < r\rho_1$, $|\lambda_2(t) - \gamma_2(t)| < r\rho_2$, $\alpha \le t \le \beta$, then (12.7.4) is valid.)
3. Prove Proposition 12.7.3. (*Hint*: The problem boils down to proving that one of the approximating closed paths for γ_0 given by Proposition 12.7.2 is O-homologous to one of those for γ_1. Use suitably small values of r in Proposition 12.7.2 together with the argument used to prove Theorem 12.7.1.)

12.8 Characterizations of a Simply Connected Region

The reader who has studied connectedness from the purely topological viewpoint is doubtless aware that there are various equivalent formulations of the concept of simple connectivity for arcwise connected sets. Frequently in topology, the definition of a simply connected region in $\mathbf{C}$ or $\mathbf{R}^2$ is taken to be that it is a region B (that is, an open connected set) with the property that every closed curve is null-homotopic in B. We have taken as our basic definition the statement that a region B is simply connected if $\hat{\mathbf{C}} \sim B$ is connected in $\hat{\mathbf{C}}$, which is equivalent to saying that B does not separate the extended plane. This is perhaps less intuitively appealing than the homotopy definition, but it seems to be more convenient to use in developing the Cauchy theory rigorously.

This section has two main purposes. The first is to present certain consequences of Theorem 12.6.1 which generalize results in the Cauchy theory proved for starlike regions in earlier chapters, and which provide at the same time necessary and sufficient conditions that a region shall be simply connected in $\mathbf{C}$. The second main purpose is to formulate a list of equivalent definitions of simple connectivity in $\mathbf{C}$ which contains some standard topological definitions of the concept, and which illustrates the close connection between the concept of simple connectivity and complex integration theory. Another purpose is to provide a review, in a new setting, of a number of important results in complex analysis which are established in previous sections. The program is contained in the following theorem and its proof.

Theorem 12.8.1 *Let B be a region in $\mathbf{C}$. The following statements involving B are equivalent.*

(i) *B is simply connected in the sense that $\hat{\mathbf{C}} \sim B$ has only one component, and so is connected in $\hat{\mathbf{C}}$.*
(ii) $\operatorname{Ind}_\gamma(a) = 0$ *for every closed path γ with graph in B and every $a \in \hat{\mathbf{C}} \sim B$.*
(iii) *For every function f analytic in B and every closed path γ with graph in B, $\int_\gamma f(z)\,dz = 0$.*
(iv) *Every function analytic in B has a primitive in B.*
(v) *Every function analytic and with no zeros in B has an analytic logarithm in B.*
(vi) *Every function analytic and with no zeros in B has an analytic square root in B.*
(vii) *B is homeomorphic to the open* disk $\Delta(0, 1)$.
(viii) *Every closed path is null-homotopic in B.*

Proof. The plan is to show that (i) ⇒ (ii) ⇒ (iii) ⇒ (iv) ⇒ (v) ⇒ (vi) ⇒ (vii) ⇒ (viii) ⇒ (i). The implication (vi) ⇒ (vii) is essentially the Riemann mapping theorem, the proof of which is deferred to the next chapter.

(i) ⇒ (ii). Recall that $\operatorname{Ind}_\gamma(a)$ is the complex line integral of the function $1/[(2\pi i)(z - a)]$ along the path γ. With $a \in \mathbf{C} \sim B$, this function is analytic in B, so (ii) follows at once from Theorem 12.6.1(iii).

(ii) ⇒ (iii). [Theorem 12.6.1(i)].

(iii) ⇒ (iv). Let $f: B \to \mathbf{C}$ be analytic. Take any point $z_0 \in B$ and hold it fixed. Let z be any other point in B, and let γ_z be any path with graph in B joining z_0 to z. By (iii), the integral $F(z) = \int_{\gamma_z} f(\zeta)\,d\zeta$ is independent of the path and so represents a well-defined single-valued function of z. Since B is open, there exists for each $z_1 \in B$ a disk $\Delta(z_1, R) \subset B$, $r = r(z_1) > 0$. Take any point $z_2 \in \Delta(z_1, r)$, and let the path joining z_0 to z_2 be $\gamma_{z_1} + [z_1, z_2]$ where γ_{z_1} is any path joining z_0 to z_1. Then $F(z_2) - F(z_1) = \int_{[z_0,z_1]} f(\zeta)\, d\zeta$. The remainder of the proof is exactly the same as the last part of the proof of Theorem 8.5.2, starting with equation (8.5.2).

(iv) $\Rightarrow$ (v). Let $f: B \to \mathbf{C}$ be analytic and $f(z) \neq 0$ in B. The function f'/f is analytic in B [Theorem 9.4.2], and therefore by (iv) has a primitive in B. Therefore, by Proposition 9.4.2, f has an analytic logarithm in B.

(v) $\Rightarrow$ (vi). [Exercise 9.4.9].

(vi) $\Rightarrow$ (vii). If $B = \mathbf{C}$, we do not need the hypothesis (vi). Look at the transformation given by $w = z/(1 + |z|)$. This clearly gives a continuous mapping of $\mathbf{C}$ (the z-plane) into $\{w: |w| < 1\}$. It is readily verified that the inverse function exists and is given by $z = w/(1 - |w|)$, with domain $\Delta(0, 1)$. This inverse clearly is continuous on its domain. Thus, $\mathbf{C}$ is homeomorphic to $\Delta(0, 1)$. The case in which $B \neq \mathbf{C}$ is covered by the proof of the Riemann mapping theorem given in Chapter 13.

(vii) $\Rightarrow$ (viii). Let $h: B \to \Delta(0, 1)$ be a homeomorphism from B onto $\Delta(0, 1)$, and let $z_0 = h^{-1}(0)$. Let γ be a closed path with graph in B and equation $z = \gamma(t)$, $\alpha \leq t \leq \beta$. Let the function $\phi: ([\alpha, \beta] \times [0, 1]) \to B$ be given by the equation $\phi(t, s) = h^{-1}[sh(\gamma(t))]$, $\alpha \leq t \leq \beta, 0 \leq s \leq 1$. Clearly ϕ is continuous on its domain [Exercises 3.2.8 and 3.2.10] and $\phi(t, 0) = h^{-1}(0) = z_0$, $\phi(t, 1) = h^{-1}[h(\gamma(t))] = \gamma(t)$, $\alpha \leq t \leq \beta$. Thus, by definition, γ is null-homotopic in B.

(viii) $\Rightarrow$ (i). Suppose that $\hat{\mathbf{C}} \sim B$ were not connected in $\hat{\mathbf{C}}$. We show that in that case there would exist a closed path γ with graph in B such that $\text{Ind}_\gamma (a) \neq 0$ for some $a \in B^c = \mathbf{C} \sim B$. Thus, this path would not be B-homologous to zero, in contradiction to Theorem 12.7.1.

If $B = \mathbf{C}$, no proof is needed. We assume henceforth that $B^c = \mathbf{C} \sim B \neq \varnothing$. Let $\hat{B}^c = \hat{\mathbf{C}} \sim B$. The assumption that $\hat{B}^c$ is not connected implies that there exist disjoint sets O_1 and O_2 open in $\hat{\mathbf{C}}$ and such that $O_1 \cap \hat{B}^c$ and $O_2 \cap \hat{B}^c$ is a separation of $\hat{B}^c$. Since $\infty \in \hat{B}^c$, one of the sets O_1, O_2 must contain ∞; let it be O_2. Since O_2 contains a disk with center at ∞, O_1 must be bounded, and then so is $O_1 \cap \hat{B}^c$. We now show that $O_1 \cap \hat{B}^c$ is closed and therefore is compact. If $O_2 \cap \hat{B}^c = \{\infty\}$, then $O_1 \cap B^c = B^c$, which is closed in $\mathbf{C}$, since B is open in $\mathbf{C}$. If $O_2 \cap \hat{B}^c$ contains points other than ∞, then $(O_2 \sim \{\infty\}) \cap B^c$ and $O_1 \cap B^c$ form a separation in $\mathbf{C}$ of the closed set B^c, and so by Exercise 5.3.14, both sets in the separation are closed in $\mathbf{C}$. Thus, in particular, $O_1 \cap B^c$ is a compact subset of the open set O_1.

This sets the stage for an application of Theorem 12.2.1, with O_1 playing the role of the O in that theorem and $O_1 \cap B^c$ playing the role of F. Theorem

12.2.1 assures the existence of a finite set of line segments $\gamma_1, \gamma_2, \ldots, \gamma_n$ with graphs in $O_1 \sim (O_1 \cap B^c) = O_1 \cap B \subset B$ such that for every function $f\colon O_1 \to \mathbf{C}$ analytic in O_1,

$$f(z) = \sum_{j=1}^{n} \frac{1}{2\pi i} \int_{\gamma_j} \frac{f(\zeta)\, d\zeta}{\zeta - z}, \tag{12.8.1}$$

Moreover, according to Propositions 12.2.1 and 12.2.2, the segments γ_j can be chosen so that they form one or more closed polygonal paths, which we here designate by $\gamma^1, \gamma^2, \ldots, \gamma^p$. In (12.8.1), we let $f(z) = 1$ for every $z \in O_1$ and take z to be a point a in $O_1 \cap B^c$. Then (12.8.1) becomes

$$\begin{aligned} 1 &= \frac{1}{2\pi i} \int_{\gamma^1} \frac{d\zeta}{\zeta - a} + \frac{1}{2\pi i} \int_{\gamma^2} \frac{d\zeta}{\zeta - a} + \cdots + \frac{1}{2\pi i} \int_{\gamma^p} \frac{d\zeta}{\zeta - a} \\ &= \operatorname{Ind}_{\gamma^1}(a) + \operatorname{Ind}_{\gamma^2}(a) + \cdots + \operatorname{Ind}_{\gamma^p}(a). \end{aligned} \tag{12.8.2}$$

Since their sum is not zero, at least one of the topological indices in (12.8.2) must be different from zero, so the corresponding closed path (of which the graph lies in B) is not B-homologous to zero. Q.E.D.

13

Conformal Mapping

13.1 The Conformal Property of the Mapping Given by an Analytic Function

Of the various special properties possessed by an analytic function f, one of the most remarkable and valuable is the fact that at a point z_0 in the domain of f at which $f'(z_0) \neq 0$, the local mapping given by the function preserves angles between curves intersecting at z_0. This section is devoted to making this statement precise and then proving it. The proof is extremely simple, once an exact formulation has been achieved.

A *ray* in **C** with *initial point* z_0 and *direction* p, with $p \in \mathbf{C}$ and $|p| = 1$, is defined to be the ordered point set R: $\{z: z = z_0 + pt, 0 \leq t < \infty\}$. Let the rays R_1 and R_2 have the same initial point z_0 and have respective directions p_1 and p_2. The *angle from* R_1 *to* R_2 is defined to be the number $\theta = \arg p_2 - \arg p_1$, where the determinations of the arguments are chosen so that $0 \leq \theta < 2\pi$. (This implies that if $p_1 = p_2$, then $\theta = 0$. In the intuitive geometric sense, our definition of θ gives the angle at z_0 between R_1 and R_2 measured in a counterclockwise sense from R_1.)

We now extend the concept of the angle between two rays to that of the angle between two intersecting arcs. To avoid unnecessary complications, we deal only with *simple* arcs. A simple arc (also called a Jordan arc) is a *one-to-one* continuous mapping of an interval $[\alpha, \beta]$ into **C**. Geometrically speaking, such an arc does not intersect itself. (The related concept of a simple closed curve was introduced in the remarks in Sec. 12.6.) Let γ be such an arc, with parametric equation $z = \gamma(t)$, $\alpha \leq t \leq \beta$, $\beta > \alpha$. Take any real numbers t_0 and h such that $\alpha \leq t_0 < t_0 + h \leq \beta$. An equation of the ray with initial point $\gamma(t_0)$ and passing through $\gamma(t_0 + h)$ is $z = \gamma(t_0) + p(t_0, h)\tau$, $0 \le \tau < \infty$, in which the direction is given by

$$p(t_0, h) = \frac{\gamma(t_0 + h) - \gamma(t_0)}{|\gamma(t_0 + h) - \gamma(t_0)|}. \tag{13.1.1}$$

If $\lim_{h\to 0} p(t_0, h) = p(t_0)$, $h > 0$, exists, then $p(t_0)$ is called the *right direction* of γ at $\gamma(t_0)$ and the ray with equation $z = \gamma(t_0) + p(t_0)\tau$, $0 \le \tau < \infty$, is called the *right half-tangent* to γ at $\gamma(t_0)$. If $\gamma'(t_0)$ exists and is not zero, then by dividing the numerator and denominator of (13.1.1) by h it is easily verified that $p(t_0)$ does exist and equals $\gamma'(t_0)/|\gamma'(t_0)|$. (In advanced calculus texts, $p(t_0)$, treated as a located vector in $\mathbf{R}^2$ with initial point $\gamma(t_0)$, is often called the "unit tangent vector to γ at $\gamma(t_0)$ in the direction of γ.")

Now let γ_1 be a simple arc with parametric equation $z = \gamma_1(t)$, $\alpha_1 \le t \le \beta_1$, and let γ_2 be another simple arc with equation $z = \gamma_2(u)$, $\alpha_2 \le u \le \beta_2$. Furthermore, suppose that the graphs of these arcs intersect at the point $z_0 = \gamma_1(t_0) = \gamma_2(u_0)$, where $\alpha_1 \le t_0 < \beta_1$, $\alpha_2 \le u_0 < \beta_2$. If the respective right directions $p_1(t_0)$ and $p_2(u_0)$ of γ_1 and γ_2 both exist, we define the angle θ_{12} from γ_1 to γ_2 at z_0 to be the angle from the right half-tangent of γ_1 at z_0 to the right half-tangent of γ_2 at z_0; that is, $\theta_{12} = \arg p_2(u_0) - \arg p_1(t_0)$, where the arguments are determined so that $0 \le \theta_{12} < 2\pi$.

Example Let γ_1 be given by $z = t + it^2$, $0 \le t \le 4$ and γ_2 by $z = u + i(1 - u)^2$, $0 \le u \le 4$. The graphs are segments of familiar parabolas with equations $y = x^2$, $y = 1 - x^2$. They intersect at $z_0 = 1/\sqrt{2} + i/2$, at which $t = t_0 = 1/\sqrt{2}$ and $u = u_0 = 1/\sqrt{2}$. Then $p_1(t_0) = \gamma_1'(1/\sqrt{2})/|\gamma_1'(1/\sqrt{2})| = (1 + i\sqrt{2})/|1 + i\sqrt{2}|$, $p_2(u_0) = (1 - i\sqrt{2})/|1 - i\sqrt{2}|$, and $\theta_{12} = \arg(1 - i\sqrt{2}) - \arg(1 + i\sqrt{2})$. Let $\alpha = \operatorname{Arg}(1 + i\sqrt{2}) = 0.96$ radians (principal argument). Then one possible determination of $\arg(1 - i\sqrt{2})$ is $-\alpha$, but we must instead choose the determination $2\pi - \alpha$ so that $0 \le \theta_{12} < 2\pi$. Then $\theta_{12} = 2\pi - 2\alpha = 4.4$ radians. The usual computation given in the elementary calculus books yields the equation $\tan \theta_{12} = 2\sqrt{2}$, and the angle with the smallest positive measure satisfying this equation is approximately 1.23 radians. This is the angle from the right half-tangent of γ_1 at z_0 to the unit tangent vector to γ_2 at z_0 in the opposite direction from that of γ_2.

After these preliminaries, we are ready to discuss the conformal property of mappings by analytic functions. First, we present a definition. Let O be an open set in $\mathbf{C}$, and let $f\colon O \to \mathbf{C}$ be continuous on O. The mapping f is said to be *conformal* or *angle-preserving* at $z_0 \in O$ if the following conditions are all fulfilled:

(i) There exists a neighborhood $N(z_0)$ of z_0 such that f is a homeomorphism from $N(z_0)$ onto a neighborhood $N(w_0)$ of $w_0 = f(z_0)$.

(ii) Given any two simple arcs γ_1 and γ_2 with graphs in $N(z_0)$ intersecting at z_0 and possessing right directions at z_0, the arcs $\gamma_1^* = f \circ \gamma_1$ and $\gamma_2^* = f \circ \gamma_2$ also possess right directions at w_0.

(iii) The angle from γ_1 to γ_2 at z_0 equals the angle from γ_1^* to γ_2^* at w_0.

Briefly, f is conformal at z_0 if it preserves angles and their orientations.

Theorem 13.1.1 *Let B be a region in* **C** *and let $f: B \to$* **C** *be analytic and nonconstant in B. Then f is conformal at every point $z_0 \in B$ at which $f'(z_0) \neq 0$.*

Proof. The fact that f satisfies condition (i) for conformality is established by Corollary 2 of Theorem 10.4.1. Let $N(z_0)$ and $N(w_0)$ be the homeomorphic neighborhoods of z_0 and $w_0 = f(z_0)$, and let γ_1 and γ_2 be simple arcs with graphs in $N(z_0)$ intersecting at z_0. Their images γ_1^* and γ_2^* under the mapping f are again simple arcs of which the graphs intersect at w_0. Now assume that γ_1 and γ_2 possess respective right directions p_1 and p_2 at z_0. We first show that γ_1^* has a right direction at w_0. Let a parametric equation of γ_1 be $z = \gamma_1(t)$, $\alpha \leq t \leq \beta$, and let $z_0 = \gamma_1(t_0)$, $\alpha \leq t_0 < \beta$. The corresponding parametric equation of $\gamma_1^* = f \circ \gamma_1$ is, of course, $w = f[\gamma_1(t)]$, $\alpha \leq t \leq \beta$, and $w_0 = f[\gamma_1(t_0)]$. According to Exercise 10.1.1 and Proposition 10.1.2, since $f'(z_0) \neq 0$, there exists a function $g(z)$ analytic and nonzero in some neighborhood $N'(z_0)$ of z_0 and such that $f(z) - f(z_0) \equiv (z - z_0)g(z)$ for every z in $N'(z_0)$. Clearly, $g(z_0) = f'(z_0)$. With $h > 0$ chosen so that $\gamma_1(t) \in N'(z_0) \cap N(z_0)$, $t_0 \leq t \leq t_0 + h$, we have the following computation for the direction (13.1.1) of γ_1^* and w_0:

$$p(t_0, h) = \frac{f[\gamma_1(t_0 + h)] - f[\gamma_1(t_0)]}{|f[\gamma_1(t_0 + h)] - f[\gamma_1(t_0)]|}$$

$$= \frac{\gamma_1(t_0 + h) - \gamma_1(t_0)}{|\gamma_1(t_0 + h) - \gamma_1(t_0)|} \cdot \frac{g[\gamma_1(t_0 + h)]}{|g[\gamma_1(t_0 + h)]|},$$

$$\lim_{h \to 0} p(t_0, h) = p_1 \frac{f'(z_0)}{|f'(z_0)|}. \tag{13.1.2}$$

Similarly, the right direction of γ_2^* at w_0 exists and equals $p_2 f'(z_0)/|f'(z_0)|$.

We now examine the relationship between the angle $\theta_{12} = \arg p_2 - \arg p_1$ from γ_1 to γ_2 at z_0 and the angle

$$\theta_{12}^* = \arg \frac{f'(z_0)}{|f'(z_0)|} p_2 - \arg \frac{f'(z_0)}{|f'(z_0)|} p_1$$

from γ_1^* to γ_2^* at w_0. It is agreed that the arguments must all be chosen so that $\theta \leq \theta_{12}, \theta_{12}^* < 2\pi$. According to Proposition 7.5.1(i),

$$\theta_{12}^* = \arg \frac{f'(z_0)}{|f'(z_0)|} + \arg p_2 - \arg \frac{f'(z_0)}{|f'(z_0)|} - \arg p_1 + 2\pi k$$

$$= \theta_{12} + \arg \frac{f'(z_0)}{|f'(z_0)|} - \arg \frac{f'(z_0)}{|f'(z_0)|} + 2\pi k, \tag{13.1.3}$$

for some integer k. Any two arguments of $f'(z_0)/|f'(z_0)|$ differ by an integer multiple of 2π, so (13.1.2) can be rewritten in the form

$$\theta^*_{12} = \theta_{12} + 2\pi K$$

where K is some integer. But because $0 \leq \theta_{12} < 2\pi$, the only way to choose K so that $0 \leq \theta^*_{12} < 2\pi$ is to set $K = 0$. Therefore, $\theta^*_{12} = \theta_{12}$. Q.E.D.

According to Remark (iii) in Sec. 10.4 and Proposition 10.4.1, if $f: B \to \mathbf{C}$ is analytic and *univalent* in the region B, then it maps B onto a region B^* and $f'(z) \neq 0$ for every $z \in B$. Also the inverse function f^{-1} is analytic on B^*, and by the inverse function law of differentiation [Theorem 6.1.3] $(f^{-1})'$ exists an is nonzero everywhere on B^*. We have the following corollary.

Corollary Let $f: B \to \mathbf{C}$ be analytic and univalent on the region B, and let $B^* = f(B)$. Then f is conformal at every point in B, and f^{-1} is conformal at every point in the region B^*.

Remark (i). A mapping of a region B onto another region is said to be a *conformal mapping* if it is a one-to-one, and also conformal at each point of B. A synonym frequently used for a conformal mapping is a conformal transformation. Thus, a univalent analytic function is a conformal mapping or conformal transformation.

Remark (ii). There seems to be a minor lack of agreement among the authors of the classical complex analysis textbooks as to the exact meaning to be given to the terms "conformal mapping" and "conformal at a point." A typical variant is given by Hille [9, p. 92], who defines a mapping to be conformal at a point z_0 if it preserves angles without necessarily preserving their orientations, and if at the same time it is a *pure magnification* at z_0 (defined in Exercise 13.1.3 and Remark (v)). Our definition of a conformal mapping, in which it is specified than the orientation of angles is to be preserved, is sometimes said to be the definition of "directly conformal," whereas if orientations are reversed, the mapping is said to be "inversely conformal." (The mapping given by $w = \bar{z}$ is inversely conformal.) In differential geometry, a mapping is called "conformal" if it is either directly or inversely conformal. The last sentence in Remark (i) is valid for all of the usual interpretations of conformality.

Remark (iii). There is a converse to Theorem 13.1.1 which can be stated as follows: Let $w = f(z) = f(x + iy)$ be continuous in a region B together with its first partial derivatives with respect to x and y, let $(\partial f/\partial x)^2 + (\partial f/\partial x)^2$ be nonzero at every point in B, and suppose that the mapping is conformal at every point of B. Then f is analytic on B. For a proof, see [9, p. 93].

Exercises 13.1

1. Show that $w = 3z + z^3$ gives a conformal mapping of $\Delta(0, 1)$ onto some region in the complex w-plane. (*Hint*: Remark (i); check univalence by showing that if $|z_1| < 1$, $|z_2| < 1$, and $3z_1 + z_1^3 = 3z_2 + z_2^3$ then $z_1 = z_2$.)
2. Let γ be an arc with parametric equation $z = \gamma(t)$, $\alpha \leq t \leq \beta$, $\alpha < \beta$. Recall from Sec. 8.2 that a change of parameter for γ is a function $g: [\alpha_1, \beta_1] \to [\alpha, \beta]$ which is

one-to-one, continuous, strictly increasing, and such that $g(\alpha_1) = \alpha$, $g(\beta_1) = \beta$. Show that if γ is a simple arc and has a right direction at $z_0 = \gamma(t_0)$, $\alpha \le t_0 < \beta$, then the arc $\gamma_1 = \gamma \circ g$ obtained from γ by a change of parameter also is a simple arc and has a right direction at z_0, which is the same as that of γ. (*Hint*: Apply Exercise 3.2.10 to $\gamma_1 \mid [\tau_0, \beta_1]$ where $g(\tau_0) = t_0$.)

Remark (iv). The implication of the exercise is that the angle between two intersecting arcs is invariant under changes of parameter.

3. Let $f: O \to \mathbf{C}$ be continuous on the open set $O \subset \mathbf{C}$. For a given $z_0 \in O$, the *magnification* of the mapping f at z_0 in the direction $p, p \in \mathbf{C}, |p| = 1$, is defined by

$$M_p(z_0) = \lim_{h \to 0} \left| \frac{f(z_0 + hp) - f(z_0)}{z_0 + hp - z_0} \right|$$

$$= \lim_{h \to 0} \frac{|f(z_0 + hp) - f(z_0)|}{h}, \qquad h > 0,$$

provided that the limit exists. Show that if f is analytic at z_0, $M_p(z_0)$ does exist and is given by $|f'(z_0)|$ and, therefore, is independent of the direction p.

Remark (v). A mapping f for which the magnification exists and is independent of the direction at each point of the domain of f (assumed to be an open set) is sometimes called a *pure magnification*. Thus, the mapping given by any analytic function is a pure magnification.

4. The purpose of this exercise is to investigate what happens to angles between arcs in a mapping by an analytic function f when the condition $f'(z_0) \neq 0$ in Theorem 13.1.1 is replaced by the condition that $f'(z)$ has a zero of order $m - 1$ at z_0, $m \ge 2$. The notation and terminology is that of Theorem 13.1.1 and its proof. The key modification in the proof is that now according to Exercise 10.1.1 and Proposition 10.1.2, there is a local representation of the form $f(z) - f(z_0) \equiv (z - z_0)^m g(z)$, where $g(z)$ is analytic and nonzero. Show that
   (a) if p_1 is the direction of γ_1 at z_0, then the direction of $\gamma_1^* = f \circ \gamma_1$ exists and equals $p_1^m g(z_0)/|g(z_0)|$;
   (b) if the angle θ_{12} from γ_1 to γ_2 at z_0 satisfies inequality $0 \le \theta_{12} < 2\pi/m$, then the angle θ_{12}^* from γ_1^* to γ_2^* at w_0 is given by $\theta_{12}^* = m\theta_{12}$.

13.2 Conformal Equivalences and a Review of Some Relevant Results in Earlier Chapters

Two regions B and B^* in $\mathbf{C}$ are said to be *conformally equivalent* if there exists a univalent analytic function $\phi: B \to B^*$ such that $B^* = \phi(B)$. The mapping ϕ is called a *conformal equivalence*. If a function f is analytic on the region B^*, the composition $f \circ \phi$ transforms f into a function analytic on B, and if F is analytic on B, the composition $F \circ \phi^{-1}$ transforms F into a function analytic on B^* [Corollary 2 of Theorem 10.4.1]. Let $\mathscr{A}(B)$ be the family of functions analytic on B and $\mathscr{A}(B^*)$ the family of functions analytic on B^*. It is easy to verify that the transformation T: $B^* \to B$ given by $T(f) = f \circ \phi$, $f \in \mathscr{A}(B^*)$, sets up a one-to-one correspondence between the members of $\mathscr{A}(B)$ and $\mathscr{A}(B^*)$. The transformation, of course, preserves sums and products of functions. These facts are the basis of a standard technique for problem solving in complex analysis and its practical applications, which follows. If B has a simple structure (and the usual cases are that B is the unit disk or a half-plane), then certain problems concerning $\mathscr{A}(B^*)$ may usefully be carried over to $\mathscr{A}(B)$ by the composition transformation and solved in $\mathscr{A}(B)$; thereafter the solution is transferred back to $\mathscr{A}(B^*)$.

Many of the theorems and exercises in the earlier chapters of this book have relevance to the study of conformal equivalences. We review here, in the language of conformal mapping, the results of this type which seem to be particularly pertinent. The order of presentation is generally that of their first appearance in the text.

First, we list examples of special conformal equivalences. In each example the conformal equivalence is given by a function $w = \phi(z)$ which maps the indicated region B in the z-plane onto the indicated region B^* in the w-plane.

Examples

(i) $w = u + iv = [Rz + (1/Rz)]/2$, $R > 1$; $B = \{z : |z| > 1\}$, $B^* = \{(u, v) : (u/\alpha)^2 + (v/\beta)^2 > 1\}$, $\alpha = R + (1/R)$, $\beta = R - (1/R)$. Also $B = \{z : 0 < |z| < 1/R^2\}$, B^* as before [Exercise 1.14.17].

(ii) $w = -z^2$; $B = \{z : 0 < \operatorname{Re} z < p, -\infty < \operatorname{Im} z < \infty\}$, $p > 0$, $B^* = \{w = w + iv : u > -p^2, v^2 < 4p^2(u + p^2)\} \sim \{w : u \geq 0\}$ [Exercise 1.14.18].

(iii) $w = z^2$; $B = \{z = x + iy : x > p,\ y^2 < x^2 - p\}$, $p > 0$, $B^* = \{w : \operatorname{Re} w > p\}$. Also $B = \{z : x < -p, p^2 < x^2 - p\}$, B^* as before.

(iv) $w = (az + b)/(cz + d)$, $a, b, c, d \in \mathbf{C}$, $ad - bc \neq 0$. If $c = 0$, $B = \mathbf{C} = B^*$. If $c \neq 0$, $B = \mathbf{C} \sim \{-d/c\}$, $B^* = \mathbf{C} \sim \{a/c\}$ [Sec. 1.16]. In any case, the transformation is a homeomorphism from $\hat{\mathbf{C}}$ onto $\hat{\mathbf{C}}$.

(v) $w = (1 + z)/(1 - z)$; $B = \{z : |z| < 1\}$, $B^* = \{w : \operatorname{Re} w > 0\}$ [Exercise 1.16.6]. Also $B = \{z : |z| > 1\}$, $B^* = \{w : \operatorname{Re} w < 0, w \neq -1\}$ [Exercise 1.16.8].

(vi) $w = 1/(z - z_0 - r)$, $r > 0$; $B = \{z : |z - z_0| < r\}$, $B^* = \{w : \operatorname{Re} w < -1/2r\}$ [Exercise 1.16.7]. Also $B = \{z : |z - z_0| > r\}$, $B^* = \{w : \operatorname{Re} w > -(1/2)r, w \neq 0\}$ [Exercise 1.16.8].

(vii) $w = e^z$; $B = \{z = x + iy : -\infty < x < \infty,\ y_0 < y < y_0 + 2\pi\}$, y_0 real, $B^* = \mathbf{C} \sim \{w : w = te^{iy_0}, 0 \leq t < \infty\}$ [Theorem 7.5.1].

(viii) $w = i(u_0 + \pi) + \operatorname{Log} z$, u_0 real; $B = \mathbf{C} \sim (-\infty, 0]$, $B^* = \{w = u + iv : -\infty < u < \infty, u_0 < u < u_0 < 2\pi\}$ [Exercise 7.5.9].

(ix) $w = z^\alpha$, $0 < \alpha < 1$ (principal logarithm); $B = C \sim (-\infty, 0]$, $B^* = \{w = u + iv : u > 0, -\alpha\pi < \arg w < \alpha\pi\}$ [Exercise 7.6.11].

We now present some theoretical results relevant to conformal equivalence, each of which is a translation (or at least a more or less direct consequence) of previously established facts. Proofs, where needed, are easy and are deferred to the exercises.

(A) A bounded region cannot be conformally equivalent to **C**. [Theorem 9.6.1 (Liouville); see Exercise 13.2.1].

(B) A nonconstant polynomial function maps **C** onto **C**, but the only conformal equivalences which are polynomial functions and which map **C** onto **C** are the linear functions $w = az + b$, $a, b \in \mathbf{C}$. [Theorem 9.6.2, Exercise 9.6.9, Proposition 10.4.1(ii); see Exercise 13.2.2.]

(C) A conformal equivalence mapping $\Delta(0, 1)$ onto $\Delta(0, 1)$ is a linear fractional transformation of the type $\phi(z) = e^{i\alpha}(z - z_0)(z - \bar{z}_0 z)^{-1}$ [Proposition 9.8.1].

(D) A conformal equivalence ϕ mapping a region B onto $\Delta(0, 1)$ is uniquely determined to within a rotation by the condition that for a given $z_0 \in B$, $\phi(z_0) = 0$ [Proposition 9.8.2]. Either of the following two further conditions complete the unique determination: (i) for a given $z_1 \in B$, $z_1 \neq z_0$, and a given $w_1 \in \Delta(0, 1)$, $\phi(z_1) = w_1$; (ii) Arg $\phi'(z_0)$ is given; in particular, $\phi'(z_0) > 0$ [corollary of Proposition 9.8.2].

(E) A conformal equivalence mapping $\Delta(0, 1)$ onto itself and having two distinct fixed points is necessarily the identity transformation $\phi(z) = z$ [Exercise 9.8.4].

(F) If f is an analytic function mapping $\Delta(0, 1)$ onto a region B_1, and if g is a conformal equivalence mapping $\Delta(0, 1)$ onto a region B_2, and if $B_1 \subset B_2$ and $f(0) = g(0)$, then $f[\Delta(0, r)] \subset g[\Delta(0, r)]$ for every r, $0 < r < 1$ [Exercise 9.8.5].

(G) Let $\langle f_n \rangle_1^\infty$ be a sequence of conformal equivalences, in which f_n maps a fixed region B onto a region B_n, $n = 1, 2, \ldots$. If $\langle f_n \rangle$ converges almost uniformly on B to a nonconstant function f, then f is a conformal equivalence mapping B onto some region B^* [Exercise 10.3.8]. See also Exercise 13.2.3.

(H) The derivative of a conformal equivalence mapping B onto B^* has no zeros in B, and the derivative of the inverse function exists everywhere on B^* and has no zeros in B^* [Proposition 10.4.1(ii)].

(I) A conformal equivalence mapping **C** onto **C** is necessarily a polynomial [Proposition 10.7.4, Theorem 10.7.2, Exercise 13.2.4], and, therefore, by (B) it must be a linear function $w = az + b$.

Exercises 13.2

1. Show that Liouville's Theorem implies (A).
2. Show that the references cited imply (B).

3. This refers to (G). Let $B_n^* = f_n(B), n = 1, 2, \ldots$. Show that each point w of B^* lies in $B_N^* \cap B_{N+1}^* \cap B_{N+2}^* \cap \cdots$, where $N = N(w) \geq 1$. (*Hint*: Let $w_0 = f(z_0)$ be a typical point in B^*. Then $\langle f_n(z) - w_0 \rangle$ converges a.u. on B to $f(z) - w_0$. If there were to exist an infinite subsequence $\langle B_{n_j}^* \rangle_{j=1}^{\infty}$ of $\langle B_n^* \rangle$ such that $w_0 \notin B_{n_j}^*$, $j = 1, 2, \ldots$, then $f_{n_j}(z) - w_0 \neq 0$ for every $z \in B$, $j = 1, 2, \ldots$. However, $f(z_0) - w_0 = 0$. Now see what Corollary 2 of Theorem 10.3.2 implies.)
4. With reference to (I), show that a conformal equivalence mapping **C** onto **C** is necessarily a nonconstant polynomial function. (*Hint*: Let the conformal equivalence be given by $w = \phi(z)$. It has an isolated singularity at ∞, and the problem is to show that this must be a pole, for then Theorem 10.7.2 would indicate that ϕ is either a constant function or a nonconstant polynomial. Of course, the constant function case is impossible. The only other case to examine is that in which ϕ has an essential singularity at ∞. In any case, ϕ maps $\Delta(0, 1)$ in the z-plane onto a bounded region B^* in the w-plane. If ϕ were to have an essential singularity at ∞, what would Proposition 10.7.4(iii) imply concerning the mapping $\phi : \Delta(\infty, 1) \to \mathbf{C}$? Would there be image points in B^*?)
5. Show that the function $\zeta = 1/\bar{z}$, which gives a continuous one-to-one mapping of $\mathbf{C} \sim \{0\}$ onto itself, is, nevertheless, conformal at no point in its domain. Thereafter, find a function $w = f(\zeta)$ which again gives a continuous one-to-one mapping of $\mathbf{C} \sim \{0\}$ onto itself and again is conformal at no point, but which is such that $w = f(1/\bar{z})$ is a conformal equivalence mapping $\mathbf{C} \sim \{0\}$ onto itself.
6. Let Γ be a closed square path in the z-plane with vertices at $(\pm 1, \pm 1)$. Show that $w = 1/z$ gives a conformal equivalence mapping the unbounded component of Γ^c (see Sec. 8.4, Example (ii)) onto the union of four circular disks in the w-plane, each with radius $1/2$ and with respective centers at $\pm 1/2, \pm i/2$).
7. Show that $w = e^z$ gives a conformal equivalence mapping the rectangular region $\{z = x + iy : \alpha < x < \beta, -\pi < y < \pi\}$ onto a certain annular region in the w-plane from which the points on the negative real axis are deleted. (*Hint*: Use polar coordinates in the w-plane.)
8. Find a conformal equivalence mapping $\{z = x + iy : -\infty < x < \infty, -\pi < y < \pi\}$ onto $\Delta(0, 1)$ in the w-plane so that $z = 0$ goes onto $w = 0$. (*Hint*: Examples (vii), (ix), and (v).)

   Answer. $w = (e^{z/2} - 1)/(e^{z/2} + 1)$.

13.3 The Stieltjes-Vitali Convergence Theorem and Normal Families

In this section, we develop certain results concerning the convergence of sequences of analytic functions which are the basis for the proof of the Riemann mapping theorem given in the next section. The two principal theorems in this section are of importance for themselves alone and rank among the more famous propositions of classical complex function theory. They have the feature in common that a conclusion of almost uniform (a.u.) convergence of a sequence or subsequence of functions analytic in a region is obtained from apparently very weak hypotheses. (Recall that a sequence of functions is said to converge a.u. on a region B if it converges uniformly on each compact subset of B.)

Theorem 13.3.1 (*Stieltjes-Vitali*). *Let $\langle f_n \rangle_1^\infty$ be a sequence of functions, each analytic on a region $B \subset \mathbf{C}$, and let the set of functions $\{f_n\}$ be uniformly bounded on each compact subset of B. Furthermore, let $S \subset B$ be an infinite subset of B with an accumulation point in B, and let $\lim_{n\to\infty} f_n(z)$ exist for every $z \in S$. Then $\langle f_n \rangle$ converges a.u. on B.*

Remark. The term "uniformly bounded on each compact subset of B" means that if F is such a subset, then there exists $M(F) > 0$ such that $|f_n(z)| \le M(F)$ for every $z \in F$, $n = 1, 2, \ldots g$ but $M(F)$ may change with F.

Proof. We first prove a special case of the theorem contained in the following lemma. Thereafter the general case is established by a simple connectivity argument. Thus, the special case is really the heart of the theorem.

Lemma Let $\langle f_n \rangle_1^\infty$ be a sequence of functions each analytic on an open set O which contains a closed disk $\bar{\Delta}(a, 2r)$, $r > 0$. Suppose that there exists $M > 0$ such that $|f_n(z)| \le M$ for every $z \in \bar{\Delta}(a, 2r)$; $n = 1, 2, \ldots$. Let $T \subset \Delta(a, r)$ be an infinite point set of which a is an accumulation point, and let $\lim_{n\to\infty} f_n(z)$ exist at each point of T. Then $\langle f_n \rangle$ converges uniformly on $\bar{\Delta}(a, r)$.

Proof of the Lemma. (i) Let the local power series for f_n at a be $f_n(z) = \sum_{h=0}^\infty a_h(n)(z - a)^h$. It converges and represents f_n in $\bar{\Delta}(a, 2r)$. We first show that $\langle a_h(n) \rangle_{n=1}^\infty$ is a Cauchy sequence for each fixed h, $h = 0, 1, 2, \ldots$. We first consider $a_0(n) = f_n(a)$. By hypothesis, for any $z \in \Delta(a, 2r)$, $|f_n(z) - f_n(a)| \le |f_n(z)| + |f_n(a)| \le 2M$, and, of course, $f_n(z) - f_n(a) = 0$ if $z = a$. Therefore, by Schwarz's lemma in the form given in Exercise 9.8.1, $|f_n(z) - f_n(a)| \le 2M|z - a|/2r = M|z - a|/r$ for every $z \in \Delta(a, 2r)$. Now for every $z \subset T$ (indeed, or every $z \in \Delta(a, 2r)$)

$$
\begin{aligned}
|a_0(n) - a_0(m)| &= |f_n(a) - f_m(a)| \\
&\le |f_n(a) - f_n(z)| + |f_n(z) - f_m(z)| + |f_m(z) - f_m(a)| \\
&\le 2M|z - a|r + |f_n(z) - f_m(z)|. \qquad (13.3.1)
\end{aligned}
$$

Choose any $\varepsilon > 0$. First, let $z_0 \in T$ be such that $|z_0 - a| < r\varepsilon/4M$. Now the sequence $\langle f_n(z_0) \rangle_1^\infty$ is convergent by hypothesis, so it is a Cauchy sequence. Choose $N > 0$ so that for $n > m \ge N$, $|f_n(z_0) - f_m(z_0)| < \varepsilon/2$. Substituting into (13.3.1), we obtain

$$
|a_0(n) - a_0(m)| < \frac{2Mr\varepsilon}{4Mr} + \frac{\varepsilon}{2} = \varepsilon.
$$

Thus, $\langle a_0(n) \rangle_{n=1}^\infty$ is a Cauchy sequence.

(ii) We now extend this result to the other coefficients $a_h(n)$ by induction on h. We set up the following function: for $z \in \bar{\Delta}(a, 2r) \sim \{a\}$,

$$\phi_{h,n}(z) = \phi_n(z) = \frac{f_n(z) - \sum_{j=0}^{h} a_j(n)(z-a)^j}{(z-a)^h}$$
$$= a_{h+1}(n) + a_{h+2}(n)(z-a) + a_{h+3}(n)(z-a)^2 + \cdots,$$

and for $z = a$, $\phi_n(a) = a_{h+1}(n)$. This function is the sum in $\Delta(a, 2r)$ of a power series, so it is analytic in this disk, and clearly ϕ_n is continuous in $\bar{\Delta}(a, 2r)$. We now assume as the induction hypothesis that $\langle a_j(n)\rangle_{n=1}^{\infty}$ converges for $j = 0, 1, 2, \ldots, h$. In particular, this implies that there exists $A > 0$ such that $|a_j(n)| \le A$, $n = 1, 2, \ldots$; $j = 0, 1, 2, \ldots, h$ [Exercise 2.3.4]. For every $z \in C(a, 2r)$,

$$|\phi_n(z)| \le \frac{M + A\sum_{j=0}^{h}(2r)^j}{(2r)^h} = M_h.$$

Therefore, by the maximum modulus principle in the version given in Corollary 2 of Theorem 9.7.1, $|\phi_n(z)| \le M_h$ for every $z \in \bar{\Delta}(a, 2r)$. We can now repeat the proof given in paragraph (i) with f_n replaced by ϕ_n, $a_0(n)$ replaced by $a_{h+1}(n)$, and M by M_h, and thereby establish that $\langle a_{h+1}(n)\rangle_{n=1}^{\infty}$ is a Cauchy sequence.

(iii) Now we show that $\langle f_n\rangle_1^{\infty}$ has the uniform Cauchy property on $\bar{\Delta}(a, r)$ and, therefore, converges uniformly on $\bar{\Delta}(a, r)$ [Theorem 3.3.2]. The Cauchy coefficient estimates (9.5.3) yield the inequalities

$$|a_h(n)| \le \frac{M}{(2r)^h}, \qquad h = 0, 1, 2, \ldots.$$

For every $z \in \bar{\Delta}(a, r)$ and for $n, m \ge 1$, $H \ge 1$,

$$\begin{aligned}|f_n(z) - f_m(z)| &= \left|\left(\sum_{h=0}^{H} + \sum_{h=H+1}^{\infty}\right)[a_h(n) - a_h(m)](z-a)^h\right| \\ &\le \sum_{h=0}^{H}|a_h(n) - a_h(m)|r^h + \sum_{h=H+1}^{\infty}[|a_h(m)| + |a_h(n)|]r^h \\ &\le \sum_{h=0}^{H}|a_h(n) - a_h(m)|r^h + 2M\sum_{h=H+1}^{\infty}\frac{r^h}{(2r)^h} \\ &= \sum_{h=0}^{H}|a_h(n) - a_h(m)|r^h + \frac{M}{2^{H-1}}. \end{aligned} \tag{13.3.2}$$

Now take $\varepsilon > 0$. We first choose H so that $M/2^{H-1} < \varepsilon/2$ and then hold H fixed. Thereafter we use the Cauchy property of $\langle a_h(n)\rangle_{h=1}^{\infty}$ to choose $N > 0$ so that for $n > m \ge N$,

$$|a_h(n) - a_h(m)| < \frac{\varepsilon}{2r^h(H+1)}, \qquad h = 0, 1, \ldots, H.$$

Substituting into (13.3.2), we find that for every $z \in \bar{\Delta}(a, r)$ and $n > m \geq N$,

$$|f_n(z) - f_m(z)| < \sum_{h=0}^{H} \frac{\varepsilon r^h}{2r^h(H+1)} + \frac{\varepsilon}{2} = \frac{\varepsilon}{2} + \frac{\varepsilon}{2} = \varepsilon.$$

Thus, $\langle f_n \rangle$ has the uniform Cauchy property on $\bar{\Delta}(a, r)$. The proof of the lemma is complete [Theorem 3.3.1].

As already stated, we complete the proof of the theorem by a connectivity argument. Let F be a compact subset of B. Then $0 < \text{dist}\,(F, B^c) \leq \infty$ [Theorem 5.6.1]. Let d be such that $0 < 2d < \text{dist}\,(F, B^c)$. For every $z \in F$, $\bar{\Delta}(z, d) \subset B$. The family of open disks $\Omega = \{\Delta(z, d): z \in F\}$ is an open covering of the compact set F, so we can extract a finite subcovering from Ω, say $\{\Delta(z_1, d), \Delta(z_2, d), \ldots, \Delta(z_k, d)\}$. The plan is to show that $\langle f_n \rangle$ converges uniformly on each closed disk $\bar{\Delta}(z_j, d)$, $j = 1, \ldots, k$. It then follows at once that $\langle f_n \rangle$ converges uniformly on $\bigcup_{j=1}^{k} \bar{\Delta}(z_j, d)$ and so in particular on F.

We hold j fixed in what follows. Let b be the accumulation point of S referred to in the hypotheses of the theorem. If $b = z_j$, the lemma applies directly with $r = d$, $T = S \cap \Delta(z_j, d)$, and nothing more needs to be said. If $b \neq z_j$, let L be a polygonal arc joining b to z_j and contained in B. Let $3\eta = \text{dist}\,(L, B^c)$, and choose a set of points $\{b = z_0, z_1, z_2, \ldots, z_K = z_j\}$ on L in such a way that $|z_s - z_{s-1}| < \eta$, $s = 1, \ldots, K$. For each s, the center of the disk $\Delta(z_s, \eta)$ lies in disk $\Delta(z_{s-1}, \eta)$, and the disk $\bar{\Delta}(z_s, 2\eta)$ is contained in B. By hypothesis, $\langle f_n \rangle$ is uniformly bounded on each closed disk $\bar{\Delta}(z_s, 2\eta)$, $s = 1, \ldots, K$. The lemma can now be applied successively to the disks $\bar{\Delta}(b, \eta)$, $\bar{\Delta}(z_1, \eta), \ldots, \bar{\Delta}(z_K, \eta)$, and it establishes uniform convergence in each disk. In the application to $\bar{\Delta}(b, \eta)$, $a = b$, $r = \eta$, $T = S \cap \Delta(b, \eta)$. In the application to $\Delta(z_s, \eta)$ (assuming $\langle f_n \rangle$ converges uniformly on $\bar{\Delta}(z_{s-1}, \eta)$), $a = z_s$, $r = \eta$, $T = \Delta(z_{s-1}, \eta) \cap \Delta(z_s, \eta)$. When we finally reach $\Delta(z_K, \eta) = \Delta(z_j, \eta)$, we can replace this disk in the application of the lemma by the disk $\Delta(z_j, d)$, because $\bar{\Delta}(z_j, 2d) \subset B$ and $\langle f_n \rangle$ is uniformly bounded in $\bar{\Delta}(z_j, 2d)$. Q.E.D.

We now study an important subclass of the class $\mathscr{A}(B)$ of functions analytic on a given region B. An infinite subclass $\mathscr{F}$ of $\mathscr{A}(B)$ is defined to be a *normal family* in B if every infinite sequence from $\mathscr{F}$ contains a subsequence which converges a.u. on B. The limit functions may or may not be members of the normal family, but they are members of $\mathscr{A}(B)$ [Theorem 9.9.1].

Example. The functions $\{f_n(z) = nz/(z = n), n = 1, 2, \ldots\}$ constitute a normal family on the region $B = \mathbf{C} \sim \{-1, -2, -3, \ldots\}$. The limit functions are all equal to the identity function $f(z) = z$. Note that the limit function is *not* a member of this normal family.

Theorem 13.3.2 (*Montel*). *Let B be a region, and let $\mathscr{F}$ be a family of* *functions analytic in B and such that the members of $\mathscr{F}$ are uniformly bounded on each compact subset of B. Then $\mathscr{F}$ is a normal family in B.*

Proof. Take $b \in B$ and $r > 0$ such that $\bar{\Delta}(b, r) \subset B$. Let $\langle z_k \rangle_1^\infty$ be a sequence of distinct points in $\Delta(b, r)$ such that $\lim_{k \to \infty} z_k = b$. Let M be the uniform bound of the functions $f \in \mathscr{F}$ on $\bar{\Delta}(b, r)$. Finally, let $\langle f_n \rangle_1^\infty$ be a sequence from $\mathscr{F}$. The points $f_n(z_1)$, $n = 1, 2, 3, \ldots$, all lie in $\bar{\Delta}(0, M)$, so by the Bolzano-Weierstrass theorem [Exercise 5.5.4] they have an accumulation point $w_1 \in \bar{\Delta}(0, M)$. Therefore, there exists a subsequence of $\langle f_n(z_1) \rangle_1^\infty$, say $\langle f_{1m}(z_1) \rangle_{m=1}^\infty$, which converges to w_1 [Exercise 5.1.2].

Now the points $f_{1m}(z_2)$, $m = 1, 2, \ldots$, all lie in $\bar{\Delta}(0, M)$, so again by the Bolzano-Weierstrass theorem, they have an accumulation point $w_2 \in \bar{\Delta}(0, M)$. Therefore, there exists a subsequence of the sequence $\langle f_{1m}(z_2) \rangle$ which converges to w_2. We continue in this way and obtain an infinite sequence of infinite sequences $\langle f_{1,m} \rangle_1^\infty, \langle f_{2,m} \rangle_1^\infty, \langle f_{3,m} \rangle_1^\infty, \ldots, \langle f_{p,m} \rangle_1^\infty, \ldots$, each of which is a subsequence of each of the preceding sequences. The sequence $\langle f_{p,m} \rangle_1^\infty$, therefore, converges at each point at which the sequences $\langle f_{p-1,m} \rangle, \langle f_{p-2,m} \rangle, \ldots, \langle f_{1m} \rangle$ converge.

Now look at the "diagonal" sequence $f_{11}, f_{22}, f_{33}, \ldots$. Recall that the convergence of a sequence is not affected by discarding a block of initial terms. Therefore, since $\langle f_{n,n} \rangle_1^\infty$ converges at z_1, $\langle f_{n,n} \rangle_2^\infty$ converges at z_2, $\langle f_{n,n} \rangle_3^\infty$ converges at z_3, and so on, it follows that the sequence $\langle f_{n,n} \rangle_1^\infty$ converges at each of the points $\{z_k, k = 1, 2, \ldots\}$, of which $b \in B$ is an accumulation point. Thus, by Theorem 13.3.1, $\langle f_{n,n} \rangle_1^\infty$ converges a.u. in B.

But now we have demonstrated that the sequence $\langle f_n \rangle$, which was an arbitrary sequence from $\mathscr{F}$, has a subsequence which converges a.u. on B, so $\mathscr{F}$ by definition is a normal family. Q.E.D.

13.4 The Riemann Mapping Theorem

This theorem establishes the existence of conformal equivalences which map a given simply connected region onto another given simply connected region. The case in which one of the regions is the entire complex plane and the other region is bounded must be ruled out, as was seen in Sec. 13.2(A). The case in which each of the regions is the entire complex plane is covered by Sec. 13.2(B) and (I); the conformal equivalences in this case are the linear functions $w = az + b$, $a \neq 0$.

Theorem 13.4.1 *Let B be a region in* **C**, $B \neq$ **C**, *and such that every function analytic and without zeros in* **C** *has an analytic square root* [*Exercise 9.4.9*] *in B. Given* $z_0 \in B$, *there exists a conformal equivalence* ϕ *mapping B onto* $\Delta(0, 1)$ *with* $\phi(z_0) = 0$ *and* $\phi'(z_0) > 0$.

Remark (i). According to 13.2(D), the conditions $\phi(z_0) = 0$ and $\phi'(z_0) > 0$ uniquely determine the promised mapping.

Remark (ii). According to Theorem 12.8.1(vi), if a region B is simply connected in the sense that $\hat{\mathbf{C}} \sim B$ is connected in $\hat{\mathbf{C}}$, then every function analytic and without zeros in B has an analytic square root.

These two remarks together with Theorem 13.4.1 establish the following form of the famous theorem referred to in the heading of this section.

Corollary 1 (Riemann Mapping Theorem). Let B be a simply connected region which is not the entire complex plane. There exists a unique conformal equivalence ϕ mapping B onto $\Delta(0, 1)$ in such a way that for a given $z_0 \in B$, $\phi(z_0) = 0$ and $\phi'(z_0) > 0$.

Remark (iii). The proof of Theorem 13.4.1 fills an important gap in the proof of Theorem 12.8.1 given in Sec. 12.8. To establish completely the chain of equivalent statements in that theorem relating to the concept of simple connectivity, it was necessary in particular to show that if B is a region with the analytic square root property, then B is homeomorphic to $\Delta(0, 1)$. In Sec. 12.8 that was done only in the trivial case in which $B = \mathbf{C}$. Incidentally, this remark implies that in Corollary 1 to Theorem 13.4.1, it is unnecessary to be specific about which one of the various standard definitions of simple connectivity in $\mathbf{C}$ is being used.

Proof of Theorem 13.4.1. Let the region B have the analytic square root property described in the theorem, and let z_0 be a given point in B. Let $\mathscr{K}(B)$ be the class of functions which are analytic and univalent in B and such that if $g \in \mathscr{K}(B)$, then $|g(z)| < 1$ for every $z \in B$, and $g(z_0) = 0$, $g'(z_0) > 0$. The plan of the proof is to show first by a simple series of constructions that $\mathscr{K}(B)$ is not empty and, thereafter, that it contains an extremal member ϕ for which $\phi'(z_0)$ is as great as possible. Each function in $\mathscr{K}$ obviously maps B *into* $\Delta(0, 1)$, and it turns out that ϕ maps B *onto* $\Delta(0, 1)$.

Lemma 1 The class $\mathscr{K}(B)$ is not empty.

Proof. Let a be any point in B^c. Since $z - a$ is obviously analytic and never zero in B, it has an analytic square root $g_1(z)$: $[g_1(z)]^2 = z - a$, $z \in B$. We need to use the following two properties of g_1:

(i) g_1 is univalent (because if $g_1(z_1) = g_2(z_2)$, then $[g_1(z_1)]^2 = [g_2(z_2)]^2$ and $z_1 - a = z_2 - a$, so $z_1 = z_2$);

(ii) $g_1(z_1) \neq -g_1(z_2)$ for every $z_1, z_2 \in B$. (For if $g_1(z_1) = -g_1(z_2)$ with $z_1, z_2 \in B$, then as in (i) from $[g_1(z_1)]^2 = [-g_1(z_2)]^2$, we must conclude that $z_1 = z_2$. But the resulting equation, $g_1(z_1) = -g_1(z_1)$ could be true only if $g_1(z_1) = 0$, and clearly g_1 has no zeros in B.)

Let $w = g_1(z)$, $z \in B$, and let $w_0 = g_1(z_0)$. The point set $B^* = g(B)$ is open [Theorem 10.4.1], so it contains a disk $\Delta = \{w: |w - w_0| < r, r > 0\}$. Let $S = \{z: z \in B, w = g_1(z) \in \Delta\}$. (In what follows it must be kept in mind that each point $w \in \Delta$ lies in B^* and so has a preimage point in S.) We claim that $|g_1(z) + g_1(z_0)| \geq r$ for every $z \in B$. For if there were to exist $z_1 \in B$ such that $|g_1(z_1) + g_1(z_0)| = |-g_1(z_1) - g_1(z_0)| < r$, then $w^* = -g_1(z_1) \in \Delta$. But w^* would have to have a preimage $z^* \in S$ such that $g(z^*) = -g_1(z_1)$, contrary to property (ii) of g_1.

In the sequel, we use several times the facts that the linear fractional transformation $w = (a\zeta + b)/(c\zeta + d)$, $ad - bc \neq 0$, is analytic and univalent on its domain D, and that the composition of such a function with a function $\zeta = f(z)$ which is analytic and univalent in a region B and with range in D is again analytic and univalent on B.

We now present a simple sequential construction which establishes the lemma. Take any α, $0 < \alpha < 1$. Let $g_2(a) = \alpha r/[g_1(z) + g_1(z_0)]$. Since $|g_1(z) + g_1(z_0)| \geq r$ for every $z \in B$, $|g_2(z)| \leq \alpha$. Also g_2 is analytic and univalent in B. Next let $g_3(z) = [g_2(z) - g_2(z_0)]/2$. Again g_3 is analytic and univalent in B and satisfies the inequality $|g_3(z)| \leq \alpha < 1$. Also $g_3(z_0) = 0$, and, of course, $g_3'(z_0) \neq 0$ [Proposition 10.4.1(ii)]. Finally, let $g(z) = g_3(z)\{\overline{g_3'(z_0)}/|g_3'(z_0)|\}$. It is readily verified that this $g \in \mathscr{K}(B)$. The proof of Lemma 1 is complete.

Lemma 2 Let $b = \sup\{g'(z_0): g \in \mathscr{K}(B)\}$. There exists $\phi \in \mathscr{K}(B)$ such that $\phi'(z_0) = b$.

Proof. We assume that the set of positive real numbers $G = \{g'(z_0): g \in \mathscr{K}(B)\}$ is infinite, since otherwise the lemma becomes trivial. We also proceed for the time being on the assumption that b is not a member of this set G; this assumption is proved to be incorrect. The fact that $|g(z)| < 1$ for every $z \in B$ and every $g \in \mathscr{K}(B)$, and the assumption that G is an infinite set (so $\{g : g \in \mathscr{K}(B)\}$ is an infinite family of functions) have the following implications:

(i) $\{g : g \in \mathscr{K}(B)\}$ is a normal family [Theorem 13.3.2];
(ii) $0 < b < \infty$, because the Cauchy estimates (9.5.3) show that $g'(z_0) \leq 1/\rho$, where $\rho > 0$ is such that $\bar{\Delta}(z_0, \rho) \subset B$;
(iii) There exists an infinite sequence $\langle g_n \rangle$ from $\mathscr{K}(B)$ such that $\lim_{n\to\infty} g_n'(z_0) = b$ [Exercise 1.4.1].

As a consequence of (i), (iii), and the definition of a normal family, there exists an infinite subsequence of $\langle g_n \rangle$, say $\langle g_{n_k} \rangle_{k=1}^{\infty}$, which converges a.u. on B. Let ϕ be the limit function of this subsequence. We deduce the following facts concerning ϕ, which add up to the conclusion of Lemma 2:

(i) ϕ is analytic in B [Theorem 9.9.1];
(ii) $\phi(z_0) = \lim_{k\to\infty} g_{n_k}(z_0) = 0$, since $g_{n_k} \in \mathscr{K}(B)$;
(iii) $\phi'(z_0) = \lim_{k\to\infty} g_{n_k}'(z_0) = b$ [Theorem 9.9.1];

(iv) $\phi'(z)$ is not constant in B, because $\phi'(z_0) > 0$;
(v) ϕ is univalent in B [Exercise 10.3.8];
(vi) $|\phi(z)| < 1$ for every $z \in B$, because for any given $z_1 \in B$, $|\phi(z_1)| = \lim_{k\to\infty} |g_{n_k}(z_1)| \leq 1$ [Exercise 2.3.2], but if it were true that $|\phi(z_1)| = 1$ for for some $z_1 \in B$ then ϕ would be a constant function [Theorem 9.7.1].

The proof of Lemma 2 is complete.

To complete the proof of Theorem 13.4.1, it remains to show that the mapping of B given by the extremal function ϕ found in Lemma 2 is *onto* $\Delta(0, 1)$. We do this by showing that if there were to exist $w_0 \in \Delta(0, 1)$ such that $\phi(z) \neq w_0$ for every $z \in B$, then it would be possible to construct another function $f \in \mathscr{K}(B)$ for which $f'(z_0) > b$, which would contradict the definition of b.

Clearly, $w_0 \neq 0$ because $\phi(z_0) = 0$. As in the proof of Lemma 1, we give a sequential construction of the mythical f. The steps are as follows:

(i) Let $\zeta = \phi_1(z) = [\phi(z) - w_0]/[1 - \bar{w}_0\phi(z)]$. The linear fractional transformation $\zeta = [w - w_0]/[1 - \bar{w}_0 w]$ maps $\Delta(0, 1)$ n a one-to-one way onto itself so that $w = w_0$ corresponds to $\zeta = 0$ [Exercise 1.13.3]. Thus, ϕ_1 is a conformal equivalence mapping B onto some region $B^* \subset \Delta(0, 1)$, and ϕ_1 has no zeros in B.
(ii) By the hypothesis of Theorem 13.4.1, ϕ_1 has an analytic square root ϕ_2 in B. The function ϕ_2 is univalent in B, because if $\phi_2(z_1) = \phi_2(z_2)$ then $[\phi_2(z_1)]^2 = [\phi_2(z_2)]^2$ and $\phi_1(z_1) = \phi_2(z_2)$, but this implies that $\phi(z_1) = \phi(z_2)$ (the linear fractional transformation is univalent), and, therefore, that $z_1 = z_2$ because ϕ is univalent. Since ϕ_2 is univalent, $\phi_2'(z_2) \neq 0$. Also since $|\phi_1(z)| < 1$ for every $z \in B$, we have $|\phi_2(z)| < 1$ for every $z \in B$.
(iii) Let

$$f(z) = \frac{|\phi_2'(z_0)|}{\phi_2'(z_0)} \cdot \frac{\phi_2(z) - \phi_2(z_0)}{1 - \overline{\phi_2(z_0)}\phi_2(z)}. \tag{13.4.1}$$

This composition of ϕ_2 with a linear fractional transformation mapping $\Delta(0, 1)$ onto $\Delta(0, 1)$ is analytic and univalent in B; also $|f(z)| < 1$ for every $z \in B$ and $f(z_0) = 0$. A straightforward computation which we defer to the exercises reveals that $f'(z_0) = (1 + |w_0|)b/2|w_0|^{1/2}$, so $f \in \mathscr{K}(B)$. The obvious inequality $(1 - \alpha)^2 > 0$, $\alpha \neq 1$, implies that $1 + \alpha^2 > 2\alpha$, $\alpha \neq 1$, and so since $|w_0|^{1/2} < 1$, it follows that $f'(z_0) > b$. Q.E.D.

Corollary 2 Let B_1 be a simply connected region in the complex z-plane, $B_1 \neq \mathbf{C}$, and let B_2 be a simply connected region in the complex ζ-plane, with $B_2 \neq \mathbf{C}$. There exists a conformal equivalence F mapping B_1 onto B_2

in such a way that for a given $z_0 \in B_1$ and $\zeta_0 \in B_2$, $F(z_0) = \zeta_0$. The function F is uniquely determined by the further condition $F'(z_0) > 0$.

Proof. According to Corollary 1, there exist conformal equivalences ϕ_1 and ϕ_2, respectively, mapping B_1 and B_2 onto $\Delta(0, 1)$ in the w-plane and such that $\phi_1(z_0) = 0 = \phi_1(\zeta_0)$, $\phi_1'(z_0) > 0$, $\phi_2'(\zeta_0) > 0$. Let $\psi_2 = \phi_2^{-1}$ (the inverse function); by Theorem 6.1.3, $\psi_2'(0) = 1/\phi_2'(\zeta_0) > 0$. Let $F = \psi_2 \circ \phi_1$. This function is analytic and univalent on B_1 and maps B_1 onto B_2, with $F(z_0) = \zeta_0$, and $F'(z_0) = \psi_2'(0)\phi_1'(z_0) > 0$ [chain law]. It remains only to determine the uniqueness of F. Let G be any function analytic and univalent on B_1, with range B_2, and such that $G(z_0) = \zeta_0$, $G'(z_0) > 0$. Look at the functions $\phi_2 \circ G$ and $\phi_2 \circ F$. These are univalent and analytic on B_1; each one maps B_1 onto $\Delta(0, 1)$; $\phi_2(G(z_0)) = 0 = \phi_2(F(z_0))$, and

$$\left.\frac{d\phi_2(G(z))}{dz}\right]_{z=z_0} = \phi_2'(\zeta_0)G'(z_0) > 0$$

with a similar inequality for the derivative of $\phi_2 \circ F$ at z_0. But by Corollary 1, such a mapping function is unique. That means that $\phi_2(G(z)) \equiv \phi_2(F(z))$, $z \in B_1$. Since ϕ_2 is univalent, it follows that $G(z) \equiv F(z)$, $z \in B_1$. Q.E.D.

If a region in **C** is not simply connected, then according to Theorem 12.8.1(vii) there cannot exist a conformal equivalence mapping it onto a region simply connected in **C**, such as $\Delta(0, 1)$. There has been a large amount of research on conformal equivalences for multiply connected regions; a brief introduction to some of the techniques is found in [2, pp. 243–253]. A very simple example of such a conformal equivalence is given by $w = 1/z$, which maps $\Delta'(\infty, 1)$ onto $\Delta'(0, 1) = \Delta(0, 1) \sim \{0\}$. A slightly more complicated case is given in Example (i) in Sec. 13.2. These examples do not involve really notable departures from the hypotheses of the Riemann mapping theorem because in each case the mapping is that of what might he called a "region simply connected in $\hat{\mathbf{C}}$" (rather than in **C**) onto another such "region." However, such mappings play a central rôle in certain parts of complex functional approximation theory. Partly for this reason and partly to introduce a certain important class of univalent Laurent series which we examine in Sec. 13.6, we generalize these examples as follows.

Theorem 13.4.2 *Let $E \subset \mathbf{C}$ be a compact connected set containing more than one point, and such that $K = \hat{\mathbf{C}} \sim E$ is connected in $\hat{\mathbf{C}}$. There exists a unique function $\psi: \Delta(\infty, 1) \to \hat{\mathbf{C}}$ which has the following properties: ψ is univalent on $\Delta(\infty, 1)$, is analytic on $\Delta'(\infty, 1)$, has a simple pole at ∞ with $\psi'(\infty) > 0$; and ψ maps $\Delta'(\infty, 1)$ onto $K \sim \{\infty\}$. The inverse function $\phi = \psi^{-1}$ also has a simple pole at ∞ with $\psi'(\infty) = 1/\psi'(\infty) > 0$.*

Remark (i). According to Exercise 10.7.14, the existence of $\psi'(\infty) \neq \infty$ and $\phi'(\infty) \neq \infty$ is implied by the fact that the poles of ψ and ϕ at ∞ are simple. The same exercise establishes that the Laurent expansion of ψ at ∞ has the form

$$\psi(w) = \psi'(\infty)w + b_0 + \frac{b_1}{w} + \frac{b_2}{w^2} + \cdots, \qquad |w| > 1, \tag{13.4.2}$$

and that of ϕ at ∞ has the form

$$\phi(z) = \phi'(\infty)z + a_0 + \frac{a_1}{z} + \frac{a_2}{z^2} + \cdots, \tag{13.4.3}$$

which is valid in some neighborhood of ∞.

Remark (ii). Among the more important obvious implications of the theorem are that there exist conformal equivalences mapping the complement in $\hat{\mathbf{C}}$ of a simple arc, or the complement in $\hat{\mathbf{C}}$ of the closure of a simply connected region in $\mathbf{C}$, onto a punctured disk with center at ∞.

Remark (iii). The number $\psi'(\infty) = d = d(E)$ is the value at E of a set function which keeps reappearing with various names in various parts of modern–classical complex analysis. Some of the names are transfinite diameter, exterior mapping radius, Robin's constant, capacity, and Chebyshev constant. (Refer to the index of [10].)

Proof of Theorem 13.4.2. Choose a point $a \in E$. The function $\zeta = 1/(z - a)$, $z = a + (1/\zeta)$, is a homeomorphism of $\hat{\mathbf{C}}$ onto $\hat{\mathbf{C}}$, which takes $z = \infty$ onto $\zeta = 0$, and $z = a$ onto $\zeta = \infty$. The image B of $K = \hat{\mathbf{C}} \sim E$ under this homeomorphism is open in $\hat{\mathbf{C}}$ [Exercise 10.7.15] and is connected [Exercise 12.3]; B contains the point $\zeta = 0$ but not $\zeta = \infty$; and $B \neq \mathbf{C}$. Thus, B is a region in $\mathbf{C}$ which contains the origin. Moreover, B is simply connected, because the preimage of $\hat{\mathbf{C}} \sim B$ is the connected set E and, therefore, $\hat{\mathbf{C}} \sim B$ is connected in $\hat{\mathbf{C}}$ [Exercise 12.3].

And so according to the Riemann mapping theorem, there exists a unique conformal equivalence $\xi = f(\zeta)$ mapping B in the ζ-plane onto $\Delta(0, 1)$ in the ξ-plane, with $f(0) = 0$ and $f'(0) > 0$. The inverse function f^{-1} is a conformal equivalence mapping $\Delta(0, 1)$ onto B, and such that $f^{-1}(0) = 0$ and $(f^{-1})'(0) = 1/f'(0) > 0$ [Theorem 6.1.3]. The zero of f^{-1} at $\xi = 0$ is simple [Exercise 10.1.1]. The function given by $z = X(\xi) = a + [1/f^{-1}(\xi)]$ is analytic and univalent on $\Delta'(0, 1)$ and has a simple pole at $\xi = 0$ [Theorem 10.5.2]. It maps $\Delta(0, 1)$ in the ξ-plane onto K in the z-plane so that $\xi = 0$ corresponds to $z = \infty$. The reciprocation $w = 1/\xi$ carries X into a function given by $z = \psi(w) = X(1/w)$, which is analytic and univalent in $\Delta'(\infty, 1)$ in the w-plane and which has a simple pole at $w = \infty$, by the definition of a pole at infinity in Sec. 10.7. According to Exercise 10.7.14, $\psi'(\infty)$ exists and can be computed as follows:

$$\psi'(\infty) = \lim_{w\to\infty} \frac{\psi(w)}{w} = \lim_{\xi\to 0} \xi X(\xi) = \lim_{\xi\to 0} [\xi a + (\xi/f^{-1}(\xi))] = f'(0) > 0.$$

The inverse function ϕ given by $w = \psi^{-1}(z) = \phi(z)$ is analytic in $\Delta'(\infty, R)$ for some $R > 0$ and, therefore, so is $\phi(z)/z$. Since $\lim_{z\to\infty} [\phi(z)/z] =$

$\lim_{w\to\infty} [w/\psi(w)] = 1/\psi'(\infty)$, which is finite and nonzero, it follows from Proposition 10.7.3(iv) that ϕ has a simple pole at ∞. Then by Exercise 10.7.14, $\phi'(\infty) = 1/\psi'(\infty)$, as stated in the theorem.

The question of uniqueness requires attention because the sequence of mappings whereby we constructed ψ seems to involve a free parameter $a \in E$. However, suppose that $\psi_1: \Delta(0, 1) \to K$ is another function analytic and univalent on $\Delta'(\infty, 1)$, with range K, and suppose that ψ_1 has a simple pole at ∞ with $\psi_1'(\infty) > 0$. Let the inverse function be denoted by ϕ_1. Consider the function given by $\xi = F(\zeta) = 1/\phi_1(\psi(1/\zeta))$, $\zeta \in \Delta(0, 1)$. It has a removable singularity at $\zeta = 0$, and when defined suitably there, it is analytic and univalent in $\Delta(0, 1)$ and maps $\Delta(0, 1)$ in the ζ-plane onto $\Delta(0, 1)$ in the ξ-plane so that $\zeta = 0$ goes into $\xi = 0$. Then according to 13.2(C) (or Proposition 9.8.1), $F(\zeta) \equiv e^{i\alpha}\zeta$ in $\Delta(0, 1)$, where α is some real number. Thus, $\phi_1(\psi(1/\zeta)) \equiv e^{-i\alpha}/\zeta$, $\psi(1/\zeta) \equiv \psi_1(e^{-i\alpha}/\zeta)$, $\zeta \in \Delta(0, 1)$, and with $w = 1/\zeta$, $\psi(w) \equiv \psi_1(e^{-i\alpha}w)$, $|w| > 1$. By Exercise 10.7.14,

$$\psi'(\infty) = \lim_{w\to\infty} \frac{\psi(w)}{w} = \lim_{w\to\infty} \frac{\psi_1(e^{-i\alpha}w)}{e^{-i\alpha}w} e^{-i\alpha} = \psi_1'(\infty)e^{-i\alpha}.$$

Since both $\psi'(\infty)$ and $\psi_1'(\infty)$ are real and positive, and $|e^{-i\alpha}| = 1$, we have $e^{-i\alpha} = 1$ and so $\psi(w) \equiv \psi_1(w)$, $|w| > 1$. Q.E.D.

Corollary Let $E_1 \subset \mathbf{C}$ and $E_2 \subset \mathbf{C}$ be compact connected sets, each containing more than one point, and such that $K_1 = \hat{\mathbf{C}} \sim E_1$ and $K_2 = \hat{\mathbf{C}} \sim E_2$ are connected in $\hat{\mathbf{C}}$. There exists a function $F: K_1 \to K_2$ which is univalent in K_1, and which is a conformal equivalence mapping $K_1 \sim \{\infty\}$ onto $K_2 \sim \{\infty\}$; the function F has a simple pole at ∞, so $F(\infty) = \infty$. The function F is uniquely determined by the further condition $F'(\infty) > 0$.

Proof. This is an exercise, since the argument is similar to the proof of Corollary 2 of Theorem 13.4.1.

Exercises 13.4

1. Show that if $f(z)$ is given by (13.4.1), then $f'(z_0) = (1 + |w_0|)b/2|w_0|^{1/2}$.
2. It has been remarked in various places in this book (e.g., 13.2(A)) that there cannot exist a conformal equivalence mapping $\mathbf{C}$ onto a *bounded* region. Prove by reference to the Riemann mapping theorem that there can exist no conformal equivalence mapping $\mathbf{C}$ onto any simply connected region, bounded or unbounded, which is not $\mathbf{C}$ itself.
3. Let B denote a simply connected region, $B \neq \mathbf{C}$, which is symmetric in the real axis; that is, if $z \in B$ then $\bar{z} \in B$. Let ϕ be a conformal equivalence mapping B onto $\Delta(0, 1)$. Prove that the function given by $w = \psi(z) = \overline{\phi(\bar{z})}$ is another conformal equivalence mapping B onto $\Delta(0, 1)$. (*Hint*: The analyticity of ψ can

be proved by the method in the hint for Exercise 6.3.7. The univalence and "onto" character of the mapping can be deduced in an elementary way from the corresponding properties of ϕ.)

4. (Continuation of Exercise 13.4.3.). With B, ϕ, and ψ as in Exercise 10.4.3, let $z_0 \in B$ be real and suppose that $\phi(z_0) = 0$, $\phi'(z_0) > 0$. Show that under these circumstances $\psi(z) \equiv \phi(z)$ for $z \in B$. Deduce from this that any pair of complex conjugate points in B is mapped by ϕ into a pair of conjugate complex points in $\Delta(0, 1)$, that the subset S of the axis of reals in B is mapped onto the interval of reals in $\Delta(0, 1)$, and that, therefore, S is a single interval of reals. (*Hint*: The identity of the two mappings follows from the uniqueness statement in the Riemann mapping theorem, provided that one can show that $\psi'(z_0) > 0$. In calculating $\psi'(z_0)$, recall that ψ can be restricted to the real axis.)
5. Prove the corollary of Theorem 13.4.2. (*Hint*: Imitate the proof of Corollary 2 of Theorem 13.4.1, using Exercise 10.7.14 to compute derivatives at ∞.)
6. Let $\psi: \Delta(\infty, R) \to \hat{\mathbf{C}}$ be analytic on $\Delta'(\infty, R)$ with a pole at ∞. Show that if ψ is univalent on $\Delta'(\infty, R)$, the pole is simple. (*Hint*: $\psi[\Delta(\infty, R)] \neq \mathbf{C}$, for if $\psi[\Delta(\infty, R)] = \mathbf{C}$, then $z = \psi(1/\xi)$ would give a conformal equivalence mapping $\mathbf{C}$ onto the bounded region $\Delta'(0, 1/R)$, and no such mapping exists [Sec. 13.2(A)]. Then for $a \notin \psi[\Delta(\infty, R)]$, let $\chi(\xi) = 1/[\psi(1/\xi) - a]$, $\xi \in \Delta'(0, 1/R)$, and let $\chi(0) = 0$. The function χ is analytic and univalent on $\Delta(0, 1/R)$. Refer to Proposition 10.4.1(ii) and Exercise 10.1.1 to determine the order of the zero of χ at $\xi = 0$, and then use Theorem 10.5.2(iv) to draw an inference concerning ψ.)

13.5 Level Curves

We present in this section a brief discussion of the *level curves* in a conformal equivalence. They are also called *equipotential curves.* Detailed information on this topic in settings more general than that of this discussion is found in [21].

For brevity, in the sequel an arc or curve is linguistically identified with its graph when no confusion can arise by doing so. First, we review some concepts introduced in the remarks in Sec. 12.6. A *simple closed curve* or *Jordan curve* is a closed curve which is homeomorphic to a circle. In less precise terms, a simple closed curve is one which does not intersect itself. The Jordan curve theorem [14, pp. 104–106] can be phrased for present purposes as follows: *The complement* Γ^c *in* $\mathbf{C}$ *of a Jordan curve* Γ *has two and only two components, one bounded and one unbounded, and* Γ *is the complete boundary of each component of* Γ^c. Since Γ is compact [Sec. 5.5, Example (ii)], the components are regions [Corollary 2 of Theorem 5.4.1]. It follows from our definition of simple connectivity that the bounded component is a simply connected region. The unbounded region is called the *exterior region* of Γ, or more briefly, the *exterior* of Γ, and is denoted by Ext Γ. The bounded region is called the *interior region*, or *interior*, of Γ, and is denoted by Int Γ.

Frequently, in the remainder of this book, the Jordan curve theorem is assumed implicitly.

We need the following simple facts about Jordan curves.

Proposition 13.5.1 *Let Γ_1 and Γ_2 be Jordan curves.*

(i) *If* $\Gamma_2 \subset \text{Int}\,\Gamma_1$, *then* $\text{Int}\,\Gamma_2 \subset \text{Int}\,\Gamma_1$.
(ii) *If* $\Gamma_2 \subset \text{Ext}\,\Gamma_1$, *then* $\text{Ext}\,\Gamma_2 \subset \text{Ext}\,\Gamma_1$.

Proof. The terminology and notation make these statements seem intuitively obvious. We give a formal proof of (i) and defer that of (ii) to the exercises.

If $\Gamma_2 \subset \text{Int}\,\Gamma_1$, then $\Gamma_1 \cup \text{Ext}\,\Gamma_1 \subset \Gamma_2^c$. Since $\Gamma_1 \cup \text{Ext}\,\Gamma_1$ is connected [Proposition 5.3.4], it lies in a component of Γ_2 [Theorem 5.4.1]. This component is unbounded, because $\Gamma_1 \cup \text{Ext}\,\Gamma_1$ is unbounded, and, therefore, the component is $\text{Ext}\,\Gamma_2$. It follows that $(\text{Ext}\,\Gamma_2)^c = (\Gamma_2 \cup \text{Int}\,\Gamma_2) \subset (\Gamma_1 \cup \text{Ext}\,\Gamma_1)^c = \text{Int}\,\Gamma_1$. Q.E.D.

Now let Ψ be analytic and univalent on either $\Delta(0, \rho)$, $\rho > 0$, or $\Delta(\infty, \rho)$, $\rho \geq 0$. The range of Ψ is a region A [Sec. 10.4, Remark (iii)], and the inverse function $\Psi^{-1} = \Phi$ is analytic and univalent on A. A level curve C_r in this mapping is a locus in A of the type $\{z\colon |\Phi(z)| = r\} = \Psi[C(0, r)]$, where r is a fixed number, with $0 < r < \rho$ or $\rho < r < \infty$, as the case may be. A natural parametrization of C_r is given by $z = \Psi(re^{i\theta})$, $0 \leq \theta \leq 2\pi$. Notice that this parametrization assigns an orientation to C_r, which we henceforth call the "Ψ-orientation." A level curve with this orientation is said to be "Ψ-oriented." (The symbol Ψ in this terminology is replaced in various contexts by the symbols for functions playing the role of Ψ here.) Since $\partial\Psi/\partial\theta$ is continuous on $[0, 2\pi]$, and $\Psi(re^{i0}) = \Psi(re^{2\pi i})$, it follows that C_r is a closed path; and since Ψ is univalent, C_r is a simple closed path. Another implication of the univalence is that if $r' \neq r$, then $C_{r'}$ and C_r do not intersect. For any given $z_0 \in A$, there is one and only one level curve of the family $\{C_r\}$ passing through z_0; namely the curve C_r for which $r = |\Phi(z_0)|$.

It is now convenient to discuss separately the types of conformal equivalences which appear respectively in Theorems 13.4.1 and 13.4.2.

Proposition 13.5.2 *Let $B \subset \mathbf{C}$ be a simply connected region, $B \neq \mathbf{C}$, and let $\psi\colon \Delta(0, \rho) \to B$, $\rho > 0$, be a conformal equivalence mapping $\Delta(0, \rho)$ onto B so that for a given $z_0 \in B$, $\psi(0) = z_0$. Let $\{C_r = \psi[C(0, r)], 0 < r < \rho\}$ be the family of level curves in this mapping. Then:*

(i) *For each r, $0 < r < \rho$, we have $z_0 \in \text{Int}\,C_r$, and if C_r is ψ-oriented, then* $\text{Ind}_{C_r}(z) = 1$ *for every* $z \in \text{Int}\,C_r$.
(ii) *For each r, $0 < r < \rho$,* $\text{Int}\,C_r = \psi[\Delta(0, r)]$.
(iii) *For every r, r' with $0 < r' < r < \rho$,* $(C_{r'} \cup \text{Int}\,C_{r'}) \subset \text{Int}\,C_r$.

Proof. (i) First, notice that $z_0 = \psi(0) \notin \psi[C(0, r)] = C_r$. We have

$$\begin{aligned}\operatorname{Ind}_{C_r}(z_0) &= \frac{1}{2\pi i}\int_{C_r}\frac{dz}{z - z_0}\\ &= \frac{1}{2\pi i}\int_0^{2\pi}\frac{\psi'(re^{i\theta})ire^{i\theta}\,d\theta}{\psi(w) - \psi(0)}\\ &= \frac{1}{2\pi i}\int_\gamma\frac{\psi'(w)\,dw}{\psi(w) - \psi(0)},\end{aligned} \tag{13.5.1}$$

where γ is a positively oriented circle with graph $C(0, r)$. Now $\psi'(0) \neq 0$, so $\psi(w) - \psi(0)$ has a simple zero at $w = 0$ but has no other zero in $\Delta(0, r)$. Therefore, by Theorem 10.2.1(i), the value of the last integral is one.

It follows from Theorem 8.4.2 that $z_0 \in \operatorname{Int} C_r$, since $\operatorname{Ind}_{C_r}(z_0) \neq 0$. Also according to the same theorem, $\operatorname{Ind}_{C_r}(z)$ has the constant value one on the component of C_r^c in which z_0 lies, which is $\operatorname{Int} C_r$.

(ii) We base this part of the proof on a variant of Theorem 5.4.1 which we call "the connectivity principle." It runs as follows: Let S_1 and S_2 be connected subsets of a set S such that $S_1 \cap S_2 \neq \varnothing$. Then S_1 and S_2 lie in the same component of S; and if S_1 is known to be itself a component of S, then $S_2 \subset S_1$.

To apply this to the proof of (ii), we first observe that $\psi[\Delta(0, r)]$ is connected [Proposition 5.3.1]; it is a subset of C_r^c containing z_0, and $\operatorname{Int} C_r$ is also a connected subset of C_r^c containing z_0. Since $\operatorname{Int} C_r$ is a component of C_r^c, it follows from the connectivity principle that $\psi[\Delta(0, r)] \subset \operatorname{Int} C_r$.

We now show that $\operatorname{Int} C_r \subset \psi[\Delta(0, r)]$, which completes the proof of (ii). First, we establish that $\operatorname{Int} C_r$ lies in the range of ψ; that is, $\operatorname{Int} C_r \subset B$. The argument is similar to that of the proof of Proposition 13.5.1(i). By our definition of simple connectivity, all the components of $B^c = \mathbf{C} \sim B$ are unbounded. Let K be a component of B^c. Since $C_r \subset B$, we have $B^c \subset C_r^c$, and by Theorem 5.4.1, the connected set K is contained in a component of C_r^c. That component cannot be $\operatorname{Int} C_r$ because K is unbounded and $\operatorname{Int} C_r$ is bounded. It follows that $K \subset \operatorname{Ext} C_r$, so $B^c \subset \operatorname{Ext} C_r$ and $(\operatorname{Ext} C_r)^c = C_r \cup \operatorname{Int} C_r \subset B$.

Now let $\phi = \psi^{-1}$, the inverse of ψ. We have shown in the preceding paragraph that $\operatorname{Int} C_r$ lies in the domain B of ϕ. The set $\phi[\operatorname{Int} C_r]$ is connected [Proposition 5.3.1] and is, therefore, contained in some component of $[C(0, r)]^c$, by Theorem 5.4.1. From the connectivity principle, we deduce that this component is necessarily $\Delta(0, r)$, because $0 \in \Delta(0, r)$ and by part (i), $0 \in \phi(\operatorname{Int} C_r)$. Therefore $\operatorname{Int} C_r \subset \psi[\Delta(0, r)]$. The proof that $\operatorname{Int} C_r = \psi[\Delta(0, r)]$ is complete.

(iii) This follows at once from (ii); $C_{r'} \cup \operatorname{Int} C_{r'} = \psi[\bar{\Delta}(0, r')] \subset \psi[\Delta(0, r)] = \operatorname{Int} C_r$. Q.E.D.

Corollary Let B, ψ, and the family $\{C_r\}$ be as described in the proposition. Let $f: B \to \mathbf{C}$ be analytic on B, and let C_r be ψ-oriented. Then we have

$$f(z) = \frac{1}{2\pi i}\int_{C_r} \frac{f(\zeta)\, d\zeta}{\zeta - z} \tag{13.5.2}$$

for every $z \in \operatorname{Int} C_r$.

Proof. By Proposition 13.5.2(i) and (ii), $\operatorname{Int} C_r \subset B$ and $\operatorname{Ind}_{C_r}(z) = 1$ for $z \in \operatorname{Int} C_r$. Therefore, by the corollary to Theorem 12.6–(9.4.1), the Cauchy integral formula (13.5.2) is valid. Q.E.D.

Proposition 13.5.3 *Let $\psi: \Delta(\infty, \rho) \to \mathbf{C}$, $\rho \geq 0$, be univalent on $\Delta(\infty, \rho)$ and analytic on $\Delta'(\infty, \rho)$, with a simple pole at ∞. Let $E = \hat{\mathbf{C}} \sim \psi[\Delta(\infty, \rho)]$. Let $\{C_R = \psi[C(0, R)], \rho < R < \infty\}$ be the family of level curves in this mapping. Then:*

(i) *For every R, $\rho < R < \infty$, we have $E \subset \operatorname{Int} C_R$; and if C_R is ψ-oriented, then $\operatorname{Ind}_{C_R}(z) = 1$ for every $z \in \operatorname{Int} C_R$.*
(ii) *For each R, $\rho < R < \infty$, $\operatorname{Ext} C_R = \psi[\Delta'(\infty, R)]$.*
(iii) *For every pair R, R' with $\rho < R < R' < \infty$, $(C_{R'} \cup \operatorname{Ext} C_{R'}) \subset \operatorname{Ext} C_R$ and $(C_R \cup \operatorname{Int} C_R) \subset \operatorname{Int} C_{R'}$.*
(iv) *If O is an open set in $\mathbf{C}$ containing E, $O \neq \mathbf{C}$, then there exists $R^* > \rho$ such that $\operatorname{Int} C_{R^*} \subset O$ and $(C_R \cup \operatorname{Int} C_R) \subset O$ for every R, $\rho < R < R^*$, but C_{R^*} intersects O^c.*

Remark. The closed set E is not empty; that is, $\psi[\Delta'(\infty, \rho)] \neq \mathbf{C}$ even if $\rho = 0$. This may be intuitively obvious to the reader, but it is an essential fact for certain of our applications, and so we give a formal proof. According to our definition of simple connectivity, $\Delta'(\infty, \rho)$, $\rho \geq 0$, is not a simply connected region in $\mathbf{C}$. However, $\mathbf{C}$ itself is simply connected in $\mathbf{C}$ and so is homeomorphic to $\Delta(0, 1)$ [Theorem 12.8.1(vii)]. If $\psi[\Delta'(\infty, \rho)]$ were to equal $\mathbf{C}$, then $\Delta'(\infty, \rho)$ would be homeomorphic to $\mathbf{C}$, and $\Delta'(\infty, \rho)$ would then also be homeomorphic to $\Delta(0, 1)$. This is impossible, again by Theorem 12.8.1(vii) and (i).

Proof. (i) Choose R, $\rho < R < \infty$, and any $z_0 \in E$. As in the proof of (13.5.1), we have

$$\operatorname{Ind}_{C_R}(z_0) = \frac{1}{2\pi i}\int_{C_R} \frac{dz}{z - z_0} = \frac{1}{2\pi i}\int_{\gamma} \frac{\psi'(w)\, dw}{\psi(w) - z_0}, \tag{13.5.2}$$

where γ is a positively oriented circle with graph $C(0, R)$. The function $\psi(w)$ has a simple pole at ∞, and, therefore, so does $\psi(w) - z_0$. The latter function is analytic and nonzero on $\Delta'(\infty, \rho)$. By Proposition 10.7.3(v), there exists $h: \Delta(\rho, \infty) \to \mathbf{C}$, analytic on $\Delta'(\rho, \infty)$ and continuous at ∞, with $h(\infty) \neq \infty$, and such that $\psi(w) - z_0 \equiv wh(w)$, $w \in \Delta'(\rho, \infty)$. Then $\psi'(w) = (d/dw)[\psi(w) - z_0] = h(w) + wh'(w)$, so

$$\frac{\psi'(w)}{\psi(w) - z_0} = \frac{1}{w} + \frac{h'(w)}{h(w)}, \qquad z \in \Delta'(\infty, \rho).$$

The function h has a Laurent expansion at ∞ of the following type: $h(w) = c_0 + c_1 w^{-1} + c_2 w^{-2} + \cdots$. It converges a.u. on $\Delta'(\infty, \rho)$, so by the corollary of Theorem 9.9.1, the series can be differentiated term-by-term to obtain the Laurent expansion of h' at ∞, and the result is $h'(w) = 0 - c_1 w^{-2} - 2c_2 w^{-3} - \cdots$. Therefore, by Proposition 10.7.1(iii), h' has a zero of order at least 2 at ∞. Since h is obviously nonzero on $\Delta'(\infty, \rho)$, the function $(1/h)h'$ is analytic on $\Delta'(\infty, \rho)$, and by Proposition 10.7.1(ii), it has a zero of order at least 2 at ∞. Therefore, by Exercise 10.7.10, $\int_\gamma [h'(w)/h(w)]\, dw = 0$.

We return to (13.5.2) with this information and find that

$$\text{Ind}_{C_R}(z_0) = \frac{1}{2\pi i} \int_\gamma \left[\frac{1}{w} + \frac{h'(w)}{h(w)}\right] dw = 1 + 0.$$

It follows from Theorem 8.4.2. that $z_0 \in \text{Int}\, C_R$ (since according to that theorem, $\text{Ind}_{C_R}(z) \equiv 0$ in the unbounded component $\text{Ext}\, C_R$ of C_R^c). It also follows from that theorem that $\text{Ind}_{C_R}(z) = 1$ for every $z \in \text{Int}\, C_R$.

(ii) We establish this part of the proposition by a connectivity argument similar to that used to prove Proposition 13.5.2(ii). First, we show that $\psi[\Delta'(\infty, R)] \subset \text{Ext}\, C_R$. The point set C_R is bounded, so the component $\text{Ext}\, C_R$ of C_R^c contains a punctured disk $\Delta'(\infty, r_1)$, $0 < r_1 < \infty$ [Proposition 5.4.2]. Now by Exercise 12.3, $\psi[\Delta(\infty, R)]$ is a connected subset of $\hat{\mathbf{C}} \sim C_R$, open in $\hat{\mathbf{C}}$ [Theorem 10.4.1], so it contains a disk $\Delta(\infty, r_2)$, $0 < r_2 < \infty$. Thus, $\text{Ext}\, C_R \cap \psi[\Delta'(\infty, R)] \neq \varnothing$, so by the connectivity principle (see the proof of Proposition 13.5.2(ii)), $\psi[\Delta'(\infty, R)] \subset \text{Ext}\, C_R$.

To obtain inclusion in the opposite direction, we first observe that by part (i), since $E \subset \text{Int}\, C_R$, we have $\text{Ext}\, C_R \subset K = \hat{\mathbf{C}} \sim E$. Let $\phi = \psi^{-1}$, the inverse function of ψ. Then $\phi(\text{Ext}\, C_R)$ is connected [Proposition 5.3.1]; it contains no point of $C(0, R)$, so it lies entirely in a component in $\mathbf{C}$ of $[C(0, R)]^c$. This component is necessarily $\Delta'(\infty, R)$, not $\Delta(0, R)$, because $\phi(\text{Ext}\, C_R)$ is unbounded. Thus, $\phi(\text{Ext}\, C_R) \subset \Delta'(\infty, R)$ and $\text{Ext}\, C_R \subset \psi[\Delta'(\infty, R)]$. The proof of part (ii) is complete.

(iii) This follows at once from (ii), since if $R' > R > \rho$, then $C_{R'} \cup \text{Ext}\, C_{R'} = \{\psi[C(0, R')] \cup \psi[\Delta(\infty, R')]\} \subset \psi[\Delta(\infty, R)] = (\text{Ext}\, C_R) \cup \{\infty\}$. We take complements to complete the proof.

(iv) Since $E \subset O$, we have $O^c = \mathbf{C} \sim E \subset \hat{\mathbf{C}} \sim E$. The homeomorphism given by $z = \psi(w)$ takes the closed set O^c into a closed set $F \subset \Delta(\infty, \rho)$ in the w-plane [Exercise 5.2.1]. Let $\eta = \text{dist}\,[F, C(0, \rho)] > 0$ [Theorem 5.6.1]. According to the proof of Theorem 5.6.1 (see Lemma 1 in that proof), there exists $w_0 \in F$ such that $\text{dist}\,[w_0, C(0, \rho)] = \eta$. Let $R^* = |w_0|$. Then

$C(0, R^*) \cap F = \varnothing$, so $C_{R^*} \cap O^c \neq \varnothing$. The annulus $\{w: \rho < |w| < R^*\}$ contains no point of F, so $F \subset \bar{\Delta}(\infty, R^*)$ and $\psi(F) = O^c \subset \psi[\bar{\Delta}(\infty, R^*)] = C_{R^*} \cup \text{Ext } C_{R^*}$. Therefore, $\text{Int } C_{R^*} \subset O$, and by (iii), $(C_R \cup \text{Int } C_R) \subset O$ for every R, $\rho < R < R^*$. Q.E.D.

Corollary Let E, K, ψ, and the family $\{C_R\}$ be as described in the proposition. Let O be an open set containing E, let $f: O \to \mathbf{C}$ be analytic on O, and let C_R be ψ-oriented. Then for every $R > \rho$ such that $(C_R \cup \text{Int } C_R) \subset O$, the Cauchy integral formula

$$f(z) = \frac{1}{2\pi i} \int_{C_R} \frac{f(\zeta)\, d\zeta}{\zeta - z} \tag{13.5.4}$$

is valid for $z \in \text{Int } C_R$ and in particular for $z \in E$.

Proof. By part (iv) of the proposition, there is an open interval (ρ, R^*) such that for every $R \in (\rho, R^*)$ we have $(C_R \cup \text{Int } C_R) \subset O$. Choose any R in this interval and let R' be such that $R < R' < R^*$. By referring to various parts of the proposition, we obtain $E \subset (C_R \cup \text{Int } C_R) \subset \text{Int } C_{R'} \subset O$. Now $\text{Int } C_{R'}$ is a simply connected region on which f is analytic, and $\text{Ind}_{C_R}(z) = 1$ for every $z \in \text{Int } C_R$. These facts, taken with the Jordan curve theorem, permit us to cite the corollary of Theorem 12.6–(9.4.1) to obtain the conclusion of this corollary. Q.E.D.

One implication of Proposition 13.5.3(iv) is that for any compact connected set E with $\hat{\mathbf{C}} \sim E$ connected in $\hat{\mathbf{C}}$, it is possible to find level curves C_R of the map of the complement given by ψ such that $E \subset \text{Int } C_R$ and $\text{dist}(C_R, E)$ is arbitrarily small. (For any $\varepsilon > 0$, the set O in that part of the proposition can be taken to be the union of a family of open disks covering E with centers in E and radii ε.) The following proposition, which is needed in our discussion of polynomial approximations later in this chapter, falls into the same general category of results concerning the approximation of point sets by level curves.

Proposition 13.5.4 *Let $\psi: \Delta(\infty, \rho) \to \hat{\mathbf{C}}$, $\rho \geq 0$, be analytic on $\Delta'(\infty, \rho)$ and univalent on $\Delta(\infty, \rho)$ and have a simple pole at ∞. For each $r > \rho$, let C_r denote the level curve $\psi[C(0, r)]$. Let the level curve C_{R^*} and the compact set E be such that $E \subset \text{Int } C_{R^*}$. Then there exists R_0, $\rho < R_0 < R^*$, such that each member of the family of level curves $\{C_R: R_0 \leq R\}$ has the property that $E \subset \text{Int } C_R$.*

Proof. The curve C_{R^*} is a compact set [Theorem 5.5.2], so $d = \text{dist}(E, C_{R^*})$ is positive [Theorem 5.6.1]. For any R_1 such that $\rho < R_1 < R^*$, ψ is continuous on the set $A = \{w: R_1 \leq |w| \leq R^*\}$. Since A is compact, ψ is uniformly continuous on A [Theorem 5.5.9]. Thus, there exists $\delta > 0$

such that if $R_1 \leq r \leq R^*$ and $|r - R^*| \leq \delta$, then $|\psi(re^{i\theta}) - \psi(R^*e^{i\theta})| < d$ for every θ, $0 \leq \theta \leq 2\pi$. Now adjust δ downward, if necessary, so that $R^* - \delta > \rho$, and let $R_0 = R^* - \delta$. The set $B = \{\psi(w): R_0 \leq |w| \leq R^*\}$ has these properties: $B = (C_{R_0} \cup \text{Ext } C_{R_0}) \cap (C_{R^*} \cup \text{Int } C_{R^*})$ [Proposition 13.5.3(ii)]; and B contains no point of E because for any $z \in B$, dist $(z, C_{R^*}) < d$. Therefore, $E \subset B^c = \text{Int } C_{R_0} \cup \text{Ext } C_{R^*}$. But no point of E lies in Ext C_{R^*}, so $E \subset \text{Int } C_{R_0} \subset \text{Int } C_R$ for every $R \geq R_0$ [Proposition 13.5.3(iii)]. Q.E.D.

An arc or closed curve γ with parametric equation $z = \gamma(t)$, $\alpha \leq t \leq \beta$, is said to be *analytic* if and only if for each $t_0 \in [\alpha, \beta]$ there exists a power series $\sum c_k(t_0)(t - t_0)^k$ convergent to $\gamma(t)$ in $(t_0 - h, t_0 + h) \cap [\alpha, \beta]$, where $h > 0$ depends on t_0 (see Exercise 6.5.3), and $\gamma'(t) \neq 0$ for every $t \in [\alpha, \beta]$. We close this section by showing that the level curves in a conformal equivalence are analytic closed curves.

Proposition 13.5.5 *Let $A \subset \mathbf{C}$ be a region and let $\phi: A \to \mathbf{C}$ be a conformal equivalence mapping A onto either $\Delta(0, \rho)$ or $\Delta'(\infty, \rho)$, $\rho > 0$. Then each level curve $C_r = \{z: |\phi(z)| = r\}$, $0 < r < \rho$ or $\rho < r < \infty$, is an analytic Jordan curve.*

Proof. Let $\psi = \phi^{-1}$, the inverse mapping function, and parametrize C_r by the equation $z = \psi(re^{i\theta})$, $0 \leq \theta \leq 2\pi$. In what follows, r is held fixed on $(0, \rho)$ or (ρ, ∞), as the case may be. Since $\psi'(w) \neq 0$ for every w in the domain of ψ [Proposition 10.4.1], by the chain law, we have $d\psi(re^{i\theta})/d\theta = \psi'(re^{i\theta})ire^{i\theta} \neq 0$, $0 \leq \theta \leq 2\pi$. Take any $\theta_0 \in [0, 2\pi]$, and let $w_0 = re^{i\theta_0}$. The domain of ψ is, of course, open, so there exists $\eta_0 > 0$ such that $\psi(w)$ is analytic in $\Delta(w_0, \eta_0) = \{w: |w - w_0| < \eta_0\}$. We know that there is a power series involving powers of $w - w_0$ which represents $\psi(w)$ in $\Delta(w_0, \eta_0)$; the problem is to show that there is also a series of powers of $\theta - \theta_0$ which represents $\psi(re^{i\theta})$ in a neighborhood in $\mathbf{R}$ of θ_0.

To do this we consider the composition of ψ with the function given by $w = re^{i\zeta}$. The latter function is analytic in the entire ζ-plane, and $re^{i\zeta} - w_0$ has a simple zero at $\zeta = \theta_0$. Therefore, by Proposition 10.1.2, there exists a function g which is continuous and nonzero on a certain closed disk D with center at $\zeta = \theta_0$, and with the property that $re^{i\zeta} - w_0 = (\zeta - \theta_0)g(\zeta)$ for every $\zeta \in D$. Let $M = \max |g(\zeta)|$, $\zeta \in D$; obviously $M > 0$. Then for every $\zeta \in D$ such that $|\zeta - \theta_0| < \eta_0/M$, we have $|re^{i\zeta} - w_0| = |\zeta - \theta_0||g(\zeta)| < (\eta_0/M)M = \eta_0$. Thus, if $\zeta \in \Delta = \Delta(\theta_0, \eta_0/M) \cap D^0$, where D^0 denotes the interior of D, we have $w = re^{i\zeta} \in \Delta(w_0, \eta_0)$; and then the composition $\psi(re^{i\zeta})$ is an analytic function of ζ on the disk Δ. It has a power series representation in Δ as follows: $\psi(re^{i\zeta}) = \psi(w_0) + \sum_1^\infty c_k(\zeta - \theta_0)^k$. In particular, with ζ real, say $\zeta = \theta$, this series converges to $\psi(re^{i\theta})$ on $\Delta \cap [0, 2\pi]$. Q.E.D.

Exercises 13.5

1. According to Exercise 1.14.17, the function given by $z = (1/2)[Rw + (1/Rw)]$ gives a one-to-one mapping of $\Delta(\infty, 1)$ onto the exterior region of an ellipse with major axis of length $R + (1/R)$ and minor axis of length $R - (1/R)$. The function and its inverse ϕ are clearly conformal equivalences of the type which appears in Theorem 13.4.2. Show that each level curve $\{z : |\phi(z)| = r\}$, $r > 1$, is also an ellipse. (*Hint*: Substitute $w = \rho e^{i\theta}$ and look at the real and imaginary parts.)
2. According to Example (v) in Sec. 13.2, the linear fractional transformation $z = (1 + w)/(1 - w)$, $w = \phi(z) = (z - 1)/(z + 1)$, is a conformal equivalence mapping the half-plane $\{z : \text{Re } z > 0\}$ onto $\Delta(0, 1)$ in the w-plane. Describe the typical level curve $C_r = \{z : |\phi(z)| = r\}$, $0 < r < 1$, in this mapping.
3. Let E, K, ψ, and $\{C^R\}$ be as described in Proposition 13.5.3. Show that there exists a function $G(x, y)$ harmonic but unbounded for $z = x + iy \in K \sim \{\infty\}$, and having the value zero for every point on C_R. (*Hint*: Theorem 13.4.2 and Exercise 9.4.11.) (*Note*: G is called Green's function for Ext C_R with pole at ∞.) Also show that given any point $z_0 \notin C_R$, there exists a function G_0 harmonic on $K \sim (\{z_0\} \cup \{\infty\})$ such that $G_0(z) = -\ln|z - z_0|$ for every $z \in C_R$.

13.6 Introduction to the Classes $\mathcal{S}$ and $\mathcal{U}$

We define the class $\mathcal{S}$ to be the family of functions f, each of which is analytic and univalent on $\Delta(0, 1)$, with $f(0) = 0$ and $f'(0) = 1$. Thus, each $f \in \mathcal{S}$ has a power series representation of the type

$$f(w) = w + a_2 w^2 + a_3 w^3 + \cdots, \qquad |w| < 1. \tag{13.6.1}$$

We define the class $\mathcal{U}$ to be the family of functions F, each of which is analytic and univalent on $\Delta'(\infty, 1)$, and has a simple pole at ∞ with $F'(\infty) = 1$. (Notice that Theorem 13.4.2 ensures that $\mathcal{U}$ is not empty.) Each $F \in \mathcal{U}$ has a Laurent expansion at ∞ of the type

$$F(w) = w + b_0 + \frac{b}{w} + \frac{b_2}{w_2} + \cdots, \qquad |w| > 1. \tag{13.6.2}$$

If $f \in \mathcal{S}$, then $f[\Delta(0, 1)]$ is a region in $\mathbf{C}$ [Sec. 10.4, Remark (iii)] which, furthermore, is simply connected [Theorem 12.8.1(vii)], but $f[\Delta(0, 1)] \neq \mathbf{C}$ [Sec. 13.2(A)]. Similarly if $F \in \mathcal{U}$, then $F[\Delta'(\infty, 1)]$ is a region K in $\mathbf{C}$. This region K cannot be $\mathbf{C}$ itself, because if it were, then the function given by $z = F(1/\zeta)$ would be a conformal equivalence mapping the bounded region $\Delta'(0, 1)$ onto $\mathbf{C}$, and according to Sec. 13.2(A), no such conformal equivalence exists. (Another argument showing that $K \neq \mathbf{C}$ was presented in the remark which follows the statement of Proposition 13.5.3.) As a matter of fact, $K^c = \mathbf{C} \sim K$ must contain at least two points. This is brought out in the remark following the proof of the next proposition.

Proposition 13.6.1

(i) *If* $F \in \mathscr{U}$, *then for any* $a \notin F[\Delta(\infty, 1)]$, $a \in \mathbf{C}$, *the function given by* $z = 1/[F(1/\xi) - a]$ *is a member of* $\mathscr{S}$.
(ii) *If* $f \in \mathscr{S}$, *then the function given by* $z = 1/f(1/\xi)$ *is a member of* $\mathscr{U}$.

Proof. (i) The function with values $F(1/\xi) - a$ is analytic, univalent, and nonzero on $\Delta(0, 1) \sim \{0\}$ and has a simple pole at $\xi = 0$. Therefore, by Theorem 10.5.2(iv), the analytic univalent function given by $z = f(\xi) = 1/[F(1/\xi) - a]$ has a removable singularity at $\xi = 0$, which becomes a simple zero after removal. To compute $f'(0)$, we use in succession the definition of $f'(0)$, the reciprocation $w = 1/\xi$, and Exercise 10.7.14:

$$f'(0) = \lim_{\xi \to 0} \frac{f(\xi)}{\xi} = \lim_{\xi \to 0} \frac{1}{\xi[F(1/\xi) - a]}$$
$$= \lim_{w \to \infty} \left[\frac{1}{F(w)/w - a/w}\right] = \frac{1}{F'(\infty) - 0} = 1.$$

Therefore $f \in \mathscr{S}$.

The proof of (ii) is very similar and is deferred to the exercises.

Remark (i). The linear fractional transformation $z = 1/(\zeta - a)$, $a \in \mathbf{C}$, maps $\hat{\mathbf{C}} \sim a$ onto $\mathbf{C}$. Thus, if a were the only point in the complement K^c of $F[\Delta(\infty, 1)]$ with respect to $\mathbf{C}$, the function appearing in Proposition 13.6.1(i) would be a conformal equivalence mapping $\Delta(0, 1)$ onto $\mathbf{C}$, and according to Sec. 13.2(A), no such mapping exists. Therefore, K^c contains more than one point.

Remark (ii) (Continuation.) We can further assert that K^c is connected in $\mathbf{C}$. For according to the discussion preceding the proposition, the function given by $z = 1/[F(1/\xi) - a]$, $a \in K^c$, is a univalent mapping of $\{\xi : |\xi| < 1\}$ onto a *simply connected* region $B \subset \mathbf{C}$. The homeomorphism from $\hat{\mathbf{C}}$ onto $\hat{\mathbf{C}}$ given by $z = 1/(\zeta - a)$ maps K^c onto $\hat{\mathbf{C}} \sim B$, and by our definition of simple connectivity, $\hat{\mathbf{C}} \sim B$ is connected in $\hat{\mathbf{C}}$. Therefore, K^c (which does not contain $\{\infty\}$) is connected in $\mathbf{C}$, by Exercise 12.3.

Example (i) The function f given by

$$z = f(w) = \frac{w}{(1 - w)^2} = w\frac{d}{dw}\left(\frac{1}{1 - w}\right) = \sum_1^\infty nw^n$$

is a member of $\mathscr{S}$. To check the univalence, suppose that $w_1, w_2 \in \Delta(0, 1)$, $w_1 \neq w_2$, and $f(w_1) = f(w_2)$. This equation implies that $w_1 - 2w_1w_2 + w_1w_2^2 = w_2 - 2w_1w_2 + w_1^2w_2$, $w_1w_2(w_2 - w_1) = w_2 - w_1$. The latter equation can be true only if $w_2 = w_1$, because if $w_2 \neq w_1$, then the equation implies that $|w_1||w_2| = 1$, and this in turn implies that w_1 or w_2 lies outside $\Delta(0, 1)$. For future reference we also note that $-1/4 \notin f[\Delta(0, 1)]$, for by elementary algebra we find that the only solution of $f(w) = -1/4$ is $w = -1$. Finally, notice that for this function, in the notation of (13.6.1), we have $|a_n| = n$. The famous Bieberbach conjecture, to which we return later in this section, is that for each $f \in \mathscr{S}$, $|a_n| \leq n$, $n = 2, 3, \ldots$. Therefore, this example is an extremal function for the Bieberbach conjecture and shows that no "tighter" inequality for $|a_n|$, $n = 2, 3, \ldots$, can be valid for every $f \in \mathscr{S}$.

Theorem 13.6.1 *If $f \in \mathscr{S}$, then for each positive integer n there exists $g \in \mathscr{S}$ such that $[g(w)]^n \equiv f(w^n)$, $w \in \Delta(0, 1)$.*

Proof. The function f is nonzero on $\Delta(0, 1)$ except for the simple zero at $w = 0$. Define a function h by $h(w) = f(w)/w$, $w \in \Delta(0, 1) \sim \{0\}$, and let $h(0) = 1$. This function is analytic and nonzero on $\Delta(0, 1)$, so by the corollary of Proposition 9.4.2, it has an analytic logarithm L on $\Delta(0, 1)$, which by definition satisfies the identity $e^{L(w)} \equiv h(w)$, $w \in \Delta(0, 1)$. Notice that $L(0) = 0$. Next define a function k by $k(w) \equiv e^{L(w)/n}$, $w \in \Delta(0, 1)$, where n is any positive integer. This function k is analytic and nonzero on $\Delta(0, 1)$; $[k(w)]^n \equiv h(w)$ and $k(0) = e^0 = 1$. We have $f(w) = w[k(w)]^n$, $f(w^n) = w^n[k(w^n)]^n$. Finally, define g by $g(w) \equiv wk(w^n)$; this function is analytic on $\Delta(0, 1)$; $g(w) \neq 0$, $w \neq 0$; $g(0) = 0$, and $g'(0) = \lim_{w \to 0} wk(w^n)/w = k(0) = 1$.

Clearly $[g(w)]^n = f(w^n)$. To prove that $g \in \mathscr{S}$, it remains only to show that g is univalent on $\Delta(0, 1)$. Suppose that w_1, $w_2 \in \Delta(0, 1)$ are such that $g(w_1) = g(w_2)$. As usual in our proofs of univalence, we obtain from this the implication that $w_1 = w_2$. We have $[g(w_1)]^n = [g(w_2)]^n$ and $f(w_1^n) = f(w_2)$. Since f is univalent, it follows that $w_1^n = w_2^n$. This implies that for some integer H, $1 \leq H \leq n$, we have $w_1 = w_2 e^{2\pi i H/n}$. Then $g(w_1) = w_1 k(w_1^n) = w_2 e^{2\pi i H/n} k(w_2^n) = g(w_2) e^{2\pi i H/n} = g(w_1) e^{2\pi i H/n}$. This equation, taken with $g(w_1) = g(w_2)$, leads to one of two conclusions. The first is that $g(w_1) = 0 = g(w_2)$, but this can happen only if $w_1 = w_2 = 0$. The second is that $g(w_1) \neq 0$ and $H = n$. But then again $w_1 = w_2 e^{2\pi i} = w_2$. Q.E.D.

Theorem 13.6.2 (*The Area Theorem*). *Let F be a member of $\mathscr{U}$ with the Laurent expansion at ∞ as follows*:

$$F(w) = w + b_0 + \frac{b_1}{w} + \frac{b_2}{w^2} + \cdots, \qquad |w| > 1.$$

Then $\sum_{n=1}^{\infty} n|b_n|^2 \leq 1$.

Remark. The classical proof of this theorem is based on a special case of Green's theorem which appears in Exercise 8.3.8, and which states that the area of the interior region of a simple closed path γ is given by $A = -\text{Im}[(1/2)\int_\gamma \bar{z}\, dz]$. Of course, this formula gives a *positive* value to A (which is crucial for the present application) only if γ is properly oriented. The curve γ in the present application would be the level curve $C_R: z = F(Re^{i\theta})$, $0 \leq \theta \leq 2\pi$, $R > 1$. Apart from the relative inaccessibility (at the level of this book) of rigorous analytical proofs of Green's theorem, there remains a question, which seems to be difficult to resolve directly, as to whether the F-orientation of C_R is the one which gives a positive value to A. We present here a proof of the theorem which avoids such problems and which is adapted from an argument used by Rudin in [18, pp. 305–307].

Proof of the Theorem. We first give the proof for a function $\psi \in \mathscr{U}$ with a Laurent expansion at ∞ as follows:

$$\psi(w) = w + \frac{\alpha}{w} + \Sigma_2^\infty \frac{c_n}{w^n}, \qquad |w| > 1, \tag{13.6.3}$$

where α is is real and nonnegative.

For any $R > 1$, let $C_R = \psi[C(0, R)]$, and let $A_{\rho,R} = \{w: \rho < |w| < R\}$, where $1 < \rho < R$. According to Proposition 13.5.3(ii), $\text{Ext}\, C_R = \psi[\Delta(\infty, R)]$, so $\psi(A_{\rho,R}) \subset \text{Int}\, C_R$.

We write

$$z = x + iy = \psi(w) = w + \frac{\alpha}{w} + \chi(w), \qquad |w| > 1.$$

Let $w = R(\cos\theta + i\sin\theta)$. Then

$$x = \left(R + \frac{\alpha}{R}\right)\cos\theta + \text{Re}\,\chi(w) \tag{13.6.4}$$

$$y = \left(R - \frac{\alpha}{R}\right)\sin\theta + \text{Im}\,\chi(w). \tag{13.6.5}$$

Let $a = R + (\alpha/R)$, $b = R - (\alpha/R)$, and assume that $R > \sqrt{2\alpha}$, from which it follows that $b > R/2$. We divide (13.6.4) by a and (13.6.5) by b, square the equations, and add, obtaining

$$\frac{x^2}{a^2} + \frac{y^2}{b^2} = \cos^2\theta + \sin^2\theta + 2\frac{\cos\theta}{a}\,\text{Re}\,\chi(w)$$
$$+ \frac{2\sin\theta}{b}\,\text{Im}\,\chi(w) + \left(\frac{\text{Re}\,\chi(w)}{a}\right)^2 + \left(\frac{\text{Im}\,\chi(w)}{b}\right)^2. \tag{13.6.6}$$

Now $\chi(w) = \sum_2^\infty c_n w^{-n}$, and if χ is not identically zero on $\Delta(\infty, 1)$, it has a zero of order at least 2 at ∞ [Proposition 10.7.1(iii)]. In either case, by Proposition 10.7.1(ii), there exists $M \geq 0$ such that $|\chi(w)| \leq MR^{-2}$, $|w| \geq R$. Notice that $1/a < 1/R$ and $1/b < 2/R$. We apply the triangle inequality to the right-hand member of (13.6.6) and obtain

$$\frac{x^2}{a^2} + \frac{y^2}{b^2} \leq 1 + \frac{2M}{R^3} + \frac{4M}{R^3} + \frac{M^2}{R^6} + \frac{4M^2}{R^6} \leq 1 + \frac{\mu}{R^3} \tag{13.6.7}$$

where $\mu = 6M + 5M^2$. Let E be the ellipse with equation

$$\frac{x^2}{a^2(1 + \mu R^{-3})} + \frac{y^2}{b^2(1 + \mu R^{-3})} = 1.$$

The inequality (13.6.7) implies that $C_R \subset E \cup \text{Int}\, E$. Therefore, by Proposition 13.5.1(i), $\text{Int}\, C_R \subset E \cup \text{Int}\, E$, and we have $\psi(A_{\rho,R}) \subset \text{Int}\, C_R \subset E \cup \text{Int}\, E$.

The area of $E \cup \text{Int}\, E$ is

$$|E \cup \text{Int } E| = \pi ab\left(1 + \frac{\mu}{R^3}\right) = \pi\left(R^2 - \frac{\alpha^2}{R^2}\right)\left(1 + \frac{\mu}{R^3}\right)$$
$$\leq \pi R^2\left(1 + \frac{\mu}{R^3}\right).$$

According to Exercise 9.4.12, the area $|\psi(A_{\rho,R})|$ exists, ane since $\psi(A_{\rho,R}) \subset E \cup \text{Int } E$, we have $|\psi(A_{\rho,R})| \leq |E \cup \text{Int } E|$. (See [6, Theorem 8.2.4].) Again referring to Exercise 9.4.12, we obtain

$$R^2\left(1 + \frac{\mu}{R^3}\right) \geq |\psi(A_{\rho,R})| = \int_\rho^R r\left[\int_0^{2\pi} |\psi'(re^{i\theta})|^2 \, d\theta\right] dr. \qquad (13.6.8)$$

According to the corollary of Theorem 9.9.1, we can differentiate the Laurent series for ψ term-by-term to obtain the Laurent expansion at ∞ for ψ', which turns out to be

$$\psi'(w) = 1 - \frac{\alpha}{w^2} - \frac{2c_2}{w^3} - \frac{3c_3}{w^4} - \cdots, \qquad |w| > 1.$$

Also

$$\overline{\psi'(w)} = 1 - \frac{\alpha}{\overline{w}^2} - \frac{\overline{2c_2}}{\overline{w}^3} - \frac{\overline{3c_3}}{\overline{w}^4} - \cdots, \qquad |w| > 1.$$

These series converge uniformly on $C(0, r)$, for each r, $\rho \leq r \leq R$, and their respective sum functions are bounded on $C(0, r)$. Therefore, by Exercise 3.3.9 and Proposition 8.3.1, the following computation is valid, with $w = re^{i\theta}$:

$$\int_0^{2\pi} |\psi'(w)|^2 \, d\theta = \int_0^{2\pi} \psi'(w)\overline{\psi'(w)} \, d\theta$$
$$= \int_0^{2\pi} \lim_{N\to\infty}\left[\left(1 - \frac{\alpha}{w^2} - \sum_{n=2}^N \frac{nc_n}{w^{n+1}}\right)\left(1 - \frac{\alpha}{\overline{w}^2} - \sum_{n=2}^N \frac{\overline{nc_n}}{\overline{w}^{n+1}}\right)\right] d\theta$$
$$= \lim_{N\to\infty} \int_0^{2\pi} \left[\left(1 - \frac{\alpha}{w^2} - \sum_{n=2}^N \frac{nc_n}{w^{n+1}}\right)\left(1 - \frac{\alpha}{\overline{w}^2} - \sum_{n=1}^N \frac{\overline{nc_n}}{\overline{w}^{n+1}}\right)\right] d\theta$$
$$= 2\pi\left[1 + \frac{\alpha^2}{r^4} + \sum_{n=2}^\infty \frac{n^2|c_n|^2}{r^{2n+2}}\right]. \qquad (13.6.9)$$

In the last step we used the fact that with $w = re^{i\theta}$,

$$\int_0^{2\pi} \frac{1}{w^n\overline{w}^m} \, d\theta = \begin{cases} 0, & n \neq m. \\ 2\pi r^{-2n}, & n = m. \end{cases}$$

The series in the last member of (13.6.9) is a series of positive terms, so for any fixed r, $\rho \leq r \leq R$, its sequence of partial sums is monotonically increasing. We return to (13.6.8) with this information and obtain, with $N \geq 2$,

$$\pi R^2\left(1 + \frac{\mu}{R^3}\right) \geq |\psi(A_{\rho,R})| \geq \int_\rho^R 2\pi\left(r + \frac{\alpha^2}{r^3} + \sum_{n=2}^N \frac{n^2|c_n|^2}{r^{2n+1}}\right) dr$$
$$= \pi\left[R^2 - \rho^2 + \alpha^2\left(\frac{1}{\rho^2} - \frac{1}{R^2}\right) + \sum_{n=2}^N n|c_n|^2\left(\frac{1}{\rho^{2n}} - \frac{1}{R^{2n}}\right)\right].$$

We divide this equation by π and then subtract R^2 from each side and add ρ^2 to each side. The result is

$$\rho^2 + \frac{\mu}{R^3} \geq \alpha^2\left(\frac{1}{\rho^2} - \frac{1}{R^2}\right) + \sum_{n=2}^N n|c_n|^2\left(\frac{1}{\rho^{2n}} - \frac{1}{R^{2n}}\right).$$

This inequality is valid for every ρ, R, and N with $1 < \rho < R$ and $N \geq 2$. The functions on either side are continuous functions of ρ on the interval $1 \leq \rho < \infty$, so the inequality persists if we let $\rho = 1$. For a similar reason, the inequality remains valid in the limit as $R \to \infty$. We obtain

$$1 \geq \alpha^2 + \sum_{n=2}^\infty n|c_n|^2, \qquad N \geq 2.$$

Finally, by Exercise 2.3.2 (or Theorem 2.4.3), the inequality remains valid as $N \to \infty$.

This completes the proof of the area theorem for any member of $\mathscr{U}$ which has a Laurent expansion of the type shown in (13.6.3). Now let $F \in \mathscr{U}$ have the more general type of expansion displayed in the statement of the theorem. Let $\lambda = e^{-i(\text{Arg } b_1)/2}$; then $|\lambda| = 1$ and $\lambda^2 b_1 = e^{-i \text{ Arg } b_1} b_1 = |b_1|$. The function $\lambda F(w/\lambda) - b_0$ is univalent and analytic on $\Delta(\infty, 1)$ and has the Laurent series

$$\lambda F(w/\lambda) - b_0 = w + \frac{|b_1|}{w} + \sum_{n=2}^N \frac{\lambda^{n+1} b_n}{w^n}, \qquad |w| > 1.$$

So this function is a member of $\mathscr{U}$ which has a Laurent expansion like (13.6.3). We therefore have $\sum_1^\infty n|\lambda^{n+1} b_n|^2 = \sum_1^\infty n|b_n|^2 \leq 1$. Q.E.D.

Corollary With F as described in the theorem, $|b_1| \leq 1$.

Example (ii) This example presents an extremal function F for the theorem and its corollary, as well as for Proposition 13.6.2. Let $f(w) = w/(1 - w)^2$ as in Example (i). By Proposition 13.6.1(ii), the function F with values $1/f(1/\xi) = \xi - 2 + \xi^{-1}$ is a member of $\mathscr{U}$ for which obviously $b_1 = 1$ and $\sum_1^\infty n|b_n|^2 = 1$.

Theorem 13.6.3 *Let f be a member of $\mathscr{S}$ with the power series representation*

$$f(w) = w + a_2 w^2 + a_3 w^3 + \cdots, \qquad |w| < 1.$$

Then $|a_2| \leq 2$.

Proof. According to Theorem 3.6.1, there exists $g \in \mathscr{S}$ such that

$$[g(w)]^2 = f(w^2) = w^2 + a_2 w^4 + a_3 w^6 + \cdots, \qquad |w| < 1.$$

Let g have the power series representation

$$g(w) = w + c_2 w^2 + c_3 w^3 + \cdots, \qquad |w| < 1. \tag{13.6.10}$$

This series converges absolutely for $|w| < 1$, so the Cauchy product of the series for g and g converges to $[g(w)]^2$, by Propositions 2.7.2 and 2.7.3. We have

$$\begin{aligned} w^2 + a_2 w^2 + \cdots &\equiv (w + c_2 w^2 + c_3 w^3 + \cdots)\cdot(w + c_2 w^2 + c_3 w^3 + \cdots) \\ &= w^2 + 2c_2 w^3 + (c_3 + c_2^2 + c_3)w^4 + \cdots. \end{aligned}$$

Thus, $c_2 = 0$, $2c_3 = a_2$, and $c_3 = a_2/2$.

Now according to Proposition 13.6.1(ii), the function with value $1/g(1/\xi)$ is a member of $\mathscr{U}$. Let its Laurent expansion at ∞ be $\xi + b_0 + b_1\xi^{-1} + b_2\xi^{-2} + \cdots$, $|\xi| > 1$. This series converges absolutely for $|\xi| > 1$, and so does the series (13.6.10) with $w = 1/\xi$. Formation of a Cauchy product series is thereby again justified, and we have, for $|\xi| > 1$,

$$\begin{aligned} 1 &= g\left(\frac{1}{\xi}\right)[\xi + b_0 + b_1\xi^{-1} + b_2\xi^{-2} + \cdots] \\ &= \left(\xi^{-1} + 0 + \frac{a_2}{2}\xi^{-3} + c_4\xi^{-4} + \cdots\right)\cdot(\xi + b_0 + b_1\xi^{-1} + b_2\xi^{-2} + \cdots). \end{aligned} \tag{13.6.11}$$

The third term in the Cauchy product series is $[b_1 + 0 + (a_2/2)]\xi^{-2}$; the coefficient of ξ^{-2} in the left-hand member of (13.6.11) is zero, so we have $a_2 = -2b_1$. Then by the corollary of Theorem 13.6.2, we have $|a_2| \le 2$. Q.E.D.

Notice that Example (i) is an extremal $f \in \mathscr{S}$ for this theorem with $a_2 = 2$. The theorem contains various implications for the class $\mathscr{U}$, of which we give only some very simple examples.

Proposition 13.6.2 *Let $F \in \mathscr{U}$ have the Laurent expansion at ∞ given by (13.6.2), and suppose that $F(w) \neq 0$ for every $w \in \Delta(\infty, 1)$. Then $|b_0| \le 2$.*

Proof. Here we can let $a = 0$ in Proposition 13.6.1(i), and then the function with values $1/F(1/\xi)$ is a member of $\mathscr{S}$. Let its power series expansion be $\xi + a_2\xi^2 + \cdots$, $|\xi| < 1$. The Cauchy product technique used in the proof of Theorem 13.6.3 is again available, and we have for $|\xi| > 1$,

$$\begin{aligned} 1 &\equiv F\left(\frac{1}{\xi}\right)[\xi + a_2\xi^2 + a_3\xi^3 + \cdots] \\ &\equiv \left(\frac{1}{\xi} + b_0 + b_1\xi + b_2\xi^2 + \cdots\right)(\xi + a_2\xi^2 + a_3\xi^3 + \cdots). \end{aligned} \tag{13.6.12}$$

The second term of the Cauchy product series is $(a_2 + b_0)\xi$; the coefficient of ξ in the left-hand member of the identity (13.6.12) is zero; so $b_0 = -a_2$ and $|b_0| = |a_2| \le 2$, by Theorem 13.6.3. Q.E.D.

Notice that Example (ii) is an extremal function for this proposition.

Corollary Let $F \subset \mathscr{U}$ have the Laurent expansion at ∞ given by (13.6.2), and let E be the complement in $\mathbf{C}$ of $F[\Delta(\infty, 1)]$. Then diam $E \le 4$.

Proof. For $z_0 \in E$, the function given by $F(w) - z_0$ is analytic, univalent, and nonzero for $w \in \Delta(\infty, 1)$, and has the following Laurent expansion at ∞: $F(w) = w + (b_0 - z_0) + b_1 w^{-1} + b_2 w^{-2} + \cdots$, $|w| > 1$. Therefore, by the proposition, $|b_0 - z_0| \le 2$. Now let z_1 and z_2 be any two points in E. We have $|z_1 - z_2| = |z_1 - b_0 + b_0 - z_2| \le |b_0 - z_1| + |b_0 - z_2| \le 2 + 2 = 4$. Q.E.D.

Example (iii) The function given by $z = w + (1/w)$, $|w| > 1$, is an extremal for this corollary of Proposition 13.6.2. The image of $\Delta(\infty, 1)$ is the complement of the interval of reals $[-2, 2]$, (This is verified in Exercise 13.6.2.)

Theorem 13.6.3 also contains the following implication for the class $\mathscr{S}$.

Proposition 13.6.3 *For any* $f \in \mathscr{S}$, *the map* $f[\Delta(0, 1)]$ *contains the* disk $\Delta(0, 1/4)$.

Proof. Let z_0 be such that $z_0 \notin f[\Delta(0, 1)]$. We prove that every such z_0 satisfies the inequality $|z_0| \ge 1/4$, and therefore $\{z: |z| < 1/4\} \subset f[\Delta(0, 1)]$.

Define a function $h: \Delta(0, 1) \to \mathbf{C}$ by

$$h(w) = \frac{f(w)}{1 - [f(w)/z_0]}.$$

Since $f(w)/z_0 \ne 1$ for $w \in \Delta(0, 1)$, it follows that h is analytic on $\Delta(0, 1)$. Also $h(0) = 0$ and $h'(0) = \lim_{w \to 0} h(w)/w = f'(0)/(1 - 0) = 1$. It is easily verified that h is univalent on $\Delta(0, 1)$. (By elementary algebra, it can be shown that the equation $h(w_1) = h(w_2)$ implies $f(w_1) = f(w_2)$, and f is univalent.) Thus, $h \in \mathscr{S}$ and has a power series representation on $\Delta(0, 1)$, which we write partially as follows: $h(w) = w + C_2 w^2 + \cdots$. Let the corresponding series for f be $f(w) = w + a_2 w^2 + \cdots$, $|w| < 1$; then the corresponding power series for $1 - [f(w)/z_0]$ is $1 - (w/z_0) - a_2(w^2/z_0) - \cdots$. We have the identity

$$\left(1 - \left[\frac{f(w)}{z_0}\right]\right) h(w) \equiv \left(1 - \frac{w}{z_0} - \frac{a_2 w^2}{z_0} - \cdots\right) \cdot (w + C_2 w^2 + \cdots)$$
$$\equiv f(w) \equiv w + a_2 w^2 + \cdots, \qquad |w| < 1.$$

All these series converge absolutely on $\Delta(0, 1)$, so the substitution of the Cauchy product series in the second member does not disturb the identity. Only the coefficients of w^2 are of interest, and by equating these, we obtain $(-1/z_0) + C_2 = a_2$, so $1/z_0 = C_2 - a_2$. Now both h and f are members of $\mathscr{S}$, so by Theorem 13.6.3, we have $|C_2| \le 2$ and $|a_2| \le 2$. Therefore, $|1/z_0| \le |C_2| + |a_2| \le 4$, and $|z_0| \ge 1/4$. Q.E.D.

Notice that Example (i) is an extremal for this proposition, for it was pointed out there that the mapping of $\Delta(0, 1)$ given by the function in that example omits the point $z = -1/4$.

The inequality $|a_2| \le 2$ is consistent with the Bieberbach conjecture of 1916 that for every $f \in \mathscr{S}$, $|a_n| \le n$, $n = 2, 3, \ldots$. An enormous amount of mathematical research has been poured into attempts to verify this conjecture, and there has been a substantial fall-out into other areas of mathematical research. For each $n > 3$ for which the conjecture has been verified, the proof has been deep and complicated.† We conclude the text of this section with an important special case for which the proof is relatively simple.

Proposition 13.6.4 (*Dieudonné, 1931*). *For any $f \in \mathscr{S}$ such that the coefficients of the power series (13.6.1) are real numbers, we have $|a_n| \le n$, $n = 1, 2, \ldots$.*

Proof. Let $f(w) = U(w) + iV(w)$, where U and V are real valued. We have for $|w| < 1$,

$$f(w) = U(w) + iV(w) = a_1 w + \Sigma_2^\infty a_n w^n,$$
$$\overline{f(\overline{w})} = U(\overline{w}) - iV(\overline{w}) = \overline{a_1 \overline{w} + \Sigma_2^\infty a_n \overline{w}^n} = f(w),$$

and $a_1 = 1$. Thus $U(w) = U(\overline{w})$ and $V(w) = -V(\overline{w})$; and with $w = re^{i\theta}$, $0 < r < 1$, we have $V(re^{i\theta}) = -V(re^{-i\theta})$. We refer now to Corollary 2 of Proposition 9.4.1, in which the following formula is established:

$$a_n = \frac{i}{\pi r^n} \int_{-\pi}^{\pi} V(re^{i\theta})(\cos n\theta - i \sin n\theta)\, d\theta, \qquad n = 1, 2, \ldots.$$

Since a_n is real here, the imaginary part of the right-hand member is zero for each n, and we have

$$a_n = \frac{1}{\pi r^n} \int_{-\pi}^{\pi} V(re^{i\theta}) \sin n\theta\, d\theta, \qquad n = 1, 2, \ldots.$$

Also since $V(re^{i\theta}) \sin n\theta$ is an even function of θ, we can write

$$a_n = \frac{2}{\pi r^n} \int_{0}^{\pi} V(re^{i\theta}) \sin n\theta\, d\theta, \qquad n = 1, 2, \ldots. \tag{13.6.12}$$

It is necessary now to show that $V(re^{i\theta}) > 0$, $0 < \theta < \pi$. We know that $V(w)$ is continuous on $H = \Delta(0, 1) \cap \{w\colon \operatorname{Im} w > 0\}$. If V were to change

†Statements as to the status quo of this work tend to become obsolete quickly. As of the end of 1972, the conjecture had been proved for $n = 2, 3, 4, 5, 6$ and it was known that $|a_n| < (7/6)^{1/2}\, n < 1.081n$ for every $n \ge 1$. See [12, review 453] and [13, review 557] for references to various research papers which are landmarks in the history of the problem.

sign on H, then by the intermediate value theorem [Proposition 5.3.2], there would exist $w_0 \in H$ such that $V(w_0) = 0$. This would imply that $V(\overline{w}_0) = -V(w_0) = 0$, and since $U(w_0) = U(\overline{w}_0)$, we would have $f(w_0) = f(\overline{w}_0)$. But the last equation would contradict the fact that f is univalent on $\Delta(0, 1)$. Therefore, V has a constant sign on H. To find out what it is, we set $n = 1$ in (13.6.12) and use the fact that $a_1 = 1 > 0$. Therefore, $V(w) > 0$ on H.

The remainder of the proof of the proposition follows readily from the following inequality, which is valid for $0 \leq \theta \leq \pi$:

$$\begin{aligned} |\sin n\theta| &= \sin\theta\left|\frac{e^{in\theta} - e^{-in\theta}}{e^{i\theta} - e^{-in\theta}}\right| \\ &= \sin\theta\left|\sum_{k=1}^{n} e^{ik\theta}e^{-i(n-k-1)\theta}\right| \\ &\leq (\sin\theta)\sum_{k=1}^{n} 1 = n\sin\theta. \end{aligned}$$

From (13.6.12), we have

$$\begin{aligned} |a_n| &\leq \frac{2}{\pi r^n}\int_0^\pi V(re^{i\theta})|\sin n\theta|\, d\theta \\ &\leq \frac{2n}{\pi r^n}\int_0^\pi V(re^{i\theta})\sin\theta\, d\theta \\ &= \frac{n}{r^{n-1}}\left[\frac{2}{\pi r}\int_0^\pi V(re^{i\theta})\sin\theta\, d\theta\right] \\ &= \frac{n}{r^{n-1}}a_1 = \frac{n}{r^{n-1}}, \qquad n = 1, 2, 3, \ldots. \qquad (13.6.13) \end{aligned}$$

This inequality holds for every r, $0 < r < 1$, and the last member is continuous on $(0, 1]$, so the inequality remains valid with $r = 1$. Q.E.D.

The function in Example (i) is an extremal for this proposition. The Bieberbach conjecture is also known to be true if $f[\Delta(0, 1)]$ is starlike, but the proof is considerably more difficult [10, pp. 356–358].

Exercises 13.6

1. Prove Proposition 13.6.1(ii).
2. Prove that for the mapping $z = f(w)$ in Example (iii), $\{f[\Delta(\infty, 1)]\}^c = \{z = x + iy: -2 \leq x \leq 2, y = 0\}$.
3. Let $f: \Delta(0, 1) \to \mathbf{C}$ be analytic and have the power series representation $f(w) = w + a_2w^2 + a_3w^3 + \cdots$, $|w| < 1$. Furthermore, suppose that $\sum_2^\infty n|a_n| < 1$. Show that f can be extended onto $C(0, 1)$ so as to be continuous on $\overline{\Delta}(0, 1)$, and that f so extended is univalent on $\overline{\Delta}(0, 1)$. (*Hint*: Use Proposition 4.1.2 to establish the continuous extension. Then for w, $w_0 \in \overline{\Delta}(0, 1)$, we have $f(w) - f(w_0) = (w - w_0)[1 + \sum_2^\infty a_n(w^n - w_0^n)/(w - w_0)]$. Therefore, if $f(w) - f(w_0) = 0$, then unless $w = w_0$, the term in square brackets is necessarily zero. Use the identity

$(w^n - w_0^n)/(w - w_0) = w^{n-1} + w^{n-2}w_0 + \cdots + w_0^n$ to show that under the hypothesis $\sum_2^\infty n|a_n| < 1$, the term in square brackets is nonzero for $w, w_0 \in \Delta(0, 1)$.)

13.7 Continuation of Analytic Functions and Conformal Equivalences Onto and Across Boundaries

Let $B \subset \mathbf{C}$ be a region, and let $\phi: B \to \mathbf{C}$ be a function which is analytic on B, so $\phi(B)$ is another region [Sec. 10.4, Remark (iii)]. The behavior of such a function ϕ and its inverse in neighborhoods of the boundary ∂B of B and the boundary $\partial\phi(B)$ of $\phi(B)$ has received a considerable amount of attention. In this section, we consider two basic questions in this area of complex analysis. The first is that of whether $\phi(z)$ tends to a limit in some sense as $z \in B$ approaches the boundary ∂B of B. The second is that of the possibility of extending the definition of ϕ across ∂B so as to provide a conformal equivalence which maps a region B_1 containing $\bar{B}$ onto a region containing $\phi(\bar{B})$. Because of considerations of space, it is necessary to give only a superficial treatment of the first question, but we look into the second question at some length. Certain proofs which are largely of a topological character are deferred to an appendix at the end of this chapter.

In Sec. 13.5, we defined a Jordan curve, or simple closed curve, to be a closed curve which is homeomorphic to a circle. A fundamental theorem relating to the first question is as follows.

Theorem 13.7.1 *Let C be a Jordan curve, and let $\psi: \Delta(0, 1) \to \mathbf{C}$ be a conformal equivalence which maps $\Delta(0, 1)$ onto* Int C. *Then ψ can be extended onto $C(0, 1)$ in such a way that the extended function is a homeomorphism from $\bar{\Delta}(0, 1)$ onto $C \cup$* Int C.

Corollary Let C be a Jordan curve. Let $\psi_0: \Delta(\infty, 1) \to \mathbf{C}$ be analytic and univalent on $\Delta'(\infty, 1)$, with a simple pole at ∞, and let ψ_0 map $\Delta'(\infty, 1)$ onto Ext C. Then ψ_0 can be extended onto $C(0, 1)$ in such a way that the extended function is a homeomorphism from $\bar{\Delta}'(\infty, 1)$ onto $C \cup$ Ext C.

The proof of the theorem unfortunately requires rather extensive preparations involving real analysis which are outside of the scope of this book, so we omit it. (A proof is given in [10, Sec. 17.5].) The corollary can be referred to the theorem by a composition of transformations used previously in the proofs of Theorem 13.4.2 and Proposition 13.6.1. First, notice that with $a \in$ Int C, the linear fractional transformation $L: \zeta = 1/(z - a)$ is a homeomorphism from $\hat{\mathbf{C}}$ onto $\hat{\mathbf{C}}$, which carries C (homeomorphic to a circle) in the z-plane onto a Jordan curve C^* in the ζ-plane. It takes the connected set Int C onto a connected set S which lies in $(C^*)^c$, so S must be entirely contained in a component of $(C^*)^c$ [Theorem 5.4.1], and since S is unbounded, this component of $(C^*)^c$ is necessarily Ext C^*. By pursuing this line of reasoning a little further, as we did in the proofs in Sec. 13.5, we can easily show that L carries Ext C onto Int C^*. We now define an analytic univalent function ψ from $\Delta(0, 1)$ in the ξ-plane onto Int C^* by $\zeta = \psi(\xi) = 1/[\psi_0(1/\xi) - a]$, $\xi \neq 0$, and $\psi(0) = 0$. We apply the theorem to $\psi(\xi)$ and then validate the conclusion of the corollary via the equation $\psi_0(w) = a + 1/\psi(1/w)$.

A much weaker result can be derived for a general region B by adopting the following definition for approach to the boundary ∂B. A sequence $\langle z_n \rangle_1^\infty$ from B is said to approach ∂B if given any compact set $E \subset B$, there exists $N = N_E$ such that for every $n \geq N$, $z_n \in E^c \cap B$. We symbolize such a limit process by $z_n \to \partial B$. We have this simple result with a purely topological content.

Proposition 13.7.1 *Let $\phi: B \to B^*$ be a homeomorphism from the region B onto B^*. If the sequence $\langle z_n \rangle$ from B is such that $z_n \to \partial B$, then $\phi(z_n) \to \partial B^*$.*

The proof is deferred to the exercises.

In the remainder of this section, we present a series of propositions which are relevant to the problem of extending an analytic function (and, in particular, a conformal equivalence) across the boundary of its domain. First, we need some further definitions. Recall that in Sec. 5.3, an arc in **C** was defined to be a continuous mapping of a closed interval of reals into **C**. We now define a *simple arc*, or *Jordan arc*, as an arc in **C** which is a homeomorphism from the closed interval $[\alpha, \beta]$ of reals into **C**. An *open arc* is a bounded continuous mapping of an open interval (α, β) into **C**, and the arc is *simple* if the mapping is one-to-one. Recall from Sec. 13.5 that in the interest of brevity, in this chapter we frequently use the same notation and nomenclature for an arc and its graph (or a closed curve and its graph) when no confusion can thereby arise.

If B is a region and ∂B denotes its boundary, and if $\Gamma \subset \partial B$ is an arc, we say that Γ is a *free boundary arc* of B if and only if Γ is an open simple arc, and for each point $z \in \Gamma$ there exists a disk $\Delta(z, \rho_z)$, $\rho_z > 0$, such that $\Delta(z, \rho_z) \cap \partial B = \Delta(z, \rho_z) \cap \Gamma$. That is, for each point in a free boundary arc, there is a neighborhood which intersects ∂B only in points of Γ. An example of an open boundary arc of a region which is *not* a free boundary arc is given by the interval $(0, 1)$ in the boundary of $B = \{z : 0 < \text{Re}\, z < 1, 0 < \text{Im}\, z < 1\} \sim \bigcup_{n=2}^\infty \{z : 0 < \text{Re}\, z < (1/2), \text{Im}\, z = 1/n\}$. (See Figure 1.)

In the affirmative mode we have the following proposition.

Proposition 13.7.2 *Let C be a Jordan arc or curve, and let Γ be an open arc contained in C. Then Γ is a free boundary arc of C^c. In particular, if C is a Jordan curve, then Γ is a free boundary arc of both* Int C *and of* Ext C.

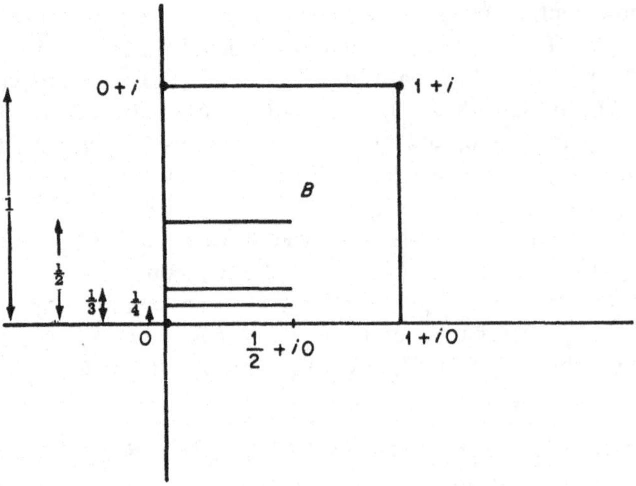

Fig. 1

Proof. According to the Jordan curve theorem, if C is a Jordan curve, then it is simultaneously the boundary of Int C and of Ext C. If C is a Jordan arc, it is shown in point-set topology that C is the boundary of C^c. (In topological language, the basic proposition here is that a simple closed curve or arc in $\mathbf{R}^n$ is nowhere dense in $\mathbf{R}^n$ [14, pp. 66–67].) It remains only to prove that Γ in either case is a *free* boundary arc of C^c. Let γ be a continuous mapping of the interval $[\alpha, \beta]$ onto C, with γ univalent on (α, β), and with $\gamma(\alpha) = \gamma(\beta)$ if C is a closed curve. Let $\Gamma = \gamma(\alpha', \beta')$, $(\alpha', \beta') \subset [\alpha, \beta]$. If Γ were not a free boundary arc of C^c, there would exist $z_0 \in \Gamma$ such that in each disk $\Delta(z_0, 1/n)$, $n = 1, 2, \ldots$, there would be a point $z^{(n)} \in C \sim \Gamma$. Clearly, $\lim_{n\to\infty} z^{(n)} = z_0 \in \Gamma$. By Theorem 5.5.2, $C \sim \Gamma$ is closed in $\mathbf{C}$ because it is the image, in the continuous mapping γ, of the set $[\alpha, \beta] \cap (\alpha', \beta')^c$, which is compact in $\mathbf{R}$. Therefore, $\lim_{n\to\infty} z^{(n)} = z_0 \in C \sim \Gamma$; that is, $z_0 \in \Gamma$ and $z_0 \in C \sim \Gamma$. Then the inverse image of z_0 under the mapping γ would contain a point in each of the two disjoint sets $[\alpha, \beta] \cap (\alpha', \beta')^c$ and (α', β'), which would be in conflict with the univalence of γ. Therefore, Γ is a free boundary arc of C^c. Q.E.D.

Our first result concerning the extension of an analytic function across the boundary of its domain is one of several variants of the Schwarz reflection principle.

Theorem 13.7.2 *Let $B \subset \mathbf{C}$ be a region in the upper half-plane $H = \{z : \operatorname{Im} z > 0\}$, such that the open interval of reals (α, β) is a free boundary arc* *of B. Let* $B^* = \{z : \bar{z} \in B\}$ *(the reflection of B in the real axis). Let* $f: (B \cup (\alpha, \beta)) \to \mathbf{C}$ *be a function which is analytic on B, continuous on* $B \cup (\alpha, \beta)$*, and real valued on* (α, β)*. Then:*

(i) *There exists a function F with domain* $D = B \cup (\alpha, \beta) \cup B^*$*, which is analytic on D and which is given by*

$$F(z) = \begin{cases} f(z), & z \in B \cup (\alpha, \beta); \\ \overline{f(\bar{z})}, & z \in B^*. \end{cases} \tag{13.7.1}$$

(ii) *If f is univalent on* $B \cup (\alpha, \beta)$ *and if the only real values of* $f(z)$ *for* $z \in B \cup (\alpha, \beta)$ *occur for* $z \in (\alpha, \beta)$*, then F is univalent on D.*

Proof. (i) According to Exercise 6.3.7, F is analytic on B^*, and since $\overline{f(\bar{z})} \equiv f(z)$, $z \in (\alpha, \beta)$, it remains only to show that F is analytic at each point of (α, β). Take $x_0 \in (\alpha, \beta)$. Since (α, β) is a free boundary arc of B, there exists a disk $\Delta(x_0, 2\rho)$, $\rho > 0$, such that $\Delta(x_0, 2\rho) \cap \partial B = (x_0 - 2\rho, x_0 + 2\rho) \in (\alpha, \beta)$. The upper half-disk $\Delta(x_0, 2\rho) \cap H$ contains points of B, but no points of B^*. In fact, $\Delta(x_0, 2\rho) \cap H$ contains no point of $B^c = \bar{B}^c \cup \partial B$ which is not in the interval $(x_0 - 2\rho, x_0 + 2\rho)$. (The reason is that such a point would necessarily lie in the open set $\bar{B}^c$, since it cannot lie in ∂B, and then $\bar{B}^c \cup B$ would be arcwise connected and, therefore, connected [Theorem 5.3.1]; but this is impossible by the definition of connectivity.) Therefore, $(\Delta(x_0, 2\rho) \cap H) \subset B$. Similarly, we have $\Delta(x_0, 2\rho) \cap H^* \subset B^*$, where $H^* = \{z : \operatorname{Im} z < 0\}$.

Now consider the circle $C(x_0, \rho) \subset \Delta(x_0, 2\rho)$. Let γ_1 denote the positively oriented upper half of $C(x_0, \rho)$; let γ_2 be the oriented interval $\{z = x + iy : y = 0, x_0 - \rho \le x \le x_0 + \rho\}$; and let γ_3 denote the positively oriented lower half of $C(x_0, \rho)$. Notice that F is analytic on $\Delta(x_0, \rho) \sim (x_0 - \rho, x_0 + \rho)$ and continuous on $\bar{\Delta}(x_0, \rho)$. The closed paths $\gamma_1 + \gamma_2$ and $\gamma_3 - \gamma_2$ are of the type described in Proposition 9.4.1, as the reader can easily verify. Therefore, according to that proposition, we have the Cauchy integral formulas

$$F(z) = \begin{cases} \dfrac{1}{2\pi i} \displaystyle\int_{\gamma_1 + \gamma_2} \dfrac{F(\zeta)\, d\zeta}{\zeta - z}, & z \in \Delta(x_0, \rho) \cap H; \\ \dfrac{1}{2\pi i} \displaystyle\int_{\gamma_3 - \gamma_2} \dfrac{F(\zeta)\, d\zeta}{\zeta - z}, & z \in \Delta(x_0, \rho) \cap H^*. \end{cases} \tag{13.7.2}$$

For every z in the unbounded component of the complement of the path $\gamma_1 + \gamma_2$, which contains $\Delta(x_0, \rho) \cap H^*$, the first integral in (13.7.2) has the value zero [Proposition 9.3.3], and for a similar reason, the second integral has the value zero in $\Delta(x_0, \rho) \cap H$. We can use this information together with a cancellation along the path γ_2 to rewrite (13.7.2) in the form

$$F(z) = \frac{1}{2\pi i}\int_{\gamma_1+\gamma_3} \frac{F(\zeta)\,d\zeta}{\zeta - z}, \qquad z \in \Delta(x_0, \rho) \sim (x_0 - \rho, x_0 + \rho). \quad (13.7.3)$$

But by Theorem 8.4.1, this integral represents a function analytic at every point z not on the path of integration. Therefore, in particular, the right-hand member of 13.7.3 is analytic on $\Delta(x_0, \rho)$. But the left-hand member as defined in (13.7.1) is continuous on $\Delta(x_0, \rho)$, so the equation (13.7.2) is valid for every $z \in \Delta(x_0, \rho)$. Thus, F is analytic at x_0 and, therefore, on (α, β).

(ii) If f is univalent on $B \cup (\alpha, \beta)$, then clearly $F|(B \cup (\alpha, \beta))$ and $F|(B^* \cup (\alpha, \beta))$ are each univalent on the indicated domain. It remains to show that F is globally univalent on D. Suppose that for $z_1 \in B$ and $z_2 \in B^*$, we were to have $F(z_1) = F(z_2)$. We would have $f(z_1) = \overline{f(\bar{z}_2)}$, with $\bar{z}_2 \in B$. Let $f(z) = u(z) + iv(z)$, where $u(z)$ and $v(z)$ are real. The equation $f(z_1) = \overline{f(\bar{z}_2)}$ implies that $v(z_1) = -v(\bar{z}_2)$. Since v is continuous on the connected set B, by the intermediate value theorem [Proposition 5.3.2], there would exist $z_3 \in B$ such that $v(z_3) = 0$, and then $f(z_3)$ would be real. This is ruled out by hypothesis, so the conclusion is that F is globally univalent on D. Q.E.D.

Another variant of the Schwarz reflection principle is the following corollary.

Corollary Let W be a region in $\Delta(0, 1)$ in the complex w-plane of which the arc $A = \{w: w = e^{i\theta}, \tau < \theta_1 < \theta < \theta_2 < \tau + 2\pi\}$ is a free boundary arc. Let $W^* = \{w: 1/\bar{w} \in W\}$ ($W^* \subset \hat{C}$ is the reflection of W in the circle $C(0, 1)$). Let $W^{**} = W^* \sim \{\infty\}$. Let $\psi: (W \cup A) \to \mathbf{C}$ be a function which is analytic on W, continuous on $W \cup A$, and real valued on A. Then:

(i) There exists a function G with domain $R = W \cup A \cup W^{**}$ which is analytic on R and which is given by

$$G(w) = \begin{cases} \psi(w), & w \in W \cup A; \\ \overline{\psi}\left(\dfrac{1}{\bar{w}}\right), & w \in W^{**}. \end{cases}$$

(ii) If ψ is univalent on $W \cup A$ and if the only real values of ψ for $w \in (W \cup A)$ occur on A, then G is univalent on R.

Proof. (i) We show that the corollary is a consequence of Theorem 13.7.2 by using the conformal equivalence given by the linear fractional transformation

$$z = i\frac{e^{i\tau} + w}{e^{i\tau} - w} = L(w), \qquad w = e^{i\tau}\left(\frac{z - i}{z + i}\right) = L^{-1}(z).$$

(See Exercises 1.16.6 and 1.16.8.) The following facts concerning this mapping are easily verified. If $w = L^{-1}(z)$, then $1/\bar{w} = L^{-1}(\bar{z}) = L^{-1}(\overline{L(w)})$; $L[\Delta(0, 1)] = H = \{z: \text{Im } z > 0\}$; $L[\Delta(\infty, 1)] = H^* = \{z: \text{Im } z < 0\}$, $L[\Delta'(\infty, 1)] = H^* \sim \{-i\}$; $L(A) = (\alpha, \beta)$, an open interval on the real axis (the explicit formulas for α and β in terms of θ_1, θ_2, τ are not needed here); $B = L(W)$ is a region in H; $B^* = L(W^*)$ is the region in H^* which is the reflection of B^* in the real axis; $B^* \sim \{-i\} = L(W^{**})$; (α, β) is a free boundary arc of B; $\psi(L^{-1}(z))$ is analytic on B, continuous on $B \cup (\alpha, \beta)$, and real valued on (α, β). Therefore, by Theorem 13.7.2, the function F given by

$$F(z) = \begin{cases} \psi(L^{-1}(z)), & z \in B \cup (\alpha, \beta), \\ \overline{\psi(L^{-1}(\bar{z}))}, & z \in B^* \sim \{-i\}, \end{cases}$$

is analytic on $B \cup (\alpha, \beta) \cup (B^* \sim \{-i\})$. Now let G be defined by

$$G(w) = F(L(w)) = \begin{cases} \psi(L^{-1}(L(w))) = \psi(w), & w \in W \cup A; \\ \bar{\psi}(L^{-1}(\overline{L(w)})) = \bar{\psi}\left(\dfrac{1}{\bar{w}}\right), & w \in W^{**}. \end{cases}$$

We know that with $z = L(w)$, dF/dz and dz/dw exist for every $w \in W \cup A \cup W^{**} = R$. Therefore by the chain law [Theorem 6.1.2], $dG/dw = dG/dz \cdot dz/dw$ exists everywhere on R.

(ii) The proof of this part is practically the same as that of part (ii) of Theorem 13.7.2, and we therefore omit it. Q.E.D.

We come now to a proposition which contains the essence of the technique which we use for extending the definition of an analytic univalent function across a free boundary arc.

Proposition 13.7.3 (a) *Let T be a bounded region in the complex t-plane symmetric in the real axis, and let the intersection of T with the real axis be the interval (α, β). Let $H = \{t: \text{Im } t > 0\}$, $H^* = \{t: \text{Im } t < 0\}$, $T_0 = T \cap H$, and $T_0^* = T \cap H^*$. Let $z = F(t)$ be analytic, bounded, and univalent on T. Then $\Omega = F(T)$ is a bounded region in the z-plane [Sec. 10.4, Remark (iii)] with the following properties:*

(i) *It is the union of the following disjoint sets: $\Omega_0 = F(T_0)$ and $\Omega_0^* = F(T_0^*)$, both of which are regions, and the arc $\Gamma = F((\alpha, \beta))$.*
(ii) *Γ is a free boundary arc of Ω_0, and also a free boundary arc of Ω_0^*.*

(b) *Let the function $\phi: \bar{\Omega}_0 \to \mathbf{C}$ be analytic and nonzero on Ω_0, continuous and univalent on $\bar{\Omega}_0 = \Omega_0 \cup \partial\Omega_0$; and let ϕ map Ω_0 onto a region $W_0 \subset \Delta(0, 1)$ and Γ onto a set $A \subset C(0, 1)$. Let $\psi = \phi^{-1}$, and let $W_0^* = \{w: 1/\bar{w} \in W_0\}$. Then:*

(i) *A is a free boundary arc of W_0.*
(ii) *The function h from $W = W_0 \cup A \cup W^*$ onto Ω defined by*

$$z = h(w) = \begin{cases} \psi(w), & w \in (W_0 \cup A), \\ F\left[\overline{F^{-1}\left(\psi\left(\frac{1}{\bar{w}}\right)\right)}\right], & w \in W_0^*, \end{cases} \tag{13.7.4}$$

is analytic and univalent on W.

(iii) *The inverse function h^{-1} is given by*

$$w = h^{-1}(z) = \begin{cases} \phi(z), & z \in (\Omega_0 \cup \Gamma), \\ \dfrac{1}{\overline{\phi[F(\overline{F^{-1}(z)})]}}, & z \in \Omega_0^*, \end{cases} \tag{13.7.5}$$

and is analytic and univalent on Ω.

Proof. (a) (i). The fact that T_0 and T_0^* are regions follows from Exercise 5.5.6. Therefore, $\Omega_0 = F(T_0)$ and $\Omega_0^* = F(T_0^*)$ are regions [Sec. 10.4, Remark (iii)].

(a) (ii). Recall that F is a homeomorphism from T onto Ω and F^{-1} is analytic on Ω [Proposition 10.4.1(ii)]. Since Γ is homeomorphic to (α, β) and F is bounded on T, by definition Γ is an open simple arc. We now prove that Γ is a boundary arc of both Ω and Ω^*. Let z_0 be any point in Γ. Since Γ is contained in the open set Ω, there are disks $\Delta(z_0, \rho)$, contained in Ω, and for each such disk $F^{-1}[\Delta(z_0, \rho)]$ is a neighborhood of $t_0 = F^{-1}(z_0)$ contained in T. Such a neighborhood in turn contains a disk $\Delta(t_0, \rho')$ which is the union of the sets $\Delta(t_0, \rho') \cap T_0$, $\Delta(t_0, \rho') \cap T_0^*$, and the interval $(t_0 - \rho', t_0 + \rho')$. Thus, $F[\Delta(t_0, \rho')] \subset \Delta(z_0, \rho)$ intersects both Ω_0 and Ω_0^*, so z_0 is a boundary point of both Ω_0 and Ω_0^*.

To see that Γ is a free boundary arc of Ω_0, first notice that $[(\partial\Omega_0) \sim \Gamma] \subset (\Omega_0^c \sim \Gamma) = \Omega^c \cup \Omega_0^*$, and since Ω_0^* is open, $[(\partial\Omega_0) \sim \Gamma] \subset \Omega^c$. Now for any $z_0 \in \Gamma$, there is a disk $\Delta(z_0, \rho) \subset \Omega$, $\rho > 0$, and $\Delta(z_0, \rho) \cap \Gamma$ contains no point of Ω^c, so it contains no point of $\partial\Omega_0 \sim \Gamma$. Thus, Γ is a free boundary arc of Ω_0. A similar argument shows that Γ is a free boundary arc of Ω_0^*.

(b) (i). We first observe that ϕ is a homeomorphism from $\overline{\Omega}_0$ onto $\overline{W}_0$ which maps $\partial\Omega_0$ onto ∂W_0. (A formal argument follows. First, $\phi(\overline{\Omega}_0)$ is a closed set [Exercise 5.2.1]. It contains W_0, so it also contains $\overline{W}_0$ [Exercise 5.1.6]. The same authorities ensure that $\overline{\Omega}_0 \subset \psi(\overline{W}_0)$, which implies that $\phi(\overline{\Omega}_0) \subset \overline{W}_0$. Therefore, $\phi(\overline{\Omega}_0) = \overline{W}_0$, and since $\phi(\Omega_0) = W_0$, it further follows that $\phi(\partial\Omega_0) = \partial W_0$.) Thus, since $\Gamma \subset \partial\Omega_0$, we have $A \subset \partial W_0$. Also, A is an open arc because $A = \phi[F((\alpha, \beta))]$. It remains to show that A is a *free* boundary arc of W_0.

Let w_0 be any point in A, and suppose that in each disk $\Delta(w_0, 1/n)$, $n = 1, 2, \ldots$, there were to exist $w^{(n)} \in (\partial W \sim A)$. Then $z^{(n)} = \psi(w^{(n)}) \in (\partial\Omega_0 \sim \Gamma)$, $n = 1, 2, \ldots$. Now $\lim_{n\to\infty} w^{(n)} = w_0$, so by continuity, $\lim_{n\to\infty} \psi(w^{(n)}) = \lim_{n\to\infty} z^{(n)} = z_0 = \psi(w_0) \in \Gamma$. Thus, in any neighborhood of $z_0 \in \Gamma$, there would exist points $z^{(n)} \in (\partial\Omega_0 \sim \Gamma)$, and this would contradict the fact that Γ is a free boundary arc of Ω_0. Thus, A is a free boundary arc of W_0.

(b) (ii). The function $t = F^{-1}[\psi(w)]$ is analytic on W_0, continuous and univalent on $W_0 \cup A$, and is real valued for $w \in A$ because $(\alpha, \beta) = F^{-1}[\psi(A)]$. Since ϕ is nonzero on Ω, we have $0 \notin W_0$, and $\infty \notin W_0^*$. Then by the corollary of Theorem 13.7.2, the function G given by the composition

$$t = G(w) = \begin{cases} F^{-1}[\psi(w)], & w \in W_0 \cup A, \\ \overline{F^{-1}}\left[\psi\left(\dfrac{1}{\bar{w}}\right)\right], & w \in W_0^*, \end{cases}$$

is analytic on $W = W_0 \cup A \cup W_0^*$. The range of this function is the region T of part (a). Now define $h: W \to \mathbf{C}$ by

$$z = h(w) = F(t) = F[G(w)] = \begin{cases} F(F^{-1}[\psi(w)] = \psi(w), & w \in W_0 \cup A. \\ F\left(\overline{F^{-1}}\left[\psi\left(\dfrac{1}{\bar{w}}\right)\right]\right), & w \in W_0^*. \end{cases}$$

This composition of analytic functions is analytic on W. Its range is Ω under the indicated restrictions on w. We now look into the question of the univalence of h. The function $t = G(w)$ maps W_0 onto T_0, and maps W_0^* onto T_0^*, so the only real values of this function occur for $w \in A$. Therefore, by part (ii) of the corollary of Theorem 13.7.2, G is univalent on W. Now F is univalent on T, so the composition h of F with G is univalent on W.

Then by Proposition 10.4.1(ii), h has a univalent analytic inverse with domain Ω. The reader should have no difficulty in inverting the formula (13.7.4) to obtain (13.7.5). Q.E.D.

Remark. Since Γ is a free boundary arc of Ω_0^* as well as of Ω_0, in part (b) of the proposition it is possible to interchange Ω_0 and Ω_0^* whenever they appear. More specifically, if we interchange Ω_0 and Ω_0^* in part (b), and let $W_0 = \phi(\Omega_0^*)$, and define W_0^* by $W_0^* = \{w: 1/\bar{w} \in W_0\}$ as before, then the conclusions (i), (ii), and (13.7.4) remain unaltered; and (iii) and (13.7.5) are valid with Ω_0 and Ω_0^* interchanged.

The definition of an analytic arc and an analytic curve is given in Sec. 13.5. We rephrase it here as follows: An arc or closed curve in $\mathbf{C}$ is analytic if and only if there exists a parametrization $z = \gamma(t)$, $\alpha \le t \le \beta$, such that for each $\tau \in [\alpha, \beta]$ there exists a power series $\sum c_h(\tau)(t - \tau)^h$ convergent to $\gamma(t)$ in $(\tau - \delta(\tau), \tau + \delta(\tau)) \cap [\alpha, \beta]$, with $\delta(\tau) > 0$; and $\gamma'(t) \neq 0$ for every $t \in [\alpha, \beta]$. An important property of a simple analytic arc is revealed in the following proposition.

Proposition 13.7.4 *Let Γ be a simple analytic arc with the analytic parametrization $z = \gamma(t)$, $\alpha \le t \le \beta$. There exist a function F and a region T in the complex t-plane with the following properties: T is bounded, symmetric in the real axis, and contains $[\alpha, \beta]$; F is bounded, analytic, and univalent on T; and $F(t) \equiv \gamma(t)$, $t \in [\alpha, \beta]$, so $\Gamma = F([\alpha, \beta])$.*

The proof of this proposition requires rather careful attention to certain elementary topological details which seem to lie apart from the mainstream of thought in this chapter, so we defer the proof to the appendix to this chapter. It is shown there that T and $F(T)$ can be made to satisfy some further specifications, to the end that it is then possible to give an accurate proof of the following fundamental theorem. Once again the proofs are largely topological in character.

Theorem 13.7.3 *Let C be a Jordan curve in the complex z-plane containing an analytic arc Γ_1, and let Γ be an open subarc of Γ_1. Let ϕ be a conformal equivalence which maps* Int C *onto* $\Delta(0, 1)$. *Then there exist a region B and a function $f: B \to \mathbf{C}$ with these properties*: $(\Gamma \cup \text{Int } C) \subset B$, $B \cap \text{Ext } C$ *is a region of which Γ is a free boundary arc, f is analytic and univalent on B, and* $f(z) \equiv \phi(z)$, $z \in \text{Int } C$.

Finally, using a slightly sharper form of this theorem which appears as Theorem 13.7.3A in the appendix, we prove the following statements, which have been the ultimate goal of this study of the posibility of extending a conformal equivalence across an analytic boundary curve.

Theorem 13.7.4 *Let C be an analytic Jordan curve in the complex z-plane.*

(a) *Let ϕ be a conformal equivalence which maps* Int C *onto* $\Delta(0, 1)$. *Then there exist a region B and a function $f: B \to \mathbf{C}$ with these properties*:

(i) $(C \cup \text{Int } C) \subset B$ *and* $B \cap \text{Ext } C$ *is a region.*
(ii) *f is analytic and univalent on B.*
(iii) $f(z) \equiv \phi(z)$, $z \in \text{Int } C$.
(iv) *There exists $\rho > 1$ such that the inverse function f^{-1} is analytic and univalent on* $\Delta(0, 1)$.

(b) *Let $\psi: \Delta(\infty, 1) \to \mathbf{C}$ be analytic on $\Delta'(\infty, 1)$, univalent on $\Delta(\infty, 1)$, with a simple pole at ∞, and, furthermore, let ψ map $\Delta'(\infty, 1)$ onto* Ext C. *Then there exists r, $0 < r < 1$, such that ψ can be extended across* $C(0, 1)$ *so as to be analytic and univalent on* $\Delta'(\infty, r)$.

Exercises 13.7

1. Prove Proposition 13.7.1. (*Hint*: By Proposition 5.3.1, B^* is a region. If $E^* \subset B^*$ is a compact set, then so also is $\phi^{-1}(E^*) \subset B$ [Theorem 5.5.2].)
2. Let B be a region, and let $\langle z_n \rangle$ be a sequence from B. Ahlfors [2, p. 224] defines the limit process $z_n \to \partial B$ by the requirement that for every $z \in B$, there exist $\varepsilon_z > 0$ and N_z such that for every $n \geq N_z$, we have $|z_n - z| \geq \varepsilon_z$. Show that this definition is equivalent to the one in this section. (*Hint*: It can be assumed that for each $z \in B$, $\Delta(z, \varepsilon_z) \subset B$. The disks $\{\Delta(z, \varepsilon_z), z \in B\}$ form an open covering of

Ω. Let $E \subset B$ be compact, and use the definition of a compact set given in Sec. 5.5.)

3. This is a variant of the corollary of Theorem 13.7.2. Let W, A, W^*, W^{**} be defined as in that corollary, with the further restriction that W is simply connected. Let $\psi: (W \cup A) \to \mathbf{C}$ be analytic on W, continuous and nonzero on $W \cup A$, and such that $\psi(A) = \{z = e^{i\lambda}, -\pi < \lambda_1 < \lambda < \lambda_2 < \pi\}$. (That is, ψ maps A onto an arc of a circle.) Then ψ can be continued across A into W^{**} so as to be analytic on $W \cup A \cup W^{**}$. The continuation is given by $\psi^*(w) \equiv 1/\overline{\psi}(1/\overline{w})$, $w \in W^{**}$. (*Hint*: By Theorem 12.8.1, ψ has an analytic logarithm on W, so there exists a function $L(w)$ analytic on W and such that $e^{iL(w)} \equiv \psi(w)$, $w \in W$. Show that L can be extended onto A so as to be continuous on $W \cup A$, and that L is real valued on A. Then apply the corollary of Theorem 13.7.2 to L.)

13.8 Applications of Conformal Mapping to the Approximation of Analytic Functions by Polynomials

According to Theorem 12.5.1(ii), if $O \subset \mathbf{C}$ is an open set such that $\hat{\mathbf{C}} \sim O$ is connected in $\hat{\mathbf{C}}$, then for any function $f\colon O \to \mathbf{C}$ analytic on O there exists a sequence of polynomials $\langle P_n \rangle_1^\infty$ converging a.u. to f on O. The sequence can be replaced by an infinite series $\sum_1^\infty p_n(z)$ by letting $p_1 = P_1$ and $p_n = P_n - P_{n-1}$, $n = 2, 3, \ldots$. In remarks following that theorem, it is pointed out that various explicit constructions of such sequences and series have been developed for special domains O. Two very simple examples in which O is a circular disk appeared in Exercises 12.4.2 and 12.4.4. In the remainder of this chapter, we discuss certain polynomial approximation processes for cases in which O is the interior region of a Jordan curve.

The first of these involves the Faber polynomials, which were introduced in Exercises 10.7.11, 10.7.12, and 10.7.13. We recapitulate here the facts relevant to this application. Let $\psi\colon \Delta(\infty, \rho) \to \hat{\mathbf{C}}$, $\rho \geq 0$, be analytic on $\Delta'(\infty, \rho)$ and univalent on $\Delta(\infty, \rho)$, with a simple pole at ∞. By the remark following Proposition 13.5.3, $\psi[\Delta'(\infty, \rho)] \neq \mathbf{C}$, even if $\rho = 0$. For $z \in \mathbf{C} \sim \psi[\Delta'(\infty, r)]$, $\rho \leq r < \infty$, the function

$$\frac{w\psi'(w)}{\psi(w) - z} = 1 + \sum_{n=1}^{\infty} p_n(z) w^{-n}, \qquad |w| > r, \tag{13.8.1}$$

is a generating function for the Faber polynomials p_n, $n = 1, 2, \ldots$, belonging to ψ. They appear here as the coefficients in the Laurent expansion at ∞ of the function of w given by the left-hand member of the equation. The polynomials p_n are independent of r, $\rho \leq r < \infty$, and since, for any $z \in \mathbf{C}$, r can be adjusted so that $z \notin \psi[\Delta'(\infty, r)]$, (13.8.1) gives an unambiguous definition of each p_n for every $z \in \mathbf{C}$. Notice that if $\psi(w) = w + b$, with $|z - b| < r$ and $|w| > r$, (13.8.1) becomes

$$\frac{w\psi'(w)}{\psi(w) - z} = \frac{1}{1 - (z - b)/w} = 1 + \sum_{n=1}^{\infty} (z - b)^n w^{-n}.$$

Thus, in this special case, the nth Faber polynomial is merely the nth power of $z - b$, $n = 1, 2, \ldots$.

We continue with our summary of the information in Exercises 10.7.11 through 10.7.13. With $r > \rho$, let C_r denote the level curve $\psi[C(0, r)]$. There exists a constant $M > 0$ such that for every $z \in C_r$.

$$|p_n(z)| \le Mr^n, \qquad n = 1, 2, \ldots. \tag{13.8.2}$$

By the maximum modulus principal [Corollary 2 of Theorem 9.7.1], (13.8.1) remains valid for every $z \in \text{Int } C_r$. Also with $z = \psi(w)$, $|w| = r > \rho$,

$$\lim_{n\to\infty} |p_n(z)|^{1/n} = |w| \tag{13.8.3}$$

uniformly for $z \in C_r$, $|w| = r$. Finally, with $z = \psi(w)$, $|w| > \rho$, there is a Laurent expansion at ∞ of the following type for each p_n as a function of w, $n = 1, 2, \ldots$:

$$p_n(z) = p_n[\psi(w)] = w^n + \sum_{k=1}^{\infty} \alpha_{nk} w^{-k}. \tag{13.8.4}$$

The series converges a.u. on $\Delta(\infty, \rho)$.

An infinite series $a_0 + a_1 p_1(z) + a_2 p_2(z) + \cdots$ in which $p_1, p_2, \ldots$ are Faber polynomials belonging to the same mapping function is called a Faber series. The family of Faber series is the natural generalization of the family of power series when the disks of convergence of the power series are replaced by interior regions of analytic Jordan curves. We show above that the family of Faber series includes the family of power series as a special case.

We start with the analog of Theorem 4.1.1, which was our starting point for the theory of convergence of power series.

Theorem 13.8.1 *Let $\psi: \Delta(\infty, \rho) \to \hat{\mathbb{C}}$, $\rho \ge 0$, be analytic on $\Delta'(\infty, \rho)$ and univalent on $\Delta(\infty, \rho)$, with a simple pole at ∞. Let $\langle p_n \rangle_1^\infty$ be the sequence of Faber polynomials belonging to ψ. Let the Faber series $a_0 + \sum_1^\infty a_n p_n(z)$ be such that for some $z_0 \in \psi[\Delta(\infty, \rho)]$, the inequality $|a_n p_n(z_0)| \le \mu$, is valid for some $\mu > 0$ and for $n = 1, 2, \ldots$. Let $z_0 = \psi(w_0)$ and $|w_0| = R_0$. Then this Faber series converges absolutely and almost uniformly on* Int C_{R_0}.

Proof. We prove absolute and uniform convergence on $C_R \cup \text{Int } C_R$, where R is any number such that $\rho < R < R_0$. Then according to Proposition 13.5.4, if E is any compact set with $E \subset \text{Int } C_{R_0}$, it is possible to adjust R, $\rho < R < R_0$, so that $E \subset C_R \cup \text{Int } C_R$. The plan is to set up a Weierstrass M-test for the series with $z \in C_R \cup \text{Int } C_R$. We have, for z so restricted and $p_n(z_0) \ne 0$,

$$|a_n p_n(z)| = |a_n|\,|p_n(z_0)| \cdot \frac{|p_n(z)|}{|p_n(z_0)|} \le \mu \frac{|p_n(z)|}{|p_n(z_0)|}.$$

According to (13.8.2), there exists $M > 0$ such that $|p_n(z)| \le MR^n$, $z \in C_R \cup \operatorname{Int} C_R$. Also, by (13.8.3), we have $\lim_{n\to\infty} |p_n(z_0)|^{1/n} = R_0$. Choose $\varepsilon > 0$ so that $R_0 - \varepsilon > R$. Then there exists $N > 0$ such that for every $n \ge N$, we have $|p_n(z_0)|^{1/n} \ge R_0 - \varepsilon$ and $|p_n(z_0)| \ge (R_0 - \varepsilon)^n > R^n$. Thus, for $n \ge N$, we have

$$|a_n p_n(z)| \le \frac{\mu M R^n}{(R_0 - \varepsilon)^n}.$$

Let $M_n = \max |a_n p_n(z)|$, $z \in C_R \cup \operatorname{Int} C_R$, $n = 1, 2, \ldots, N-1$, and $M_n = \mu M R^n/(R_0 - \varepsilon)^n$, $n = N, N+1, \ldots$. The series $\sum_1^\infty M_n$ converges, since $\sum_{n=N}^\infty M_n$ is the product of a constant and a convergent geometric series, so by Theorem 3.4.1 our Faber series converges absolutely and a.u. on $C_R \cup \operatorname{Int} C_R$. Q.E.D.

Corollary With ψ, z_0, R_0, and $\langle p_n \rangle$ as described in the proposition, if $\sum_0^\infty a_n p_n(z)$ converges for $z = z_0$, then it converges absolutely and a.u. for $z \in \operatorname{Int} C_{R_0}$.

Theorem 13.8.2 *Let $\psi: \Delta(\infty, \rho) \to \hat{\mathbf{C}}$ be analytic on $\Delta'(\infty, \rho)$, $\rho \ge 0$, univalent on $\Delta(\infty, \rho)$, with a simple pole at ∞. Let $\langle p_n \rangle$ be the sequence of Faber polynomials belonging to ψ. Let C_R denote the level curve $\psi[C(0, R)]$, for each $R > \rho$. Let f and C_{R^*} be such that f is analytic on* Int C_{R^*}. *Then:*

(i) *There exists a Faber series representation of f, $f(z) = a_0 + a_1 p_1(z) + a_2 p_2(z) + \cdots$, convergent absolutely and a.u. on* Int C_{R^*}.
(ii) *If f and C_{R^*} are such that f cannot be extended onto C_{R^*} so as to be analytic on $C_{R^*} \cup$* Int C_{R^*}, *then the series in (i) diverges for each $z \in$* Ext C_{R^*}.
(iii) *For any R, $\rho < R < R^*$.*

$$a_n = \frac{1}{2\pi i} \int_{\gamma_R} f[\psi(w)] w^{-n-1}\, dw, \qquad n = 0, 1, 2, \ldots, \tag{13.8.5}$$

and a_n is independent of R, $\rho < R < R^*$; here γ_R is the positively oriented circle with graph $C(0, R)$.

(iv) *Under the additional hypothesis in (ii), we have $\overline{\lim_{n\to\infty}} |a_n|^{1/n} = 1/R^*$.*
(v) *If $\rho = 1$, then for each R, $1 < R < R^*$, there exists $\mu(R) > 0$ such that for every $z \in E = \hat{\mathbf{C}} \sim \psi[\Delta(\infty, 1)]$,*

$$\left| f(z) - \left[a_0 + \sum_{n=1}^{N} a_n p_n(z) \right] \right| \le \frac{\mu(R)}{R^N}, \qquad N = 1, 2, \ldots, \tag{13.8.6}$$

where $\mu(R)$ depends on R but not on z and N.

(vi) *The Faber series in (i) is unique in the sense that if a Faber series* $A_0 + \sum_1^\infty A_n p_n(z)$ *with* p_n *belonging to* ψ, $n = 1, 2, \ldots$, *converges uniformly to* f *on a level curve* C_R, $\rho < R < R^*$, *then* $A_n = a_n$, $n = 0, 1, 2, \ldots$.

Proof. (i) and (ii). Choose r and R, $\rho < r < R < R^*$. Let C_R be ψ-oriented, and let γ_R denote the positively oriented circle with graph $C(0, R)$. We use the corollary of Proposition 13.5.3 and the uniform convergence of the Laurent series in (13.8.1) as authorities for the following calculation, with $z \in (C_r \cup \text{Int } C_r) \subset \text{Int } C_R$ [Proposition 13.5.3(iii)]:

$$\begin{aligned} f(z) &= \frac{1}{2\pi i}\int_{C_R} \frac{f(\zeta)\, d\zeta}{\zeta - z} \\ &= \frac{1}{2\pi i}\int_0^{2\pi} \frac{f[\psi(Re^{i\theta})]\psi'(Re^{i\theta})\,iRe^{i\theta}\, d\theta}{\psi(Re^{i\theta}) - z} \\ &= \frac{1}{2\pi i}\int_{\gamma_R} \frac{f[\psi(w)]w\psi'(w)\, dw}{w[\psi(w) - z]} \\ &= \frac{1}{2\pi i}\int_{\gamma_R} f[\psi(w)]\left[\frac{1}{w} + \sum_{n=1}^{\infty} \frac{p_n(z)}{w^{n+1}}\right] dw \\ &= a_0 + \sum_{n=1}^{\infty} a_n p_n(z), \end{aligned} \tag{13.8.7}$$

where a_n is given by (13.8.5), $n = 0, 1, 2, \ldots$.

Since the series in the last member of (13.8.7) converges on C_r, by the corollary of Theorem 13.8.1, it converges absolutely and a.u. on Int C_r. If E is any compact set such that $E \subset \text{Int } C_{R^*}$, then by Proposition 13.5.4, r can be adjusted so that $E \subset \text{Int } C_r \subset \text{Int } C_{R^*}$. Thus, the series in (13.8.7) converges absolutely and uniformly on such a set E. We have now established absolute and a.u. convergence on Int C_{R^*}.

To complete the proof of (ii), it remains to show that a_n in (13.8.5) is independent of R for $\rho < R < R^*$. Let R' be any number such that $\rho < R' < R^*$. The function $f[\psi(w)]w^{-n-1}$ is analytic on the annulus $\{w: \rho < |w| < R^*\}$, so by Theorem 12.6-(10.6.1), with $\gamma_{R'}$ denoting the positively oriented circle with graph $C(0, R')$, we have

$$\begin{aligned} a_n &= \frac{1}{2\pi i}\int_{\gamma_R} f[\psi(w)]w^{-n-1}\, dw \\ &= \frac{1}{2\pi i}\int_{\gamma_{R'}} f[\psi(w)]w^{-n-1}\, dw, \qquad n = 1, 2, \ldots. \end{aligned}$$

(iii). Suppose that the series in (13.8.7) were to converge at a point $z_0 \in \text{Ext } C_{R^*}$. Let w_0 be such that $z_0 = \psi(w_0)$; then $|w_0| > R^*$ [Proposition 13.5.3(ii)]. According to the corollary of Theorem 13.8.1, the series converges a.u. on Int $C_{|w_0|}$, which contains $C_{R^*} \cup \text{Int } C_{R^*}$ [Proposition 13.5.3(iii)].

The sum function of the series is analytic on Int $C_{|w_0|}$ [corollary of Theorem 9.9.1], and provides an analytic extension of f onto C_{R^*}, which is prohibited by the hypothesis of this part of the theorem. Therefore, the series diverges for each $z \in \text{Ext } C_{R^*}$.

(iv). According to the nth root test [Theorem 2.6.1], we have $\overline{\lim}\, |a_n p_n(z)|^{1/n} \le 1$ for every $z \in \text{Int } C_{R^*}$. Now choose any R with $\rho < R < R^*$. By (13.8.3), for $z \in C_R$ we have $\lim_{n\to\infty} |p_n(z)|^{1/n} = R$. Thus, for $z \in C_R$, by Exercise 3.4.10, we have $\overline{\lim}\, |a_n p_n(z)|^{1/n} = R\, \overline{\lim}\, |a_n|^{1/n} \le 1$ and $\overline{\lim}\, |a_n|^{1/n} \le 1/R$. This inequality holds for every R, $\rho < R < R^*$, so by the continuity of $1/R$ it must persist when $R = R^*$. Now suppose that $\overline{\lim}\, |a_n|^{1/n} = \delta < 1/R^*$. Then $(1/\delta) > R^*$. Choose any R_0 such that $(1/\delta) > R_0 > R^*$. Then $R_0\, \overline{\lim}\, |a_n|^{1/n} < (1/\delta)\, \overline{\lim}\, |a_n|^{1/n} = 1$, and $1 > R_0\, \overline{\lim}\, |a_n|^{1/n} = \overline{\lim}\, |a_n p_n(z)|^{1/n}$ for $z \in C_{R_0}$, by (13.8.3). The nth root test would indicate convergence of the Faber series in part (i) for $z \in C_{R_0}$. But $C_{R_0} \subset \text{Ext } C_{R^*}$ [Proposition 13.5.3(iii)], so we have reached a contradiction to the divergence statement in part (ii). Thus, $\overline{\lim}\, |a_n|^{1/n} = 1/R^*$.

(v) Let R_1 and R_2 be such that $1 < R_1 < R_2 < R^*$, and let $M(f; R_2) = \max |f(z)|$, $z \in C_{R_2}$. With $R = R_2$ in (13.8.5), we obtain

$$|a_n| \le \frac{1}{2\pi} M(f; R_2) R_2^{-n-1} \cdot 2\pi R_2 = \frac{M(f; R_2)}{R_2^n}, \qquad n = 1, 2, \ldots.$$

According to (13.8.2), there exists $M(R_1) > 0$ such that $|p_n(z)| \le M(R_1) R_1^n$ for $z \in C_{R_1}$, $n = 1, 2, \ldots$. Thus, for every $z \in C_{R_1}$, we have

$$|a_n p_n(z)| \le M(f; R_2) M(R_1) \left(\frac{R_1}{R_2}\right)^n, \qquad n = 1, 2, \ldots,$$

and

$$\begin{aligned} \left| f(z) - \left[a_0 + \sum_{n=1}^{N} a_n p_n(z) \right] \right| &= \left| \sum_{N+1}^{\infty} a_n p_n(z) \right| \le \sum_{N+1}^{\infty} |a_n p_n(z)| \\ &\le \sum_{N+1}^{\infty} M(f; R_2) M(R_1) \left(\frac{R_1}{R_2}\right)^n \\ &= M(f; R_2) M(R_1) \left(\frac{R_1}{R_2}\right)^N \frac{R_2}{R_2 - R_1}, \qquad N = 1, 2, \ldots. \end{aligned} \tag{13.8.8}$$

By the maximum modulus principal, this inequality is valid for $z \in C_{R_1} \cup \text{Int } C_{R_1}$, and by Proposition 13.5.3(i), $E \subset \text{Int } C_{R_1}$. Now suppose that R is any number such that $1 < R < R^*$. Choose $R_1 = (R^* + 1)/(R + 1)$ and

$R_2 = R(R^* + 1)/(R + 1)$. It is easily verified that $1 < R_1 < R_2 < R^*$ and $R_1/R_2 = 1/R$. Let $\mu(R) = M(f; R_2)M(R_1)R_2/(R_2 - R_1)$, with R_1 and R_2 chosen as indicated. With this substitution, the right-hand member of (13.8.8) assumes the form promised in part (v) of the theorem, and the proof of part (v) is complete.

(vi) The hypothesis of (vi), taken with part (i), implies that the series $A_0 - a_0 + \sum_1^\infty (A_n - a_n)p_n(z)$ converges to zero uniformly on C_R. Let γ_R be the positively oriented circle with graph $C(0, R)$ in the w-plane. We use the Laurent expansion of $p[\psi_n(w)]$ at ∞ given in (13.8.4) to obtain the following formula for any nonnegative integer h and $n = 0, 1, 2, \ldots$ ($p_0 = 1$):

$$\int_{\gamma_R} p_n[\psi(w)]w^{-h-1}\,dw = \begin{cases} 0, & h \neq n; \\ 2\pi i, & h = n. \end{cases} \tag{13.8.9}$$

The integration of the series in (13.8.4) term-by-term is justified by the uniform convergence on $C(0, R)$. The series $\sum_0^\infty (A_n - a_n)p_n[\psi(w)]w^{-h-1}$ also converges uniformly on $C(0, R)$, and the sum function is zero. With h a nonnegative integer, we integrate this series term-by-term along γ_R and use (13.8.9). The result is $2\pi i(A_h - a_h) = 0$ for each h, $h = 0, 1, 2, \ldots$. Q.E.D.

Now let C be an analytic Jordan curve, and let $\psi: \Delta(\infty, 1) \to \hat{\mathbf{C}}$ be the unique conformal equivalence mapping $\Delta'(\infty, 1)$ onto Ext C, and such that ψ has a simple pole at ∞ with $\psi'(\infty) > 0$ [Theorem 13.4.2]. According to Theorem 13.7.4(b), there exists r, $0 < r < 1$, such that ψ can be extended across $C(0, 1)$ so as to be analytic and univalent on $\Delta(\infty, r)$. Let ρ be the greatest lower bound of the set of such values of r; it is easy to see that this set must be the interval $\{r: r \geq \rho\}$ because $\Delta(\infty, r') \subset \Delta(\infty, r)$ for $r' > r$. We call such a curve C a ρ-analytic curve.

Corollary 1 Let C be a ρ-analytic Jordan curve, let f: Int $C \to \mathbf{C}$ be analytic and such that no analytic extension of f onto C is possible, and let $\psi: \Delta(0, 1) \to \{\infty\} \cup$ Ext C be the mapping function for Ext C described in Theorem 13.4.2. Then:

(i) f can be expanded in a Faber series $f(z) = a_0 + \sum_1^\infty a_n p_n(z)$ in which the Faber polynomials p_n belong to ψ, and which converges a.u. on Int C but diverges for $z \in$ Ext C.

(ii) For any R, $\rho < R < 1$,

$$a_n = \frac{1}{2\pi i}\int_{\gamma_R} f[\psi(w)]w^{-n-1}\,dw, \qquad n = 0, 1, 2, \ldots,$$

where γ_R is a positively oriented circle with graph $C(0, R)$; the coefficients a_n are independent of R.

(iii) $\overline{\lim}\, |a_n|^{1/n} = 1$.

Proof. In the light of the discussion of the meaning of the term ρ-analytic which immediately precedes the corollary, the corollary is seen to be merely a restatement of Theorem 13.8.1(i), (ii), (iii), and (iv), in which $R^* = 1$. Q.E.D.

According to Corollary 1 of Theorem 12.4.1, a function which is analytic on a compact set E such that $\hat{\mathbf{C}} \sim E$ is connected in $\hat{\mathbf{C}}$ can be approximated arbitrarily closely by polynomials. Our next result here is that if E is connected, then the polynomials can be taken to be the nth partial sums of a Faber series. It is remarked in Sec. 12.1 that in stating a proposition of this type, we might as well start with the hypothesis that the given function is analytic on an open set containing E.

Corollary 2 Let E be a compact connected set in $\mathbf{C}$ such that $\hat{\mathbf{C}} \sim E$ is connected in $\hat{\mathbf{C}}$, and let f be analytic on an open set O containing E. There exists a Faber series representation of f, $f(z) = a_0 + \sum_1^\infty a_n p_n(z)$, convergent a.u. in a simply connected region containing E; and there exists $R^* > 1$ such that for each R, $1 < R < R^*$, we have for every $z \in E$,

$$\left| f(z) - \left[a_0 + \sum_{n=1}^{N} a_n p_n(z) \right] \right| < \frac{M(R)}{R^N}, \qquad N = 1, 2, \ldots, \tag{13.8.10}$$

where $M(R)$ is independent of N.

Proof. If E consists of only a single point, say b, then f can be expanded in series of powers of $(z - b)$ convergent a.u. on some disk $\Delta(b, \rho)$ [Theorem 9.4.2]. It is shown earlier in this section that such a series is a special Faber series. The sequence of inequalities (13.8.10) for this case can easily be deduced from the Cauchy coefficient estimates; we defer the details to an exercise.

If E contains more than one point, then according to Theorem 13.4.2, there exists a function $\psi: \Delta(\infty, 1) \to \hat{\mathbf{C}}$, analytic on $\Delta'(\infty, 1)$ and univalent on $\Delta(\infty, 1)$ with a simple pole at ∞, and such that $\psi[\Delta(\infty, 1)] = \hat{\mathbf{C}} \sim E$. For each $R > 1$, let C_R denote the level curve $\psi[C(0, R)]$ in this mapping. By Proposition 13.5.3(iv), there exists $R^* > 1$ such that $(C_R \cup \operatorname{Int} C_R) \subset \operatorname{Int} C_{R^*} \subset O$ for every R, $1 < R < R^*$. Then by Theorem 13.8.2, there exists a Faber series representation of f, with the polynomials in it belonging to ψ, convergent a.u. on Int R^*. The series converges uniformly on E because $E \subset \operatorname{Int} C_{R^*}$ [Proposition 13.5.3(i)]. The inequalities (13.8.10) are authorized by part (v) of Theorem 13.8.2. Q.E.D.

We conclude this section with a generalization of the important inequality (13.8.2) for the Faber polynomials. In one form or another, the following result, sometimes known as Bernstein's inequality, appears repeatedly in the theory of polynomial approximation.

Theorem 13.8.3 *Let $E \subset \mathbf{C}$ be a compact connected set containing more than* *one point, and such that* $\hat{\mathbf{C}} \sim E$ *is connected in* $\hat{\mathbf{C}}$. *Let* $P(z)$ *be a polynomial of degree* $\leq n$, *and let* $M = \max|P(z)|$, $z \in E$. *Let* $\psi: \Delta(\infty, 1) \to \mathbf{C}$ *be analytic on* $\Delta'(\infty, 1)$, *univalent on* $\Delta(\infty, 1)$, *with a simple pole at* ∞, *and such that* $\psi[\Delta(\infty, 1)] = \mathbf{C} \sim E$; *and let* C_R *denote the level curve* $\psi[C(0, R)]$ *for each* $R > 1$. *Then* $|P(z)| \leq MR^n$ *for every* $z \in C_R \cup \text{Int } C_R$.

Remark. The theorem apparently dates back to 1912, when S. N. Bernstein published it for the special case in which E is a line segment. Various inequalities for real polynomials and trigonometric polynomials go under Bernstein's name.

Proof of the Theorem. The existence of M follows from the fact that $|P(z)|$ is continuous on the compact set E [Corollary 1 of Theorem 5.5.2]. Another consequence of this continuity is that for any $\varepsilon > 0$, for each $\zeta \in E$ there exists $\delta_\zeta > 0$ such that $|P(z)| < M + \varepsilon$ for every $z \in \Delta(\zeta, \delta_\zeta)$. Let $O = \bigcup_{\zeta \in E} \Delta(\zeta, \delta_\zeta)$; this is an open set in $\mathbf{C}$ and $E \subset O$. By Proposition 13.5.3(iv), there exists $R^* > 1$ such that $\text{Int } C_{R^*} \subset O$ and $C_{R_0} \subset \text{Int } C_{R^*} \subset O$ for each R_0, $1 < R_0 < R^*$. We have $|P(z)| < M + \varepsilon$ for every $z \in \text{Int } C_{R^*}$.

The polynomial $P(z)$ has a pole of order not greater than n at ∞, and $\psi(w)$ has a simple pole at ∞. By using Proposition 10.7.3(v) twice, it is easily shown that $P[\psi(w)]/w^n$ is analytic for $w \in \Delta'(\infty, 1)$ and has a removable singularity at ∞. We complete the definition at ∞ properly and make the transformation $w = 1/\xi$. This carries $P[\psi(w)]/w^n$ into a function $\chi(\xi) = \xi^n P[\psi(1/\xi)]$ analytic on $\Delta(0, 1)$, and it also carries C_{R_0} into $C(0, 1/R_0) \subset \Delta(0, 1)$. On the circle $C(0, 1/R_0)$, $|\chi(\xi)| < (M + \varepsilon)/R_0^n < M + \varepsilon$, and then by the maximum modulus principle in the form of Corollary 2 of Theorem 9.7.1, $|\chi(\xi)| < M + \varepsilon$ for every ξ, $|\xi| < 1/R_0$. Since R_0 can be chosen arbitrarily near to 1, and $\varepsilon > 0$ is also arbitrary, we have $|\chi(\xi)| \leq M$ for every $\xi \in \Delta(0, 1)$. Therefore $|P[\psi(w)]/w^n| \leq M$ for every $w \in \Delta(\infty, 1)$, and the conclusion of the proposition follows at once from this inequality and the maximum modulus principle. Q.E.D.

Exercises 13.8

1. Let $\psi: \Delta(\infty, \rho) \to \hat{\mathbf{C}}$, $\rho \geq 0$, be analytic on $\Delta'(\infty, \rho)$, univalent on $\Delta(\infty, \rho)$, and have a simple pole at ∞. Let $\langle p_n \rangle_1^\infty$ be the sequence of Faber polynomials belonging to ψ. Show that if $\langle a_n \rangle_0^\infty$ is a sequence of complex numbers such that $\overline{\lim}\, |a_n|^{1/n} = 1/R$, where $R > \rho$, and if $C_R = \psi[C(0, R)]$, than the Faber series $a_0 + \sum_{n=1}^\infty a_n p_n(z)$ converges absolutely and a.u. on $\text{Int } C_R$ to an analytic function, and diverges for every $z \in \textit{Ext } C_R$. (*Hint*: See the proof of Theorem 13.8.2(iv).)
2. Complete the proof of Corollary 2 of Theorem 13.8.2 for the case in which E consists of a single point. (*Hint*: Recall the Cauchy estimates in Sec. 9.5.)
3. Let $E \subset \mathbf{C}$ be a compact connected set containing more than one point, and such that $\hat{\mathbf{C}} \sim E$ is connected in $\hat{\mathbf{C}}$. Let $\langle P_n(z) \rangle_{n=1}^\infty$ be a sequence of polynomials in which P_n is of degree at most n, and which converges on E to a function

$f: E \to \mathbf{C}$. Furthermore, let the sequence and f satisfy the inequality $|P_n(z) - f(z)| < M/R^n$, $n = 1, 2, \ldots$, for every $z \in E$, where $M > 0$ and $R > 1$. Show that the sequence converges uniformly on a region containing E. (*Hint*: Let ψ be the mapping function for $\hat{\mathbf{C}} \sim E$ described in Theorem 13.4.2, and let $C_r = \psi[C(0, r)]$, $r > 1$. With $1 < r < R$, show by Bernstein's inequality that $\langle P_n(z) \rangle$ is a Cauchy sequence on C_r.)

13.9 Continuation of Section 13.8; Interpolation in the Fekete Points and in the Fejér Points

This section consists of a discussion of certain modern-classical approximation processes which involve interpolation with polynomials to analytic functions. First, we recapitulate in a slightly revised form some facts about interpolation polynomials. In Exercise 12.4.3, a formula was presented for a polynomial $L(z)$ of degree at most $n - 1$ which assumed n given values $w_1, w_2, \ldots, w_n$ at respective distinct points $\zeta_1, \zeta_2, \ldots, \zeta_n$ in the z-plane. The points ζ_k are called *nodes*. Here it is convenient to change n to $n + 1$ and to let the subscripts on the points and values run from 0 to n. The formula becomes

$$L_n(z) = \sum_{k=0}^{n} w_k \frac{\omega(z)}{(z - \zeta_k)\omega(\zeta_k)}, \tag{13.9.1a}$$

$$\omega(z) = (z - \zeta_0)(z - \zeta_1) \cdots (z - \zeta_n). \tag{13.9.1b}$$

This is called the Lagrange interpolation formula. The kth fraction (starting the count with $k = 0$) in the summation in (13.9.1a) has a removable singularity at the point ζ_k, and the appropriate completion of its definition at this point is given by

$$\begin{aligned}\lim_{z \to \zeta_k} \frac{\omega(z)}{(z - \zeta_k)\omega'(\zeta_k)} &= \lim_{z \to \zeta_k} \frac{\omega(z) - \omega(\zeta_k)}{(z - \zeta_k)\omega'(\zeta_k)} \\ &= \frac{\omega'(\zeta_k)}{\omega'(\zeta_k)} = 1, \qquad k = 0, 1, 2, \ldots, n.\end{aligned}$$

Thus, $L_n(\zeta_k) = 0 + 0 + \cdots + 0 + w_k + 0 + \cdots + 0 = w_k$. For each k, the term in the summation in (13.9.1a), after cancellation of $z - \zeta_k$ in the numerator and denominator, is clearly a polynomial in z of degree exactly n, so the sum of these polynomials is a polynomial of degree at most n. If $P_n(z)$ is a polynomial of degree at most n and such that $P_n(\zeta_k) = w_k$, $k = 0, 1, 2, \ldots, n + 1$, then $P_n(z) - L_n(z)$ is a polynomial of degree at most n which has zeros at the $n + 1$ points $\zeta_0, \zeta_1, \ldots, \zeta_n$. Therefore, by the corollary of Theorem 9.6.2, $P_n - L_n$ is a zero polynomial and $P_n(z) \equiv L_n(z)$, $z \in \mathbf{C}$. If the numbers $w_0, w_1, \ldots, w_n$ are the respective values of a given function f at $\zeta_0, \zeta_1, \ldots, \zeta_n$, then we use the expanded notation $L_n(f; z)$ for $L_n(z)$. Thus, if $Q(z)$ is any polynomial of degree at most n, we have for every $z \in \mathbf{C}$,

$$L_n(Q; z) = \sum_{k=0}^{n} Q(\zeta_k) \frac{\omega(z)}{(z - \zeta_k)\omega'(\zeta_k)} = Q(z). \tag{13.9.2}$$

Another formula for $L_n(f; z)$ was given in Exercise 12.6.3 for the case in which f is analytic on a simply connected region B. Let $\zeta_0, \zeta_2, \ldots, \zeta_n$ be a set of numbers, not necessarily distinct, such that the corresponding points lie in B. Let γ be a simple closed path with graph $\Gamma \subset B$ such that $\{\zeta_0, \zeta_1, \ldots, \zeta_n\} \subset \operatorname{Int} \Gamma$, and also let γ have the further property that $\operatorname{Ind}_k(z) = 1$ for $z \in \operatorname{Int} \Gamma$. Then if the points ζ_k are distinct, we have, with $\omega(z)$ given by (13.9.1b),

$$L_n(f; z) = \frac{1}{2\pi i} \int_\gamma \frac{f(\zeta)}{\zeta - z} \left(1 - \frac{\omega(z)}{\omega(\zeta)}\right) d\zeta \tag{13.9.3}$$

for every $z \in \operatorname{Int} \Gamma$. If the nodes ζ_k are not distinct, the right-hand member of (13.9.3) is a polynomial of degree at most n which satisfies $n + 1$ interpolation conditions, but now these conditions involve coincidence with derivatives of f at repeated nodes. Further details are given in Exercise 12.6.3.

The first of the two polynomial interpolation processes which we discuss involves nodes which are known as Fekete points. To study interpolation in these points, it is convenient to introduce still another formula for the unique polynomial $L_n(f; z)$ which assumes the values $f(\zeta_k)$ at the distinct points $z = \zeta_k$, $k = 0, 1, 2, \ldots, n$. Let $\mathbf{V}_n$ denote the following Vandermonde matrix:

$$\mathbf{V}_n = \begin{pmatrix} 1 & \zeta_0 & \zeta_0^2 & \cdots & \zeta_0^n \\ 1 & \zeta_1 & \zeta_1^2 & \cdots & \zeta_1^n \\ \vdots & & & & \\ 1 & \zeta_n & \zeta_n^2 & \cdots & \zeta_n^n \end{pmatrix},$$

and let $\mathbf{V}_{nk}(z)$ denote the matrix obtained from $\mathbf{V}_n$ be substituting the row-vector $(1, z, z^2, \ldots, z^n)$ for the row-vector $(1, \zeta_k, \zeta_k^2, \ldots, \zeta_k^2)$, $k = 0, 1, \ldots, n$. For any square matrix $\mathbf{A}$, let $\det \mathbf{A}$ denote the determinant of $\mathbf{A}$. The expansion of $\det \mathbf{V}_{nk}(z)$ according to the minors of the kth row reveals $\det \mathbf{V}_{nk}(z)$ to be a polynomial of degree at most n. Also $\det \mathbf{V}_{nk}(\zeta_h) = 0$, $h \neq k$, because $\mathbf{V}_{nk}(\zeta_h)$ has two identical rows. From these facts and the fundamental theorem of algebra (in the form of the corollary of Theorem 9.6.2), it can be deduced that

$$\det \mathbf{V}_n = \prod_{0 \le h < k \le n} (\zeta_h - \zeta_k), \tag{13.9.4}$$

and thus $\det \mathbf{V}_n \neq 0$ when the points ζ_k are distinct. With this information the representation

$$L_n(f; z) = \sum_{k=0}^{n} f(\zeta_k) \frac{\det \mathbf{V}_{nk}(z)}{\det \mathbf{V}_n} \tag{13.9.5}$$

is seen to be valid, since the polynomial of degree at most n in the right-hand member satisfies the same interpolation conditions that $L_n(f; z)$ satisfies, and such an interpolation polynomial is unique. In particular, if $Q(z)$ is any polynomial of degree at most n, then we obtain from (13.9.5) and (13.9.2) the identity

$$\sum_{k=0}^{n} Q(\zeta_k) \frac{\det \mathbf{V}_{nk}(z)}{\det \mathbf{V}_n} \equiv L_n(Q; z) \equiv Q(z), \qquad z \in \mathbf{C}. \tag{13.9.6}$$

Now let E be a compact infinite point set in $\mathbf{C}$. To make use of facts established in Sec. 5.5 relevant to the behavior of $|\det \mathbf{V}_n|$ with the $n + 1$ points $\zeta_0, \zeta_1, \ldots, \zeta_n$ restricted to E, we temporarily regard E as a compact set in $\mathbf{R}^2$. It can be shown by elementary topological considerations that the $(n + 1)$-fold Cartesian product $E^n = E \times E \times \cdots \times E$ is compact in $\mathbf{R}^{2n+2}$. Let $\zeta_k = \sigma_k + i\tau_k$, where σ_k, τ_k are real, $k = 0, 1, \ldots, n$. The function $\det \mathbf{V}_n$ clearly is a complex-valued polynomial in the $2n + 2$ real variables σ_k, τ_k, $k = 0, 1, \ldots, n$. Such a polynomial is known to be continuous on $\mathbf{R}^{2n+2}$, and from this we can infer by using a trivial extension of Exercise 3.2.5 that $|\det \mathbf{V}_n|$ is also continuous on $\mathbf{R}^{2n+2}$. Then in particular, $|\det \mathbf{V}_n|$ is continuous on the compact set E^n, so according to Corollary 1 of Theorem 5.5.2, $|\det \mathbf{V}_n|$ attains a maximum value on E^n. We translate this fact back to the situation in $\mathbf{C}$ and assert that $|\det \mathbf{V}_n|$ attains a maximum value at a set of points $F_n = \{\hat{\zeta}_{n0}, \hat{\zeta}_{n1}, \ldots, \hat{\zeta}_{nn}\} \subset E$. We denote the corresponding matrices $\mathbf{V}_n$ and $V_{nk}(z)$ by $\hat{\mathbf{V}}_n$ and $\hat{\mathbf{V}}_{nk}(z)$. Equation (13.9.4) shows that $|\det \hat{\mathbf{V}}_n| > 0$.

A set of $n + 1$ points $F_n \subset E$ which maximizes $|\det \mathbf{V}_n|$ on E is called a set of Fekete points belonging to E. Such a set need not be unique. Fekete introduced these points in 1923 in connection with an algebraic problem. He showed among other things that if E satisfies the hypotheses of Theorem 13.4.2, then his points are connected with conformal mapping and various other branches of complex analysis. In particular, he derived the equation $\lim_{n \to \infty} |\det \hat{\mathbf{V}}_n|^{2/n(n+1)} = d(E)$, where $d(E)$ is the transfinite diameter of E which we introduced briefly in Remark (iii) of Sec. 13.4. (See [10, pp. 268 ff.] for proofs of this and various related results.) Here we discuss only an application of the Fekete points to polynomial approximations to analytic functions. It is contained in the following theorem.

Theorem 13.9.1 *Let E be an infinite connected compact set in $\mathbf{C}$ such that*

$\hat{\mathbf{C}} \sim E$ is connected in $\mathbf{C}$, *let O be an open set containing E, and let $f: O \to \mathbf{C}$ be analytic on O. For each n, $n = 1, 2, \ldots$, let $F_n = \{\zeta_{n0}, \zeta_{n1}, \ldots, \zeta_{nn}\}$ be a set of Fekete points belonging to E, and let $\hat{L}_n(f; z)$ be the polynomial of degree at most n which coincides with f(z) at the points F_n. Then:*

(i) $\lim_{n\to\infty} \hat{L}_n(f; z) = f(z)$ *a.u. on a simply connected region containing E.*

(ii) *There exists $R^* > 0$ such that for each R, $1 < R < R^*$, we have $|f(z) - \hat{L}_n(f, z)| \le M(R)/R^n$ for every $z \in E$, where M(R) depends on R but not on z and n.*

Proof. (i) Since $|\det \hat{\mathbf{V}}_n|$ is the maximum value of $|\det \mathbf{V}_n|$ for $\{\zeta_0, \zeta_1, \ldots, \zeta_n\} \subset E$, it follows that $|\det \hat{\mathbf{V}}_{nk}(z)| \le |\det \hat{\mathbf{V}}_n|$ for every $z \in E$, $k = 0, 1, 2, \ldots, n$. Let $\psi: \Delta(\infty, 1) \to \hat{\mathbf{C}}$ be the function described in Theorem 13.4.2, which is analytic on $\Delta'(\infty, 1)$, univalent on $\Delta(\infty, 1)$, with a simple pole at ∞, and such that $\psi[\Delta(\infty, 1)] = \hat{\mathbf{C}} \sim E$. For each $R > 1$, let C_R be the level curve $\psi[C(0, R)]$ in this mapping. By Proposition 13.5.3(iv), there exists $R^* > 1$ such that $(C_R \cup \text{Int } C_R) \subset \text{Int } C_{R^*} \subset O$ for every R, $1 < R < R^*$. The function $\lambda_{nk}(z) = \det \hat{\mathbf{V}}_{nk}(z)/\det \mathbf{V}_n$ is a polynomial in z of degree at most n, with $\max |\lambda_{nk}(z)| = 1$, $z \in E$, $k = 0, 1, 2, \ldots, n$, so by Theorem 13.8.3 (Bernstein's inequality), $|\lambda_{nk}(z)| \le R^n$ for every $z \in C_R$ and every $R > 1$, $k = 0, 1, \ldots, n$.

By Theorem 13.8.2, there exists a Faber series representation of f which converges to f a.u. on Int C_{R^*} and has the property that the nth partial sum $P_n(z)$ (which is a polynomial in z of degree at most n) satisfies the following inequality for $z \in E$:

$$|f(z) - P_n(z)| \le \frac{\mu(R_1)}{R_1^n}, \qquad n = 1, 2, \ldots, \tag{13.9.7}$$

for every R_1, $1 < R_1 < R^*$, where $\mu(R_1)$ depends on R_1 but not on z and n. We use (13.9.5), (13.9.6), (13.9.7), and Bernstein's inequality, in the following computations:

$$\begin{aligned} f(z) - \hat{L}_n(f; z) &= f(z) - P_n(z) + P_n(z) - \hat{L}_n(f; z) \\ &= f(z) - P_n(z) + \sum_{k=0}^{n} [P_n(\zeta_k) - f(\zeta_k)]\lambda_{nk}(z) \end{aligned} \tag{13.9.8}$$

and for every $z \in C_R \cup \text{Int } C_R$, $1 < R < R_1 < R^*$,

$$\begin{aligned} |f(z) - \hat{L}_n(f; z)| &\le |f(z) - P_n(z)| + \sum_{k=0}^{n} |P_n(\zeta_k) - f(\hat{\zeta}_k)|\,|\lambda_{nk}(z)| \\ &\le |f(z) - P_n(z)| + (n + 1)\mu(R_1)\frac{R^n}{R_1^n}. \end{aligned} \tag{13.9.9}$$

If E_0 is any compact subset of Int C_{R^*}, then by Proposition 13.5.4, R and

R_1 can be adjusted so that $E_0 \subset \operatorname{Int} C_R$ and $1 < R < R_1 < R^*$. We have $\lim_{n\to\infty} (R/R_1)^n = 0$, $\lim_{n\to\infty} (nR^n/R_1)^{1/n} = (\lim_{n\to\infty} n^{1/n})(R/R_1) = R/R_1$ [Sec. 2.4, Example (iv)], so $\lim_{n\to\infty} (nR^n/R_1) = 0$. Also $\lim_{n\to\infty} P_n(z) = f(z)$ a.u. on Int R^*, by Theorem 13.8.2(i). It follows from (13.9.9) that $\lim_{n\to\infty} \hat{L}_n(f; z) = f(z)$ a.u. on Int R^*. The proof of part (i) is complete.

(ii) Let R and R_1 again be such that $1 < R < R_1 < R^*$. Notice again that $|\lambda_{nk}(z)| \leq 1$ for $z \in E$, $k = 0, 1, \ldots, n$. From (13.9.7) and (13.9.9), with $z \in E$, we obtain

$$|f(z) - \hat{L}_n(f; z)| \leq \frac{\mu(R_1)}{R_1^n} + (n+1)\frac{\mu(R_1)}{R_1^n}$$
$$= \frac{1}{R^n}\left[\frac{(n+2)\mu(R_1)R^n}{R_1^n}\right], \qquad n = 1, 2, \ldots.$$

As indicated previously, $\lim_{n\to\infty} (R^n/R_1) = 0 = \lim_{n\to\infty} nR^n/R_1^n$. Thus, the limit of the term in square brackets here as $n \to \infty$ is zero. Therefore, there exists N such that for every $n > N$, this term is less than one. Let $M(R)$ in part (ii) of the theorem be $\max\{1, (n+1)(R_1)R^n/R_1^n, n = 1, 2, \ldots, N\}$. Q.E.D.

The second interpolation process which we discuss was published by L. Fejér in 1918. He was interested in providing a solution to a problem which had attracted the interest of a number of mathematicians around the turn of this century; namely, that of finding a sequence of polynomials $p_1(z), p_2(z), \ldots$, which "belong" to a given region, in the sense that any function f analytic in the region can be expanded in a convergent series $a_0 + \sum_1^\infty a_k p_k(z)$ in which the coefficients a_k, but not the polynomials p_k, depend on f. George Faber published a solution in 1903 which we have presented in various forms in Theorem 13.8.1 and the corollaries. It is characteristic of his solution that the given region to which his polynomials "belong" is bounded by an analytic Jordan curve. (A number of recent attempts have been made to lift this restriction on the curve in the Faber approximation theory, but none of them seem to be entirely satisfactory from an esthetics point of view. See [5] for some details and references.) In Fejér's solution, the given region is bounded by an *arbitrary* Jordan curve, but his method—which we present in the remainder of this section—requires that the function f be analytic on the closure of this region.

Let Γ be a Jordan curve in the z-plane. The set $\Gamma \cup \operatorname{Int}\Gamma$ is compact and connected in $\mathbf{C}$, and $(\operatorname{Ext}\Gamma) \cup \{\infty\}$ is connected in $\hat{\mathbf{C}}$. Let $\psi: \Delta(\infty, 1) \to (\operatorname{Ext}\Gamma) \cup \{\infty\}$ be the mapping function described in Theorem 13.4.2; ψ is analytic on $\Delta'(\infty, 1)$, univalent on $\Delta(\infty, 1)$ with a simple pole at ∞ at which $c = \psi'(\infty) > 0$. By the corollary of Theorem 13.7.1, ψ has a continuous univalent extension onto $C(0, 1)$. We assume implicitly in the sequel that ψ has been so extended.

For each n, $n = 1, 2, \ldots$, let U_n denote a set of $n + 1$ distinct numbers $\{\theta_{n0}, \theta_{n1}, \theta_{n2}, \ldots, \theta_{nn}\} \subset [0, 2\pi)$. The sequence $S = \langle U_n \rangle_{n=1}^{\infty}$ is said to be *uniformly distributed* on $[0, 2\pi]$ if and only if for every α, β, such that $0 \le \alpha < \beta \le 2\pi$, we have $\lim_{n\to\infty} N_n[\alpha, \beta)/(n + 1) = (\beta - \alpha)/2\pi$, where $N_n[\alpha, \beta)$ is the number of points in U_n which are contained in the interval $[\alpha, \beta)$. When S is uniformly distributed on $[0, 2\pi]$, we say that the sequence of sets $\langle\{w_{nk} = e^{i\theta_{nk}}, k = 0, 1, \ldots, n\}_{n=1}^{\infty}$ is uniformly distributed on $C(0, 1)$ and the sequence of sets $\langle\{z_{nk} = \psi(w_{nk}), k = 0, 1, \ldots, n\}\rangle_{n=1}^{\infty}$ is uniformly ψ-distributed on Γ, where ψ is the mapping function referred to in the preceding paragraph. (A simple example of a uniformly distributed sequence of points on $[0, 2\pi]$ is given by the sequence in which the set U_n consists of the equally spaced points $\theta_{nk} = 2\pi k/(n + 1)$, $k = 0, 1, .\ \ , n$. We defer the proof to an exercise.)

Fejér's method is based on the following proposition, which is reminiscent of equation (13.8.3) for the Faber polynomials.

Proposition 13.9.1 *Let the sequence of sets* $\langle\{z_{nk} = \psi(w_{nk}), k = 0, 1, \ldots, n\}\rangle_{n=1}^{\infty}$ *be uniformly* ψ*-distributed on the Jordan curve* Γ, *where* ψ *is the exterior mapping function described in Theorem 13.4.2 with* $\psi'(\infty) = c > 0$. *Let* $\omega_n(z) = (z - z_{n0})(z - z_{n1}) \cdots (z - z_{nn})$, $n = 1, 2, \ldots$. *Then for each* $R > 1$,

$$\lim_{n\to\infty} |\omega_n(z)|^{1/(n+1)} = cR \tag{13.9.10}$$

uniformly for $z \in C_R = \psi[\Delta(\infty, R)]$.

Proof. The plan of the proof is to show that

$$\lim_{n\to\infty} \frac{1}{n + 1} \ln |\omega_n(z)| = \frac{1}{2\pi} \int_0^{2\pi} \ln |z - \psi(e^{i\theta})|\, d\theta \tag{13.9.11}$$

uniformly for $z \in C_R$, and thereafter to evaluate the integral in (13.9.11) by using a variant of the Gauss mean value theorem. Let $w_{nk} = e^{i\theta_{nk}}$, $k = 0, 1, \ldots, n$; $n = 1, 2, \ldots$. Then

$$\frac{1}{n + 1} \ln |\omega_n(z)| = \frac{1}{n + 1} \sum_{k=0}^{n} \ln |z - \psi(e^{i\theta_k})|. \tag{13.9.12}$$

It is a known property of uniformly distributed points that a sequence of sums such as the sum in the right-hand member of (13.9.12), when formed for a continuous function, behaves like a sequence of Riemann sums [21, pp. 164–166]. However, in the present application, it is essential also to establish *uniform* convergence. To this end we use the following lemma.

Lemma 1 Let the sequence $\langle U_n = \{\theta_{nk}, k = 0, 1, \ldots, n\}\rangle_1^\infty$ be uniformly distributed on $[0, 2\pi]$, and let the complex valued function $f(\sigma, \theta)$ be continuous on $X = \{(\sigma, \theta): 0 \le \sigma \le 2\pi, 0 \le \theta \le 2\pi\}$. Then

$$\lim_{n\to\infty} \frac{1}{n+1} \sum_{k=0}^{n} f(\sigma, \theta_{nk}) = \frac{1}{2\pi} \int_0^{2\pi} f(\sigma, \theta)\, d\theta$$

uniformly for $\sigma \in [0, 2\pi]$.

Proof. The set $X = [0, 2\pi] \times [0, 2\pi]$ is compact in $\mathbf{R}^2$, so by Theorem 5.5.3, f is uniformly continuous on this square. Choose any $\varepsilon > 0$. There exists δ such that if $|\theta - \theta'| < \delta$ with $\theta, \theta' \in [0, 2\pi]$ then $|f(\sigma, \theta) - f(\sigma, \theta')| < \varepsilon/3$ for every $\sigma \in [0, 2\pi]$. Impose a partition on $[0, 2\pi]$ denoted by $0 = \theta_0 < \theta_1 < \theta_2 < \cdots < \theta_{N+1} = 2\pi$, and chosen so that $\max_K |\theta_{K+1} - \theta_K| < \delta$, $K = 0, 1, \ldots, N$. Then for every $\sigma \in [0, 2\pi]$ and $\theta \in [\theta_K, \theta_{K+1}]$, we have $|f(\sigma, \theta) - f(\sigma, \theta_K)| < \varepsilon/3$, $K = 0, 1, \ldots, N$.

We now assign new indices to the points θ_{nk}, $k = 0, 1, \ldots, n$, of U_n so that the subset of U_n which is contained in the interval $[\theta_K, \theta_{K+1})$ is denoted by $\{\theta_{n1}^{(K)}, \theta_{n2}^{(K)}, \ldots, \theta_{np_K}^{(K)}\}$, $K = 0, 1, 2, \ldots, N$. Notice that $p_K = N_n[\theta_K, \theta_{K+1})$, $K = 0, 1, 2, \ldots, N$, and that $\sum_{K=0}^N p_K = n + 1$, the total number of points in U_n. We have

$$\begin{aligned}
D_n &= \left| \frac{1}{n+1} \sum_{k=0}^{n} f(\sigma, \theta_{nk}) - \frac{1}{2\pi} \int_0^{2\pi} f(\sigma, \theta)\, d\theta \right| \\
&\le \left| \frac{1}{n+1} \sum_{K=0}^{N} \sum_{k=1}^{p_k} f(\sigma, \theta_{nk}^{(K)}) - \frac{1}{n+1} \sum_{K=0}^{N} f(\sigma, \theta_K) N_n[\theta_K, \theta_{K+1}) \right| \\
&\quad + \left| \frac{1}{n+1} \sum_{K=0}^{N} f(\sigma, \theta_K) N_n[\theta_K, \theta_{K+1}) - \frac{1}{2\pi} \sum_{K=0}^{N} f(\sigma, \theta_K)(\theta_{K+1} - \theta_K) \right| \\
&\quad + \left| \frac{1}{2\pi} \sum_{K=0}^{N} f(\sigma, \theta_K)(\theta_{K+1} - \theta_K) - \frac{1}{2\pi} \int_0^{2\pi} f(\sigma, \theta)\, d\theta \right| \\
&= D_n^{(1)} + D_n^{(2)} + D_n^{(3)}.
\end{aligned}$$

Then

$$\begin{aligned}
D_n^{(1)} &= \left| \frac{1}{n+1} \sum_{K=0}^{N} \sum_{k=0}^{p_k} [f(\sigma, \theta_{nk}^{(K)}) - f(\sigma, \theta_K)] \right| \\
&< \frac{1}{n+1} \sum_{K=0}^{N} \left(p_K \frac{\varepsilon}{3} \right) = \frac{\varepsilon}{3}.
\end{aligned}$$

Also,

$$D_n^{(2)} \le \sum_{K=0}^{N} |f(\sigma, \theta_K)| \left| \frac{N_n[\theta_K, \theta_{K+1})}{n+1} - \frac{\theta_{K+1} - \theta_K}{2\pi} \right|.$$

We can assume that $f \not\equiv 0$ on $X = \{(\sigma, \theta): 0 \le \sigma \le 2\pi, 0 \le \theta \le 2\pi\}$, because otherwise the lemma is trivially true. Let $M = \max |f(\sigma, \theta)| > 0$ for $(\sigma, \theta) \in X$. Let $n_K > 0$ be such that for every $n \geq n_K$, we have

$$\left|\frac{N_n[\theta_K, \theta_{K+1})}{n+1} - \frac{\theta_{K+1} - \theta_K}{2\pi}\right| < \frac{\varepsilon}{3M(N+1)},$$

$K = 0, 1, \ldots, N$; and let $\mu_\varepsilon = \max n_K$, $K = 0, 1, \ldots, N$. Then for every $n \geq \mu_\varepsilon$, we have

$$D_n^{(2)} < \sum_{K=0}^{N} M \cdot \frac{\varepsilon}{3M(N+1)} = \frac{\varepsilon}{3}.$$

Finally,

$$\begin{aligned} D_n^{(3)} &\leq \sum_{K=0}^{N} \frac{1}{2\pi} \int_{\theta_K}^{\theta_{K+1}} |f(\sigma, \theta_K) - f(\sigma, \theta)|\, d\theta \\ &< \sum_{K=0}^{N} \left(\frac{1}{2\pi}\right)\left(\frac{\varepsilon}{3}\right)(\theta_{K+1} - \theta_K) = \frac{2\pi}{2\pi}\frac{\varepsilon}{3} = \frac{\varepsilon}{3}. \end{aligned}$$

Thus, for any $n \geq \mu_\varepsilon$, $D_n < (\varepsilon/3) + (\varepsilon/3) + (\varepsilon/3) = \varepsilon$. The inequality is valid for every $\sigma \in [0, 2\pi]$, so the proof of Lemma 1 is complete.

Now in (13.9.12), let $z = \psi(Re^{i\sigma})$, $\sigma \in [0, 2\pi]$, $R > 1$. The function $\chi(\sigma, \theta) = \psi(Re^{i\sigma}) - \psi(e^{i\theta})$ is continuous and nonzero on $X = \{(\sigma, \theta)\colon 0 \leq \sigma \leq 2\pi, 0 \leq \theta \leq 2\pi\}$, so $|\chi(\sigma, \theta)|$ is also continuous and nonzero on X, and then by the lemma of Theorem 7.7.2, the composition of $\ln(\cdot)$ with $|\chi|$ is continuous on X. Thus, the hypotheses of Lemma 1 are satisfied with $\ln|\psi(Re^{i\sigma}) - \psi(e^{i\sigma})|$ playing the role of $f(\sigma, \theta)$ in the lemma, so equation (13.9.11) is valid uniformly for $z \in C_R$.

To evaluate the integral in (13.9.11) we use the following version of the Gauss mean value theorem (9.5.2).

Lemma 2 Let $g\colon \bar{\Delta}(\infty, 1) \to \hat{\mathbb{C}}$ be analytic on $\Delta'(\infty, 1)$, continuous, and nonzero on $\bar{\Delta}(\infty, 1)$, and let $g(\infty) = b$, where $b \neq 0$, $b \neq \infty$. Then $\int_0^{2\pi} \ln|g(e^{i\theta})|\, d\theta = 2\pi \ln|b|$.

Proof of Lemma 2. We first show that $\int_0^{2\pi} \ln|g(Re^{i\theta})|\, d\theta = 2\pi \ln|b|$ for every $R > 1$. The function $h(\xi) = g(1/\xi)$ is analytic and nonzero on $\Delta(0, 1) \sim \{0\}$, and has a removable singularity at zero. We remove the singularity by letting $h(0) = b$. Then by Lemma 1 for Theorem 11.3.2, we have, with $r = 1/R$, $R > 1$,

$$\begin{aligned} 2\pi \ln|b| &= \int_0^{2\pi} \ln|h(re^{i\theta})|\, d\theta \\ &= \int_0^{2\pi} \ln\left|g\left(\frac{e^{-i\theta}}{r}\right)\right| d\theta \\ &= \int_0^{2\pi} \ln|g(Re^{i\theta})|\, d\theta. \end{aligned} \tag{13.9.13}$$

The function $\ln |z|$ is continuous on $\Delta'(\infty, 0)$, and the function $z = |g(Re^{i\theta})|$ is continuous on $S = \{(R, \theta): 1 \le R < \infty, 0 \le \theta \le 2\pi\}$, so $\ln |g(Re^{i\theta})|$ is continuous on S. Therefore, by Exercise 8.3.7, the last integral in (13.9.13) is a continuous function of R for $1 \le R < \infty$, so (13.9.13) remains true with $R = 1$. This completes the proof of Lemma 2.

We can now evaluate the integral in (13.9.11). We choose $z \in C_R$, $R > 1$, and let $z = \psi(\xi)$, where $|\xi| = R$; ξ is held fixed in the ensuing calculation. We have

$$\begin{aligned}\frac{1}{2\pi}\int_0^{2\pi} &\ln |\psi(\xi) - \psi(e^{i\theta})|\, d\theta \\ &= \frac{1}{2\pi}\int_0^{2\pi} \ln\left|\frac{\psi(\xi) - \psi(e^{i\theta})}{\xi - e^{i\theta}}\right| d\theta + \frac{1}{2\pi}\int_0^{2\pi} \ln |\xi - e^{i\theta}|\, d\theta \\ &= I_1 + I_2.\end{aligned}$$

By Lemma 3 for Theorem 11.3.2, $I_2 = \ln |\xi| = \ln R$. To study I_1, we define g as a function of w by

$$g(w) = \begin{cases} \dfrac{\psi(\xi) - \psi(w)}{\xi - w}, & |w| > 1,\ w \ne \xi. \\ \psi'(\xi), & w = \xi. \end{cases}$$

This function is analytic on $\Delta'(\infty, 1)$, continuous and nonzero on $\bar{\Delta}'(\infty, 1)$, and has a removable singularity at ∞, where the appropriate value of g is

$$g(\infty) = \lim_{w \to \infty} \frac{\psi(\xi)/w - \psi(w)/w}{(\xi/w) - 1} = \frac{0 - \psi'(\infty)}{0 - 1} = c.$$

Therefore, by Lemma 2, $I_1 = \ln |c|$. Thus, the value of the integral in (13.9.11) for $z \in C_R$ is $\ln |c| + \ln R = \ln |cR| = \ln cR$, since both c and R are positive.

Finally, to show that (13.9.11) (with $\ln cR$ substituted for the right-hand member) implies (13.9.10), we restate (13.9.11) in arithmetic terms as follows: For any $\varepsilon > 0$ there exists N such that for every $z \in C_R$ and every $n \ge N$, we have $-\varepsilon/cRe < \ln |\omega_n(z)|^{1/(n+1)} - \ln cR < \varepsilon/cRe$, which implies that $0 < |\omega_n(z)|^{1/(n+1)}/cR < e^{\varepsilon/cRe}$. (Here e is the number introduced at the beginning of Chapter 7.) It is easy to see by using the mean value theorem of differential calculus that if x is such that $0 < x \le 1$, then $e^x < ex + 1$. We adjust ε, if necessary, so that $\varepsilon \le cRe$. Then we have, for $n > N_\varepsilon$ and every $z \in C_R$, $0 < |\omega_n(z)|^{1/(n+1)}/cR < e(\varepsilon/cRe) + 1 = 1 + \varepsilon/cR$, and $0 < |\omega_n(z)|^{1/(n+1)} - cR < cR\varepsilon/cR = \varepsilon$. Q.E.D.

Remark (i). Let $\phi = \psi^{-1}$ be the inverse of the mapping function in Proposition 13.9.1. By using a variant of Lemma 1 it is possible to prove that under the hypotheses of the proposition, we have

$$\lim_{n\to\infty} |\omega_n(z)|^{1/(n+1)} = c|\phi(z)| \tag{13.9.14}$$

a.u. on Ext Γ.

Remark (ii). L. Kalmar in 1926 published a converse to Proposition 13.9.1 as modified by Remark (i). Specifically, it runs as follows: Let $Z = \langle\{z_{nk}, k = 0, 1, \ldots, n\}\rangle_{n=1}^{\infty}$ be a sequence of subsets of a Jordan curve Γ such that with $\omega_n(z) = (z - z_{n0})(z - z_{n1})\cdots(z - z_{nn})$, equation (13.9.14) holds a.u. on Ext Γ. Then Z is uniformly ψ-distributed. For a proof, see [21, pp. 167–170].

We are now ready to state and prove Fejér's Theorem of 1918.

Theorem 13.9.2 *Let Γ be a Jordan curve in the z-plane, let O be an open set containing $\Gamma \cup \operatorname{Int}\Gamma$, and let $f\colon O \to \mathbf{C}$ be analytic on O. Let $\psi\colon \Delta(\infty, 1) \to (\operatorname{Ext}\Gamma) \cup \{\infty\}$ be the exterior mapping function for $\Gamma \cup \operatorname{Int}\Gamma$ described in Theorem 13.4.2, extended onto $C(0, 1)$ so as to be a homeomorphism from $\bar{\Delta}(\infty, 1)$ onto $\Gamma \cup (\operatorname{Ext}\Gamma) \cup \{\infty\}$. For each $R > 1$, let C_R denote the level curve $\psi[C(0, R)]$. Let the sequence of sets $\langle F_n = \{z_{n0}, z_{n1}, \ldots, z_{nn}\}\rangle_{n=1}^{\infty}$ be uniformly ψ-distributed on Γ, let the points in F_n be distinct, and let $L_n(f; z)$ be the polynomial of degree at most n which satisfies $L_n(f; z_{nk}) = f(z_{nk})$, $k = 0, 1, \ldots, n$; $n = 1, 2, \ldots$. Then:*

(i) *There exists $R^* > 1$ such that* $\operatorname{Int} C_{R^*} \subset O$ *and* $\lim_{n\to\infty} L_n(f; z) = f(z)$ *a.u. on* $\operatorname{Int} C_{R^*}$.

(ii) *For each R, $1 < R < R^*$, and for every $z \in (\Gamma \cup \operatorname{Int}\Gamma)$, we have*

$$|L_n(f; z) - f(z)| \le \frac{K}{R^n}, \qquad n = 1, 2, \ldots, \tag{13.9.15}$$

where K depends on R but not on z and n.

Proof. Propositions 13.5.3(i) and (iv) ensure the existence of $R^* > 1$ with the properties that for every R, $1 < R < R^*$, we have $(\Gamma \cup \operatorname{Int}\Gamma) \subset (C_R \cup \operatorname{Int} C_R) \subset \operatorname{Int} C_{R^*} \subset O$. Now choose R_1 and R_2 so that $1 < R_1 < R_2 < R^*$, and let $\omega_n(z) = (z - z_{n0})(z - z_{n1})\cdots(z - z_{nn})$, where $z_{nk} \in F_n$, $k = 0, \ldots, n$. According to Proposition 13.5.3(i), for any $R > 1$, we have $\operatorname{Ind}_{C_{R2}}(z) = 1$ for every $z \in \operatorname{Int} C_{R_2}$, so we can use the formula (13.9.3) to represent $L_n(f; z)$ for $z \in \operatorname{Int} C_{R_2}$:

$$L_n(f; z) = \frac{1}{2\pi i}\int_{C_{R_2}} \frac{f(\zeta)}{\zeta - z}\left(1 - \frac{\omega_n(z)}{\omega_n(\zeta)}\right) d\zeta. \tag{13.9.16}$$

Then for $z \in C_{R_1}$, using the corollary of Proposition 13.5.3, we obtain

$$|L_n(f; z) - f(z)| \le \frac{1}{2\pi}\int_{C_{R_2}} \left|\frac{f(\zeta)}{\zeta - z}\right|\left|\frac{\omega_n(z)}{\omega_n(\zeta)}\right| d\zeta. \tag{13.9.17}$$

Let $d = \operatorname{dist}(C_{R_1}, C_{R_2})$ and $M(f; R_2) = \max|f(z)|$, $z \in C_{R_2}$. According to Proposition 13.9.1, the sequence $\langle|\omega_n(z)|^{1/(n+1)}\rangle_1^{\infty}$ converges uniformly to cR_1 for $z \in C_{R_1}$ and converges uniformly to cR_2 for $z \in C_{R_2}$. Let r_1 and r_2 be any numbers which satisfy $1 < R_1 < r_1 < r_2 < R_2$. Because of the uniform convergence, there exists N such that for every $n \geq N$ and every $z \in C_{R_1}$, we have $|\omega_n(z)|^{1/(n+1)} < cr_1$, so $|\omega_n(z)| < (cr_1)^{n+1}$; and for every $n \geq N$ and every $z \in C_{R_2}$, we have $|\omega_n(z)|^{1/(n+1)} > cr_2$, so $|\omega_n(z)| > (cr_2)^{n+1}$ From (13.9.17) and (8.3.3) we obtain, for $n \geq N$,

$$|L_n(f; z) - f(z)| \leq \frac{1}{2\pi} \frac{M(f; R_2)}{d} \left(\frac{cr_1}{cr_2}\right)^{n+1} l(C_{R_2}) \qquad (13.9.18)$$

for $z \in C_{R_1}$, where $l(C_{R_2})$ is the length of the path C_{R_2}. By the maximum modulus principle in the form of Corollary 2 of Theorem 9.7.1, the inequality (13.9.18) is valid for every $z \in \text{Int } C_{R_1}$. Since $\lim_{n \to \infty} (r_1/r_2)^n = 0$, and since R_1 can be adjusted within the indicated restrictions so that any given compact set $E \subset \text{Int } C_{R^*}$ lies in $\text{Int } C_{R_1}$ [Proposition 13.5.4], part (i) of the theorem has now been established.

(ii) For this part, we proceed as in the proof of Theorem 13.8.2(v) to juggle the various R's and r's in the proof of part (i) so as to obtain an inequality of the type in (13.9.15). Specifically, let R be any number such that $1 < R < R^*$. Then let $r_1 = (R^* + 1)/(R + 1)$, $r_2 = R(R^* + 1)/(R + 1)$; so $r_1/r_2 = 1/R$ and $1 < r_1 < r_2 < R^*$. Thereafter, choose R_1 and R_2 so that $1 < R_1 < r_1 < r_2 < R_2 < R^*$. In (13.9.17), let $K = (2\pi d)^{-1} M(f; R_2)(r_1/r_2)l(C_{R_2})$. With these substitutions, (13.9.18) assumes the appearance of (13.9.15). Since (13.9.18) is valid for every $z \in \text{Int } C_{R_1}$, the equality is, of course, valid on $(\Gamma \cup \text{Int } \Gamma) \subset \text{Int } C_{R_1}$. Q.E.D.

Remark (iii). Notice that the only property of $L_n(f; z)$ used in the proof is that L_n can be represented by the formula (13.9.16). In the light of this fact, the conclusion of the theorem remains true if the hypothesis that for each n, the points in F_n be distinct is dropped, and $L_n(f; z)$ is then defined by (13.9.16) for each n and each R_2, $1 < R_2 < R^*$, instead of by the interpolation conditions $L_n(f; z_{nk}) = f(z_{nk})$, $k = 0, 1, \ldots, n$.

Remark (iv). The sequence of sets of nodes $\langle F_n \rangle$ is sometimes called a sequence of Fejér points on Γ.

Remark (v). As we have pointed out previously, the hypothesis concerning the function f in the theorem can be restated without explicit reference to an open set O by requiring that f be analytic on $\Gamma \cup \text{Int } \Gamma$, because the definition of analyticity at a point entails analyticity in a neighborhood of the point. For each $n, n = 1, 2, \ldots$, let $\Phi_n = \{\zeta_{nk} = \psi(e^{i\theta_{nk}}); k = 0, 1, \ldots, n\}$, where $\theta_{n0}, \theta_{n1}, \ldots, \theta_{nn}$ are any distinct points in $(0, 2\pi)$. Let $L_n(f; z)$ be the polynomial of degree at most n which is determined by the conditions $L_n(f; \zeta_{nk}) = f(\zeta_{nk})$, $k = 0, 1, \ldots, n$, where f is analytic on $\Gamma \cup \text{Int } \Gamma$. The theorem implies that a sufficient condition on the sequence $\langle \Phi_n \rangle_1^\infty$ for the conclusion that $\lim_{n \to \infty} L_n(f; z) = f(z)$ uniformly on $\Gamma \cup \text{Int } \Gamma$ for *every* function f analytic on $\Gamma \cup \text{Int } \Gamma$ is that $\langle \Phi_n \rangle_1^\infty$ be uniformly ψ-distributed on Γ. In the paper referred to in Remark (ii), Kalmár proved that this condition on $\langle\Phi_n\rangle$ is also necessary. In fact, it is known [4] that if $\langle\Phi_n\rangle$ is such that for just a *single* point $z_0 \in \text{Int}\,\Gamma$, $\overline{\lim}_{n\to\infty} |L_n(1/(z-z_0); z_0|^{1/n} \le 1$ for every $t \in \text{Ext}\,\Gamma$, then the sequence $\langle\Phi_n\rangle$ is necessarily ψ-distributed on Γ.

We now return to the problem which interested Fejér; namely, that of finding a sequence of polynomials $p_1, p_2, p_3, \ldots$, which "belong" to a given region B in the sense that every function f analytic on B can be expanded on B in a series $a_0 + \sum_1^\infty a_k p_k(z)$ in which the coefficients a_k depend on f but the polynomials do not. To apply Fejér's interpolation method to this problem, it is useful to know that there exist sequences of sets $S = \langle U_n = \{\theta_{nk}, k = 0, \ldots, n\}\rangle_{n=1}^\infty$ which are uniformly distributed on $[0, 2\pi]$, and are such that for each n, the points in U_n are distinct, and θ_{nk} depends on k but not on n, $k = 0, 1, \ldots, n$. That is, the structure of such a sequence is of this type: $U_1 = \{\theta_0, \theta_1\}$, $U_2 = \{\theta_0, \theta_1, \theta_2\}, \ldots$, $U_{n-1} = \{\theta_0, \theta_1, \ldots, \theta_{n-1}\}$, $U_n = \{\theta_0, \theta_1, \ldots, \theta_{n-1}, \theta_n\}, \ldots$, where $\langle\theta_k\rangle_0^\infty$ is a infinite sequence of distinct points in $[0, 2\pi)$. A classical example is $\langle U_n = \{2\pi k\theta - 2\pi[k\theta]$, $k = 0, 1, 2, \ldots, n\}\rangle_1^\infty$ where θ is any positive irrational real number, $0 < \theta < 1$, and $[\alpha]$ denotes the greatest integer in the nonnegative real number α. (See [16, pp. 70–73] for a proof and further interesting information concerning uniform distributions.)

Corollary With Γ and ψ as described in the theorem, let $\langle F_n = \{z_0, z_1, \ldots, z_n\}\rangle_1^\infty$ be uniformly ψ-distributed on Γ, and let the points in F_n be distinct, $n = 1, 2, \ldots$. Then:

(i) Any function f analytic on $\Gamma \cup \text{Int}\,\Gamma$ has a series representation

$$f(z) = a_0 + \sum_{k=1}^{\infty} a_k \omega_{k-1}(z),$$

$$\omega_k(z) = (z - z_0)(z - z_1)\cdots(z - z_k), \tag{13.9.19}$$

which converges a.u. on a region containing $\Gamma \cup \text{Int}\,\Gamma$.

(ii) For each n, $n = 0, 1, 2, \ldots$, the nth partial sum $S_n(f; z) = a_0 + \sum_{n=1}^\infty a_k\omega_{k-1}(z)$ is the polynomial of degree at most n which assumes the values $f(z_k)$ at the points z_k, $k = 0, 1, \ldots, n$.

(iii) The coefficients a_k are given by

$$a_k = \frac{1}{2\pi i}\int_{C_R} \frac{f(\zeta)}{\omega_k(\zeta)}\,d\zeta, \qquad k = 0, 1, \ldots,$$

where C_R is any level curve of the mapping ψ such that f is analytic on $C_R \cup \text{Int}\,C_R$.

Proof. For each n, $n = 0, 1, 2, \ldots$, let $L_n(f; z)$ be the polynomial of degree at most n which assumes the values $f(z_k)$ at the points z_k, $k = 0$,

$1, \ldots, n$. The function f is analytic on an open set O containing $\Gamma \cup \operatorname{Int}\Gamma$. According to Theorem 13.9.2(i), there exists $R^* > 1$ such that $\operatorname{Int} C_{R^*} \subset O$, where $C_{R^*} = \psi[C(0, R^*)]$, and $\lim_{n\to\infty} L_n(f; z) = f(z)$ a.u. on $\operatorname{Int} C_{R^*}$. We have

$$\begin{aligned} L_n(f; z) = L_0(f; z) &+ [L_1(f; z) - L_0(f; z)] \\ &+ [L_2(f; z) - L_1(f; z)] \\ &+ \cdots + [L_n(f; z) - L_{n-1}(f; z)]. \end{aligned} \tag{13.9.20}$$

Choose any R, $1 < R < R^*$, and recall that $\operatorname{Ind}_{C_R}(z) = 1$ for $z \in \operatorname{Int} C_R$ [Proposition 13.5.3(i)]. Therefore, the formula (13.9.3) is available to represent L_n, $n = n = 0, 1, 2, \ldots$, and we obtain

$$\begin{aligned} L_k(f; z) - L_{k-1}(f; z) &= \frac{1}{2\pi i}\int_{C_R} \frac{f(\zeta)}{\zeta - z}\left[\frac{\omega_{k-1}(z)}{\omega_{k-1}(\zeta)} - \frac{\omega_k(z)}{\omega_k(\zeta)}\right] d\zeta \\ &= \frac{1}{2\pi i}\int_{C_R} \frac{f(\zeta)}{\zeta - z}\frac{\omega_{k-1}(z)}{\omega_{k-1}(\zeta)}\left[1 - \frac{z - z_k}{\zeta - z_k}\right] d\zeta \\ &= \omega_{k-1}(z)\cdot\frac{1}{2\pi i}\int_{C_R} \frac{f(\zeta)}{\omega_k(\zeta)}\, d\zeta, \qquad k = 1, 2, \ldots. \end{aligned}$$

Also,

$$\begin{aligned} L_0(f; z) &= \frac{1}{2\pi i}\int_{C_R} \frac{f(\zeta)}{\zeta - z}\left[1 - \frac{z - z_0}{\zeta - z_0}\right] d\zeta \\ &= \frac{1}{2\pi i}\int_{C_R} \frac{f(\zeta)}{\zeta - z}\, d\zeta. \end{aligned}$$

From this computation and (13.9.20), it is apparent that the partial sum $S_n(f; z)$ in part (ii) of the corollary, with the coefficients given in part (iii), is identically equal to $L_n(f; z)$. Q.E.D.

The results in this section and the preceding section are among the older modern classics of the field of approximation by polynomials in **C**. They have been widely generalized and extended in various ways, particularly by J. L. Walsh [21] and various Russian writers. A relatively up-to-date survey of Russian work is contained in [20].

Exercise 13.9

Prove that the sequence of roots of unity

$$\langle\{w_{nk} = e^{2\pi i k/(n+1)},\quad k = 0, 1, \ldots, n\}\rangle_{n=1}^{\infty}$$

is uniformly distributed on $C(0, 1)$. (*Hint*: Show that if $N_n(\alpha, \beta)$ denotes the number of points of the set $\{\theta_{nk},\ k = 0, 1, \ldots, n\}$ which lie on (α, β), then $(\beta - \alpha)/2\pi \le N_n[\alpha, \beta)/(n + 1) < 1/(n + 1) + (\beta - \alpha)/2\pi$.)

Appendix to Chapter 13

Proofs of Proposition 13.7.4 and Theorems 13.7.3 and 13.7.4

Proposition 13.7.4 *Let Γ be a simple analytic arc with the analytic parametrization $z = \gamma(t)$, $\alpha \leq t \leq \beta$. There exist a function F and a region T in the complex t-plane with these properties: T is bounded, symmetric in the real axis, and contains $[\alpha, \beta]$; F is bounded, analytic, and univalent on T, and $F(t) \equiv \gamma(t)$, $t \in [\alpha, \beta]$, so $F([\alpha, \beta]) = \Gamma$.*

Proof. The power series $\sum c_k(\tau)(t - \tau)^k$, which represents $\gamma(t)$ in a neighborhood of $\tau \in [\alpha, \beta]$, converges on an interval $I(\tau) = (\tau - \delta(\tau), \tau + \delta(\tau))$. Then by the corollary of Theorem 4.1.1, with t complex, this series converges on the disk $\Delta(\tau, \delta(\tau))$, and the sum function $f_\tau(t)$ is analytic on this disk [Theorem 6.4.1]. Clearly $f_\tau(t) \equiv \gamma(t)$, $t \in I(\tau) \cap [\alpha, \beta]$. Since $\gamma'(\tau) \neq 0$, it follows from the definition of the complex derivative that $f'_\tau(\tau) \neq 0$. Thus, the zero of $f_\tau(t) - f_\tau(\tau)$ at $t = \tau$ is simple [Exercise 10.1.1], and by the open mapping theorem [Sec. 10.4, Remark (iv)] there exists a disk $\Delta^*(\tau) = \Delta^*(\tau, \delta^*(\tau))$ with $0 < \delta^*(\tau) \leq \delta(\tau)$, which is such that f_τ is analytic and univalent on $\Delta^*(\tau)$. This function gives a one-to-one mapping of the interval $I^*(\tau) = (\tau - \delta^*(\tau), \tau + \delta^*(\tau)) \cap [\alpha, \beta]$ onto a subarc of Γ, say $\Gamma(\tau) = \gamma[I^*(\tau)]$. If $\Gamma \sim \Gamma(\tau) \neq \varnothing$, then $\Gamma \sim \Gamma(\tau)$ is compact in $\mathbf{C}$ because it is the image under the continuous mapping $\gamma\colon [\alpha, \beta] \to \mathbf{C}$ of the set $[\alpha, \beta] \cap [I^*(\tau)]^c$, which is compact in $\mathbf{R}$ [Theorem 5.5.2]. A point $t \in \Delta^*(\tau)$ such that $f_\tau(t) \in \Gamma(\tau)$ necessarily lies in $I^*(\tau)$.

For reasons which are made clear later on, we further restrict f_τ to a disk $\Delta^{**}(\tau) \subset \Delta^*(\tau)$ which has the property that if for a point $t \in \Delta^{**}(\tau)$, we have $f_\tau(t) \in \Gamma$, then $t \in I^*(\tau)$, which is equivalent to the condition that $f_\tau(t) \in \Gamma(\tau)$. If $[\alpha, \beta] \subset \Delta^*(\tau)$, then no further adjustment of $\Delta^*(\tau)$ is needed. Otherwise we use the continuity of f_τ as follows to show that such a disk exists. Assume the contrary; then each disk $\Delta(\tau, \delta^*(\tau)/n)$, $n = 1, 2, \ldots,$ would contain a point t_n such that $f_\tau(t_n) \in \Gamma \sim \Gamma(\tau)$. Since $\Gamma \sim \Gamma(\tau)$ is compact, it follows from the Bolzano-Weierstrass theorem [Exercise 5.5.4] that the set $\{f_\tau(t_n)\}$ would have an accumulation point a in $\Gamma \sim \Gamma(\tau)$; and then there would be a subsequence of $\langle f_\tau(t_n)\rangle$, say $\langle f_\tau(t_m)\rangle$, such that $\lim_{m\to\infty} f_\tau(t_m) = a$ [Exercise 5.1.2]. Now $\lim_{m\to\infty} t_m = \tau$ and f_τ is continuous on $\Delta^*(\tau)$, so we have $\lim_{m\to\infty} f_\tau(t_m) = f_\tau(\tau) = \gamma(\tau) = a$. But $\gamma(\tau) \in \Gamma(\tau)$ and $a \notin \Gamma(\tau)$, so we have a contradiction.

For greater clarity in the ensuing construction, it seems useful to replace disks by squares. A square with center τ on the real axis and edges parallel to the real and imaginary axes is by definition a region of the type $S(\tau, k) = \{t\colon \tau - k < \operatorname{Re} t < \tau + k, \tau - k < \operatorname{Im} t < \tau + k\}$ where, of course, $k > 0$. Its boundary is the closed polygonal path with vertices $\tau \pm k \pm ik$, and its intersection with the real axis is the open interval $I(\tau, k) = (\tau - k, \tau + k)$. Two squares $S(\tau, k)$ and $S(\tau', k')$ intersect if and only if $I(\tau, k) \cap I(\tau', k')$ is a nonempty open interval; and $S(\tau, k) \subset S(\tau', k')$ if and only if $I(\tau, k) \subset I(\tau', k')$.

Now for each $\tau \in [\alpha, \beta]$, let $k = k(\tau)$ be chosen so that $\overline{S(\tau, k)} \subset \Delta^{**}(\tau)$. (This is obviously possible; for example, take k equal to 1/2 of the radius of the disk.) The family $\{S(\tau, k)\colon \tau \in [\alpha, \beta]\}$ is an open covering of the compact set $[\alpha, \beta]$, so there exists a finite subcovering Σ of $[\alpha, \beta]$ from this family. We now reduce Σ as follows to a finite open Σ-covering of $[\alpha, \beta]$, say Σ_0, which is possibly simpler than Σ. First, we discard any square in Σ which is contained in another square in Σ; we denote the thereby reduced subcovering by Σ'. Next, we choose a square in Σ' which contains α, an endpoint of $[\alpha, \beta]$. If this square also contains β, we let Σ_0 consist of this square alone. If this square does not contain β, we denote the square by S_1, its center by τ_1, and its intersection with the real axis by $(\tau_1 - k_1, \tau_1 + k_1)$. There must exist a square $S(\tau, k(\tau))$ in Σ' with $\tau + k(\tau) > \tau_1 + k_1$, but then $\tau - k(\tau) > \tau_1 - k_1$ (because otherwise $S_1 \notin \Sigma'$). Denote this new square by S_2, the center by τ_2, and the real-axis intersection by $(\tau_2 - k_2, \tau_2 + k_2)$. If $\beta \in S_2$, we let $\Sigma_0 = \{S_1, S_2\}$. If $\beta \notin S_2$, again there must exist a square $S(\tau, k(\tau))$ in Σ' with $\tau + k(\tau) > \tau_2 + k_2$, $\tau - k(\tau) > \tau_2 - k_2$; let this square be denoted by S_3 with center τ_3. We continue this construction until β is contained in a square in the sequence $S_1, S_2, \ldots,$ say S_N. (The sequence is certainly finite because Σ' is a finite subcovering of $[\alpha, \beta]$.) Then we let $\Sigma_0 = \{S_1, S_2, \ldots, S_N\}$. See Figure A1.

Now let $f_j = f_{\tau_j}|\overline{S}_j$, $j = 1, 2, \ldots, N$. The function f_j has the following

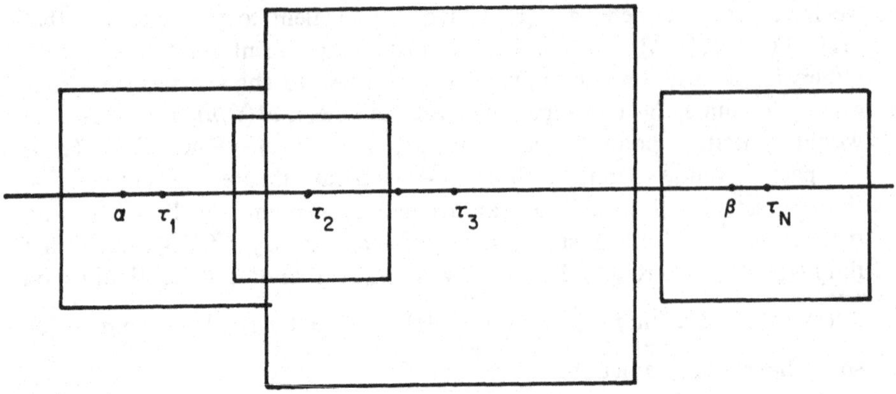

Fig. A1

properties, which we recall from the earlier part of this proof. It is analytic on S_j, continuous, and univalent on $\bar{S}_j$; $f_j(t) \equiv \gamma(t)$, $t \in \bar{S}_j \cap [\alpha, \beta]$; and last but not least, since $\bar{S}_j = \overline{S(\tau_j, k_j)} \subset \Delta^{**}(\tau_j)$, if with $t \in \bar{S}_j$ we have $f_j(t) \in \Gamma$, then $t \in [\alpha, \beta]$.

Since $f_1(t) \equiv \gamma(t) \equiv f_2(t)$ on the interval $[\alpha, \beta] \cap S_1 \cap S_2$, which contains an open interval of reals, by Corollary 2 of Theorem 6.5.1 we have $f_1(t) \equiv f_2(t)$, $t \in S_1 \cap S_2$. Let F_2 be defined by

$$F_2(t) = \begin{cases} f_1(t), & t \in S_1; \\ f_2(t), & t \in S_2. \end{cases}$$

Then F_2 is well defined (that is, single valued), bounded, and analytic on $S_1 \cup S_2$ and $F_2(t) \equiv \gamma(t)$, $t \in [\alpha, \beta] \cap (S_1 \cup S_2)$. By construction, $[\alpha, \beta] \cap (S_1 \cup S_2) \cap S_3$ contains an open interval of reals, and $f_3(t) \equiv \gamma(t) \equiv F_2(t)$ on this interval. Then as before we have $f_3(t) \equiv F_2(t)$, $t \in (S_1 \cup S_2) \cap S_3$. Let F_3 be defined by

$$F_3(t) = \begin{cases} F_2(t), & t \in S_1 \cup S_2; \\ f_3(t), & t \in S_3. \end{cases}$$

Then F_3 is well defined, bounded, and analytic on $S_1 \cup S_2 \cup S_3$. We continue in this way until S_N is reached, and thereby obtain a function $F = F_N$ which is bounded and analytic on $T_0 = S_1 \cup S_2 \cup \cdots \cup S_N$. It has these other properties: (i) $F(t) \equiv \gamma(t)$, $t \in [\alpha, \beta]$, so $F([\alpha, \beta]) = \Gamma$ and F is univalent on $[\alpha, \beta]$; (ii) $F(t) \equiv f_j(t)$, $t \in S_j$, so F has a continuous and univalent extension onto $\bar{S}_j$ given by the function f_j, $j = 1, 2, \ldots, N$; and (iii) the set $F(T_0)$ is open [Theorem 10.4.1]. We do not attempt to define F everywhere on $\bar{T}_0 = \bar{S}_1 \cup \bar{S}_2 \cup \cdots \cup \bar{S}_N$ because of possible ambiguities at common boundary points of the squares. (Again see Figure A1.)

We now prove that there exists an open set $T_1 \subset T_0$ with $[\alpha, \beta] \subset T_1$ and such that F is univalent on T_1. Let z' be any point in Γ, and let $t' = \gamma^{-1}(z') = F^{-1}(z')$. The point t' must be contained in at least one of the squares S_j, say S. We claim that there exists a disk $\Delta[z', \rho(z')] \subset F(T_0)$, $\rho(z') > 0$, such that the corresponding neighborhood of t', $N(t') = F^{-1}(\Delta[z', \rho(z')])$ lies entirely in S. Assume the contrary; then in each disk $\Delta(z', 1/n)$, $n = 1, 2, \ldots$, there would be a point z_n such that one of the preimages of z_n in the inverse image $F^{-1}(z_n)$, say t_n, would lie in S^c. The set $\overline{T}_0 \cap S^c$ is compact in $\mathbf{C}$ and $\{t_n; n = 1, 2, \ldots\}$ is contained in this compact set, so by the Bolzano-Weierstrass theorem, there exists an accumulation point b of $\{t_n\}$ in $\overline{T}_0 \cap S^c$ and a subsequence $\langle t_m \rangle$ from $\langle t_n \rangle$ such that $\lim_{m \to \infty} t_m = b$. The point b is in at least one of the closed squares $\overline{S}_j$, and it might lie on a common boundary of two or more squares. However, there is only a finite number of these squares in T_0, and $\langle t_m \rangle$ is an infinite sequence of points from $T_0 = S_1 \cup S_2 \cup \cdots \cup S_N$, so there necessarily is an infinite subsequence of $\langle t_m \rangle$, say $\langle t_p \rangle$, such that the set $\{t_p, p = 1, 2, \ldots\}$ lies in just one of the squares in T_0, say S_P; and $\lim_{p \to \infty} t_p = b$. Then $b \in \overline{S}_P$, and by the continuity of the function f_P on $\overline{S}_P$, we have $\lim_{p \to \infty} f_P(t_p) = f(b)$. Now clearly we also have $\lim_{n \to \infty} z_n = z' \in \Gamma$, and $F(t_n) = z_n$, so after the renumbering of the sequence $\langle t_n \rangle$ to designate the subsequence $\langle t_p \rangle$, we have $z_p = f_P(t_p)$, and $z' = \lim_{p \to \infty} z_p = \lim_{p \to \infty} f_P(t_p) = f(b)$. Thus, b is a point in $\overline{S}_P$ such that $z' = f_P(b) \in \Gamma$. Recall now that one of the properties of f_P is that if $f_P(b) \in \Gamma$ with $b \in \overline{S}_P$, then $b \in [\alpha, \beta]$. No member of the set $\{t_n, n = 1, 2, \ldots\}$ lies in S, so $b \neq t'$. But now we have reached a contradiction with the univalence of γ, because $f(b) = \gamma(b) = z'$ and $\gamma(t') = z'$.

To summarize: We have now shown that for any $z' \in \Gamma$, and $t' = \gamma^{-1}(z')$, there exists a disk $D(z')$ such that its preimage $N(t') = F^{-1}[D(z')]$ is contained entirely in the square in T_0 which contains t'. The restriction of F to each of the squares in T_0 is univalent on that square, so we conclude that the set of preimages of any point $z \in D(z')$ consists of one and only one point in T_0. Let $G = \bigcup_{z' \in \Gamma} D(z')$. Then each $z \in G$ has one and only one preimage point in the mapping $F: T_0 \to \mathbf{C}$. Now let $T_1 = F^{-1}(G)$. Then $T_1 \subset T_0$, $[\alpha, \beta] \subset T_1$, and F is bounded, analytic, and univalent on T_1.

We conclude the proof by replacing T_1 with a region $T \subset T_1$ which is symmetric in the real axis. Since F is continuous on T_0 and G is obviously open, it follows that $T_1 = F^{-1}(G)$ is open [Theorem 5.2.1]. Let $d = \text{dist}([\alpha, \beta], T_1^c) > 0$ [Theorem 5.6.1], and let $T = \bigcup_{\tau \in [\alpha, \beta]} \Delta(\tau, d)$. Then T is clearly open and arcwise connected, so T and $F|T$ have the properties promised in the statement of the proposition. Q.E.D.

We now develop a more detailed form of Proposition 13.7.4 for the case in which Γ is a subarc of a Jordan curve. The result is correlated with the hypotheses of Proposition 13.7.3(a) to facilitate proving Theorems 13.7.3 and 13.7.4.

Proposition 13.7.5 *Let C be a Jordan curve in the complex z-plane, and let Γ_1 be a subarc of C such that $C \sim \Gamma_1 \neq \varnothing$ and such that Γ_1 is an analytic arc with analytic parametrization $z = \gamma(t)$, $\alpha \le t \le \beta$. Let Γ be the corresponding open arc $\gamma((\alpha, \beta))$. Then there exists a region T° in the complex t-plane and a function $F\colon T^\circ \to \mathbf{C}$ with these properties:*

(i) *T° is bounded and symmetric in the real axis and its intersection with the real axis is (α, β).*
(ii) *F is analytic, bounded, and univalent on T, and $F(t) \equiv \gamma(t)$, $t \in (\alpha, \beta)$.*
(iii) *The region $\Omega = F(T^\circ)$ is such that $\Omega \cap C = \Gamma$.*
(iv) *The subregions $\Omega_0 = F(T^\circ \cap H)$ and $\Omega_0^* = F(T^\circ \cap H^*)$, where $H = \{t\colon \operatorname{Im} t > 0\}$ and $H^* = \{t\colon \operatorname{Im} t < 0\}$, lie in different components of C^c (that is, if $\Omega_0 \subset \operatorname{Int} C$ then $\Omega_0^* \subset \operatorname{Ext} C$ and vice versa); and Γ is a free boundary arc of both Ω_0 and Ω_0^*.*
(v) *If E is a given closed subset of* Int C, *then $E \subset \Omega$.*

Remark. In the statement of this proposition and elsewhere in the remainder of this appendix, we take for granted without repeated back-references the fact that a nonconstant complex-valued function analytic on a region maps the region onto another region. [The specific authority is Sec. 10.4, Remark (iii).]

Proof of Proposition 13.7.5. The arc Γ_1 is simple, since it is contained in a simple closed curve C and $C \sim \Gamma_1 \neq \varnothing$. Then according to Proposition 13.7.4, there exist a region T symmetric to the real axis in the complex t-plane, with $[\alpha, \beta] \subset T$, and a function $F\colon T \to \mathbf{C}$, which is analytic and univalent on T, such that $F(t) \equiv \gamma(t)$, $t \in [\alpha, \beta]$. Now $F(T)$ is open [Theorem 10.4.1] and $\Gamma \subset F(T)$. According to Proposition 13.7.2, Γ is a free boundary arc of C^c. Therefore, for each point $z \in \Gamma$, there exists a disk $\Delta(z, \rho_z) \subset F(T)$, $\rho_z > 0$, such that this disk contains no point of $C \sim \Gamma$. Furthermore, if $E \subset \operatorname{Int} C$ is a closed set (and we now refer to part (v)), then since dist $(E, C) > 0$ [Theorem 5.6.1], we choose $\rho_z <$ dist (E, C) for each $z \in \Gamma$, and then for each such z, $\Delta(z, \rho_z)$ does not intersect E. Thus, for each $z \in \Gamma$, the disk $\Delta(z, \rho_z) \subset F(T)$ does not intersect $E \cup (C \sim \Gamma)$.

Let $z \in \Gamma$ for the moment be held fixed, and let $t = F^{-1}(z) \in (\alpha, \beta)$. The open set $N(t) = F^{-1}[\Delta(z, \rho_z)]$ lies in the region T, and there exists a disk $\Delta(t, \delta_t) \subset N(t)$. Of course, $F[\Delta(t, \delta_t)] \subset \Delta(z, \rho_z)$, so $F[\Delta(t, \delta_t)]$ does not intersect $E \cup (C \sim \Gamma)$.

Now let $T^\circ = \bigcup_{t \in (\alpha, \beta)} \Delta(t, \delta_t)$, and let $\Omega = F(T^\circ)$. The set T° clearly is a bounded region which intersects the real axis only in (α, β); the region $\Omega = F(T^\circ) = \bigcup_{t \in (\alpha, \beta)} F[\Delta(t, \delta_t)]$ contains Γ but does not intersect $E \cup$

$(C \sim \Gamma)$, so $\Omega \cap C = \Gamma = F((\alpha, \beta))$. Thus, T° and $F|T^\circ$ satisfy the specifications for T° and F in parts (i), (ii), (iii), and (v) of Proposition 13.7.5, and it remains only to prove that (iv) is satisfied by the subsets Ω_0, Ω_0^* of Ω there defined.

Notice that T° and $F|T^\circ$ (which we henceforth denote simply by F) satisfy the conditions on T and F in the hypothesis of Proposition 13.7.3(a), so according to that proposition, $\Omega_0 = F(T^\circ \cap H)$ and $\Omega_0^* = F(T^\circ \cap H^*)$ are regions, and Γ is a free boundary arc of both Ω_0 and Ω_0^*. Neither one of these regions intersects C. They are connected sets, so each one lies in one and only one component of C^c [Theorem 5.4.1]. The components of C^c are Int C and Ext C [Jordan curve theorem]. Suppose now that both Ω_0 and Ω_0^* were to be contained in Ext C. There exists a disk Δ with center on Γ such that $\Delta \subset \Omega$ and every point of Δ is a point of one of the three disjoint sets Ω_0, Ω_0^*, and Γ. This implies, under the assumption that $\Omega_0, \Omega_0^* \subset$ Ext C, that if $z \in \Delta$, $z \notin \Gamma$, then $z \in$ Ext C. But Γ is a boundary arc of Int C and so Δ contains points of Int C. Thus, we have a contradiction, and it follows that at least one of the regions Ω_0, Ω_0^* lies in Int C. A similar argument shows that both Ω_0 and Ω_0^* cannot lie in Int C. Q.E.D.

We now turn to the proofs of Theorems 13.7.3 and 13.7.4. With a view toward clarifying the construction used in our proof of Theorem 13.7.4, we formulate here a more detailed version of Theorem 13.7.3 in which the situation in the range of the conformal equivalence ϕ is emphasized.

Theorem 13.7.3A *Let C be a Jordan curve in the complex z-plane, let ϕ be a conformal equivalence mapping* Int C *onto $\Delta(0, 1)$ in the w-plane, and let ϕ be extended onto C so as to be a homeomorphism from $C \cup$* Int C *onto $\bar{\Delta}(0, 1)$ [Theorem 13.7.1]. Let ϕ^{-1} be the inverse function. For a given $z_0 \in$* Int C, *let $\phi(z_0) = 0$. Let Γ_1 be an analytic subarc of C such that $C \sim \Gamma_1 \neq \varnothing$, and let $\Gamma \subset \Gamma_1$ be the corresponding open arc. Let $\phi(\Gamma_1) = A_1 = \{w = e^{i\theta}: \theta_1 \leq \theta \leq \theta_2, 0 < \theta_2 - \theta_1 < 2\pi\} \subset C(0, 1)$, and let A be the corresponding open subarc of $C(0, 1)$. Then there exist a region R in the w-plane and a function $h: R \to C$ with these properties:*

(i) $[A \cup \Delta(0, 1)] \subset R$.
(ii) $R \cap \Delta(\infty, 1)$ *is a region contained in the infinite sector* $S = \{w = re^{i\theta}: 0 < r < \infty, \theta_1 < \theta < \theta_2\}$.
(iii) *h is analytic and univalent on R and* $h(w) \equiv \phi^{-1}(w)$, $w \in [A \cup \Delta(0, 1)]$.
(iv) $[\Gamma \cup \text{Int } C] \subset h(R) = B$, *which is a region.*
(v) $B \cap$ Ext C *is a region.*

Proof. The existence of a region R which has all of the listed properties except (ii) is a fairly immediate consequence of Propositions 13.7.3 and 13.7.5. We identify C, Γ_1, and Γ in this Theorem 13.7.3A with the curve and arcs denoted by the same symbols in the hypothesis of Proposition 13.7.5, and z_0 with E in part (v) of the proposition. The region T°, the interval (α, β), the function F, and the regions $\Omega, \Omega_0, \Omega^*$ in the conclusions of Proposition 13.7.5 fit the respective descriptions of T, (α, β), $F, \Omega, \Omega_0, \Omega_0^*$ in Proposition 13.7.3(a). Notice that $z_0 \notin \Omega$.

Now suppose that $\Omega_0 \subset \operatorname{Int} C$. Then according to part (iv) of Proposition 13.7.5, $\Omega_0^* \subset \operatorname{Ext} C$, and Γ is a common free boundary arc of Ω_0 and Ω_0^*. The extended function ϕ is analytic and nonzero on Ω_0, continuous and univalent on $\overline{\Omega}_0 \subset (C \cup \operatorname{Int} C)$; it maps Ω_0 onto a region $W_0 \subset \Delta(0, 1)$ in the w-plane, and by hypothesis $\phi(\Gamma) = A$. Then with $W_0^* = \{w: 1/\overline{w} \in W_0\}$, Proposition 13.7.3(b) (ii) ensures the existence of a function $z = h(w)$ from $D = \Delta(0, 1) \cup A \cup W_0^*$ onto $(\operatorname{Int} C) \cup \Omega = (\operatorname{Int} C) \cup \Gamma \cup \Omega_0^*$ which is analytic and univalent on D, and is such that $h(w) \equiv \phi^{-1}(w)$, $w \in [\Delta(0, 1) \cup A]$.

At this point, the proof of Theorem 13.7.3 as stated in the text of this chapter is essentially complete if we identify f in that theorem with h^{-1} and B in that theorem with $h(D) = \operatorname{Int} C \cup \Omega$. For both $\operatorname{Int} C$ and Ω are regions, and they intersect in the region Ω_0, so $\operatorname{Int} C \cup \Omega$ is a region. Also $(\operatorname{Int} C \cup \Omega) \cap \operatorname{Ext} C = \Omega_0^*$, and this is a region. According to Proposition 13.7.5(iv), Γ is a free boundary arc of Ω_0^*.

However, Theorem 13.7.3A implies that D is to be replaced by a subregion $R \subset D$, which still satisfies (i), (ii), (iv), (v), and also is contained in the sector S. We now construct such a region R.

Since $\Omega \cup \operatorname{Int} C$ is a region, it follows that $D = h^{-1}[\Omega \cup \operatorname{Int} C]$ is also a region. Also $W_0^* = h^{-1}(\Omega_0^*)$ is a region. Consider the open set $W^* \cap S$. This intersection of regions is not necessarily a region itself, but we construct a subset U^* which is a region having A as a boundary arc and is such that $R = \Delta(0, 1) \cup A \cup U^*$ is a region. Notice that at each point $z \in A$ there exists a disk $\Delta(z, \rho(z))$, $\rho(z) > 0$, contained in both D and S. Let $U = \bigcup_{z \in A} \Delta(z, \rho(z))$, and let $U^* = U \cap (W_0^* \cap S)$. Clearly, U is a region containing A. Since A is a boundary arc of $\Delta(0, 1)$, each disk in U contains points of $\Delta(0, 1)$, so the open set $R = \Delta(0, 1) \cup U = \Delta(0, 1) \cup A \cup U^*$ is a region. It remains to show that $U = U \cap \Delta(\infty, 1) = R \cap \Delta(\infty, 1)$ is a region, for then it would follow at once that $B \cap \operatorname{Ext} C = h[R \cap \Delta(\infty, 1)]$ is a region. We state and prove a relevant result formally as follows.

Lemma Let $O \subset \mathbf{C}$ be a region which intersects both $\Delta(0, 1)$ and $\Delta(\infty, 1)$, and such that the intersection of O with $C(0, 1)$ is an open arc A of $C(0, 1)$. Then both $O \cap \Delta(0, 1)$ and $O \cap \Delta(\infty, 1)$ are regions.

Proof of the Lemma. Let $A = \{w = e^{i\theta}: \theta_1 < \theta < \theta_2, 0 < \theta_2 - \theta_1 < 2\pi\}$. The proof can be given directly by an adaptation of the argument suggested for Exercise 5.5.6, or it can be referred to Exercise 5.5.6 by means of the transformation

$$z = i\frac{e^{i\theta_1} + w}{e^{i\theta_1} - w} = L(w)$$

used in the proof of the corollary of Theorem 13.7.2. We elect to travel the latter route. Note that according to the description of the properties of L given in that proof, $K = L(O)$ is a region which intersects both the upper open half-plane H and its reflection H^* in the real axis, and the intersection of K with the real axis is an open interval. We have $L[\Delta(\infty, 1)] = H^*$, so $L[O \cap \Delta(\infty, 1)] = K \cap H^*$. According to Exercise 5.5.6, $K \cap H^*$ is a region, so $O \cap \Delta(\infty, 1) = L^{-1}(K \cap H^*)$ is also a region. If we replace $\Delta(\infty, 1)$ with $\Delta(0, 1)$ in this argument, H^* is replaced by H, and the same reasoning shows that $O \cap \Delta(0, 1)$ is a region. The proof of the lemma is complete.

To use the lemma to complete the proof of Theorem 13.7.3A, we have only to identify the region U with O in the lemma and U^* with $O \cap \Delta(\infty, 1)$. Q.E.D.

To prove Theorem 13.7.4(a), it is convenient to prove part (iv) first, which we restate.

Theorem 13.7.4(a) (iv) *Let C be an analytic Jordan curve in the complex z-plane, and let ϕ be a conformal equivalence which maps* Int C *onto* $\Delta(0, 1)$ *in the w-plane. Let ϕ^{-1} denote the inverse function. Then there exist a number $\rho > 1$ and a function $h: \Delta(0, \rho) \to \mathbf{C}$ such that h is analytic and univalent on $\Delta(0, \rho)$ and $h(w) \equiv \phi^{-1}(w)$, $w \in \Delta(0, 1)$.*

Proof. We assume that ϕ has been extended onto C in accordance with Theorem 13.7.1 so as to be a homeomorphism from $C \cup \text{Int}\, C$ onto $\bar{\Delta}(0, 1)$. Let $A(\theta_1, \theta_2)$ denote the open arc $\{w = e^{i\theta}: \theta_1 < \theta < \theta_2, 0 < \theta_2 - \theta_1 < 2\pi\}$, and let $S(\theta_1, \theta_2)$ denote the open infinite sector $\{w = re^{i\theta}: 0 < r < \infty, \theta_1 < \theta < \theta_2\}$. Let Γ^1 be an open subarc of C such that $\phi(\Gamma^1) = A(0, 3\pi/2)$, and let Γ^2 be such that $\phi(\Gamma^2) = A(\pi, 5\pi/2)$. By Theorem 13.7.3A, there exist regions R_1 and R_2 in the complex w-plane, and corresponding functions $h_1: R_1 \to \mathbf{C}$ and $h_2: R_2 \to \mathbf{C}$, which are analytic and univalent on their respective domains, with the following properties: $[A(0, 3\pi/2) \cup \Delta(0, 1)] \subset R_1$; $[R_1 \cap \Delta(\infty, 1)]$ is a region contained in $S(0, 3\pi/2)$; $h_1(w) \equiv \phi^{-1}(w)$, $w \in [A(0, 3\pi/2) \cup \Delta(0, 1)]$; $[A(\pi, 5\pi/2) \cup \Delta(0, 1)] \subset R_2$; $R_2 \cap \Delta(\infty, 1)$ is a region contained in $S(\pi, 5\pi/2)$; and $h_2(w) \equiv \phi^{-1}(w)$, $w \in [A(\pi, 5\pi/2) \cup \Delta(0, 1)]$. The general idea of the proof now is to construct a subregion R of the region $R_1 \cup R_2$ such that $\bar{\Delta}(0, 1) \subset R$, R is a region, and h_1 and h_2 can be "fitted together" on R so as to form a single analytic univalent function. The major difficulty facing this program is that the open set $R_1 \cap R_2$ is not necessarily a region and may have various components which contain no subsets on which $h_1(w) = h_2(w)$.

So we proceed as follows. (See Figure A2.) Choose $d > 0$ so that $d < 1/3$ and d also is less than the least of these distances: dist $(\{1\}, \partial R_2)$, dist $(\{i\}, \partial R_j)$, dist $(\{-1\}, \partial R_1)$, and dist $(\{-i\}, \partial R_2)$. (As usual, ∂R_j denotes the boundary of R_j.) Construct circular disks $\Delta^1, \Delta^2, \Delta^3, \Delta^4$ with respective centers at $1, i, -1, -i$, and each with radius d. (The condition $d < 1/3$ is imposed merely so that the closed disks do not intersect.) We have $\bar{\Delta}^1 \subset R_2$, $\bar{\Delta}^2 \subset R_1$, $\bar{\Delta}^3 \subset R_1$, $\bar{\Delta}^4 \subset R_2$. At each point w of $A^1 \subset A(0, \pi/2) \subset (R_1 \cap R_2)$ there exists a disk $\Delta(w, \rho(w))$ such that $\Delta(w, \rho(w)) \subset (R_1 \cap R_2 \cap S(0, \pi/2))$. Let $D^1 = \bigcup_{w \in A^1} \Delta(w, \rho(w))$. Clearly, D^1 is a region containing A^1, and we have $h_1(w) \equiv h_2(w) \equiv \phi^{-1}(w)$, $w \in A^1$, so by Corollary 2 of Theorem 6.5.1, we have $h_1(w) \equiv h_2(w)$, $w \in D^1$. At each point w of $A^2 = A(\pi/2, \pi)$ there exists a disk $\Delta(w, \rho(w))$ such that $\bar{\Delta}(w, \rho(w)) \subset (R_1 \cap S(\pi/2, \pi))$. Let $D^2 = \bigcup_{w \in A^2} \Delta(w, \rho(w))$; this is a region containing A^2. At each point w of $A^3 = A(\pi, 3\pi/2)$, there exists a disk $\Delta(w, \rho(w))$ such that $\Delta(w, \rho(w)) \subset (R_1 \cap R_2 \cap S(\pi, 3\pi/2))$. Let $D^3 = \bigcup_{w \in A^3} (w, \rho(w))$. Clearly, D^3 is a region containing A^3, and we have $h_1(w) \equiv h_2(w) \equiv \phi^{-1}(w)$, $w \in A^3$, so (as on D^1) we have $h_1(w) \equiv h_2(w)$, $w \in D^3$. Finally, at each point w of

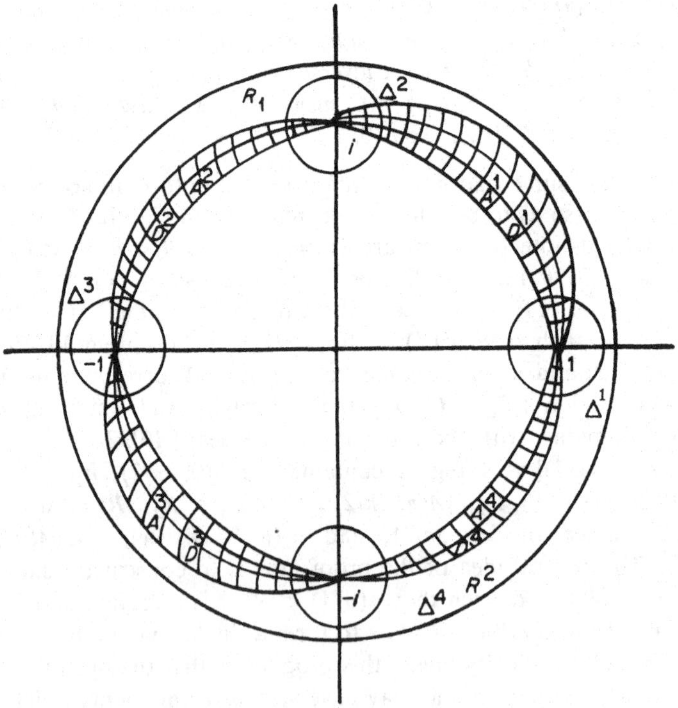

Fig. A2

$A^4 = A(3\pi/2, 2\pi)$ there exists a disk $\Delta(w, \rho(w))$ such that $\bar{\Delta}(w, \rho(w)) \subset (R_2 \cap S(3\pi/2, 2\pi))$. Let $D^4 = \bigcup_{w \in A^4} \Delta(w, \rho(w))$; this is a region containing A^4.

Now let $\Sigma = \Delta(0, 1) \cup \Delta^1 \cup D^1 \cup \Delta^2 \cup D^2 \cup \Delta^3 \cup D^3 \cup \Delta^4 \cup D^4$. This open set contains $C(0, 1)$, so it is a region. We define a function $h: \Sigma \to \mathbf{C}$ as follows

$$h(w) = \begin{cases} h_1(w), & w \in R^1 = D^1 \cup \Delta^2 \cup D^2 \cup \Delta^3 \cup D^3 \cup \Delta(0, 1); \\ h_2(w), & w \in R^2 = D^3 \cup \Delta^4 \cup D^4 \cup \Delta^1 \cup D^1 \cup \Delta(0, 1). \end{cases}$$

Notice that this formula apparently assigns to h the two values $h_1(w)$ and $h_2(w)$ for every $w \in R^1 \cap R^2 = D^1 \cup D^3 \cup \Delta(0, 1)$, but this does not involve ambiguity because $h_1(w) \equiv h_2(w)$, $w \in R^1 \cap R^2$.

It is now possible to choose a disk $\Delta(0, \rho)$ with $\rho > 1$ such that $\Delta(0, \rho) \subset \Sigma$, and if we did so, the proof would be complete *provided* that the function h were not required to be univalent on $\Delta(0, \rho)$. (In fact, in that case a simpler construction of Σ suffices.) But since a univalent function h is indeed promised in the conclusion of the theorem, we show now that there exists a subregion R of Σ such that $h|R$ is univalent on R and R contains $\Delta(0, 1)$.

Clearly, h has some local univalence properties on Σ. Specifically, since h_1 is univalent on R_1 and h_2 is univalent on R_2, and $R^1 \subset R_1$, $R^2 \subset R_2$, it follows that $h|R^1$ is univalent and so is $h|R^2$. But there is no assurance that h is univalent on $W_1 \cup W_2$, where $W_1 = \Delta^2 \cup D^2 \cup \Delta^3$ and $W_2 = \Delta^4 \cup D^4 \cup \Delta^1$. Let $A[\theta_1, \theta_2]$ denote the arc $\{w = e^{i\theta}: \theta_1 \le \theta \le \theta_2\}$. We show how to construct regions W_1^* and W_2^* with these properties: $A[\pi/2, \pi] \subset W_1^* \subset W_1 \subset R^1$, $A[3\pi/2, 2\pi) \subset W_2^* \subset W_2 \subset R^2$, and $h(w_1) \neq h(w_2)$ for every pair w_1, w_2 with $w_1 \in W_1^*$ and $w_2 \in W_2^*$. (See Figure A3.)

To avoid interrupting the chain of thought here, we assume for the moment that regions W_1^* and W_2^* with the indicated properties do exist, and we show that if we let $R = \Delta(0, 1) \cup D^1 \cup W_1^* \cup D^3 \cup W_2^*$, then R is a region containing $\bar{\Delta}(0, 1)$ and $h|R$ is univalent on R. In the first place, since $A[\pi/2, \pi] \subset W_1^*$, it follows that W_1^* contains a subarc of $C(0, 1)$ which also lies in D^1 and so $W_1^* \cap D^1$ is nonempty. Similarly, $W_1^* \cap D^3$, $D^3 \cap W_2^*$, and $W_2^* \cap D^4$ are each nonempty. Also, of course, $D^1 \cap \Delta(0, 1)$ is nonempty. It follows from the corollary of Proposition 5.4.1 that the open set R is a region. Clearly, $\bar{\Delta}(0, 1) \subset R$.

The function $h|R$ is defined by

$$h(w) = \begin{cases} h_1(w), & w \in B_1 = D^1 \cup W_1^* \cup D^3 \cup \Delta(0, 1); \\ h_2(w), & w \in B_2 = D^3 \cup W_2^* \cup D^1 \cup \Delta(0, 1). \end{cases}$$

Notice that $B_1 \subset R^1$ and $B_2 \subset R^2$. Now suppose that w_1, $w_2 \in R$, and we have $h(w_1) = h(w_2)$. The only different cases to examine are (a) w_1, $w_2 \in B_1$, (b) w_1, $w_2 \in B_2$, (c) $w_1 \in W_1^*$ and $w_2 \in W_2^*$ (or vice versa). In cases (a) and

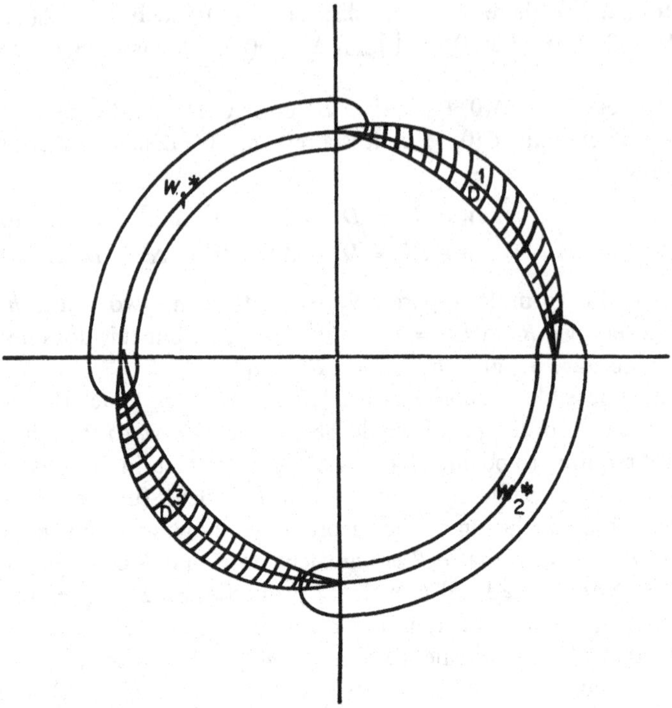

Fig. A3

(b), the conclusion is that $w_1 = w_2$ because of the local univalence of h, respectively, on R^1 and on R^2. In case (c), since $h(w_1) \neq h(w_2)$ for $w_1 \in W_1^*$ and $w_2 \in W_2^*$, it follows that we have either $w_1, w_2 \in W_1^* \subset R^1$ or $w_1, w_2 \in W_2^* \subset R^2$, and once again the conclusion is that $w_1 = w_2$. Thus, $h|R$ is univalent on the region R. The conclusion of the theorem itself now follows when we let $\delta = \text{dist}\,[\bar{\Delta}(0, 1), R^c] > 0$ [Theorem 5.6.1] and $\rho = \delta + 1$.

It remains to show how to construct W_1^* and W_2^* so that $h(w_1) \neq h(w_2)$ for every $w_1 \in W_1^*$ and $w_2 \in W_2^*$. This is done by exploiting the continuity properties of h. Recall that the construction of Σ is such that the closed disks $\bar{\Delta}^2$ and $\bar{\Delta}^3$, and the closure of each disk in the union D^2, are all contained in R_1. The function h is continuous (in fact, analytic) on each of these closed disks, so it is uniformly continuous on each one individually. The family of corresponding open disks form an open covering of the compact set $A[\pi/2, \pi]$, so there is a finite open subcovering from this family which includes Δ^2 and Δ^3. Let K_1 be the union of the disks in this finite subcovering. Then $\bar{K}_1$ is the union of the corresponding closed disks, and, therefore, h is uniformly continuous on $\bar{K}_1$. Also, $K_1 \subset W_1$ and $A[\pi/2, \pi] \subset K_1$.

Now let $\Gamma_1 = h(A[\pi/2, \pi]) = \phi^{-1}(A[\pi/2, \pi])$ and $\Gamma_2 = h(A[3\pi/2, 2\pi]) = \phi^{-1}(A[3\pi/2, 2\pi])$. These are disjoint subarcs of the Jordan curve C; they are compact in $\mathbf{C}$ [Theorem 5.5.2]; so $2\varepsilon = \text{dist}(\Gamma_1, \Gamma_2) > 0$ [Theorem 5.6.1]. Because of the uniform continuity of h on $\overline{K}_1$, there exists $\eta > 0$ such that for every pair $w', w'' \in K_1$ with $|w' - w''| < \eta$, we have $|h(w') - h(w'')| < \varepsilon$. Now choose $\eta_1 = \min(\eta, \text{dist}(A[\pi/2, \pi], K_1^c)$ and construct a disk $\Delta(w, \eta_1)$ for each $w \in A[\pi/2, \pi]$. Let W_1^* be the union of these disks. Then W_1^* has these properties: W_1^* is a region; $A[\pi/2, \pi] \subset W_1^* \subset W_1 \subset R^1$; and for each $w \in W_1^*$, we have $\text{dist}(h(w), \Gamma_1) < \varepsilon$, because for each $w \in W_1^*$ there exists $w' \in A[\pi/2, \pi]$ such that $|w - w'| < \eta_1$. It follows that for each $w \in W_1^*$, we have $\text{dist}(h(w), \Gamma_2) > \varepsilon$. (The reasoning in detail is this: By Lemma 1 for Theorem 5.6.1, with w held fixed in W_1^*, there exist points $z_1 \in \Gamma_1$ and $z_2 \in \Gamma_2$ such that $\text{dist}(h(w), \Gamma_1) = |h(w) - z_1|$ and $\text{dist}(h(w), \Gamma_2) = |h(w) - z_2|$. We have $2\varepsilon \le |z_1 - z_2| \le |z_1 - h(w)| + |h(w) - z_2| < \varepsilon + |h(w) - z_2|$, so $|h(w) - z_2| > \varepsilon$.)

We now start all over again with Δ^4, Δ^1, and D^4 substituted for Δ^2, Δ^3, and D^2, and by the same method construct a region W_2^* such that $A[3\pi/2, 2\pi] \subset W_1^* \subset W_1 \subset R^1$, and for each $w \in W_2^*$, $\text{dist}(h(w), \Gamma_2) < \varepsilon$, $\text{dist}(h(w), \Gamma_1) > \varepsilon$. (See Figure A3.) Clearly, for every pair of points w_1, w_2 with $w_1 \in W_1^*$ and $w_2 \in W_2^*$, we have $h(w_1) \neq h(w_2)$, because $h(w_1)$ and $h(w_2)$ satisfy conflicting inequalities. Q.E.D.

The other conclusions of Theorem 13.7.4 in the text of Sec. 13.7 follow easily from part (a) (iv), which we have just now proved. For the rest of part (a), we let $f = h^{-1}$ and $B = h[\Delta(0, \rho)]$. Since $\Delta(0, \rho)$ and $\Delta(0, \rho) \sim \overline{\Delta}(0, 1)$ obviously are regions, it follows that B is a region which contains $C \cup \text{Int}\, C = h(\overline{\Delta}(0, 1)) = \phi^{-1}(\overline{\Delta}(0, 1))$, and $B \cap \text{Ext}\, C = h[\Delta(0, \rho) \sim \overline{\Delta}(0, 1)]$ is also a region.

We now prove part (b) of Theorem 13.7.4 by the mapping technique used several times in this chapter, and specifically in the proof of the corollary of Theorem 13.7.1. With $a \in \text{Int}\, C$, the linear fractional transformation $L: \zeta = 1/(z - a)$ is a homeomorphism from $\hat{\mathbf{C}}$ onto $\hat{\mathbf{C}}$ which carries the Jordan curve C onto a Jordan curve C^* in the ζ-plane and Ext C onto Int C^*. Let $z = \gamma(t)$, $\alpha \le t \le \beta$, be an analytic parametrization of C. Then (see the beginning of the proof of Proposition 13.7.4) at each point $\tau \in [\alpha, \beta]$ there exists a disk $\Delta(\tau, \delta(\tau))$ and a function $f_\tau(t)$ analytic in the complex variable t on $\Delta(\tau, \delta(\tau))$, with $f_\tau(t) \equiv \gamma(t)$, $t \in \{\Delta(\tau, \delta(\tau)) \cap [\alpha, \beta]\}$ and $f_\tau'(\tau) \neq 0$. Let $\gamma^*(t) = 1/(\gamma(t) - a)$; then $\zeta = \gamma^*(t)$, $\alpha \le t \le \beta$, is a parametric equation for C^*. We claim that γ^* is an analytic parametrization of C^* with a nonzero derivative, and, therefore, C^* is an analytic Jordan curve. For let g_τ be given by $g_\tau(t) = 1/[f_\tau(t) - a]$, $t \in \Delta(\tau, \delta(\tau))$. This composition of analytic functions is analytic on $\Delta(\tau, \delta(\tau))$, and it coincides in value with $\gamma^*(t)$ on $\Delta(\tau, \delta\tau) \cap [\alpha, \beta]$. Therefore, γ^* is represented by a series of powers of $t - \tau$

in a neighborhood of each $\tau \in [\alpha, \beta]$. Also $[\gamma^*(\tau)]' = g_\tau'(\tau) = -f_\tau'(\tau)/[f_\tau(\tau) - a]^2 \neq 0$. Therefore, C^* is an analytic Jordan curve.

Now let ψ be the mapping function described in Theorem 13.7.4(b). We define an analytic univalent function ψ^* from $\Delta(0, 1)$ in the ξ-plane onto Int C^* in the ζ-plane by

$$\zeta = \psi^*(\xi) = \begin{cases} \dfrac{1}{\psi(1/\xi) - a}, & \xi \neq 0; \\ 0, & \xi = 0. \end{cases}$$

By Theorem 13.7.4(a), there exist $\rho > 0$ and a function h^* from $\Delta(0, \rho)$ in the ξ-plane into the ζ-plane such that h^* is analytic and univalent on $\Delta(0,\infty)$ and $h^*(\xi) \equiv \psi^*(\xi)$, $\xi \in \Delta(0, 1)$. Since $h^*(0) = 0$, we have $h^*(\xi) \neq 0$ for $\xi \in \Delta(0, \rho) \sim \{0\}$. Let $h\colon \Delta'(\infty, 1/\rho) \to \mathbf{C}$ be given by $h(w) = a + 1/h^*(1/w)$. Then h is analytic and univalent on $\Delta'(\infty, 1/\rho)$ and $h(w) \equiv \psi(w)$, $w \in \Delta'(\infty, 1)$.

Q.E.D.

Some Notation Listed by Chapter and Section Where They First Appear

Section 1.1

$\in$	is a member of
$\notin$	is not a member of
$\supset$	contains
$\subset$	is a part of
$\{x: P(x)\}$	the set of x such that $P(x)$
$\{a, b, c, \ldots\}$	set with elements (or members) $a, b, c, \ldots$
$\{a, b\}, \{b, a\}$	set with members a and b
(a, b)	ordered pair; also open interval (see Section 1.6)
$\varnothing$	the empty set
$X \times Y$	Cartesian product set of the sets X and Y: $\{(x, y): x \in X, y \in Y\}$
$\cup$	union
$\cap$	intersection
$\bigcup$	union (for a family of sets)
$\bigcap$	intersection (for a family of sets)
$\sim$	difference of sets
E^c	complement of the set E

Section 1.2

$\mathscr{F}$	field
$-a$	additive inverse of a

a^{-1}	multiplicative inverse of a
$\sum_{k=m}^{m+p} a_k$	the sum of the elements a_k, k ranging from m to $m+p$
$\mathbf{R}$	the real number field

Section 1.3

$\mathscr{P}$	positive part of a field
$>$	greater than
$<$	less than
$\geq$	not less than
$\leq$	not greater than

Section 1.4

$\sup E$	least upper bound of the set E
$\inf E$	greatest lower bound of E

Section 1.6

$\infty, -\infty$	plus and minus infinity in the extended real number system
$\hat{\mathbf{R}}$	extended real number system
(α, β)	open interval $\{x: \alpha < x < \beta\}$
$[\alpha, \beta]$	closed interval $\{x: \alpha \leq x \leq \beta\}$
$[\alpha, \beta)$	$\{x: \alpha \leq x < \beta \leq \infty\}$
$(\alpha, \beta]$	$\{x: -\infty \leq \alpha < x \leq \beta\}$

Section 1.7

$f: A \to Y$	f maps A into Y
$f\|C$	the restriction of f to C
$f(C)$	the image of C with respect to f
$f^{-1}(D)$	$\{x: f(x) \in D\}$, the preimage of D with respect to f
$f(x) \equiv b, x \in A$	f is a function with the constant value b for every $x \in A$
f^{-1}	the inverse function for the function f
$g \circ f, g(f)$	the composition of the function g with the function f

Section 1.8

$\mathbf{C}$	the complex number system

Section 1.9

$\mathbf{R}^2$	$\mathbf{R} \times \mathbf{R} = \{(x, y), x \in \mathbf{R}, y \in \mathbf{R}\}$

Section 1.10

i	the number in $\mathbf{C}$ corresponding to the 2-vector (0, 1) in $\mathbf{R}^2$
$x + iy$	classical representation of a complex number

Section 1.11

$\operatorname{Re} z$	real part of the complex number z
$\operatorname{Im} z$	imaginary part of the complex number z
$\bar{z}$	the conjugate complex number of z

Section 1.12

$\lvert z \rvert$	the absolute value of the complex number z

Section 1.15

$\hat{\mathbf{C}}$	the extended complex number system
∞	the element called the point at infinity in $\hat{\mathbf{C}}$
$\mathbf{R}^3$	$\mathbf{R} \times \mathbf{R} \times \mathbf{R}$

Section 2.1

$\mathbf{Z}$	the real integers
$\langle x_k \rangle_{k=n}^{\infty}$	the sequence $x_n, x_{n+1}, x_{n+2}, \ldots$
$\langle x_k \rangle$	the sequence $\langle x_k \rangle_{k=0}^{\infty}$ unless the domain is otherwise indicated
$\{x_k\}$	the range (set of values) of the sequence $\langle x_k \rangle$
$\langle x_k \rangle + \langle y_k \rangle$	the sequence $\langle x_k + y_k \rangle$ (used when the ranges lie in a field)
$\langle x_k \rangle \cdot \langle y_k \rangle$	the sequence $\langle x_k \cdot y_k \rangle$ (used when the ranges lie in a field)
$\langle x_k \rangle / \langle y_k \rangle$	the sequence $\langle x_k / y_k \rangle$ (used when the ranges lie in a field)

Section 2.2

(X, d)	metric space with metric d
$\Delta(a, r)$	open disk $\{z : \lvert z - a \rvert < r\}$ in $\mathbf{C}$
$C(a, r)$	circle $\{z : \lvert z - a \rvert = r\}$ in $\mathbf{C}$

Section 2.3

$\lim_{k \to \infty} x_k = x$	indicates that the sequence $\langle x_k \rangle_n^{\infty}$ has a limit which is x

$x_k \to x$ abbreviation for $\lim_{k\to\infty} x_k = x$

Section 2.4

$\overline{\lim}\, x_k$, $\limsup x_k$ the upper limit of the sequence $\langle x_k\rangle_n^\infty$

$\underline{\lim}\, x_k$, $\liminf x_k$ the lower limit of the sequence $\langle x_k\rangle_n^\infty$

Section 2.6

$\sum_m^\infty z_p$ the infinite series defined by the sequence $\langle z_m + z_{m+1} + \cdots + z_n\rangle_{n=m}^\infty$

Section 3.1

$|f|$ modulus function of the function f, defined by $|f|(z) = |f(z)|$

Section 4.1

$\bar{\Delta}(a, r)$ the closed disk $\Delta(a, r) \cup C(a, r)$ in $\mathbf{C}$

Section 5.1

$B(a, r)$, $B_X(a, r)$ open ball in metric space (X, d): $B(a, r) = B_X(a, r) = \{x: x \in X, d(a, x) < r\}$

$\bar{E}$ closure of the set E

O an open set in a given metric space

∂E frontier or boundary of the set E; $\partial E = \bar{E} \cap \overline{E^c}$

Section 5.6

dist (x, E) the distance between an element x of a metric space (X, d) and a nonempty subset $E \subset X$

dist (E_1, E_2) the distance between the two nonempty subsets E_1, E_2 of the metric space (X, d)

diam (E) the diameter of the nonempty set E

Section 6.1

$\delta_a(z)$ the difference quotient of a function f at a

$f'(a)$, $\left.\frac{df}{dz}\right]_{z=a}$ the derivative of the function f at a

$F_x(x_0, y_0)$, $F_y(x_0, y_0)$ partial derivatives of $F(x, y)$ with respect to x and y at $x = x_0$, $y = y_0$

$\bar{f}$ the function with the same domain as that of the

complex-valued function f, and with conjugate complex values

Section 7.1

e	$1 + \sum_{k=1}^{\infty} (1/k!)$; base of natural logarithms
E	function with values $E(z) = 1 + z + (z^2/2!) + \cdots$; the exponential function
e^z	conventional notation for $E(z)$

Section 7.2

$\cos z$	value of the cosine function at z
$\sin z$	value of the sine function at z
$\tan z$	$\sin z/\cos z$
$\sec z$	$1/\cos z$
$\operatorname{cosec} z$	$1/\sin z$
$\sinh z$	value of the hyperbolic sine function at z
$\cosh z$	value of the hyperbolic cosine function at z

Section 7.3

π	area of a circular disk with radius $1 = 3.14159\ldots$

Section 7.5

$\ln u$	natural logarithm of u, given by $\ln u = E^{-1}(u)$, $0 < u < \infty$
$\log w$	a logarithm of w
$\arg w$	an argument of w
$\operatorname{Arg} w$	principal argument of w, the restriction of range of $\arg w$ to $(-\pi,\pi]$
$\operatorname{Log} w$	principal logarithm of w

Section 7.6

w^z	general exponential function, defined by $w^z = e^{z \log w}$

Section 8.1

$\int_\alpha^\beta f(t)\, dt$	Riemann integral of the complex-valued function over the interval of reals $[\alpha, \beta]$

Section 8.2

$\int_\gamma f(z)\, dz$	complex line integral of f along the arc (or closed curve) γ

$\gamma_1 + \gamma_2$	sum of the arcs γ_1 and γ_2, as here defined
$-\gamma$	opposite arc of γ
$\|\pi\|$	norm of the partition π (the maximum subinterval length)
$V(\gamma; \pi)$	variation of the arc γ over the partition π of a given interval $[\alpha, \beta]$
$T(\gamma)$	total variation of the arc γ over all partitions π of a given interval $[\alpha, \beta]$

Section 8.3

$L(\gamma)$	length of the path γ
$[a, b]$	the path γ given by $z = a + t(b - a)$, $0 \leq t \leq 1$, $a, b \in \mathbf{C}$; also any equivalent path; also the graph of γ, which is a line segment joining a to b

Section 8.4

$\mathrm{Ind}_\gamma (z)$	topological index of the point z with respect to the arc (or closed curve) γ (Also called the winding number of γ with respect to z.)
$\Delta'(\infty, R)$	$\{z: \|z\| > R\}$ in $\mathbf{C}$

Section 9.2

T^*	"triangle" determined by a set T of three points (T^* is the convex hull of T.)

Section 9.6

$p^*(z)$	paraconjugate polynomial corresponding to the polynomial $p(z)$

Section 9.9

a.u.	almost uniformly (A sequence or series of functions of which the common domain is an open set $O \subset \mathbf{C}$ converges a.u. on O if and only if it converges uniformly on each compact set contained in O.)

Section 9.10

$L(F; z)$	Laplace transform of the function $F: [0, \infty) \to \mathbf{R}$

Section 10.7

$\Delta(\infty, r)$	$\Delta'(\infty, r) \cup \{\infty\}$ in $\hat{\mathbf{C}}$

Res $(f; a)$	Residue of the function f at the point $a \in \mathbf{C}$
Res $(f; \infty)$	Residue of the function f at ∞

Section 11.3

$\int_u^{\tau-} \phi(t)\,dt$	abbreviation for $\lim_{\substack{s\to\tau \\ u<s<\tau}} \int_u^s \phi(t)\,dt$
$\int_{\tau+0}^u \phi(t)\,dt$	$\lim_{\substack{s\to\tau \\ u<s<\tau}} \int_s^u \phi(t)\,dt$

Section 11.4

$PV\int_{-\infty} \phi(t)\,dt$	principal value of the improper integral $\int_{-\infty}^{\infty} \phi(t)\,dt$, as defined by (11.4.2)

Section 12.3

(X, Φ)	topological space of which Φ is the family of open sets

Section 12.5

int E	interior of E, the maximal open set contained in E (concept, but not the symbol, was introduced in Section 5.1)

Section 13.5

Ext Γ	exterior region of the Jordan curve Γ
Int Γ	interior region of the Jordan curve Γ

Section 13.7

$z_n \to \partial B$	indicates that the sequence $\langle z_n \rangle$ from a region B approaches the boundary ∂B in a special sense defined in this section

Section 13.9

det $\mathbf{A}$	determinant of the square matrix $\mathbf{A}$

Appendix to Chapter 13

$S(\tau, k)$	square region $\{t: \tau - k < \operatorname{Re} t < \tau + k, \tau - k < \operatorname{Im} t < \tau + k\}$

References

[1] M. Abramowitz and L. A. Stegun, eds., *Handbook of Mathematical Functions*, National Bureau of Standards, Applied Mathematics Series, No. 55, 1964.

[2] L. V. Ahlfors, *Complex Analysis*, 2nd ed., McGraw-Hill, New York, 1966.

[3] T. J. I'A Bromwich, *An Introduction to the Theory of Infinite Series*, 2nd ed. revised, MacMillan, New York, 1926.

[4] J. H. Curtiss, *Annals of Mathematics*, **42**, 634–646 (1941).

[5] J. H. Curtiss, *American Mathematical Monthly*, **78**, 577–596 (1971).

[6] A. Devinatz, *Advanced Calculus*, Holt, Rinehart and Winston, New York, 1968.

[7] J. Duncan, *The Elements of Complex Analysis*, Wiley, London, 1968.

[8] M. Heins, *Complex Function Theory*, Academic, New York, 1968.

[9] E. Hille, *Analytic Function Theory*, Vol. 1, Ginn, Boston, 1959.

[10] *E. Hille, Analytic Function Theory*, Vol. 2, Ginn, Boston, 1962.

[11] S. MacLane and G. Birkhoff, *Algebra*, MacMillan, New York, 1967.

[12] *Mathematical Reviews*, Vol. 47, American Mathematical Society, Providence, R.I. (1974).

[13] *Mathematical Reviews*, Vol. 49, American Mathematical Society, Providence, R.I. (1975).

[14] M. H. A. Newman, *Elements of the Topology of Plane Sets of Points*, 2nd ed., Cambridge University Press, London, 1951.

[15] L. I. Pennisi, *Elements of Complex Analysis*, Holt, Rinehart, and Winston, New York, 1963.

[16] G. Pólya and G. Szegö, *Aufgaben und Lehrsätze aus der Analysis*, 3rd ed., Vol. 1 (Grund. Math. Wiss. No. 19), Springer, Berlin, 1964.

[17] R. M. Redheffer, *American Mathematical Monthly*, **76**, 778–787 (1969).

[18] W. Rudin, *Real and Complex Analysis*, 2nd ed., McGraw-Hill, New York, 1974.

[19] S. Saks and A. Zygmund, *Analytic Functions* (translated by E. J. Scott), *Monografje Matematyczne*, Vol. 28, 2nd ed., Warsaw, 1965.

[20] V. I. Smirnov and N. A. Lebedev, *Functions of a Complex Variable* (translated by Scripta Technica Ltd.), MIT Press, Cambridge, Massachusetts, 1968.

[21] J. L. Walsh, *Interpolation and Approximation by Rational Functions in the Complex Domain*, 2nd ed., American Mathematical Society, Providence, R.I., 1956.

References

Index

G

H

I

J

L

M

V

W

Z